GROWTH FACTORS AND CYTOKINES
IN HEALTH AND DISEASE

A Multi-Volume Treatise

Volume 2A • 1997

CYTOKINES

GROWTH FACTORS AND CYTOKINES IN HEALTH AND DISEASE

A Multi-Volume Treatise

CYTOKINES

Editors: DEREK LEROITH
Diabetes Branch
NIDDK
National Institutes of Health
Bethesda, Maryland

CAROLYN BONDY
Developmental Endocrinology Branch
NICHD
National Institutes of Health
Bethesda, Maryland

VOLUME 2A • 1997

JAI PRESS INC.

Greenwich, Connecticut *London, England*

55 Old Post Road No. 2
Greenwich, Connecticut 06836

JAI PRESS LTD.
38 Tavistock Street
Covent Garden
London WC2E 7PB
England

ISBN: 0-7623-0117-1

~~*Manufactured in the United States of America*~~

Transferred to digital printing in 2008.

CONTENTS

CONTENTS

LIST OF CONTRIBUTORS

Bharat B. Aggarwal	Cytokine Research Laboratory The University of Texas, Houston
Erika A. Bach	Center for Immunology Washington University School of Medicine
Jacques Banchereau	Schering-Plough Laboratory for Immunological Research Dardilly, France
Russell Basser	Centre for Developmental Cancer Therapeutics Royal Melbourne Hospital
Heinz Baumann	Department of Molecular and Cellular Biology Roswell Park Cancer Institute Buffalo, New York
Jonathan Cebon	Centre for Developmental Cancer Therapeutics Royal Melbourne Hospital
Richard R. Dickason	Departments of Medicine, Microbiology, and Immunology Baylor College of Medicine
Anand S. Dighe	Center for Immunology Washington University School of Medicine St. Louis, Missouri

Alexander H. Enk — Department of Dermatology
University of Mainz
Mainz, Germany

Mark A. Goldsmith — Gladstone Institute of Virology
and Immunology
University of California,
San Francisco

Warner C. Greene — Gladstone Institute of Virology
and Immunology
University of California,
San Francisco

Andrew C. Greenlund — Center of Immunology
Washington University School
of Medicine
St. Louis, Missouri

Matthias Grell — Institute of Cell Biology
and Immunology
University of Stuttgart
Stuttgart, Germany

Akihisa Harada — Department of Hygiene
School of Medicine
Kanazawa University
Kanazawa, Japan

Douglas J. Hilton — The Cooperative Research Center
for Cellular Growth Factors
Parkville, Australia

David P. Huston — Departments of Medicine,
Microbiology, and Immunology
Baylor College of Medicine

Chun-fai Lai — Department of Molecular and
Cellular Biology
Roswell Park Cancer Institute
Buffalo, New York

Judith E. Layton — Ludwig Institute for Cancer
Research
Melbourne Tumour Biology Branch

Michael T. Lotze — Human Gene Therapy Program
University of Pittsburgh
Medical Center

David H. Lynch — Department of Immunobiology
Immunex Corporation
Seattle, Washington

Kouji Matsushima — Department of Pharmacology
Cancer Research Institute
Kanazawa University
Kanazawa, Japan

Naofumi Mukaida — Department of Pharmacology
Cancer Research Institute
Kanazawa University
Kanazawa, Japan

Sandra Nicholason — Ludwig Institute for Cancer Research
Melbourne Tumour Biology Branch

Nicos A. Nicola — The Walter and Eliza Hall
Institute of Medical Research
Parkville, Australia

Sidney Pestka — Department of Molecular Genetics
and Microbiology
University of Medicine and
Dentistry of New Jersey

Michel Revel — Department of Molecular Genetics
and Virology
Weizmann Institute of Science
Rehovot, Israel

Sem Saeland — Schering-Plough
Laboratory for Immunological
Research
Dardilly, France

Peter Scheurich — Institute of Cell Biology
and Immunology
University of Stuttgart
Stuttgart, Germany

John W. Schrader	The Biomedical Research Centre University of British Columbia
Robert D. Schreiber	Center for Immunology Washington University School of Medicine St. Louis, Missouri
Sanjaya Singh	Cytokine Research Laboratory University of Texas, Houston
Neil Stahl	Regeneron Pharmaceuticals Tarrytown, New York
Kiyoshi Takatsu	Department of Immunology Institute of Medical Science University of Tokyo
George D. Yancopoulos	Regeneron Pharmaceuticals Tarrytown, New York

PREFACE

Advances in molecular technology in recent years have catalyzed an explosive growth of information about intercellular peptide messengers and their receptors. For example, 10 years ago the only neurotrophin characterized at the molecular level was nerve growth factor (NGF) and the only recognized neurotrophin receptor was the p75 NGF receptor. At present, the number of described neurotrophic peptides approaches 30 and the number of receptors is increasing apace. Just six years ago, the characterized interleukins numbered about three while now there are at least 16. Because many of these new peptide ligands and receptors were identified by "reverse genetic" techniques the understanding of their biological roles lags behind the knowledge of their molecular structures. Over the past few years, however, a new era of functional studies has begun because recombinant proteins have become available for clinical studies. In addition, animal models have been and are being developed using recombinant DNA techniques. Both the clinical studies and studies of transgenic and target deleted mice will allow for further physiologic elucidation of the biological roles of these messenger peptides and their receptors.

This series of growth factors and cytokines is divided into three main sections: Growth Factors (Volumes 1A and 1B), Cytokines (Volumes 2A and 2B), and Systems (Volumes 3A and 3B). Although Volumes 1 and

2 are separate the distinction between "growth factors" and "cytokines" is probably more historical or pragmatic than indicative of differences in function. The term "growth factors" refers to a wide variety of locally or systemically produced proteins with pleiotropic actions on tissue growth and differentiation. The term "cytokines" describes a group of proteins identified primarily within the immune and hematopoietic systems, although it is likely that such a narrow view of cytokines will not survive for long. For example, it appears that some interleukins and interleukin receptors are expressed by neuroepithelial cells *in vivo* suggesting that these interleukins may have intrinsic roles within the nervous system. Furthermore, tumor necrosis factor (TNF) has been identified as a potential adipose tissue regulatory factor which is both produced and acts locally. The third volume titled *Systems* deals more directly with the role of these factors in both normal physiology and the disease processes resulting from the deficiency or excess of growth factors/cytokines and their receptors.

This second volume deals with cytokines and their receptors. The field is advancing at a rapid pace and we have attempted to cover as many of the cytokines as possible. Interleukins 11-15, in fact, will be discussed under "The Immune System" in the third volume. While we have attempted to be as comprehensive and inclusive as possible, there will always be some regrettable omissions. At the publishing date we recognize that a few growth factors and cytokines have not been included in this review. These new discoveries will for certain be reviewed in similar pages in the future.

Derek LeRoith
Carolyn Bondy
Editors

INTERLEUKIN-2

Michael T. Lotze

Growth Factors and Cytokines in Health and Disease
Volume 2A, pages 1-47.

ISBN: 0-7623-0117-1

I. INTRODUCTION

In studies conducted 30 years ago, soluble factors derived from mixed lymphocyte culture (MLC) were shown to cause DNA synthesis in freshly isolated lymphocytes (Kasakura and Lowenstein, 1965; Gordon and MacLean, 1965) This mitogenic factor was identified (Morgan et al., 1976) as a T cell growth factor (TCGF) and subsequently termed interleukin 2 (IL-2). IL-2 is now one of several identified T cell growth factors whose receptors appear to share the common chain of the IL-2 receptor (Taniguchi and Minami, 1993; Aversa et al., 1993) now including IL-2, IL-4, IL-7, IL-9, and IL-15 (Table 1). Another set of factors which appear to function by altering apoptotic cell death (tumor necrosis factor, interferon gamma, IL-10, and IL-12) also modulate T cell growth and function (Table 1).

Ten years ago, based on murine studies (Donohue et al., 1984a, 1984b) IL-2 entered clinical trials (Lotze et al., 1984, 1985a, 1985b, 1986a, 1986b, 1987; Rosenberg et al., 1985; Rosenberg and Lotze, 1986) The rationale for IL-2 administration at the time of initiation of clinical trials was primarily to induce cells with lymphokine activated killer (LAK) activity and to support the growth of adoptively transferred lym-

Table 1. T Cell Growth Factors

Proliferative Signals-Gamma Common Chain Receptors	*Antiapoptotic-Receptors*
IL-2, IL-4, IL-7, IL-9, IL-15	TNFα, IL-10, IL-12, IFN-γ

phoidcells. In addition, during the early clinical trials of IL-2, it became apparent that antitumor activity was observed in murine tumor models. This allowed a series of studies designed to assess, first in mice and then in clinical trials, the critical relationship of individual elements including immune cells, cytokines and the genetic background of individuals and then subsequently to test those principles in patients. Ultimately all of these studies were designed to allow improvement in the efficacy of treatment or to alter the toxicity without compromising efficacy. The administration of IL-2 is only associated with at best a 20-50% response rate in patients with certain tumors (renal cell carcinoma, melanoma, and lymphoma). Recently, Rosenberg et al. summarized the results of my early studies at the NIH (Rosenberg et al., 1994) and concluded that only a 7% complete response is observed but that, surprisingly, the duration of these responses was quite durable lasting for as long as 10 years. Such dramatic effects in a subset of patients with these disseminated tumors suggests that immunotherapies can be quite effective and that means to understand the basis for these results and extend the therapeutic benefits to others could be quite important.

Coadministration of other cytokines (TNF, α interferon, γ interferon) has not substantially improved the response rates in patients treated with these regimens. Although adoptive transfer of tumor infiltrating lymphocytes has been said to increase response rates in up to 35% of patients (Rosenberg et al., 1988, Forthcoming), this has yet to be confirmed in other series. Adoptive transfer models in mice have been quite helpful in defining the relevant populations of cells in the immune response to tumors and have clearly identified the role for tumor-specific T cells capable of releasing cytokines such as GM-CSF, TNF, and interferon γ following tumor stimulation.

The mechanism for the antitumor effect of IL-2 has been studied in murine models. In animals bearing 10-day established tumors the antitumor effect of IL-2 is abrogated by irradiation of the animal (presumed to be due to elimination of radiosensitive lymphoid cells), by elimination of CD8+ cells, which mediate recognition of peptide antigens in the context of Class I Major Histocompatibility (MHC) molecules, and partially by elimination of CD4+ cells, which mediate recognition of peptide antigens presented by MHC Class II molecules (Mule et al., 1987). Interestingly, elimination of cells with NK phenotype had no adverse effects on the antitumor effects of IL-2 administration but rather was associated with decreased toxicity. Likewise elimination of toxicity was observed with exogenous administration of steroids (Papa et al.,

1986; Vetto et al., 1987; Morse et al., 1990). This also led to the abrogation or lessening of the antitumor effect. IL-2 antitumor effects are limited in nude mice as well.

Collectively this information suggests that the antitumor effect of IL-2 is mediated by lymphoid effectors, presumably both CD4+ and CD8+ T cells. Gratifyingly in human tumors one observes infiltration by macrophages, T cells and widespread expression of Class II molecules, presumably an epiphenomenon related to gamma interferon production subsequent to widespread immune activation (Cohen et al., 1987; Rubin et al., 1989).

Just as the search for reasons for failure to respond to chemotherapy led to identification of factors involved with drug resistance including the multidrug resistance gene, it seems likely that a critical analysis of the factors associated with IL-2 "nonresponsiveness" should also lead to important leads important for extending the results of immunotherapy. This in practice has been quite difficult. In humans significant het-

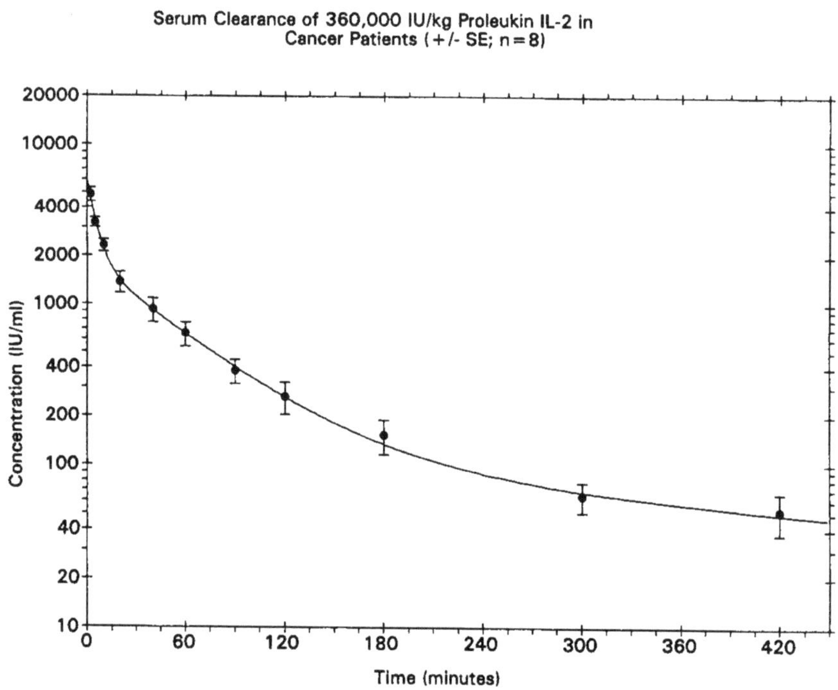

Figure 1. Serum ckearebce if 360m000 IU/kg of IL-2 in patients treated by bolus interferon. Results shown are from eight patients treated with the associated SEM of the mean blood level at each time point.

Table 2. Association Between Clinical and First Course Treatment Variables and Response to IL-2

Factor	*p2*[1]
Pre-tx. Platelets/mm^3 x 10^{-3}	0.055
Min. Platelets (cycle 1)/mm^3 x 10^{-3}	0.083
WBC (max - pre-tx.)/mm^3 x 10^{-3}	0.096
Pre-tx. Creatinine (Renal ca) (mg%)	0.17
Pre-tx. Creatinine (Melanoma) (mg%)	0.25
Max. Creatinine (Renal ca) (mg/%)	0.13
Max. Creatinine (Melanoma) (mg/%)	0.18
Min. Platelets/mm^3 x 10^{-3}	0.15
Max. Lymphocytes (%)	0.25
Max. Lymphocytes/mm^3 x 10^{-3}	0.28
Pre-tx. Weight (Kg)	0.15
Max. Weight (Kg)	0.19
Pre-tx. WBC/mm^3 x 10^{-3}	0.36
Prior Radiation (No)	0.78
Prior Surgery (Yes)	0.45
Age (50+)	0.12
Oliguria (Yes)	0.97
Pulmonary Tox. (Gr. 3 or 4)	0.30
Sepsis (Yes)	0.67
Confusion (Yes)	0.35

Note: [1] Not corrected for the large number of comparisons performed. Associations were made based on the 164 patients who actually received at least one dose of IL-1. From Rosenberg et al. (1993).

erogenneity of the disease stage, the presence of a recently defined associations of response with the polymorphic major histocompatibility loci (Marincola et al., 1992; Rubin et al., 1995), the individual loss of various molecules in the tumor or the responding T cells or macrophages suggest that treatments will be best developed when appreciating the immense complexity of the immune response. In animal models it appears clear that increasing doses of IL-2 are associated with enhanced antitumor effects. In humans IL-2 is cleared from the circulation quite quicly (Figure 1 and Table 2) and means to enhance its retention such as pegylation have been fraught with enhanced toxicities. Unfortunately many patients stop short of the "maximum" amount of IL-2 because they have underlying medical problems which preclude further administration or alternatively bothersome side effects lead either the physician, the patient or the patient's family to stop treatment. These factors make defining the relationship between any one factor and response almost impossible. Below we will deal with some of the known issues related to IL-2 production and responsiveness and then turn again to findings from clinical trials.

II. PRODUCTION OF IL-2—CRITICAL ELEMENTS FOR IMMUNE RESPONSES

The IL-2 gene exists as a single copy on human chromosome 4q26-28 (Goldsmith et al., 1994; Kuziel and Green, 1991; Lotze, 1991a, 1991b). 5' regulatory sequences included in the enhancer include binding sites for at least seven nuclear proteins including NFAT-1, NFkB, AP-1, and certain octamer proteins (Goldsmith et al., 1994). IL-2 production is also regulated by altering message stability through AU-rich motifs of the 3' untranslated region. IL-2 is normally produced following T cell engagement with cognate antigen expressed as peptides present within Class I or Class II molecules. Subsequent to antigen or mitogen stimulation a variety of cellular events including production of IL-2 message are initiated early (Figure 2). T cell receptor activation is mediated through activation of a cascade of molecules including the TCRξ or FcεRγ chain, phosphorylation of $p59^{fyn}$ ZAP 70, and then of phospholipase C (Nghiem

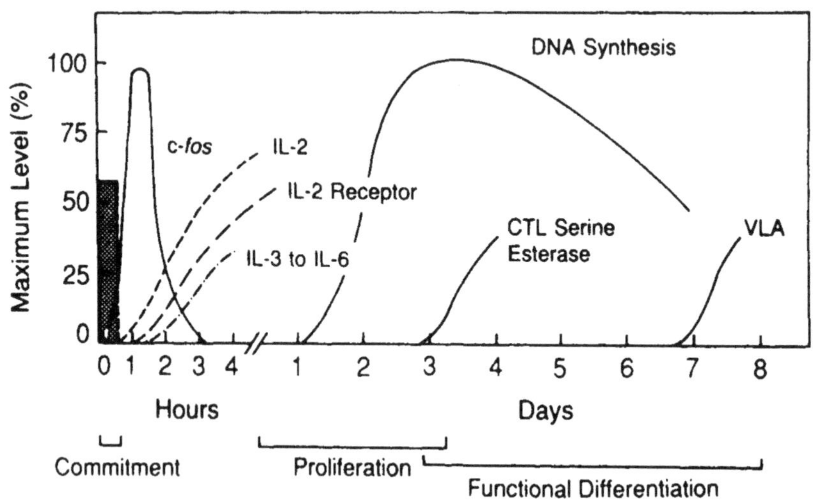

Figure 2. Sequential steps following T cell activation. After mitogenic signaling a series of programmed steps are observed. IL-2 message production is observed within an hour folowed by production of other cytokines, DNA synthesis, enhanced lytic activity, and expression of the very late antigens (VLA).

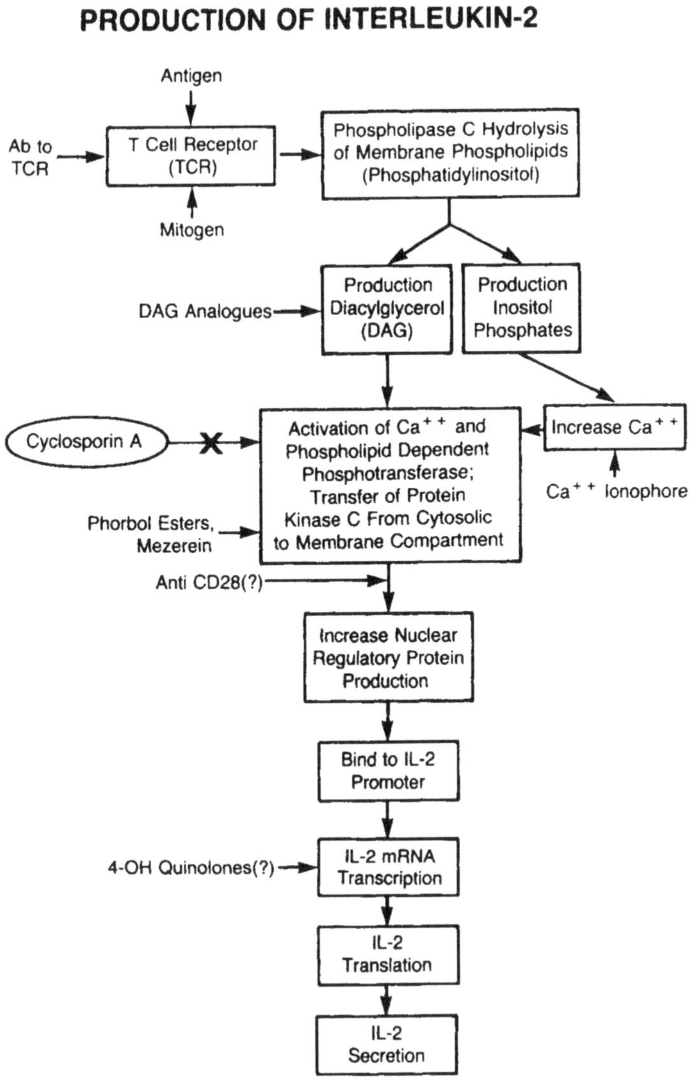

Figure 3. Production of interleukin 2. IL-2 production occurs distal to T cell receptor signaling with the subsequent PI turnover and production of diacylglycerol. Cyclosporin A mediates its effects by binding to cyclophylin, calcineurin, and calmodulin forming a complex which prevents the dephosporylation of NFAT and decreasing in turn its ability to translocate to the nucleus. NFAT binding of an associated protein in the nucleus allows induction of IL-2 message transcription. Hydroxyquinolones (such as ciprofloxacin) have been associated with enhanced stability of the IL-2 message which may be related to their DNA gyrase activity.

et al., Forthcoming; Cornellusen et al., 1994; Tamir and Isakov, 1994; Paliogianni et al., 1993a). Activation following T cell receptor ligation (Figure 3) is accompanied by *de novo* phosphoinositol production, phosphoinositol hydrolysis by phospholipase Cγ1, generation of inositol phosphate and diacylglycerol, increased [Ca^{2+}] activation of protein kinase C, and finally proliferation (Tamir and Isakov, 1994). The increase in Ca^{2+} leads to activation of the T lymphocyte multifunctional Ca^{2+}/calmodulin-dependent protein kinase (CaM kinase) and, with additional stimulation by factors such as phorbol myristic acetate (PMA), to activation of calcineurin, a Ca^{2+}/calmodulin-dependent 2B heterodimeric protein phosphatase. CaM kinase activation predominantly leads to apoptosis or programmed cell death and calcineurin to activation (Nghiem et al., Forthcoming).

Ca^{2+} loaded CaM kinase directly inhibits many of the basic-helix-loop-helix transcription factors, preventing them from binding to DNA (Cornellussen et al., 1994). Several of the immuno-modulators including other cytokines and immunosuppressive agents such as cyclosporin, FK506 and rapamycin mediate their functions by interfering with this pathway (Thomson and Starzl, 1993). cAMP inhibits PI turnover, inhibition of PLC-1 tyrosine phosphorylation and inhibition of the phosphorylation of the gamma and epsilon T-cell receptor chains. PKC induction of transcription of members of the *fos* and *jun* family are also inhibited by CAMP (Paliogianni et al., 1993a). Prostaglandin E2 increases cAMP which in turn activates protein kinase A by binding to the inactive tetrameric holoenzyme and dissociating its two regulatory and catalytic subunits. It appears that direct phosphorylation of calcineurin by cAMP leads to inhibition of IL-2 gene expression. IL-4 can inhibit many of the steps leading to IL-2 or interferon gamma secretion (Schwarz et al., 1993) presumably by inducing cAMP and downregulating the transcription factor complex NFIL-2B which binds to a region in the IL-2 enhancer. Signaling through CD28 ligation (the physiologic ligand is B7-2) can overcome the IL-4 induced suppression by stabilizing IL-2 transcripts or by induction of the nuclear factor CD28RC which binds to a site immediately adjacent to the NFIL-2B site.

NF-AT is also sensitive to cyclosporin A and FK506 which bind the immunophilins cyclophilin A and FKBP12 respectively. Their major function is to bind calcineurin and inhibit its phosphatase activity (Thomson and Starzl, 1993), perhaps leading to blocking of the assembly of functional NF-AT, preventing the cytosolic component from being translocated to the nucleus. If appears that NF-ATc may be a substrate of calcineurin, its dephosphorylation being the proximate event prior to translocation to the nucleus. Glucocorticoids appear to have their major role as immunosuppressives by inhibiting the binding of the nuclear transcription factor AP-1 (Paliogianni et al., 1993b) and, thus, leading to decreased IL-2 production.

Significant levels of IL-2 are only rarely found in the peripheral blood (Licinio et al., n.d.). Increases have been found primarily in inflammatory disorders including cerebral malaria, allograft rejection, primary biliary cirrhosis, rheumatoid arthritis, multiple sclerosis, and in patients with periodontal disease. Recently, episodic fluctuations in normal healthy women of low serum levels of IL-2 suggesting diurnal variability (Licinio et al., n.d.). has been reported. These may represent "spillover" since most of the IL-2 has been detected in ELISAs and is presumably not biologically active at the concentrations measured. Most of the available evidence would suggest that IL-2 is produced locally during an immune response and its short half-life of 2-10 mins in rodents and man suggest that rapid means for its elimination exist, presumably by binding to high affinity receptors on cells but also by elimination through binding to proteins in the kidney such as uromodulin (Hession et al., 1987). Although numerous examples of ways that microorganisms subvert the immune system, primarily focusing on the processing and presentation of peptides as well as the "effector" cytokines, IL-1, TNFalpha, and interferon gamma, little evidence for direct interaction of the IL-2/IL-2R signaling system and such microorganisms exist. Interestingly proteases of both Pseudomonas species and Legionella pneumophila inactivate IL-2 selectively (Mintz et al., 1993). It is possible that IL-2 might have a role in disorders caused by these microbes.

III. THE IL-2/IL-2R SIGNALING SYSTEM

IL-2 interacts with cell surface IL-2 receptors (IL-2R; Figure 4), mediating its biologic effects. IL2R's have been identified on T cells, B cells, macrophages, and NK cells (Lotze, 1991a, 1991b). IL-2Rs have also been observed on oligodendroglial cells and certain tumors (Lin et al., 1993). The response to IL-2 varies depending on the maturational and activation state of the cell. Activation can occur subsequent to a number of different signals including T cell receptor and/or CD28 ligation as well as by several cytokines including IL1, IL-2, IL-4, IL-6, IL-7, TNF,

SCHEMATIC DRAWING SHOWING POSSIBLE MODE OF IL-2 BINDING TO TWO DISTINCT RECEPTORS

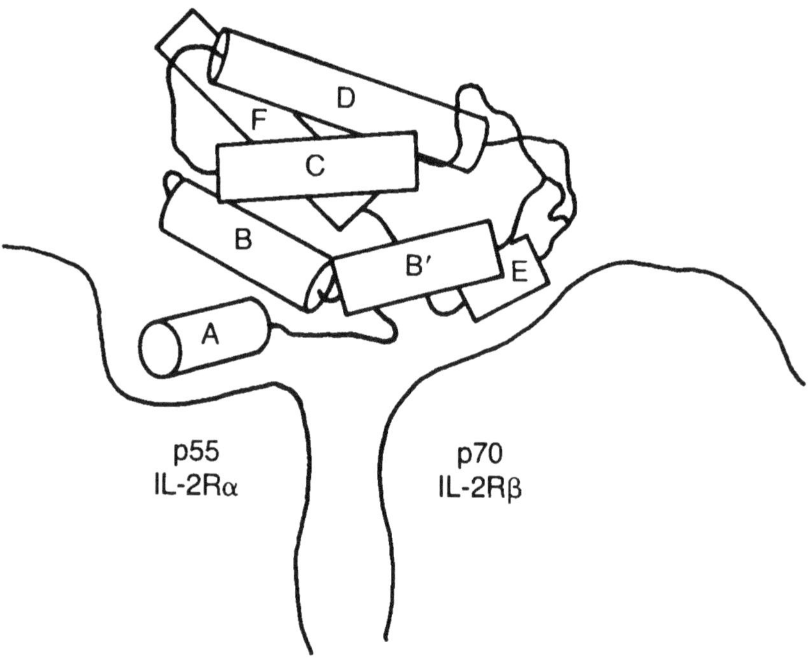

Figure 4. IL-2 interacts with three chains of the IL-2R, two of which are shown. Schematicially, the interaction of IL-2 in its tertiary configuration mediates the activation and induction of IL-2 signaling. Currently the IL-2 receptor is known to consist of various combinations of three chains. The affinity of IL-2 for the various receptors markedly varies with as much as two orders of magnitude between the intermediate (present on unactivated cells) and high affinity receptors (seen on activated cells due to antigenic stimulation or exposure to IL-2 causing up-regulation of the alpha chain). The alpha chain of the IL-2R has a short cytoplasmic tail and is thought to be most important for increasing the affinity of IL-2 for its receptor and a rapid "on-rate." Recently, heterodimerization of the beta and gamma subunit cytoplasmic domains has been shown to be necessary and sufficient for the tyrosine phosphorylation and cellular proliferation induced by IL-2 signaling.

GM-CSF, ADF/thioredoxin, and the recently described Sezary T cell activating factor (SAF; Abrams et al., 1993).

Response of cells to IL-2 is mediated by the expression of various components of what is now known as the IL-2R complex. This is composed of an α chain (p55, Tac), a β chain (p75), and a γ chain (p64). Each provides a specialized function to the complex (Kuziel et al., 1993) and may be expressed singly or in combination (Stahl and Yancopoulos, 1993; Kishimoto et al., 1993). The α chain is expressed after activation (interferon γ stimulation on macrophages, IL-2 stimulation of T cells, and T cell stimulation of B cells). It allows rapid association kinetics ($t_{1/2}$ of 1.7 seconds) even though it binds IL-2 only weakly by itself ($k_d = 1.4 \times 10^{-8}$ M). This is thought to be caused by conformational changes in the β chain (Kuziel et al., 1993). The presence of this chain on T cells or NK cells suggests recent activation even though it expresses an extremely short cytoplasmic tail and is devoid of any signalling capacity. Upwards of 30-40% of peripheral blood lymphocytes express low levels of the p55 chain of the IL-2 receptor. Coexpression of both the p55 and p75 chains is found on only 1-6% of cells, predominantly CD4+ memory cells expressing the CD45RO isoform (Sheldon et al., 1993).

The β chain is expressed constituitively on circulating monocytes and NK cells and explains the relatively high concentrations of IL-2 required for activation of these cells when expressed alone ($k_d = 2.5 \times 10^{-7}$). Its cytoplasmic domain is the largest of the IL-2Rs and signals in part through the *src* family protein tyrosine kinases (Hata Keyman et al., 1991) including $p56^{lck}$. Although it binds more strongly, its association kinetics are slow ($t_{1/2}$ =46 mins).

The third chain of the receptor, the γ_c chain, has been more recently described and is expressed as a necessary component for high affinity binding and internalization of the receptor. It appears to be expressed in association with the β chain on only approximately 12% of the circulating NK cells following IL-2 therapy, (Vos et al., 1992) suggesting that its downregulation may be important for the suppression of immune responsiveness of these cells following treatment. Its importance has recently also been recognized with the finding that it is mutated in the X-linked severe combined immunodeficiency. It serves as a common receptor for the T cell growth factors noted above including IL-4, IL-7, IL-9, and IL-15 (Kondo et al., 1993; Boussiotis et al., 1994). Heterodimerization of the β and γ chains is necessary for IL-2 signaling through the receptor (Miyazaki et al., 1994; Nelson et al., 1994) and appears to prevent anergy gollowing antigen stimulation (Nakamura et al., 1994) with associated Jak3 phosphorylation. Some of the biologic activities of IL-2 may also represent inadvertent crosstalk with IL-15 receptors which have a separate alpha chain but common beta and gamma chains (Grabstein and Gillis, personal communication). Transformed epithelial cells of the gut express IL-15α receptors and many of the gastrointestinal side effects observed with high dose IL-2 administration may be due to crypt cell stimulation.

Interaction of the IL-2/IL-2R complex on activated cells allows G1 progression to the S phase and is modulated by several additional factors (Figure 2). Tyrosine-specific and serine/threonine specific phosphorylation occurs very rapidly after IL-2R activation (Watts et al., 1993). $p56^{lck}$ associates with the beta subunit of the IL-2R and its kinase activated (by phosphorylation at serine residue 59) following IL-2 ligation. This can be demonstrated to occur *in vitro* by adding purified mitogen activated protein kinase (MAPK). IL-2 induces the rapid activation of the two MAPK isoforms, $p42^{erk\text{-}2}$ and $p44^{erk\text{-}1}$, which presumably phosphorylate $p56^{lck}$ and down-regulate its activation following IL-2 signaling. Two other protein tyrosine kinases also appear to be activated after IL-2 stimulation, $p59^{fyn}$ and $p53/56^{lyn}$, although the mechanism of subsequent signaling by these molecules is unclear. Several protooncogenes are expressed after IL2R activation including c-*fos*, c-*myc*, c-*raf*-1, and c-*myb*. Beadling et al. (1993) have recently identified eight separate immediate early genes induced by IL-2, termed CR 1-8. Two of them have been previously identified including CR-4, as a matrix associated DNA-binding protein, SATB1 and CR-7, as *pim*-1 which encodes a 33-kDa cytoplasmic kinase. It is possible that many of these immediate early genes may also be induced following ligation of other cytokine receptors in the common family shared with IL-2.

Other factors modulate IL-2 signaling. Interestingly, crosslinking of DP and DR on T cells but not DQ enhances the mitogenic response to IL-2 (but not IL-4). This appears to be mediated in part by activation of PLC-γ1 and up-regulation of the IL-2R. Class II expression on T cells is induced by CD2 or CD3 signaling and is maximal 3-7 days after ligation of the TCR. IL-2 receptors have been visualized using injection of ^{123}iodine-interleukin 2 into rats rejecting renal allografts (Abbs et al., 1993). This has also been taken advantage of by administering the IL-2-diphtheria fusion toxin DAB_{486} IL-2 to eliminate reactive cells in the setting of transplant rejection and IL-2R positive hematologic malignancies (Bacha et al., 1988; Kelley et al., 1988; Lorberboum-Galski et al., 1988; Tepler et al., 1994; Le Maistre et al., 1993).

IV. CONSEQUENCES OF TARGETED DISRUPTION OF THE IL-2 GENE

The IL-2 gene was disrupted using targeting vectors introduced into embryonic stem cells (Figure 5) in 1991. Animals that can no longer make IL-2 have normal numbers of thymocytes and peripheral T cells (Schorle et al., 1991). Decreased ability to secrete IgM after crosslinking with antibody and increased levels of IgG1 and IgG2a (presumably due to alterations in T cell help) were noted. The response of T cells to many mitogens was suppressed. This suggested that the redundancy of the T cell growth factors was compensated for the loss of IL-2. With longer followup, these animals developed an ulcerative colitis-like disease associated with colon expression of Class II molecules, development of anti-colon antibodies, large numbers of activated T cells and B cells,

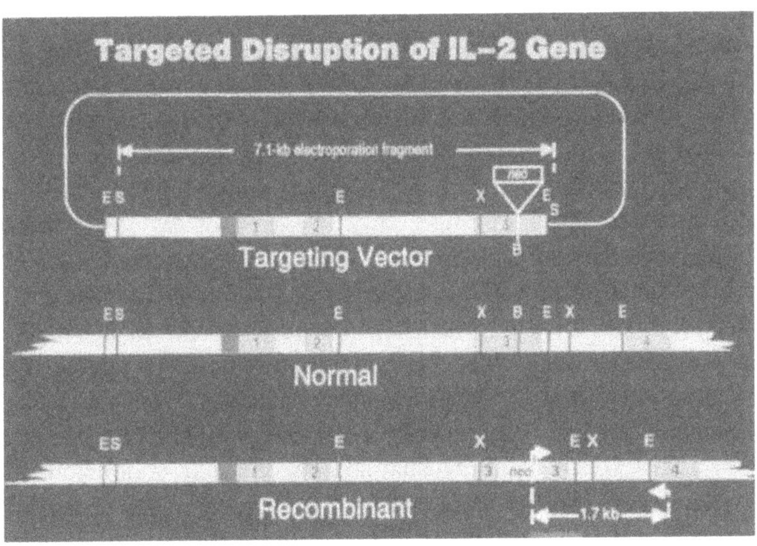

Figure 5. Targetted disruption of the IL-2 gene. One means to determine the role of IL-2 in the immune response is to cause targetted disruption of the IL-2 gene in embryonic stem cells which are then chimerized in early blastocysts. Creation of the targetting vector required introduction of the neomycin resistence gene into the third exon of the genomic IL-2 gene and homologous recombination with the wild type "normal" sequence. Interruption of the third exon causes disruption of the IL-2 protein and rapid degradation.

relative reduction of $\gamma\delta$ TCR+ cells and of $\alpha\alpha$CD8/aβ TCR+ cells in the GI tract, and destruction of the crypt architecture with prominent regeneration of the epithelia (Sadlack et al., 1993). More than half of the animals die by 4-9 weeks of age. Similar lesions have been observed in animals deficient for TCR and a panenteritis develops in animals deficient for IL-10. Dysregulation of IL-2 production may alter the fine interplay of IL-2, IL-7 and IL-15 which occurs normally in the colon.

One criticism of such "knock-out" mice has been that a complex interplay of occurs during development of the immune response. Many of these steps may be critically affected by IL-2 availability. Transgenic mice have been made such that cells constitutively express the HSV 1-thymidine kinase driven by an IL-2 promoter (Minasi et al., 1993). Cells induced to produce IL-2 can be specifically destroyed by administering ganciclovir. When CD4+ cells from transgenic mice are stimulated with the staphylococcal enterotoxin A (SEA), the relevant CD4+Vbeta3 cells are eliminated but only a proportion of the cells activated with the mitogen concanavalin A, suggesting that alternative pathways other than IL-2 are induced.

Animals lacking both IL-2 and IL-4 were also created (Sadlack et al., 1994). These mice have normal T cell and B cell subsets, suggesting that these two cytokines are not critical for the development of the immune response. Proliferation of T cells is enhanced in the IL-2 and IL-2/IL-4 knockout mice, suggesting that IL-2 either causes the production of antiproliferative factors (such as TGFβ) or alternatively that at some phase of the immune response, IL-2 (Lenardo et al., 1991) induces apoptotic cell death of some cells. Evidence for the former comes from studies of macrophages stimulated by IL-2. Stimulation of macrophages with IL-2 causes them to release large quantities of TGFβ, an antiproliferative signal for many cell types and primarily immunosuppressive.

The notion that IL-2 exposure induces memory T cells to undergo apoptosis when they subsequently encounter antigen (Lenardo et al., 1991), may explain many unusual aspects of the IL-2 deficient mice described above. The poorly understood phenomenon of IL-2 treatment of animals preventing graft versus host disease and permitting alloengraftment (Sykes et al., 1990). When IL-2 is given at high doses early, protection from GVH is observed, but after seven days, GVH is increased. Oral tolerance to an antigen inducing uveitis can also be enhanced by systemic administration of IL-2 suggesting that route of antigen administration also is critical for IL-2 effects (Rizzo et al., 1994). Conversely, IL-2 administration (Andreu-Sanchoz et al., 1991) breaks tolerance in neonatally thymectomized mice and induces autoimmune disease. Provision of costimulation is required for the initial encounter of antigen by T cells in order to be activated to produce either IL-4 (TH2) or interferon gamma (TH1). Traditionally this has been thought to be mediated by ligation of CD28, present on T cells. Alternatively IL-2 may provide exogenous help for maturation of the helper response (Seder et al., 1994). A synthesis of the available data would suggest that *effective* antigenic stimulation should precede IL-2 treatment. Whether the tumor is adequately being presented at the time of IL-2 administration is questionable, except in tumors such as melanoma and renal cell carcinoma which are associated with T cell infiltrates. The purposeful immunization of patients prior to IL-2 treatment would be reasonable in future protocols. This would allow new thymic emigrants to encounter antigen prior to receiving an IL-2 stimulus. Mice appear to respond best to IL-2 when bearing immunogenic tumors when the IL-2 is given 7-10 days after tumor inoculation.

V. BIOLOGIC EFFECTS OF IL-2 *IN VITRO* AND *IN VIVO*

A. Effects on T Cells

High levels of IL-2 induce nonspecific LAK-like activity from many T cells especially when provided at high levels (Lotze et al., 1980a, 1980b, 1981; Weil-Hillman, 1991). Following activation, T cells proliferate in the presence of IL-2 and can be cultured with cycles of antigenic stimulation followed by delivery of IL-2 for many weeks in culture. In tumor models the major effector mediating antitumor effects following IL-2 treatment appears to be the T cell. Depletion of T cells abrogates the antitumor effects. Humans treated with IL-2 develop acute cutaneous and systemic anergy (Weibke, 1988), as well as depressed cytolytic activity (Weil-Hillman et al., 1991), that is transient. Responding lesions from patients treated with IL-2 (Cohen et al., 1987; Rubin et al., 1989) demonstrate large numbers of infiltrating activated T cells. Similarly, in experimental infectious diseases (Murray et al., 1993), the effect of IL-2 appears to be primarily mediated by T cells. Many of these effects are presumably due to release of interferon gamma by specific T cells at the site of inflammation. The graft versus host and graft versus leukemia effects (Cohen et al., 1993) in the setting of bone marrow transplantation appear to be also mediated by T cells and not NK cells. Adoptive transfer of T cells expanded in IL-2 can successfully cause tumor regression in murine models (Eberlein et al., 1982) and in humans (Rosenberg et al., 1994).

B. Effects on NK Cells

An early and important observation, that IL-2 activated endogenous NK cells to lyse fresh syngeneic natural killer-resistent murine (Rosenstein et al., 1984) and human tumor cells (Lotze et al., 1980b, 1981) generated much subsequent excitement about the possible application of these cells to the treatment of patients with cancer. The activity induced by IL-2 was termed "lymphokine activated killer" or LAK activity and is largely mediated by NK cells although some T cells can also mediate this activity when exposed to high concentrations of IL-2. Interestingly many of the T cell growth factors including IL-4, IL-7, IL-10, IL-12, and IL-15 (Lotze et al., 1992; Rosenberg and Lotze, 1986) mediate induction of LAK activity as well, suggesting that the early term, based on IL-2 activation, was well founded! IL-2 induces rapid increases in adherence of these cells, up-regulation of the alpha chain of the IL-2R and enhances interferon gamma and TNF release. During IL-2 treatment, the number and activity of NK cells appears to decrease, presumably due to disappearance of these nonrecirculating cells into the periphery. (Lotze et al., 1985b). Following discontinuation of IL-2 marked increase in the number and activity of cells are observed (Caligiura et al., 1991). Unfortunately, in murine models the major apparent effects of the adoptive transfer of IL-2 activated cells mediating LAK activity appear to be in animals bearing early tumors (less than 3 days) and depletion of NK cells in IL-2 treatment models suggests an actual beneficial effect in animals bearing ten day tumors, probably by abrogating some of the NK mediated toxicity. Randomized prospective trials comparing the administration of IL-2 and IL-2 with the adoptive transfer of "LAK" cells have failed to demonstrate major benefit of transfer of such cells (Rosenberg et al., 1993). Since these are "end cells" which appear unlikely to recirculate using NK cells activated with IL-2 to deliver endogenously produced cytokines such as interferon gamma and TNF or genes introduced into them such as IL-12 seems to be reasonable. Efforts to directly infuse such cells directly into the liver (Keilholz et al., 1992), extremity (Belli et al., 1992) or brain (Blancher et al., 1993) has met with limited success. Alternative approaches include means to enhance the activity of such cells by subsetting the cells which are capable of adhering to plastic (Vitolo et al., 1993) following IL-2 activation (so-called ANK cells) or by providing the appropriate reducing environment for optimal activation (Yamaauchi and Bloom, 1993) seem to be reasonable next steps.

C. Effects on Macrophages

Studies of macrophages stimulated by IL-2 (Sadlack et al., 1994) were noted above. Macrophages rapidly remove IL-2 from culture through binding of the intermediate affinity (beta chain) receptor present at approximately 4,000 sites per cell. The alpha chain of the IL-2 receptor is induced on macrophages by treatment with interferon gamma or LPS. Stimulation of macrophages with IL-2 causes them to release large quantities of TGFβ, IL-1β, TNFα, IL-6, and IL-8 (Espinoza-Delgado et al., 1995) and to enhance their tumor cytocidal activity.

D. Effects on B cells

Germinal center (GC) B cells express no *Bcl*-2 and rapidly die in culture unless this molecule is induced by (1) cross-linking of sIg, (2) stimulation with soluble CD23 and IL-1alpha, or (3) cross-linking of surface CD40 by antibody or the CD40 ligand (Holder et al., 1993). IL-2 does not promote rescue from apoptosis of this population but does cause rapid proliferation of a subset of appropriately activated GC B cells. IL-2 serves as an essential and late acting factor for B cells making IgM but less so for B cells producing other antibodies (which generally require some additional source of T cell help). When GC B cells are cultured for prolonged periods in IL-2 they express *Bcl*-2 but apoptose when IL-2 is withdrawn. Little evidence, however, exists to support a major role for B cells in the immune response to tumors.

E. Effects on Other Cells

Multiple other cells types have been reported to express the beta chain of the IL-2 receptor. These include oligodendroglial cells, possibly crypt cells in the gut and some tumor cells. Much of the effect of IL-2 on other cell types is mediated by the production of other cytokines subsequent to IL-2/IL-2R interaction. For example, the local effect of IL-2 production by transfected tumor cells is associated with neutrophilic infiltrates (see following). IL-2 treatment also induces alteration in circulating neutrophil Fc receptor expression as well as reductions in chemotactic activity (Klempner et al., 1990; Bock et al., 1990) IL-2 treatment protocols are associated with marked increases in line related infections which can be abrogated in large part with prophylactic administration of antibiotics (Hartman and Urba, 1989; Bock et al., 1990). In addition IL-2 administration causes an increase in circulating eosinophils (Bock et al., 1990) which are of an activated phenotype and apparently are also capable of killing NK susceptible targets including K562, SCLC, and many melanoma cell lines. This is apparently in part due to the induction of IL-5 production.

F. Anti-tumor Effects of Bolus IL-2

The first description of the antitumor effects of IL-2 were those of Gorelik and Rosenberg (unpublished, 1982) in which animals were pretreated with small doses of IL-2 (up to 270 U/d given IV or IP prior to tumor challenge). IL-2 could decrease the number of tumors in the lungs of animals. This effect seemed to be mediated by NK cells since antibodies to asialo GM1 abrogated the antitumor effects. Subsequently, IL-2 was shown to have anti-tumor effects in patients with melanoma (Rosenberg et al., 1985; 1987, 1989; Lotze et al., 1986a, 1986b) and reported in 1985. The FDA approved recombinant IL-2 for treatment of patients with renal cell carcinoma in 1992 (Richards and Lotze, 1992). A recent update of our 283 consecutive patients with metastatic melanoma or renal cell cancer treated with high-dose bolus IL-2 (Rosenberg et al., 1994) revealed a 7% complete response (CR) rate and a 10-13% partial response (PR). Of the 19 patients with a CR, 15 remained in response 7-91 months after treatment. The mean duration of PR was 12 months for patients with melanoma and 18 months for those with renal cell cancer (contrast with a 1% CR, 9% PR rate in 149 patients treated with RCC and high dose alpha interferon; Minasian et al., 1993). Mortality of 1.1% (3 patients) was observed early in the experience with this agent. Other tumors for which IL-2 holds promise include acute myelogenous leukemia (Mirannchi et al., 1991; Foa, 1993), lung cancer (Clamon et al., 1993; Fisher et al., 1990), childhood neuroblastoma (Rosenberg et al., 1989a, 1989b; Sigal et al., 1991; Fowler et al., 1991; Neagrier et al., 1991), and thymomas (Sander et al., 1993). Further development of IL-2 treatment for patients with these neoplasms will require means to enhance activity and decrease toxicity. For example murine models of neuroblastoma predict that IL-2 therapy would have a role but the therapy in children was extremely toxic (Negrier et al.,

1991) and without clinical benefit in a limited number of patients. Discerning the relevant factor predicting response (see below) would appear to be critical to allowing more successful therapy.

G. Combination Therapy with Interferon

Alpha interferon can be successfully applied in patients with melanoma or renal cancer. Increase in cell surface expression of Class I molecules has been demonstrated with human tumors as well as in murine models with alpha interferon treatment and coadministration with IL-2 would seem to be very reasonable. Our sequential evaluation of combinations of IL-2 and alpha interferon suggested synergistic toxicity and enhanced efficacy when applied with the highest (4.5×10^6 U/m^2 IL-2 and 6×10^6 U/m^2 interferon a thrice daily) doses possible. This particular regimen was associated in 29 patients with 8PR and 4CR (41%). In this instance response was associated with the highest dose of IL-2 able to be applied (Rosenberg et al., 1989a, 1989b). Subsequent studies (Atkins et al., 1993; Sparano et al., 1993; Keilholz et al., 1993; Jablons et al., 1990; Vogelzang et al., 1993; Culver et al., 1992; Krager et al., 1990; Mittelman et al., 1991; Huberman et al., 1991) failed to show marked benefit even when compared in randomized trials, I suspect related to the lower doses of IL-2 employed. An update of our earlier trial failed to show longterm benefit (Marincola et al., 1995). The only notable study reproducing the 40% response rate was that of Keilholz et al. (1993) who employed very high doses of IL-2 by continuous infusion which was rapidly deescalated in a decrescendo protocol to limit the amount of TNF generated by the IL-2 treatment (see following). Certainly this response rate appears to be the upper limit of that attainable in this cohort of patients. Most responders had pulmonary and nodal disease.

H. Other Means to Enhance Efficacy of IL-2

Combination therapy with chemotherapy has attracted much attention in the last few years. There appears to be little to recommend it in patients with renal cancer (Kuebler et al., 1993). In melanoma, however, combination of cisplatin with IL-2 and possibly alpha interferon (Khayat et al., 1993; Dorval et al., 1994; Kasid et al., 1990; Rixo et al., 1994) are associated with improved response rates up to 51%. The only factor associated with improvement in response was number of sites (< 2) of disease ($p = 0.007$) and the presence of vitiligo ($p = 0.007$). The addition of tamoxifen failed to enhance efficacy in these European trials. Randomized studies designed to compare IL-2 with or without cisplatin are in order. Similarly, combination with radiation therapy has been associated with improvement in murine models but not as of yet in human clinical trials (Cameron et al., 1990; Hunter et al., 1992; Lange et al., 1992). Combination with anti-CD3 treatment in patients with RCC or MM has yet to improve efficacy (Sosman et al., 1993). Increased levels of soluble IL-2R (Tac) were noted in patients treated at the highest doses of the two agents. Although surgical resection of renal primaries has been advocated prior to initiating IL-2 therapy, no randomized clinical trial has yet been performed to document enhanced responsiveness in those subjected to nephrectomy prior to IL-2 treatment. In April, 1991 patients with RCC, spontaneous regression of metastases was noted (Marcus et al., 1993), suggesting the possibility that autocrine growth factors produced by the tumor may be required in rare instances with the maintenance of tumors at metastatic sites. In a review of records from 399 patients with RCC from 14 different institutions, the 11 patients made surgically NED had no evidence of recurrence and comparable survival to those that obtained a CR with IL-2 treatment alone (Lotze, 1991). Each of them had substantially better survival than the patients who had partial response and in whom surgical resection was either not possible or not feasible. For this cohort of patients, as is true for patients with limited RCC without IL-2 treatment, resection seems to be associated with prolonged survival.

I. Other Means to Deliver IL-2

IL-2 has been given a number of different ways. Many of them are designed to decrease the toxicity of IL-2. An early development in IL-2 delivery was administration by slow continuous intravenous infusion (Dillman et al., 1991; Lotze, 1991a; Gratama et al., 1993). This allows delivery of only about 10-15% of the IL-2 able to be delivered by intermittent systemic bolus every eight hours. Although evidence for activation of the immune system is present (Gratama et al., 1993)—with a dose-dependent increase in lymphocyte number, percentage of CD56+/CD3- cells, percentage of CD3+ cells which are DR+, number of IL-2Ra+ cells, number of T cells which are of the "memory" (CD45RO+) phenotype, and ability to lyse the NK resistent Burkitt cell line, Daudi—in general less of a response is observed than with intermittent high dose bolus administration of IL-2. IL-2 has also been given subcutaneously (Caruso et al., 1992) and intravenously as a polyethylene glycol formulation to prolong its half-life from about eight minutes to eight hours. Recently, PEG-IL-2 was administered twice weekly intradermally to patients with HIV1 infection (Teppler et al., 1993). Even though no substantive toxicity was observed, only minimal erythema and induration at the site of injection, evidence for increased CD4+ cell numbers and enhanced proliferative responses were noted. Followup trials are now being conducted at the NIAID by Lane and colleagues.

In animal models, and more recently in clinical trials, IL-2 (Anderson et al., 1990) has been given as part of a liposome formulation consisting of multilamellar vesicles with a 7:3 ration of DMPC (dimysristoylphosphatidylcholine) and DMPG (dimyristoylphosphatidylglycerol). With a lipid:protein ration of 300:1, 10-50% of the IL-2 could be entrapped. When administered to mice enhanced antitumor effects and survival were observed compared to IL-2 or liposomes alone. Similarly slow release from polymer implants or from ethiodol, a commercially available derivative of poppyseed oil are under study. Ethiodol has been used for years as an imaging agent for lymphangiograms and is now being tested to deliver chemotherapeutic agents or IL-2 intraarterially to patients with hepatic metastases or large renal primaries (Rubin and Lotze, unpublished data). Systemic delivery of IL-2 in ethiodol will be tested at the Pittsburgh Cancer Institute.

An alternative approach to altering the toxicity of IL-2 is to create analogues of IL-2 that cause less secretion of cytokines (interleukin-1β, tumor necrosis factorα, and interferonγ) but substantial enhancement of cytotoxic activity (Heaton et al., 1993, 1994). These particular analogues (R38A and F42K) differ by only a single amino acid from the parental rhIL-2 but are associated with 25-50% of the levels of IL-2 induced cytokine secretion, presumably due to interaction with only the intermediate (and not the high affinity) receptor. No reports are available yet to discern whether such an approach is associated with enhanced antitumor activity.

VI. USING IL-2 REGIONAL THERAPY TO ACTIVATE NEW THYMIC EMIGRANTS OR ENDOGENOUS T CELLS

As noted above IL-2 may have paradoxical effects by priming unactivated T cells for apoptotic death when they subsequently have their T cell receptor crosslinked. This may represent a physiologic mechanism to prevent inadvertent expansion of autoreactive cells at the site of an immune response. For that reason delivery of IL-2 may best be accomplished regionally (Bubenik et al., 1987) directly at the site of tumor where the predominant activated cell type would be presumed to be tumor reactive. Emigration of new thymic emigrants at that inflammatory site, using the new paradigm, would necessarily see IL-2 first and undergo deletion. New thymic emigrants finding their way to the lymph node would potentially encounter antigen delivered there by activated phagocytic cells, become activated and then return to the appropriate site via the blood stream. Many of the recent applications of IL-2 are empiric and were based on intuition and detailed animal models. New applications will need to take into consideration these more recent insights.

A. Direct Injection of IL-2

Now classic studies by Bubenik (Bubenik et al., 1983; Bubenik, 1990) and subsequently Forni and colleagues (1985) suggested that local immunity to tumors could be induced by direct injection of a cytokine such as IL-2. This was actually taken into clinical trials with administration of IL-2 directly into head and neck tumors and into the bladder for the treatment of tumors arising there. These effects were IL-2 concentration-dependent and required multiple injections of peritumoral IL-2. Long lived immunity mediated by tumor specific T cells has been noted. Production of interferon appears to be critical for the antitumor effects observed (Forni et al., 1985). Subsequently, prolonged delivery of IL-2 using techniques of gene transfer have been the predominant mode of testing IL-2 locally. IL-2 administered intraperitoneally is associated with responses in patients at local and distant sites (Lotze et al., 1986b) but complicated by marked increases in ascites. Intrapleural administration of IL-2 in patients with malignant pleural effusions has been associated with 10/22 responses in one study (Astoul et al., 1993) and 5/23 patients in another study (Viallat et al., 1993). Administration is complicated by empyema in some patients. Administration by inhalation (Huland et al., 1994) in combination with systemic low dose IL-2 and interferon a was associated with clinical responses in nine of 15 patients with pulmonary metastatic renal cell cancer! Inhalation was carried out five times daily for 28 days with five inhalations of 10^5 -3 x 10^5 IU of natural IL-2/d. Prolonged therapy was possible allowing stabilization of disease as well in most of the other patients.

B. Ethiodol IL-2

We have been very enthusiastic about the application of depot formulations of IL-2 in ethiodol. Murine models suggest not only decreased growth of tumor but also induction of a long lived protective immune response in animals injected with tumor admixed with IL-2/ethiodol. Clinical trials began approximately one year ago (Bubenik et al., 1988).

C. IL-2 Gene Therapy

The first cytokine gene therapy was developed in Czechoslovakia by Bubenik and colleagues and used IL-2 delivered by a fibroblast line to treat HeLa cells growing in nude mice. (Bubenik, 1993; Bubenik et al., 1988, 1990, 1993). Lytic activity for tumor was observed in the spleen of treated animals. Similarly injection of a murine plasmacytoma cell line into normal BALB/c mice led to tumor development in 7/7 mice with inocula of as few as 10^3 cells but 0/7 mice when they were injected with as many as 10^7 cells transfected with the gene for murine IL-2. This result has now been confirmed by many groups (Fearon et al., 1990; Russell et al., 1990, 1992; Gansbacher et al., 1990; Bubenik et al., 1991, 1992; Cavallo et al., 1992; Kim et al., 1992). and shown to require T cells, primarily of the CD8+ phenotype. Interestingly, TS/A, a spontaneous "nonimmunogenic" murine mammary adenocarcinoma, induces an immune response when transfected with IL-2 that is associated with a prominent neutrophilic infiltrate that is dependent on an initial response of local CD4+ cells (Carvallo et al., 1992). Clinical trials utilizing allogeneic melanoma or renal cell cancers expressing the common restriction Class I molecule, HLA-A2 and retrovirally transfected with the gene for IL-2 have begun by Gansbacher and colleagues at the Memorial-Sloan Kettering Institute. Local responses characterized by extensive induration and erythema have so far been reported. When directly compared in experimental gene therapy studies, (Zaloom and Gallagher, 1993; Patel et al., 1993), IL-4 appears to induce and sustain a superior immune response when compared to IL-2. Therapy of established tumors with vaccines consisting of IL-2 transfected tumor cells requires multiple immunizations administered twice weekly and only cures approximately 60% of mice injected starting seven days after tumor transplantation on the same side as the tumor inoculum and, when started 14 days later, only 20% of the animals. Similar results have recently been reported in the treatment of mice with a mammary adenocarcinoma (Cavallo et al., 1993) and in mice bearing spontaneous metastases arising from the 3LL lung carcinoma (Porgador et al., Forthcoming). Use of larger viruses such as vaccinia (Hugin et al., 1993) or alternatively herpes simplex (Latchman, 1991) would be of interest.

D. Toxicities of IL-2

Many of the toxicities of IL-2 could be effectively abrogated by concomitant administration of various medications but treatment is usually stopped short of the planned therapy because of resistent side effects. Means to diminish or alter the nature of the toxic effects experienced by patients may increase the dose intensity of therapy capable of being administered and pari passu increase responses.

E. Vascular Leak Syndrome

The major dose-limiting toxicity of IL-2 is due to extravazation of fluid and protein into the interstitium (Rosenstein et al., 1986; Cotran et al., 1987; Ettinghausen et al., 1988). The vascular leak syndrome is dose related and in murine models is markedly abrogated by steroids and lymphocyte depletion which usually also decreases the antitumor efficacy (Papa et al., 1986). Elimination of NK cells using antibody to NK 1.1 markedly decreases toxicity without altering efficacy. Induction of the leak appears to be mediated by activation of the endothelium, presumably subsequent to release of IL-2 induced cytokines such as IL-1, TNF a, and interferonë. Although cultured endothelia appear to be killed by LAK effectors, activation of these cells by IL-1, TNF a, or interferon is associated with a decrease in susceptibility to lysis by LAK effectors and subsequent albumin diffusion across cultured monolayers (Rosenstein and Lotze, 1986; Cotran et al., 1987). The endothelial alterations induced by systemic administration of IL-2 have been visualized in a number of species including the rabbit (Mier et al., 1989; Goldblum et al., 1990) and include intracellular and subcellular blebs with separation of endothelial cells from the basement membrane with associated interstitial edema and endothelial adhesion and transmigration of lymphocytes into the tissue. Clinical management involves limiting fluid intake and liberal use of vasopressors. It does not seem to matter whether fluid resuscitation is done with colloid or crystalloid (Pockaj et al., 1994).

F. Endocrine Toxicity

Hypothyroidism is observed in up to 50% of treated patients and appears to be associated with response to therapy although this is still controversial (Atkins et al., 1988; Weijl et al., 1993; Kruit et al., 1993; Vassilopoulou et al., n.d.). Antimicrosomal antibodies and antithyroglobulin antibodies are observed in many patients and appears to correlate with duration of treatment and amount of IL-2 delivered (which also correlated with response ($p = 0.01$). It is difficult to ascertain whether the response to treatment is due to the common variable of IL-2 dose and duration of treatment. Other forms of endocrine disorders such as adrenal insufficiency (Vandremolan et al., 1989) only occur rarely in patients receiving IL-2.

G. Neuropsychiatric Effects

Perhaps one of the most distressing side effects of IL-2 therapy is the development of cognitive dysfunction (confusion) during IL-2 therapy (Denicoff et al., 1987, 1989; Caraceni et al., 1992). The mechanism is not clearly understood but I believe that it is due to the leaky capillary syndrome as manifested in the brain (Saris et al., 1989). Many patients are unable to receive additional IL-2 because of this toxicity and means to ameliorate it would seem to be a high priority.

H. Gastrointestinal Effects

The other major toxicity associated with IL-2 administration is nausea and vomiting as well as a secretory diarrhea. This can be particularly troublesome to manage. Droperidol is a particularly effective antiemetic and does not appear to be improved on by antagonists of the 5-hydroxytryptamine receptor (ondansetron; Kim et al., Forthcoming). The diarrhea has not been carefully evaluated but may be due to disturbances in the recently described IL-15/IL-15R interactions which share the IL-2R β and chains (Boussiotis et al., 1994).

I. Inhibitors of Toxicity

Hemodynamic toxicity (Siegel and Puri, 1991; Lee et al., 1989; Ognibene et al., 1988) of IL-2 is in part mediated by induction of TNFα

(Kasid et al., 1990) and subsequent production of nitric oxide (NO). Efforts have been made to decrease these mediators with a variety of different agents.

Nitric oxide (NO) is made constitituitively in blood vessels and in the brain and regulates most physiologic processes including vasodilation. In response to proinflammatory cytokines (Hibbs et al., 1992; Kilbourn and Belloni, 1990) such as TNF, IL-1, interferonγ, and IL-6 as well as LPS, another form of the enzyme nitric oxide synthase is induced which leads to enhanced NO production. Patients receiving IL-2 make increased amounts of NO (Ochoa et al., 1992) as reflected in enhanced byproducts found in the serum including nitrites and nitrates. N-monomethyl arginine (NMA), a competitive inhibitor of NO production was shown in dogs (Kilbourn et al., 1992) and humans, to reverse the production of NO induced by IL-2 and ameliorate some of the hypotension and reflex tachycardia observed with treatment and presumably mediated by NO. Another inhibitor of nitric oxide formation is aminoguanidine (Corbett et al., 1992; Soulis-Liparota et al., 1991; Leder et al., n.d.). We have tested aminoguanidine in murine treatment models and have demonstrated that it inhibits NO production observed with IL-2 treatment but fails to alter the IL-2 induced capillary leak syndrome. It has no effect or modestly decreases the antitumor effect of IL-2 in murine models (Leder et al., forthcoming, n.d.).

PGE_2 suppresses IL-2 production and IL-2R expression, possibly through a cAMP mediated mechanism. Suppression of PGE_2 with administration of the cyclooxygenase inhibitor indomethacin by itself or in combination with histamine receptor antagonists such as ranitidine (Mertens et al., 1993) have been suggested to mediate antitumor effects. These drugs are used extensively to ameliorate the toxicity observed with IL-2 treatment. Endogenous production of prostaglandins may decrease immune activation associated with IL-2 treatment.

The level of TNF increases progressively with IL-2 treatment in association with the cytokine IL-6 (Jablons et al., 1990). TNFα administration to humans leads to hypotension, hepatic dysfunction and alteration in the function of neutrophils. Ways to decrease TNF in animal models of IL-2 treatment using inhibitors of TNF including antibody, pentoxifylline and its derivatives, thalidomide, chimeric soluble TNFR, hydroxyzine, and steroids have had mixed results in terms of antitumor efficacy of IL-2 although clearly ameliorating some of the toxicities of IL-2 administration (Thompson et al., 1993; Edwards et al., 1991, 1992; Doherty et al., 1992; Han et al., 1990; Mier et al., 1990; Fraker et al.,

1989; Shiloni and Matzner, n.d.). Evidence that pentoxifylline can decrease the toxicity observed with IL-2 treatment without impacting on efficacy was reported by Edwards et al. in rodent models (1991, 1992) and by Thompson et al. in patients (1993). We conducted similar trials in patients undergoing IL-2 treatment and found that renal toxicity is somewhat lessened but that the development of nausea and vomiting associated with oral administration of high doses of PTX preclude its successful application in the clinic. We tested a PTX analogue termed CT1501R which is believed to be the active, and less toxic, metabolite of PTX in a double blind randomized study. No enhancement of efficacy or decreased toxicity was observed. Murine studies of the coadminstration of IL-2 and this agent that we performed have shown no apparent decrement in the antitumor effect in pulmonary metastases models.

J. Resistence to IL-2 Therapy

Most patients treated with IL-2 fail to respond. Production of counterregulatory factors in the host designed to reduce the activation of host effectors and "reduce stress" may also limit the efficacy of IL-2 treatment including the induction of endogenous glucocorticoids (Papa et al., 1986; Vetto et al., 1987; Morse et al., 1990; Shiloni and Matzner, n.d.). In murine models and subsequently in a limited number of patients administration of steroids abrogated the antitumor effects of IL-2. Subsequently, it was shown that steroids significantly decreased the vascular leak and reduced the lymph flow enhanced by IL-2 treatment ($p < 0.004$; Morse et al., 1990). Concomitant treatment with steroids (Mier et al., 1990) decreased TNF levels, the hepatic acute-phase response, and the neutrophil chemotactic defect observed with IL-2 administration. Any means designed to reduce toxicity might be problematic since it might be directly tied to response (Vetto et al., 1987). Whether endogenous production (Lotze et al., 1985a) of steroids in response to the stress of IL-2 treatment diminishes the anti-tumor response is unclear. One could administer agents such as RU486, a cortisone receptor antagonist, to block some of the effects of heightened steroid production. We initially tested such a strategy in murine models (Lotze and Yang, unpublished) with mixed results. Further testing is probably warranted.

Other antagonists can be found in the serum of IL-2 treated patients. These include IL1Ra, an inhibitor of IL-1 induced by IL-2 and IL-4 treatment, which is detectable in the serum within four hours following IL-2 treatment. It inhibits the production of additional IL-1 and TNF by activated PBMC (Yang et al., 1994). Soluble TNFRp55 is also found in the serum but its physiologic relevance is less clear. Circulating levels of neutralizing antibodies have also been observed in the serum of some patients treated with IL-2 (Tilg et al., 1994; Kirchner et al., 1991; Scharenberg et al., 1994) although this is only true in about 6% of patients when IL-2 is given intravenously and 10% of instances when given subcutaneously.

K. Mechanism of Anti-tumor Effects and Toxicity

It is possible that any attempt to decrease toxicity may lead to decreased efficacy. This is certainly true for the use of glucocorticoids and may be true for the TNF inhibitors as well. The information from the murine models suggest that: (1) an anti-tumor effect requires intact T cell immunity for established (d10 tumors) including primarily CD8+ cells, (2) more IL-2 is generally associated with enhanced anti-tumor effects, (3) toxicity is correlated with TNF production and presence of NK cells, and (4) the tumor's "immunogenicity," including the expression of relevant Class I molecules, is critically important in terms of the nature and quality of the immune response. Little additional information is available from the current set of clinical trials (Table 3). In general more treatment is associated with enhanced response (with some caveats, see below) and patients with a good performance status (ECOG 1 or 2) tend to do better than those with more limited disease. This has been examined in a number of studies (Blay et al., 1990; Deehan et al., 1994; Schwartzentruber, 1993; Mier and Atkins, 1993; Oppenheim and Lotze, Forthcoming) but unfortunately only with limited insights. In a prospective randomized trial of IL-2 compared to IL-2 and the adoptive transfer

Table 3. Pharmacokinetic Parameters After Bolus Injection or 15-Minute Infusion (Means ± Standard Deviations)

Route	*Dose IU/kg*	*AUC[1] Iuml*min*	*CL ml/kg/min*
Bolus	360,000	157,360 ±2,021	2.71 ±0.51
Bolus	540,000	286,007 ±43,871	2.22 ±0.38
Influsion	720,000	222,264 ±127,332	4.31 ±2.02

Note: [1] AUC was calculated using the trapezoidal rule and was extrapolated to infinity.

Table 4. Factors Associated with Increased Response to Treatment in Various Series

Author	*Patient Groups*	*Patient Number*	*Factor Associated with Improved Response*
Blay et al. (1990)	Renal Cell Cancer	25	Sustained production of TNF 48 h. after completing IL-2 treatment
Rosenberg et al. (1993)	Renal Cell Cancer, Melanoma	164	Pretreatment platelet count; minimal platelet count after IL-2; males; lack of prior chemotherapy; maximum WBC after therapy
Palmer et al. (1992)	Renal Cell Cancer	327	Performance Score; time from diagnosis to treatment; fewer (<2) metastatic sites
Blay et al. (1992)	Renal Cell Cancer	138	Decreased IL-6 and C reactive protein levels pretreatment; increased IL-6 posttreatment
Broom et al. (1992)	Colorectal Cancer	15	Decreased pretreatment C reactive protein
Marincola et al. (1992)	Melanoma	146	HLA-A11 phenotype
Rubin et al. (Forthcoming)	Melanoma	84	HLA-DQI phenotype
Deehan et al. (1994)	Colorectal Cancer	20	Lower pretreatment IL-6 levels and sIL-2R: increased IL-6 and no change in PGE_2 with treatment
West et al. (1987)	Various Malignancies, LAK + IL-2	40	Higher pretreatment and posttreatment lymphocyte counts
Rubin et al. (1989)	Melanoma, Renal Cell Cancer, Breast Carcinoma	48	Increased DR on tumor and infiltrating T cells in responding lesions
Aebersold et al. (1991)	Melanoma TIL Treatment	55	Increased cytolytic activity for autologous rumor of adoptively transferred TIL
Rosenberg et al. (Forthcoming)	Melanoma TIL Treatment	76	TIL decreased time in culture and doubling Tretime; increased lysis of autologous tumor; subcutaneous site of harvest of tumor
Pockoj et al.(1994)	Melanoma TIL Treatment	38	Localization of TIL at tumor sit using^{III} In oxine labelled cells
Kradin (1989)	Melanoma, Renal Cell Cancer, NSCLC	28	Pretreatment lymphocyte count

of IL-2 activated lymphocytes a number of factors appeared to possibly correlate with enhanced survival (Table 3). These were not corrected for the large number of comparisons but did include pretreatment and minimum platelet counts with therapy, the maximum WBC response, pretreatment creatinine in patients with renal cell cancer, male gender, and the lack of prior chemotherapy. A number of other variables including lower levels of pretreatment IL-6 and C-reactive protein, expression of HLA-11 in patients with melanoma, higher pretreatment lymphocyte counts, sustained TNF production, reactivity and rapidity of growth of TILs for patients receiving IL-2 treatment along with TIL's and site of harvest (skin deposits better than lymph node). These are summmarized in Table 4.

L. Mechanism of IL-2 Resistance

One can envision a number of factors that are associated with resistence to IL-2 therapy both in responsive malignancies and those not yet shown to be associated with significant response.

M. Host Factors

The critical role of T cells in murine and human studies would suggest that the major host factors are related to these cells. This would include tolerance or anergy to nominal tumor antigens, broad immunosuppression related to tumor or other factors (such as that occurring in AIDS), failure to produce other T cell and macrophage-derived factors requisite for T cell maturation, and so forth. Other trivial factors would include the inability to traffic across endothelial barriers into tumors. This might be due to increased intratumoral pressure or altered endothelium within the tumor, recently shown by investigators at the Basel Institute (Piali et al., 1995).

N. Tumor Factors

Tumors which have successfully navigated the sequential steps required for altered growth or apoptotic death, or through a process of tumor darwinism (selection of neoplastic clones able to grow in the face of adequate host immunity) can be envisioned. Good evidence for failure to express Class I molecules due to lack of adequate beta 2 microglobulin, individual allelic losses or failure to generate peptide or transport it into the endoplasmic reticulum as is required for normal Class I assembly has been identified in many tumors of different histologic subtypes. Failure to be sensitive to either cytotoxic factors such as perforin or those inducing apoptotic cell death are now being identified. Is the resistence to IL-2 therapy due to resistance to cytotoxicity? There are many factors which are important in obtaining tumor lysis. Several of these have not been dealt with here. For example, it appears that many tumors can fail to express relevant antigen as peptides in Class I or Class II molecules. This presumably occurs by alterations in one of the many steps necessary to process and present antigen (requiring the peptide transporters TAP1 and TAP2, individual class I heavy chain alleles, beta 2 microglobulin, etc) These tumors are likely to be perceptibly different from immune effectors and may have enhanced sensitivity to NK or LAK effectors. Alternatively, strategies to enhance acquiescence to programmed cell death or sensitivity to proteolysis following insertion of the perforin membrane attack complex, including use of gamma interferon or gamma interferon inducers such as IL-2 or IL-12, would seem to be in order.

REFERENCES

Abbs, I.C., Pratt, J.R., Dallman. M.J., & Sacks, S.H. (1993). Analysis of activated T cell infiltrates in rat renal allografts by gamma camera imaging after injection of 123 iodine-Interleukin-2. Transplant Immunology 1, 45-51.

Abrams, J.T., Ghosh, S.K, Z& Defreitas, E. (1993). Sezary T-cell-activating factor induces functional Interleukin-2 receptors on T-cells derived from patients with Sezary Syndrome. Cancer Research 53, 5501-5506.

Aebersold, P., Hyatt, C., Johnson, S., Hines, K., Korchak, L., Sandrs, M., Loetze, M., Topallian, S., Yang, J., & Rosenberg, S.A. (1991). Lysis of autologous melanoma cells by tumor infiltrating lyphocytes: Association with clinical response. JNCI 83, 932-937.

Aebersold, P., Hyatt, S., Johnson, S., Hinesk, Korcak, L., Sanders, M., Lotze, M., Topalian, S., Yang, J., & Rosenberg, S.A. (1991). Lysis of autologous melanoma cells by tumor infiltrating lymphocytes: Association with clinical response. Journal of the National Cancer Institute 83, 932-937.

Anderson, P.M., Katsanis, E., Leonard, A.S., Schow, D., Loeffler, C.M., Goldstein, M.B., & Ochoa, A.C. (1990). Increased local antitumor effects of interleukin 2 liposomes in mice with MCA-106 sarcoma pulmonary metastases. Cancer Research 50, 1853-1856.

Andreu-Sanchez, J.L., de Alboran, I., Marcos, M.A.R., Sanchez-Movilla, A., Martinez, A.C., & Kroemer, G. (1991). Interleukin 2 abrogates the nonresponsive state of T cells expressing a forbidden T cell receptor repertoire and induces autoimmune disease in neonatally thymectomized mice. J. Exp. Med. 173, 1323-1329.

Ardizzoni, A., Bonavia, M., Viale, M., Baldini, E., Mereu, C., Verna, A., Ferrini, S., Cinquegrana, A., Molinari, S., Mariani, G.L., Roest, G.J., Sharenberg, J., Palmer, P.A., Rosso, R., Ropolo, F., & Raso, C. (1994). Biologic and clinical effects of continuous infusion interleukin-2 in patients with non-small cell lung cancer. Cancer 73, 1353-1360.

Astoul, P., Viallat, J.R., Laurent, J.C., Brandely, M., & Boutin, C. (1993). Intrapleural recombinant IL-2 in passive immunotherapy for malignant pleural effusion. Chest 103, 209-213.

Atkins, M.B., Mier, J.W., Parkinson, D.R., et al. (1988). Hypothyroidism after treatment with Interleukin-2 and lymphokine-activated killer cells. NEJM 318, 1557-1563.

Atkins, M.B., Sparano, J., Fisher, R.I., Weiss, G.R., Margolin, K.A., Fink, K.I., Rubinstein, L., Louie, A., Mier, J.W., Gucalp, R., Sosman, J.A., Boldt. D.H., Doroshow, J.H., Aronson, F.R., & Sznol, M. (1993). Randomized Phase II trial of high dose interleukin-2 either alone or in combination with interferon alfa-2b in advanced renal cell carcinoma. J. Clin. Onc. 11, 661-670.

Aversa, G., Punnonen, J., Cocks, B.G., de Waal Malefyt, R., Vega, F., Zurawski,S.M., Zurawski, G., & de Vries, J.E. (1993). An Interleukin 4 (IL-4) mutant protein inhibits both IL-4 or IL-13 induced human immunoglobulin G4 (IgG4) and IgE Synthesis and B cell proliferation: Support for a common component shared by IL-4 and IL-13 receptors. J. Exp. Med. 178, 2213-2218.

Bacha, P., Williams, D.P., Waters, C., Williams, J.M. , Murphy, J.R., & Strom, T.B. (1988). Interleukin 2 receptor targeted cytotoxicity. Interleukin 2 receptor-mediated action of a diphtheia toxin-related interleukin 2 protein. J. Exp. Med. 167, 612-622.

Beadling, C., Johnson, K.W., & Smith, K.A. (1993). Isolation of Interleukin -2 induced immediate early genes. Proc. Natl. Acad. Sci. USA 90, 2719-2723.

Belli, F., Arienti, F., Rivotini, L., Santinami, M., Mascheroni, L., Prada, A., Ammatuna, M., Marchesi, E., Parmiani, G., Cascinelli, N., & Vaglini, M. (1992). Treatment of recurrent in transit metastases from cutaneous melanoma by isolation perfusion in extracorporeal circulation with interleukin-2 and lymphokine activated killer cells. A pilot study. Melanoma Research 2, 263-271.

Blancher, A., Roubinet, F., Grancher, A.S., Tremoulet, M., Bonate, A., Delisle, M.B., Calot, J.P., Pourreau, C., Franks, C., Ducos, J., Rascol, A., & Clanet, M. (1993). Local immunotherapy of recurrent glioblastoma multiforme by intracerebral perfusion of interleukin-2 and LAK cells. Eur. Cytokine Netw. 4, 331-341.

Blay, J-Y., Negrier, S., Combaret, V., Attali, S., Gorillot, E., Merrouche, Y., Mercatello, A., Revault, A., Tourani, M.M., Mokovtchenko, J-F., Philip, T., & Favrot, M. (1992). Serum level of interleukin-6 as a prognosis factor in metastatic renal cell carcinoma. Cancer Res. 52, 3317-3322.

Blay, J.Y., Favrot, M.C., Negier, S., Combaret, V., Chouaib, S., Mercatello, A., Kaemmerlen, P., Franks, C.R., & Philip, T. (1990). Correlation between clinical response to interleukin-2 therapy and sustained production of tumor necrosis factor. Cancer Res. 50, 2371-2374.

Bock, S.N., Lee, R.E., Fisher, B., Rubin, J.T., Schwartzentruber, D., Wei, J.P., Callender, D., Yang, J.C., Lotze, M.T., Pizzo, P.A., & Rosenberg, S.A. (1990). A prospective randomized trial evaluating prophylatic antibiotics to prevent catheter-related sepsis in patients treated with immunotherapy. J. Clin. Oncology 8, 161-169.

Boussiotis, V.A., Barber, D.L., Nakarai, T., Freeman, G.J., Gribben, J.G., Bernstein, G.M., D'Andrea, A.D., Ritz, J., & Nadler, L.M. (1994) Prevention of T cell anergy by signaling through the γc chain of the IL-2 receptor. Science 206, 1039-1044.

Broom, J., Heys, S.C., Whiting, P.H., Park, K.G.M., Strachen, A., Rothne, I., Franks, C.R., & Eremin, O. (1992). Interleukin-2 therapy in cancer, identification of responders. Br. J. Cancer 66,1185-1187.

Bubenik, J. (1989). Local immunotherapy of cancer with Interleukin 2. Immunology Lett. 21, 267-274.

Bubenik, J. (1990). Local and regional immunotherapy of cancer with interleukin 2. J. Cancer Res. Clin. Onc. 116, 1-7.

Bubenik, J. (1993). IL-2 and gene therapy of cancer (Review). Intl. J. Onc. 2, 1049-1052.

Bubenik, J., & Indrova, M. (1987). Cancer immunotherapy using local Interleukin-2 administration. Immunol. Lett. 16, 305-310.

Bubenik, J., Perlman, P., Indrova, M., Simova, J., Jandova, T., & Neuwirt, J. (1983). Growth inhibition of an MC-induced mouse sarcoma by TCGF (IL-2) containing preparations. Cancer Immunol. Immunotherapy 14, 205.

Bubenik, J., Voitenok, N.N., Kieler, J., Prassolov, V.S., Chumakov, P.M., Bubenikova, D., Simova, J., & Jandlova, T. (1988). Local administration of cells containing an inserted IL-2 gene and producing IL-2 inhibits growth of human tumours in nu/nu mice. Immunology Lett. 19, 279-282.

Bubenik, J., Simova, J., & Jandlova, T. (1990). Immunotherapy of cancer using local administration of lymphoid cells transformed by IL-2 cDNA and constitutively producing IL-2. Immunology Lett. 23, 287-292.

Bubenik, J., Lotzova, E., Zuethen, J., Bubinikova, D., Simova, J., & Indrova, M. (1991). Gene therapy of cancer: Use of Interleukin-2(IL-2) gene transfer in lcoal immunotherapy. J. Exp. Clin. Cancer Res. 10, 213-220.

Bubenik, J., Zeuthen, J., Simova, J., Jandlovam, T., Bubenikova, D., & Radzikowski, C. (1992). Activation and phenotyping of LAK generated by exposure to product of cells transformed by interleukin 2 cDNA. Anticancer Research 12, 965-968.

Bubenik, J., Simova, J., Bubenikova, D., Zeuthen, J., & Radzikowski, C. (1993). Utilization of interleukin-2 gene transfer in local immunotherapy of cancer. J. Cancer Res. Clin. Oncol. 119, 253-256.

Caligiuri, M.A., Murray, C., Robertson, M.J., Wang, E., Cochran, K., Cameron, C., Schow, P., Ross, M.E., Klumpp, T.R., Soiffer, R.J., Smith, K.A., & Ritz, J. (1991). Selective modulation of human natural killer cells *in vivo* after prolonged infusion of low dose recombinant interleukin 2. J. Clin. Invest. 91, 123-132.

Cameron, R., Spiess, P., & Rosenberg, S. (1990). Synergistic antitumor activity of tumor-infiltrating lymphocytes interleukin 2, and local tumor irradiation. J. Exp. Med. 171, 249-263.

Caraceni, A., Martini, C., Belli, F., Mascheroni, L., Rivoltini, L., Arienti, F., & Cascinelli, N. (1992). Neurophychological and neurophysiological assessment of the central effects of Interleukin-2 administration. Eur. J. Cancer 29, 1266-1269.

Caruso, B., Giavarina, D., Jirillo, A., Giavarina, M., Balli, M. (1992). Pharmacokinetics of subcutaneous recombinant Interleukin-2 administration in cancer patients. J. Immunol. Immunopharmacol. 12, 207-209.

Cavallo, F., DiPierro, F., Giovarelli, M., Gulino, A., Vacca, A., Stoppacciaro, A., Forni, M., Modesti, A., & Forni, G. (1993). Protective and curative potential of vaccination with Interleukin-2-gene-transfected cells from a spontaneous mouse mammary adenocarcinoma. Cancer Res. 53, 5067-5070.

Cavallo, F., Giovarelli, M., Gulino, A., Vacca, A., Stoppacciaro, A., Modesti, A., & Forni, G. (1992). Role of neutrophils and CD4+ T lymphocytes in the primary and memory response to nonimmunogenic murine mammary adenocarcinoma made immunogenic by IL-2 gene. J. Immunology. 149, 3627-3635.

Clamon, G., Herndon, J., Perry, M.C., Ozer, H., Kreisman, H., Maher, T., Ellerton, J., & Green, M.R. (1993). Interleukin-2 activity in patients with extensive small-cell lung cancer: A Phase II trial of cancer and leukemia group B. JNCI 85, 316-320.

Cohen, P., Vourka-Karussis, U., Weiss, L., & Slavin, S. (1993). Spontaneous and IL-2-induced anti-leukemic and anti-host effects against tumor-/and host-specific alloantigens. J. Immunology 151, 4803-4810.

Cohen, P.J., Lotze, M.T., Roberts, J.R., Rosenberg, S.A., & Jaffee, E.S. (1987). The immunopathology of sequential tumor biopsies in patients treated with interleukin-2. Correlation of response with T cell infiltration and HLA-DR expression. Amer. J. Pathol. 129, 208-216.

Corbett, J.A., Tilton, R.G., Chang, K., Hasan, K.S., Ido, Y., Wang, J.L.,, Sweetland, M.A., Lancaster, J., Williamson, J.R., & McDaniel, M.L. (1992). Aminoguanidine, a novel inhibitor of nitric oxide formation, prevents diabetic vascular dysfunction. Diabetes 41, 552-556.

Cornellussen, B., Holm, M., Waltersson, Y., Onions, J., Hallberg, B., Thornell, A., & Grundstrom, T. (1994). Calcium/calmodulin inhibition of basic-helix-loop-helix transcription factor domains. Nature 368, 760-764.

Cotran, R., Pober, J.S., Gimbrone, Jr., Springer, T.A., Wiebke, E.A., Gaspari, A.A., Rosenberg, S.A., & Lotze, M.T. (1987). Endothelial activation during interleukin-2 (IL-2) immunotherapy: A possible mechanism for the vascular leak syndrome. J. Immunol. 139,1883-1888.

Culver, K., Cornetta, K., Morecki, S., Aebersold, P., Freeman, S., Kasid, A., Morgan, R., Lotze, M.T., Rosenberg, S.A., Anderson, W.F., & Blaese, R.M. (1992). Gene therapy: Lymphocytes as cellular vehicles for gene transfer in mouse and man. Proc. Nat. Acad. Sci. 88, 3155-3159.

Deehan, D.J., Heys, S.D., Broom. J., & Eremin, O. (1994). Cytokine production in patients treated with recombinant interleukin 2 preducts response to therapy. Proc. Soc. Univ. Surgeons 55, 111.

Denicoff, K.D., Durkin, T.M., Lotze, M.T., Quinlan, P.E., Davis, C.L., Listwak, S.J., Rosenberg, S.A., & Rubinow, D.R. (1989). The neuroendocrine effects of Interleukin-2 treatment. J. Clin. Endocrinology and Metabolism 69, 402-410.

Denicoff, K.D., Rubinow, D.R., Papa, M.Z., Simpson, C., Seipp, C.A., Lotze, M.T., Chang, A.E., Rosenstein, D., & Rosenberg, S.A. (1987). The neuropsychiatric effects of interleukin 2/lymphokine activated killer cell treatment. Ann. Int. Med. 107, 293-300.

Dillman, R.O., Oldham, R.K., Tauer, K.W., Orr, D.W., Barth, N.M., Blumenschein, G., Arnold, J., Birch, R., & West, W.H. Continuous Interleukin-2 and lymphokine-activated killer cells for advanced cancer: A national biotherapy study group trial. J. Clin. Onc. 9, 1233-1240, 1991.

Doherty, G., Jensen, J., Alexander, H., et al. (1992). Pentoxifylline suppression of tumor necrosis factor gene transcription. Surgery 110, 192-198.

Donohue, J.H., Lotze, M.T., Robb, R.J., Rosenstein, M., Braziel, R.M., Jaffe, E.S., & Rosenberg, S.A. (1984b). *In vivo* administration of purified Jurkat derived Interleukin-2 in mice. Cancer Research 44, 1380-1386.

Donohue, J.H., Rosenstein, M., Chang, A.E., Lotze, M.T., Robb, R.J., & Rosenberg, S.A. (1984a). The systemic administration of purified Interleukin-2 enhances the ability of sensitized murine lymphocyte lines to cure a disseminated syngeneic lymphoma. J. Immunol. 132, 2123-2128.

Dorval, T., Negrier, S., Chevreau, C., Baume, D., Cupissol, D., Oskam, R., dePeuter, M.A., Herrera, A., Escudier, B. (1994). Results of a french multicentric randomized trial of chemoimmunotherapy (Cisplatin (P), IL-2 (Proleukin), with or without IFN (Roferon) in metastatic malignant melanoma. Proc. ASCO 13, 395 (1347).

Eberlein, T.J., Rosenstein, M., & Rosenberg, S.A. (1982). Regression of a disseminated syngeneic solid tumor by systemic transfer of lymphoid cells expanded in interleukin 2. J. Exp. Med. 156, 385-397.

Edwards, M., Abney, D., & Miller, F. (1991). Pentoxifylline inhibits interleukin-2 induced leukocyte-endothelial adherence and reduces systemic toxicity. Surgery 110, 199-204.

Edwards, M., Heniford, B., Klar, E., et al. (1992). Pentoxifylline inhibits interleukin-2 induced toxicity in C57B16 mice but preserves antitumor efficacy. J. Clin. Invest. 90, 637-641.

Espinoza-Delgado, I., Bosco, M.C., Musso, T., Gusella, G.L., Longo, D.L., & Varesio, L. (1995). Interleukin-2 and human monocyte activation. J. Leukocyte Biology 57, 13-19.

Ettinghausen, S.E., Puri, R.K., & Rosenberg, S.A. (1988). Increased vascular permeability in organs mediated by the systemic administration of lymphokine-activated killer cells and recombinant interleukin-2 in mice. JNCI 80, 177-188.

Fearon, E.R., Pardoll, D.M., Itaya, T., Golumbek, P., Levitsky, H.I., Simons, J.W., Karasuyama, H., Vogelstein, B., & Frost, P. (1990). Interleukin-2 production by tumor cells bypasses T helper function in the generation of an antitumor response. Cell 60, 397-400.

Fisher, B., Keenan, A., Garra, B.S., Steinberg, S.M., White, D.E., DeBisceglie, A.M., Hoofnagle, J.H., Yolles, P., Rosenberg, S.A., & Lotze, M.T. (1990). Interleukin-2 induces profound reversible cholestasis: A detailed analysis in treated patients. J. Clin. Oncol. 7, 1852-1862.

Foa, R. (1993). Does Interleukin-2 have a role in the management of acute leukemia? J. Clin. Onc. 11, 1817-1825.

Forni, G., Giovarelli, M., Santoni, A. (1985). Lymphokine activated tumor inhibition in vivo: 1. The local administration of interleukin 2 triggers nonreactive lymphocytes from tumor-bearing mice to inhibit tumor growth. J. Immunol. 134, 1305.

Fowler, C.L., Brooks, S.P., Squire, R., Rich, G.A., Rossman, J.E., Finegold, M.J., Allen, J.E., & Cooney, D.R. (1991). Enhanced resection and improved survival in murine neuroblastoma (C1300-NB) after preoperative immunotherapy. J. Ped. Surg. 26, 381-388.

Fraker, D.L., Langstein, H.N., & Norton, J.A. (1989). Passive immunization against tumor necrosis factor partially abrogates interleukin-2 toxicity. J. Exp. Med. 170, 1015-1019.

Gansbacher, B., Zier, K., Daniels, B., Cronin, K., Bannerji, R., & Gilboa, E. (1990). Interleukin 2 gene transfer into tumor cells abrogates tumorigenicity and induces protective immunity. J. Exp. Med. 172, 1217-1224.

Goldblum, S.E., Yoneda, K., Cibull, M., Pearson, T., Hall, C., & Marshall, M.E. (1990). Human recombinant Interleukin-2 provokes acute pulmonary vascular endothelial injury in rabbits. J. Biol. Res. Mod. 9, 127-139.

Goldsmith, M.A., & Greene, W.C. (1994). Interleukin-2. *The Cytokine Handbook*, 2nd ed. (A. Thompson, Ed.). Academic Press, San Diego, pp. 57-82.

Gordon, J., & MacLean, L.D. (1965). A lymphocyte-stimulating factor produced *in vitro*. Nature 208, 795-796.

Gratama, J.W., Bruin, R.J., Lamers, C.H.J., Oosterom, R., Braakman, E., Stotter, G., & Bolhuis, R.L.H. Activation of the immune system of cancer patients by continuous i.v. recombinant IL-2 (rIL-2) therapy is dependent on dose and schedule of rIL-2. Clin. Exp. Immunol. 92, 185-193, 1993.

Han, J., Thompson, P., & Beutler, B. (1990). Dexamethasone and pentoxifylline inhibit endotoxin-induced cachectin/tumor necrosis factor synthesis at separate points in the signaling pathway. J. Exp. Med. 172, 391-394.

Hartman, L.C., & Urba, W.J. (1989). Use of prophylactic antibiotics for prevention of intravascular catheter related infections in Interleukin-2 treated patients. JNCI 81, 1190-1192.

Hatakeyama, M,, Kono, T., Kobayashi, N., Kawahara, A., Levin, S.D., Perlmutter, R.M., & Taniguchi, T. (1991). Interaction of the IL-2 receptor with the src-family kinase $p56^{lck}$: Identification of novel intermolecular association. Science 252, 1523-1528.

Heaton, K.M., Ju, G., & Grimm, E.A. (1993). Human Interleukin-2 analogues that preferentially bind the intermediate-affinity Interleukin 2 receptor lead to reduced secondary cytokine secretion: Implications for the use of these Interleukin 2 analogues in cancer immunotherapy. Cancer Research 53, 2597-2602.

Heaton, K.M., Ju, G., & Grimm, E.A. (1994). Induction of lymphokine-activated killing with reduced secretion of Interleukin-1B, tumor necrosis factor-a, and Interferon-g by Interleukin-2 analogs. Annals of Surg. Onc. 1, 198-203.

Hession, C., Decker, J.M., Sherblom, A.P., Kumar, S., Yul, C.C., Mattaliano, R.J., Tizard, R., Kawashima, E., Schmeissner, V., & Heletky, S. (1987). Uromodulin (Tamm Horsfall glycoprotein): A renal ligand for lymphokines. Science 237, 1479-1484.

Hibbs, J.B., Westenfelder, C., Taintor, R., Vavrin, Z., Kablitz, C., Baranowski, R.L., Ward, J.H., Menlove, R.L., McMurry, M.P., Kushner, J.P., & Samlowski, W.E. (1992). Evidence for cytokine-inducible nitric oxide synthesis from L-arginine in patients receiving Interleukin-2 therapy. J. Clin. Invest. 89, 867-877.

Holder, M.J., Abbot, S.D., Milner, A.E., Gregory, C.D., Casamayor, M., Johnson, G.D., MacLennan, I.C.M., & Gordon, J.(1993). IL-2 expands and maintains IgM plasmablasts from a CD5+ subset contained within the germinal centre cell-enriched (surface IgD-/CD39-buoyant) fraction of human tonsil. Intl. Immunology 5, 1059-1066.

Huberman, M., Bering, H., Fallon, B., Tessitore, J., Sonnenborn, H., Paul, S., Zeffren, J., Levitt, D., & Groopman, J. (1991). A Phase I study of an outpatient regimen of recombinant human interleukin-2 and alpha-2a-interferon in patients with solid tumors. Cancer 68, 1708-1713.

Hugin, A.W., Flexner, C., & Moss, B. (1993). Clearance of recombinant vaccinia virus expressing IL-2: Role of local host immune responses. Cellular Immunology 152, 499-509.

Huland, E., Heinzer, H., Huland, H. (1994). Inhaled interleukin-2 in combination with low-dose systemic interleukin-2 and interferon a in patients with pulmonary metastatic renal cell carcinoma: Effectiveness and toxicity of mainly local treatment. J. Cancer Res. Clin. Oncol. 120, 221-228.

Hunter, N., Toshitake Nakayama, T., Ito, H., Woo, S., & Milas, L. (1992). Combination of interleukin-2 and irradiation in therapy of murine tumors. Clin. Exp. Metatasis 10, 431-436.

Jablons, D., Bolton, E., Mertins, S., Rubin, M., Pizzo, P., Rosenberg, S.A., & Lotze, M.T. (1990). IL-2 based immunotherapy alters circulating neutrophil Fc Receptor expression and chemotaxis. J. Immunology 144, 3630-3636.

Kasakura, S., & Lowenstein, L. (1965). A factor stimulating DNA synthesis derived from the medium of leucocyte cultures. Nature 208, 794-795.

Kasid, A., Morecki, S., Aebersold, P., Cornetta, K., Culver, K., Freeman, S., Director, E., Lotze, M.T., Blaese, R.M., Anderson, F., &Rosenberg, S.A. Human gene transfer: Characterization of human tumor infiltrating lymphocytes as vehicles for retroviral mediated gene transfer in man. Pro. Nat. Acad. Sci. 87, 473-477, 1990.

Kasid, AT., Director, E., Stovroff, M.C., Lotze, M.T., & Rosenberg, S.A. (1990). Cytokine regulation of tumor necrosis factor-a and β(lymphotoxin)- messenger RNA expression in human peripheral blood lymphocytes. Cancer Res. 50, 5072-6.

Keilholz, U., Scheibenbogen, C., Tilgen, W., Bergmann, L., Weidmann, E., Seither, E., Richter, M., Brado, B., Mitrou, P.S., & Hunstein, W. (1993). Interferon-a and interleukin-2 in the treatment of metastatic melanoma. Cancer 72, 607-614.

Keilholz, U., Schlag, P., Tilgen, W., Brado, B., Galm, F., Gorich, J., Kauffmann, G.W., Moller, P., Schneider, S. & Hunstein, W. (1992). Regional administration of lymphokine-activated killer cells can be superior to intravenous application. Cancer 69, 2172-2175.

Kelley, V.E., Bacha, P., Pankewycz, O., Nichols, J.C, Murphy, J.R., & Strom, T.B. (1988). Interleukin 2 diptheria toxin fusion protein can abolish cell mediated immunity *in vivo*. Proc. Natl. Acad. Sci. 85, 3980-3984.

Khayat, D., Borel, C., Tourani, J.M., Benhammouda, A., Antoine, E., Rixe, O., Vullemin, E., Bazex, P.A., Thill, L., Franks, R., Auclerc, G., Soubrane, C., Banzet, P., & Weil, M. (1993). Sequential chemoimmunotherapy with cisplatin, interleukin-2, and interferon alfa-2a for metastatic melanoma. J. Clin. Onc. 11 , 2173-2180.

Kilbourn, R.G., Belloni, P. (1990). Endothelial cell production of nitrogen oxides in response to interferon-gamma in combination with tumor necrosis factor, interleukin-1 or endotoxin. JNCI 82, 772-776.

Kilbourn, R.G., Owen-Schaub, L., Griffith, O.W. (1992). Interleukin-2 (IL-2) mediated hypotension in dogs is reversed by N-methyl-L-arginine (NMA), an inhibitor of nitric oxide (NO) formation. Proc. Am. Assoc. Cancer. Res. 33, 328.

Kim, B., & Louie, A.C. (1992). Surgical resection following interleukin 2 therapy for metastatic renal cell carcinoma prolongs remission. Arch. Surg. 127, 1343-1349.

Kim, H., Rosenberg, S.A., Steinberg, S.M., Cole, D.J., & Weber, J.S.(Forthcoming). A Randomized double blinded comparison of the antiemetic efficacy of ondansetron and droperidol in patients receiving high dose interleukin-2. J. Immunotherapy.

Kim, T.S., Russell, S.J., Collins, M.K.L., & Cohen, E. (1992). Immunity to B16 melanoma in mice immunized with IL-2 secreting allogeneic mouse fibroblasts expressing melanoma associated antigens. Int. J. Cancer 51, 1-7.

Kirchner, H., Korfer, A., Evers, P., Szamel, M.M., Knuver-Hopf, J., Mohr, H., Franks, C.R., Pohl, U., Resch, K., Hadam, M., Poliwoda, H., & Atzpodien, J. (1991) The development of neutralizing antibodies in a patient receiving subcutaneous recombinant and natural interleukin-2. Cancer 67, 1862-1864.

Kishimoto, T., Taga, T., & Akira, S. (1994). Cytokine signal transduction. Cell 76, 253-262.

Klempner, M.S., Noring, R., Mier, J.W., & Atkins, M.B. (1990). An acquired chemotactic defect in neutrophils from patients receiving interleukin-2 immunotherapy. NEJM 322, 959-965.

Kondo, M., Takeshita, T., Ishii, N., Nakamura, M., Watanabe, S., Arai, K., Sugamura, & K. (1993). Sharing of the Interleukin-2 (IL-2) receptor γ chain between receptors for IL-2 and IL-4. Science 262, 1874-1877.

Kradin R.L., Kurnick, J.T., Lazarus, D.S., Preffer, F.I., Dubinett, S.M., Pinto, C.E., Gifford, J., Davidson, E., Grove, B., Callahan, R.J., & Strauss, H.W. (1989). Tumor infiltrating lymphocytes and interleukin-2 in the treatment of advanced cancer. Lancet 1,577-580.

Kragel, A.H., Travis, W.D., Feinberg, L., Pittalugia, S., Striker, L.M., Roberts, W.C., Lotze, M.T., Yang, J.J., & Rosenberg, S.A. (1990). Pathologic findings associated with interleukin-2 based immunotherapy for cancer: A postmortem study of 19 patients. Human Pathology 21, 493-502.

Kruit, W.H.J., Bolhuism R.L.H., Goeym S.H., Jansen, R.L.H., Eggermont, A.M.M., Batchelorm D., Schmitz, P.I.M., & Stoter, G. (1993). J. Clin. Onc. 11, 921-924.

Kuebler, J.P., Whitehead, R.P., Ward, D.L., Hemstreet, G.P., & Bradley, E.C. (1993). Treatment of metastatic renal cell carcinoma with recombinant interleukin-2 in combination with vinblastine or lymphokine-activated killer cells. J. Urology 150, 814-820.

Kuziel, W.A. & Greene, W.C. (1991). Interleukin-2. In: *The Cytokine Handbook* (Thompson, A., Ed.). Academic Press, San Diego, pp. 83-102.

Kuziel, W.A., Ju, G., Grdina, T.A., & Greene, W.C. (1993). Unexpected effects of the Il-2 receptor a subunit on high affinity IL-2 receptor assembly and function detected with a mutant IL-2 analog. J. Immunology 150, 3357-3365.

Lange, J.R., Raubitschek, A.A., Pockaj, B.A., Spencer, W.F., Lotze. M.T., Topalian, S.L., Yang, J.C., & Rosenberg, S.A. (1992). A pilot study of combination of Interleukin-2 based immunotherapy and radiation therapy. J. Immunotherapy 12, 265-271.

Latchman, D.S. (1991). Herpes simplex virus life cycle and the design of viral vectors. In: *Methods in Molecular Biology, Vol 8: Practical Molecular Virology: Viral Vectors for Gene Expression* (Collins, M., Ed.). The Human Press Inc., Clifton, NJ.

Leder, G.H., Oppenheim, M., Rosenstein, M., Shah, N., Hoffman, R., Simmons, R., & Lotze, M.T. (n.d.). The IL-2 induced nitric oxide production but not IL-2 induced capillary leak syndrome is decreased by aminoguanidine. Submitted.

Leder, G.H., Oppenheim, M., Rosenstein, M., Shah, N., Hoffman, R., Simmons, R., & Lotze, M.T. (Forthcoming). The IL-2 induced nitric oxide production but not the IL-2 induced capillary leak syndrome is decreased by aminoguanidine. Third International Nitric Oxide Meeting.

Lee, R.E., Lotze, M.T., Skibber, J.M., Tucker, E., Bonow, R.O., Ognibene, F.P., Carrasquillo, J.A., Shelhammer, J.H., Parrillo, J.E., & Rosenberg, S.A. (1989). Cardiorespiratory effects of immunotherapy with interleukin-2. J. Clin. Oncol. 7, 7-20.

LeMaistre, C.F., Craig, F.E., Meneghetti, C., McMullin, B., Parker, K., Reuben, J., Boldt, D.H., Rosenblum , M., & Woodworth, T. (1993). Phase I trial of a 90 minute infusion of the fusion toxin DAB_{486} IL-2 in hematological cancers. Cancer Research 53, 3930-3934.

Lenardo, M.J. (1991). Interleukin-2 programs mouse ab T lymphocytes for apoptosis. Nature 353, 858-861.

Licinio, J., Wong, M.L., Altemus, M., Bongiorno, P.B., Bernat, A., Tamarkin, L., & Gold, P.W. (n.d.). Twenty-four-hour concentrations of Interleukin-2 in healthy women exhibit episodic fluctuations: Analysis of integrated basal levels and discrete pulse properties. J. Clin. Liver.

Lin, W.C., Yasumura, S., & Whiteside, T.L. (Forthcoming). Transfer of IL-2 receptor genes into squamous cell carcinoma: Modification of tumor cell growth. Arch Otolaryngol. Head Neck Surg.

Lorberboum-Galski, H., Kozak, R.W., Waldmann, T.A., Bailon, P., Fitzgerald, D.J.P., & Pastan, I. (1988). Interleukin 2 (IL-2) PE40 is cytotoxic to cells displaying either the p55 or p70 subunit of the IL-2 receptor. J. Biol. Chem. 263, 18650-18656.

Lotze, M.T. (1991a). Cancer treatment with Interleukin-2. In: *Principles and Practice of Biologic Therapy* (DeVita, V., Hellman, S., & Rosenberg, S.A., Eds.). J. Lippincott, Philadelphia.

Lotze, M.T. (1991b). Interleukin-2—Basic principles. In: *Principles and Practice of Biologic Therapy* (DeVita, V., Hellman, S., & Rosenberg, S.A., (Eds.). J. Lippincott, Philadelphia.

Lotze, M.T. (1991c). Interleukin-2: Basic principles. In: *Biologic Therapy of Cancer* (Devita, V., Helliman, S., & Rosenberg, S., Eds.). Lippincott, Philadelphia, pp. 123-141.

Lotze, M.T., & Rosenberg, S.A. (1981). *In vitro* growth of cytotoxic human lymphocytes. III. The preparation of lectin free T cell growth factor (TCGF) and an analysis of its activity. J. Immunol. 126, 2215-2220.

Lotze, M.T., Zeh, H.J., Elder, E.M., Cai, Q., Pippin, B.A., Whiteside, T.L., & Herberman, R.B. (1992). Use of T-cell growth factors (IL-2 to IL-12) in the investigation and treatment of melanoma. J. Immunotherapy 12, 212-217.

Lotze, M.T., Custer, M.C., Sharrow, S.O., Rubin, L.A., Nelson, D.L., & Rosenberg, S.A. (1987). *In vivo* administration of purified human interleukin-2. III. Development of Tac positive cells and circulating soluble IL-2 recep~tors following IL-2 administration to humans. Cancer Res. 47, 2188-2195.

Lotze, M.T, Custer, M.C., & Rosenberg, S.A. (1986a). Intraperitoneal administration of interleukin-2 in patients with cancer. Arch. Surg. 121, 1373-1379.

Lotze, M.T., Chang, A.E., Seipp, C.A., Simpson, C., Vetto, J.T., & Rosenberg, S.A. (1986b). High dose recombinant interleukin-2 in the treatment of patients with disseminated cancer: Responses, treatment related morbidity and histologic findings. JAMA. 256, 3117-3124.

Lotze, M.T., Frana, L.W., Sharrow, S.O., Robb, & R.J., Rosenberg, S.A. (1985a). *In vivo* administration of purified human interleukin-2 I. Half-life and immunologic effects of the Jurkat cell line derived IL-2. J. Immunol. 134, 157-166.

Lotze, M.T., Matory, Y.L., Ettinghausen, S.E., Rayner, A.A., Sharrow, S.O., Seipp, C.A.Y., Custer, M.C., & Rosenberg, S.A (1985b). *In vivo* administration of purified human interleukin-2. II. Half life, immunologic effects and expansion of peripheral lymphoid cells *in vivo* with recombinant IL-2. J. Immunol. 135, 2865-2875.

Lotze, M.T., Robb, R.J., Sharrow, S.O., Frana, L.W., & Rosenberg, S.A, (1984). Systemic administration of interleukin-2 in humans. J. Biol. Resp. Mod. 3, 475-482.

Lotze, M.T., Grimm, E.A., Mazumder, A., Strausser, J.L., & Rosenberg, S.A. (1981). Lysis of fresh and cultured autologous tumor by lymphocytes cultured in T cell growth factor (TCGF). Cancer Res. 41, 4420-4425.

Lotze, M.T., Strausser, J.L., & Rosenberg, S.A. (1980a). *In vitro* growth of cytotoxic human lymphocytes. II. Use of T cell growth factor (TCGF) to clone human T cells. J. Immunol. 124, 2972-2978.

Lotze, M.T., Lines, B.T., Mathisen, D.J., & Rosenberg, S.A. (1980b). The *in vivo* distribution of cultured autologous human and murine lymphoid cells grown in T cell growth factor (TCGF): Implications for the adoptive immunotherapy of tumors. J. Immunol. 125, 2972-2978.

Marannchi, D., Blaise, D., Viens, P., Brandely, M., Olive, D., Lopez, M., Sainty, D., Marit, G., Stoppa, A.M., Reiffers, J., Gratecos. N., Bertau-Perez, P., Mannoni, P., Mawas, C., Hercend, T., Sobahoun, G., & Carcassonne, Y. (1991). High-dose recombinant interleukin-2 and acute myeloid leukemias in relapse. Blood 78, 2182-2187.

Marcus, S.G., Choyke, P.L., Reiter, R., Jaffe, G.S., Alexander, R.B., Linehan, W.M., Rosenberg, S.A., & Walther, M.M. (1993). Regression of metastatic renal cell carcinoma after cytoreductive nephrectomy. J. Urology 150, 463-466.

Marincola, F.M., Venzon, D., White, D., Rubin, J.T., Lotze, M.T., Simonis, T.B., Balkisson, J., Rosenberg, S.A., & Parkinson, D.R.. (1992). HLA association with response and toxicity in melanoma patients treated with interleukin-2 based immunotherapy. Cancer Research 52, 6561-6566.

Marincola, F.M., White, D.E., Wise, A.P., Rosenberg, S.A. (1995). Combination therapy with interferon α-2a and interleukin 2 for the treatment of metastatic cancer. J. Clin. Oncol. 13, 1110-112.

Mertens, W.C., Bramwell, V.H.C., Banerjee, D., Gwadry-Sridhar, F., & Lala, P.K. (1993). Sustained Indomethacin and ranitidine with intermittent continuous infusion interleukin-2 in advanced malignant melanoma. A phase II study. Clin. Onc. 5, 107-113.

Mier, J., Vachino, G., Klempner, M., et al. (1990). Inhibition of interleukin-2 induced tumor necrosis factor release by dexamethasone: Prevention of an acquired neutrophil chemotaxis defect and differential suppression of interleukin-2 associated side effects. Blood 76, 1933-1940.

Mier, J.W., & Atkins, M.B. (1993). Mechanisms of action and toxicity of immunotherapy with cytokines. Current Opn. Onc. 5, 1067-1072.

Mier, J.W., Brandon, E.P., Libby, P., Janicka, M.W., & Aronson, F.R. (1989). Activated Endothelial cells resist lymphokine activated killer cell mediated injury. Possible role of induced cytokines in limiting capillary leak during IL-2 therapy. J. Immunology 143, 2407-2414.

Minasi, L.E., Kamogawa, Y., Carding, S., Bottomly, K., & Flavell, R.A. (1993). The selective ablation of interleukin-2 producing cells isolated from transgenic mice. J. Exp. Med., 177, 1451-1459.

Minasian, L.M., Motzer, R.J., Gluck, L., Mazumdar, M., Vlamis, V., & Krown, S.E. (1993). Interferon α-2a in advanced renal cell carcinoma: Treatment results and survival in 159 patients with long-term follow-up. J. Clin. Onc. 11, 1368-1375.

Mintz, C.S., Miller, R.D., Gutgsell, N.S., & Malek, T. (1993). *Legionella pneumophila* protease inactivates Interleukin-2 and cleaves CD4 on human T cells. Infection and Immunity 61, 3416-3421.

Mittelman, A., Puccio, G., Ahmed, T., Zeffren, J., Choudhury, A., Arlin, A. (1991). A phase II trial of Interleukin-2 by continuous infusion and interferon by intramuscular injection in patients with renal cell carcinoma. Cancer 68, 1699-1702.

Miyazaki, T., Kawahara, A., Fujii, H., Nakagawa, Y., Minami, Y., Liu, Z-J., Oishi, I., Silvennoinen, O., Witthuhn, B.A., Ihle, J.N., & Taniguchi, T. (1994). Functional activation of Jak1 and Jak3 by selective association with IL-2 receptor subunits. Science 266, 1045-1047.

Morgan, D.A., Rusetti, F.W., & Gouro, R.G. (1976). Selective *in vitro* growth of T-lymphocytes from normal bone marrow. Science 193, 1007-1008.

Morse, E.D., Gunther, R.A., Jesmok, G.J., & Hughes, K.S. (1990). Steroid pretreatment reduces interleukin-2 toxicity in sheep. Surgery 107, 639-647.

Mule, J., Yang, J., Afreniere, R., Shu, S, & Rosenberg, S.A. (1987). Identification of cellular mechanisms operational *in vivo* during the regression of established pulmonary metastases by the administration of high dose recombinant interleukin 2. J Immunology, 139, 285-294.

Murray, H.W., Miralles, G.D., Stoeckle, M.Y., & McDermott, D.F. (1993). Role and Effect of IL-2 in experimental visceral leishmaniasis. J. Immunology 151, 929-938.

Nakamura, Y., Russell, S.M., Mess, S.A., Friedmann, M., Erdos, M., Francois, C., Jacques, Y., Adelstein, S., & Leonard, W.J. (1994). Heterodimerization of the IL-2 receptor beta- and gamma-chain cytoplasmic domains is required for signalling. Nature 369, 330-333.

Negrier, S., Michon, J., Floret, D., Bouffet, E., Gentet, J.C., Philip, I., Cochat, P., Stamm, D., Costil, J., Gaspard, M., Andreu, G., Palmer, P., Franks, C.R., Zucker, J.M., Bernard, J.L., Fridman, W.H., Favrot, M., & Philip, T. (1991). Interleukin-2 and lymphokine-activated killer cells in 15 children with advanced metastatic neuroblastoma. J. Clin. Onc., 9,1363-1370.

Nelson, B.H., Lord, J.D., & Greenberg, P.D. (1994). Cytoplasmic domains of the interleukin-2 receptor beta and gamma chains mediate the signal for T-cell proliferation. Nature 369, 333-336.

Nghiem, P., Gardner, P., & Schulman, H. (Forthcoming). Interleukin-2 transcriptional block by multifunctional CaM kinase. Submitted.

Noguchi, M., Yi, H., Rosenblatt, H.M., Filipovich, A.H., Adelstein, S., Modi, W.S., McBride, O.W. & Leonard, W.J. (1993). Interleukin-2 Receptorγ chain mutation results in x-linked severe combined immunodeficiency in humans. Cell 73 147-157.

Ochoa, J.B., Curti, B., Peitzman, A.B., et al. (1992). Increased circulating nitrogen oxides after human tumor immunotherapy: Correlation with toxic hemodynamic changes. JNCI 84, 864-867.

Ognibene, F.P., Rosenberg, S.A., Lotze, M.T., Skibber, J., Parker, M.M., Shelhamer, J.H., & Parrillo, J.E. Interleukin-2 administration causes reversible hemodynamic changes and left ventricular dysfunction similar to those seen in septic shock. Chest 94, 750-754, 1988.

Oppenheim, M.H., & Lotze, M.T. (Forthcoming). Interleukin-2: Solid tumor therapy. Oncology.

Palamer, P.A., Vinke, J., Philip, T., Negrier, S., Atzpodien, J., Kirchner, H., Oskam, R., Franks, & C.R. (1992). Prognostic factors for survival in patients with advanced renal cell carcinoma treated with recombinant interleukin-2. Ann. Oncol. 3, 4750-4780.

Paliogianni, F., Kincaid, R.L., & Boumpas, D.T. (1993a). Prostaglandin E_2 and other cyclic AMP elevating agents inhibit Interleukin-2 gene transcription by counteracting calcineurin-dependent pathways. J. Exp. Med., 178, 1813-1817.

Paliogianni, F., Raptis, A., Ahuja, SS., Najjar, S.M., & Boumpas, D.T. (1993b). Negative transcriptional regulation of human Interleukin-2 (IL-2) gene by glucocorticoids through interference with nuclear transcription factors AP-1 and NF-AT. J, Clin, Investigation 92, 1481-1489.

Papa, M.Z., Vetto, J.T., Ettinghausen, S.E., et al. (1986). Effects of corticosteroids on the antitumor activity of lymphokine activated killer cells and IL-2 in mice. Cancer Res. 46, 5618-5623.

Patel, P.M., Flemming, C.L., Russell, S.J., McKay, I.A., MacLennan, K.A., Box, G.M., Eccles, S.A., & Collins, M.K.L. (1993). Comparison of the potential therapeutic effects of interleukin 2 or interleukin 4 secretion by a single tumour. Br. J. Cancer 68, 295-302.

Piali, L., Fichtel. A., Terpe. H.J., Imhof, B.A., & Gisler. R.H. (1995). Endothelial vascular cell adhesion molecule 1 expression is suppressed by melanoma and carcinoma. J. Exp. Med. 181(2), 811-816.

Pockaj, B.A., Sherry, R.M., Wei, J.P., Yannelli, J.R., Carter, C.S., Leitman, S.F., Carasquillo, J.A., Steinberg, S.M., Rosenberg, S.A., & Yang, J.C. (1994). Localization of 111 indium-labeled tumor infiltrating lymphocytes to tumor in patients receiving adoptive immunotherapy. Cancer 73, 1731-1737.

Pockaj, B.A., Yang, J.C., Lotze, M.T., Lange, J.R., Spencer, W.F., Steinberg, S.M., Topalian, S.L., Schwartzentruber, D.J., White, D.E., & Rosenberg, S.A. (1994). A prospective randomized trial evaluating colloid versus crystalloid resuscitation in the treatment of the vascular leak syndrome associated with interleukin-2 therapy. J. Immunotherapy 15, 22-28.

Porgador, A., Bannerji, R., Tzehoval, E., Gilboa, E., Feldman, M., & Eisenbach, L. (Forthcoming). Anti-metastatic vaccination of tumor-bearing mice with IL-2 gene inserted tumor cells. Intl. J. Cancer.

Richards, J.M., & Lotze, M.T. (1992). IL-2 therapy: Current status and future direction. Contemporary Oncology 2, 34-42.

Rivoltini, L., Viggiano, V., Spinazze, S., Santoro, A., Colombo, M.P., Takatsu, K., & Parmiani, G. (Forthcoming). *In vitro* anti-tumor activity of eosinophils from cancer patients treated with subcutaneous administration of interleukin 2. Role of interleukin 5. Intl. J. Cancer.

Rixo, O., Benhammouda, A., Antoine, E., Petit, T., Tourani, J.M., Borel, C., Franks, C., Kalif, B., Mousseau, M., Thomas, A., Bensfia, S., Nizri, D., Soubrane, C.L., Herrera, A., Bizzari, J.P., Auclerc, G., Weil, M., Banzet, P., & Khayat, D. (1994). Final results of a prospective multicentric study on 91 metastatic malignant melanoma (MMM) patients treated by chemo-immunotherapy (CH-IM) with cisplatin interleukin 2 (IL-2) Interferon-a (IFN). Proc. ASCO 13, 398 (1360).

Rizzo, L.V., Miller-Rivero, N.E., Chan, C-C., Wiggert, B., Nussenblatt, R.B., & Caspi, R.R. (1994). Interleukin-2 potentiates induction of oral tolerance in a murine model of autoimmunity. J. Clin. Oncol. 94:1668-1672.

Rosenberg, S.A., & Lotze, M.T. (1986). Cancer immunotherapy using interleukin-2 and interleukin-2-activated lymphocytes. Ann. Rev. Immunol. 4, 681-709.

Rosenberg, S.A., Yanelli, J.R., Yang, J.C., Topalian, S.L., Schwartzentruber, D.J., Weber, J.S., Parkinson, D.R., Seipp, C.A., Einhorn, J.H., & White, D.E. (Forthcoming). Treatment of patients with metastatic melanoma using autologous tumor infiltrating lymphocytes and interleukin-2. Submitted for publication.

Rosenberg, S.A., Yang, J.C., Topalian, S.L., Schwartzentruber, D.J., Weber, J.S., Parkinson, D.R., Seipp, C.A., Einhorn, J.H., & White, D.E. (1994). Treatment of 283 consecutive patients with metastatic melanoma or renal cell cancer using high-dose bolus interleukin 2. JAMA 271, 907-913.

Rosenberg, S.A., Lotze, M.T., Yang, J.C., Topalian, S.L., Chang, A.E., Schwartzentruber, D.J., Aebersold, P., Leitman, S., Linehan, W.M., Seipp, C.A., White, D.E., & Steinberg, S.M. (1993). Prospective randomized trial of high-dose Interleukin-2 alone or in conjunction with lymphokine-activated killer cells for the treatment of patients with advanced cancer. JNCI 85, 622-632.

Rosenberg, S.A., Aebersold, P., Cornetta, K., Kasid, A., Morgan, R.A., Moen, R., Karson, E.M., Lotze, M.T., Yang, J.C., Topalian, S.L., Merino, M.J., Culver, K., Miller, A.D., Blaese, R.M., & Anderson, W. (1990). Gene transfer into humans: Immunotherapy of patients with advanced melanoma using tumor infiltrating lymphocytes modified by retroviral gene transduction. N. Engl. J. Med. 323, 570-578.

Rosenberg, S.A., Lotze, M.T., Yang, J.C., Aebersold, P.M., Linehan, W.M., Seipp, C.A., & White, D.E. (1989a). Experience with the use of high dose interleukin-2 in the treatment of 652 patients with cancer. Ann. Surg. 210, 484-485.

Rosenberg, S.A., Lotze, M.T., Yang, J.C., Linehan, W.M., Seipp, C., Calabro, S., Karp, S.E., Sherry, R.M., Steinberg, & White, D.E. (1989b). Combination therapy with Interleukin-2 and Alpha-Interferon for the treatment of patients with advanced cancer. J. Clin. Onc. 7,1863-1874.

Rosenberg, S.A., Packard, B., Aebersold, P.M., Solomon, D., Topalian, S.L., Toy, S., Simon, P., Lotze, M.T., Yang, J.C., Seipp, C.A., Simpson, C., Carter, C., Bock, S., Schwartzentruber, D., Wei, J.P., & White, D.E. (1988). Immunotherapy of patients with metastatic melanoma using tumor infiltrating lymphocytes and Interleukin-2: Preliminary report. N. Engl. J. Med. 319, 1676-1680.

Rosenberg, S.A., Lotze, M.T., Muul, L.M., Chang, A.E., Avis, F.P., Leitmann, S., Linehan, W.M., Robertson, C.N., Lee, R.E., Rubin, J.T., Seipp, C.A., Simpson, C.G., & White, D.E. (1987). Clinical experience with the treatment of 157 patients with advanced cancer using lymphokine activated killer cells and interleukin 2 or high dose interleukin-2 alone. NEJM 316, 889-905.

Rosenberg, S.A., Lotze, M.T., Muul, L.M., Leitman, S., Chang, A.E., Ettinghausen, S.E., Matory, Y.L., Skibber, J.M., Shiloni, E., Vetto, J.T., Seipp, C.A., Simpson, C., & Reichert, C.M. (1985). Observations on the systemic administration of autologous lymphokine activated killer cells and recombinant interleukin-2 to patients with metastatic cancer. NEJM. 313, 1485-1492.

Rosenstein, M., Ettinghausen, S.E., Rosenberg, S.A. (1986). Extravasation of intravascular fluid mediated by the systemic administration of recombianant interleukin-2. J. Immunol. 137, 1735-1742.

Rosenstein, M., Yron, I., Kaufmann, Y., & Rosenberg, S.A. (1984). Lymphokine-activated killer cells: Lysis of fresh syngeneic natural killer-resistant murine tumor cells by lymphocytes cultured in interleukin 2. Cancer Res. 44, 1946-1953.

Rubin, J.T., Day, R., Duquesnoy, R., Simonis, T.B., Adams, S., & Lotze, M.T. (1995). HLA-DQ1 is associated with clinical response and prolonged survival of patients with melanoma who are treated with interleukin-2 based immunotherapy in humans. Therapeutic Immunology 2, 1-6.

Rubin, J.T., Elwood, L., Rosenberg, S.A., & Lotze, M.T. (1989). Immunohistochemical correlates of response to recombinant Interleukin-2 based immunotherapy in humans. Cancer Research 49, 7086-7092.

Russell, S.J., Brandenburger, A., Flemming, C.L., Collins, M.K.L., & Rommelaere, J. (1992). Transformation-dependent expression of interleukin genes delivered by a recombinant parvovirus. J. Virology 66, 2821-2828.

Russell, S.J., Flemming, C., Eccles, S.A., Johnson, C., & Collins, M.K.L. (1990). Transfer in and expression of the human IL-2 gene in a transplantable rat sarcoma. In: *Cellular Immunology and the Immunotherapy of Cancer* (Lotze, M.T., & Finn, O.J., Eds.). A.J. Liss and Co, pp. 275-278.

Sadlack, B. Kuhn, R., Schorle, H., Rajewsky, K., Muller, W., & Horak, I. (1994). Development and proliferation of lymphocytes in mice deficient for both interleukins 2 and 4. Eur. J. Immunol. 24, 281-284.

Sadlack, B., Merz, H., Schorle, H., Schimpl, A., Feller, A.C., & Horak, I. (1993). Ulcerative colitis-like disease in mice with a disrupted interleukin-2 gene. Cell 75, 253-261.

Sander, E., Prolla, G., Marques, A.R., Zelmanowitz, T., Zampese, M., & Schwartsmann, G. (1993). Reversal by interleukin-2 of alopecia universalis, mucocutaneous candidiasis and sexual impotence in a patient with malignant thymoma. JNCI 85, 673.

Saris, S.C., Patronas. W.J., Rosenberg, S.A., et al. (1989). The effect of intravenous interleukin-2 on brain water content. J. Neurosurg. 71, 169-174.

Scharenberg, J.G.M., Stam, A.G.M., von Blombergm B.M.E., Roest, G.J., Palmer, P.A., Franks, C.R., Meijer, C.J.L.M., & Scheperm R.J. (1994). The development of anti-interleukin-2 (IL-2) antibodies in cancer patients treated with interleukin-2. European J. Cancer 30A, 1804-1809.

Schorle, H., Holtschke, T., Hunig, T., Schimpl, A., & Horak, I. (1991). Development and function of T cells in mice rendered interleukin-2 deficient by gene targeting. Nature 352, 621-624.

Schwartzentruber, D.J. (1993). *In vitro* predictors of clinical response in patients receiving interleukin-2 based immunotherapy. Current Opn. Onc. 5, 1055-1058.

Schwarz, E.M., Salgame, P., & Bloom, B.R. (1993). Molecular regulation of human Interleukin-2 and T-cell function by Interleukin-4. Proc Natl Acad Sci. USA 90, 7734-7738.

Seder, R.A., Germain, R.N., Linsley, P.S., & Paul, W.E. (1994). CD28-mediated costimulation of interleukin 2 (IL-2) production plays a critical role in T cell priming for IL-4 and interferon γ production. J. Exp. Med. 179, 299-304.

Sheldon, A., Flego, L., & Zola, H. (1993). Coexpression of IL-2 receptor p55 and p75 by circulating blood lymphocytes. J. Leukocyte Biology 54, 161-167.

Shiloni, E., & Matzner, Y. (n.d.). Inhibition of Interleukin-2 induced tumor necrosis factor release by dexamethasone: Does it reduce the antitumor therapeutic efficacy?

Siegel, J.P., & Puri, R.J. (1991). Interleukin-2 toxicity. J. Clin. Oncol. 9, 684-704.

Sigal, R.K., Lieberman, M.D., Reynolds, J.V., Shou, J., Ziegler, M.M., & Daly, J.M. (1991). Lowdose interferon gamma renders neuroblastoma more susceptible to interleukin-2 immunotherapy. J. Ped. Surg. 26, 389-396.

Sosman, J.A., Weiss, G.R., Margolin, K.A., Aronson, F.R., Sznol, M., Atkins, M.B., O'Boyle, K., Fisher, R.I., Boldt, D.H., Doroshow, J., Ernest, M.L., Fisher, S.G., Mier, J., Vachino, G., & Caliendo, G. (1993). Phase IB clinical trial of anti-CD3 followed by high-dose bolus interleukin-2 in patients with metastatic melanoma and advanced renal cell carcinoma: Clinical and immunologic effects. J. Clin. Onc. 11, 1496-1505.

Soulis-Liparota, T., Cooper, M., Papazoglou, D., Clarke, B., & Jerums, G. (1991). Retardation by aminoguanidine of development of albuminuria, mesangial expansion, and tissue fluorescence in streptozocin-induced diabetic rat. Diabetes 1328-1334.

Sparano, J.A., Fisher, R.I., Sunderland, M., Margolin, K., Ernest, M.L., Sznol, M., Atkins, M.B., Dutcher, J.P., Micetich, K.C., Weiss, G.R., Doroshow, J.H., Aronson, F.R., Rubinstein, L.V., & Mier, J.W. (1993). Randomized Phase III Trial of treatment with high-dose interleukin-2 either alone or in combination with interferon alfa-2a in patients with advanced melanoma. J. Clin. Onc. 11, 1969-1977.

Stahl, N., & Yancopoulos, G.D. (1993). The alphas, betas, and kinases of cytokine receptor complexes. Cell 74, 587-590.

Steiger, J., Nickerson, P.W., Steurer, W., Moscovitch-Lopatin, M., Strom, T.B. (1995). IL-2 knockout recipient mice reject islet cell allografts. J. Immunol 155, 489-498.

Sykes, M., Romick, M.L., Hoyles, K.A., & Sachs, D.H. (1990). *In vivo* administration of interleukin 2 plus T cell-depleted syngeneic marrow prevents graft-versus-host disease mortality and permits alloengraftment. J. Exp. Med. 171, 645-658.

Tamir, A., & Isakov, N. (1994). Cyclic AMP inhibits phosphatidylinositol-coupled and-uncoupled mitogenic signals in T lymphocytes. J. Immunology 152, 3391-3399.

Taniguchi, T., & Minami, Y. (1993). The IL-2/IL-2 receptor system: A current overview. Cell 73, 5-8.

Tepler, I., Schwartz, G., Parker, K., Charette, J., Kadin, M.E., Woodworth, T.G., & Schnipper, L.W. (1994). Phase I trial of an interleukin-2 fusion toxin (DAB486IL-2) in hematologic malignancies: Complete response in a patient with Hodgkin's Disease refractory to chemotherapy. Cancer 73, 1276-1285.

Teppler, H., Kaplan, G., Smith, K.A., Montana, A.L., Meyn, P., & Cohn, Z.A. Prolonged immunostimulatory effect of low-dose polyethylene glycol Interleukin 2 in patients with human immunodeficiency virus type 1 infection. J. Exp. Med. 177, 483-492, 1993.

Thompson, J.A., Benyunes, M.C., Bianco, J.A., & Fefer, A. (1993). Treatment with pentoxifylline and ciprofloxacin reduces the toxicity of high-dose interleukin-2 and lymphokine-activated killer cells. Seminars Onc. 20, 46-51.

Thomson, A.W., & Starzl, T.E. (1993). New immunosuppressive drugs. Mechanistic insights and potential therapeutic advances. In: *Immunological Reviews*, Vol. 136 (Moller, G., Ed.).

Tilg, H., Shapiro, L., Vannier, E., Poutsiaka, D.D., Trehu, E., Atkins, M.B., Dinarello, C.A., & Mier, J.W. (1994). Induction of circulating antagonists to IL-1 and TNF by IL-2 administration and their effects on IL-2 induced cytokine production in vitro. J. Immunology 152, 3189-3198.

VandreMolen, L.A., Smith, J.W., Longo, D.L., & Steis, R. (1989). Adrenal insufficiency and Interleukin-2 therapy. Ann. Int. Med. 111, 185.

Vassilopoulou-Sellin, R., & Sella, A. (n.d.). Interleukin-2-induced thyroid dysfunction is correlated with treatment duration but not with tumor response. J. Clin. Onc.

Vetto, J.T., Pappa, M.Z., Lotze, M.T., et al. (1987). Reduction of toxicity of interleukin-2 and lymphokine activated killer cells in humans by the administration of corticosteroids. J. Clin. Oncol. 5, 496-503.

Viallat, J.R.,Boutin, C., Rey, F., Astoul, P., Farisse, P., & Brandely, M. (1993). Intrapleural immunotherapy with escalating doses of interleukin-2 in metastatic pleural effusions. Cancer 71, 4067-4071.

Vitolo, D.,Vvujanovic, N.L., Rabinowich, H., Schlesinger, M., Herberman, R.B., & Whiteside, T.L. (1993). Rapid IL-2 induced adherence of human natural killer cells. J. Immunology 151, 1926-1937.

Vogelzang, N.J., Lipton, A., & Figlin, R.A. (1993). Subcutaneous Interleukin-2 plus interferon alfa-2a in metastatic renal cancer: An outpatient multicenter trial. J. Clin. Onc. 11, 1809-1816.

Voss, S.D., Sondel, P.M., & Robb, R.J. (1992). Characterization of the Interleukin 2 receptors (IL-2R) expressed on human natural killer cells activated *in vivo* by IL-2: Association of the p64 IL-2R g chain with the IL-2β chain in functional intermediate-affinity IL-2R. J. Exp. Med. 176, 531-54.

Watts, J.D., Welham, M.J., Kalt, L., Schrader, J.W., & Aebersold, R. (1993). IL-2 stimulation of T lymphocytes induces equential activation of mitogen-activated protein kinases and phosphorylation of $p56^{lck}$ at serine-59. J. Immunology 151, 6862-6871.

Weber, G.F. & Cantor, H. (1994). Phosphatidylinositol synthesis is a proximal event in intracellular signaling coupled to T cell receptor ligation. J Immunology 152, 4433-4443.

Weijl, N.K., Van Der Harst, D., Brand, A., Kooy, Y., Van Luxemburg, S., Schroder, J., Lentjes, E., Van Rood, J.J., Cleton, F.J., & Osanto, S. (1993). Hypothyroidism during immunotherapy with interleukin-2 is associated with antithyroid antibodies and response to treatment. J. Clin. Onc. 11, 1376-1383.

Weil-Hillman, G., Schell, K., Segal, D.M., Hank, J.A., Sosman, J.A., & Sondel, P.M. (1991). Activation of human T cells obtained pre- and post-interleukin-2 (IL-2) therapy by anti-CD3 monoclonal antibody plus IL-2: Implications for combined *in vivo* treatment. J. Immunotherapy 10, 267-277.

West, W.H., Tauer, K.W., Yannelli, J.R., Marshall, G.D., Orr, D.W., Thurman, G.B., & Oldham, R.K. (1987). Constant infusion recombinant interleukin-2 in adoptive immunotherapy of advanced cancer. N. Engl. J. Med. 316, 898-905.

West, W.H., Tauer, K.W., Yannelli, J.R., Marshall, G.D., Orr, D.W., Thurman, G.B., & Oldham, R.F. (1987). Constant-infusion recombinant Interleukin-2 in adoptive immunotherapy of advanced cancer. NEJM 316, 898-905.

Wiebke, E., Rosenberg, S.A., & Lotze, M.T. (1988). Acute immunologic effects of Interleukin-2 therapy in cancer patients: Decreased delayed type hypersensitivity response and decreased proliferative response to soluble antigens. J. Clin. Oncol. 6, 1440-1449.

Yamauchi, A., & Bloom, E.T. (1993). Requirement of thiol compounds as reducing agents for IL-2 mediated induction of LAK activity and proliferation of human NK cells. J. Immunology 153, 5535-5544.

Yang, J.C., Topalian, S.L., Parkinson, D., Schwartzentruber, D.J., Weber, J.S., Ettinghausen, S.E., White, D.E., Steinberg, S.M., Cole, J., Kin, H.I., Levin, R., Gularia, A., MacFarlane, M.P., White, R.L., Einhorn, J.H., Seipp, C.A., & Rosenberg, S.A. (1994). A randomized comparison of high-dose and low-dose intravenous Interleukin-2 for the therapy of metastatic renal cell carcinoma—an interim report. J. Clin. Oncol. 12, 1572-1576.

Zaloom, Y., & Gallagher, G. (1993). IL-2 inhibits the induction of systemic antitumour immunity by IL-4 in the peritumoural treatment of experimental melanoma. Anticancer Res. 13, 1081-1086.

INTERLEUKIN-3

John W. Schrader

Growth Factors and Cytokines in Health and Disease
Volume 2A, pages 49-84.

ISBN: 0-7623-0117-1

I. INTRODUCTION

Interleukin-3 (IL-3) is a member of a large family of glycoproteins that are characterized by a characteristic three dimensional structure. A minority of this family, including growth hormone, prolactin, and erythropoietin, is comprised of classical endocrine hormones, but the majority are cytokines that usually act in the vicinity of their cellular sources, rather than systemically. Typically these cytokines are released during inflammatory responses—in response to the presence of microorganisms or other cytokines, or, as appears to be the case with IL-3, activation of cells of the immune system. A small minority of cytokines, which includes ciliary neurotrophic factor (CNTF), colony stimulating factor-1 (CSF-1), Steel factor (SLF), and leukemia inhibitory factor (LIF), but probably not IL-3, also function during development or normal physiological turnover of cells and are then present in normal animals. In contrast, IL-3 is released only when the T lymphocytes or mast cells of the immune system are activated. IL-3 usually acts locally, in the vicinity of its cellular source, and targets a spectrum of cells within the lymphohemopoietic system, but probably does not act outside it.

The molecule now recognized as IL-3 was in hindsight probably responsible for generation of mast cells in cultures of murine thymocytes (Ginsberg and Sachs, 1963), and the first report of the generation of nucleated hemopoietic cells in tissue culture. In that the thymocytes were not obtained from inbred mice, it is likely that interactions between allogenic T lymphocytes resulted in the production of IL-3 which stimulated the generation of mast cells from undifferentiated precursors.

Most of the early insights into the role of cytokines in hemopoiesis, however, came from analysis of other soluble factors that were required for the development of colonies of macrophages and/or neutrophils from bone-marrow cells in semi-solid culture medium (Bradley and Metcalf, 1966; Pluznik and Sachs, 1965). Metcalf and colleagues showed that generation of colonies of mature neutrophils or macrophages, or of mixtures of the two cell types, depended on the presence of a series of factors given the operational title of colony-stimulating factors (CSFs). Initially, three such CSFs were defined: granulocyte-macrophage colony-stimulating factor (GM-CSF) stimulated the growth of colonies containing both neutrophils and macrophages, granulocyte-CSF (G-CSF) stimulated the growth of colonies of neutrophils, and CSF-1 or macrophage (M)-CSF stimulated the generation of colonies of macrophages. As discussed below, IL-3 also proved to be a CSF, but differed from these CSFs in that instead of stimulating the production of cells of a particular lineage or closely-related lineages it stimulated the production of cells of a wide range of the various cell-lineages that are ultimately derived from the pluripotential hemopoietic stem cell.

This multiplicity of the targets of IL-3 resulted in its independent emergence in different laboratories that were studying a variety of immunological or hematological phenomena and cell-types. One of these cells was the mast cell. In the mouse, supernatants from cultures of activated T lymphocytes (Nabel et al., 1981; Razin et al., 1981; Schrader and Nossal, 1980; Schrader, 1981) or from cultures of a monocytic leukemia WEHI-3B, promoted the growth of homogenous populations of mast cells (Tertian et al., 1981). These mast cells were dependent on the continued presence of a specific growth factor and died in its absence. The factor was shown to be a direct product of T lymphocytes and to be distinct from other lymphokines that were shown to be produced by the same T cell clones, such as IL-2 or GM-CSF (Schrader and Nossal, 1980; Schrader, 1981). Other groups had shown that conditioned medium from cultures containing mitogen-stimulated spleen cells or WEHI-3B cells permitted the growth of factor-dependent lines of cells with characteristics of mast cells or various types of immature hemopoietic cells (Dexter et al., 1980; Hasthorpe, 1980; Greenberger 1980; Nagoa et al., 1981).

At the same time, Ihle and colleagues had been studying an activity that was present in the supernatant of cultures of activated T lymphocytes and a myelomonocytic leukemia WEHI-3B, and which stimulated the induction in populations of bone marrow cells of the enzyme 20 a hydroxy steroid dehydrogenase (20αSDH; Ihle et al., 1983; Lee et al., 1982). This enzyme was postulated to be a differentiation marker for T lymphocytes, and in recognition of the supposed role of this factor in T cell differentiation, it was termed IL-3. In keeping with this notion, Ihle's group reported that IL-3 purified from WEHI-3B conditioned medium supported the *in vitro* generation of lines of T lymphocytes (Hapel et al., 1981). This report, however, appears to have been in error, and the "T cell lines" appear to have been in fact contaminating WEHI-3B cells (Hapel, personal communication). It was subsequently recognized that 20αSDH was not restricted to T lymphocytes, and was also apparent in myeloid cells (Hapel et al., 1985). The induction of 20αSDH activity by IL-3 thus reflected its role in promoting the development of mast cells rather than T lymphocytes. As discussed further below, the role of IL-3 in the development of lymphoid cells is still being clarified. Nevertheless, using the induction of 20αSDH activity as an assay, Ihle successfully purified IL-3 to homogeneity and determined an amino-terminal amino acid sequence (Ihle et al., 1983). He established, in collaboration with laboratories using other assays, that purified IL-3 stimulated the growth and survival of primary mast cells and of factor-dependent lines of hemopoietic cells, and induced the synthesis of histamine in bone marrow cells.

Using a bioassay based on the stimulation of the proliferation and survival of mast cells derived from murine bone-marrow, our laboratory had purified a protein that included the amino-acid sequence reported by Ihle and colleagues, but had an additional five amino acids at the amino-terminus (Clark-Lewis et al., 1984). Subsequent experiments using an antibody specific for these five amino acids indicated that most, but not all, of IL-3 occurring *in vivo* had these five amino acids (Ziltener et al., 1987). We had termed this biological activity "persisting cell stimulating factor" (PSF), based on the operational name "persisting cells" that we had originally given to the homogenous populations of mast cells that developed when murine lymphohemopoietic cells were cultured for several weeks in the presence of PSF.

II. ACTIONS OF IL-3 ON HEMOPOIETIC STEM CELLS AND PROGENITOR CELLS OF MULTIPLE LINEAGES

During our biochemical characterizations, we observed that IL-3 co-purified with a variety of bioactivities. These included activities that stimulated the growth in agar-cultures of murine bone-marrow of multiple types of colonies including colonies of megakaryocytes, macrophages, neutrophils, eosinophils, as well as colonies containing mixtures of cell types. IL-3 also co-purified with an activity that induced the expression of the Thy-1 antigen on hemopoietic cells from murine bone-marrow, termed "Thy-1 inducing activity" (Schrader et al., 1982,

1988) as well as with an activity that stimulated the *in vitro* survival of bone-marrow-derived cells (CFUs) that were capable of forming colonies of myeloid and erythroid cells in the spleens of irradiated mice (Schrader et al., 1982). Experiments with purified material confirmed that IL-3 acted as a colony-stimulating factor that was capable of generating colonies of cells of multiple lineages; moreover, using single cells it was demonstrated that the ability of IL-3 to stimulate hemopoietic progenitor cells was direct and not mediated via stimulation of another cell type (Schrader et al., 1988). Homogenous preparations of IL-3 directly or indirectly stimulated the *in vitro* survival of pluripotential CFUs and also supported directly or indirectly the *in vitro* survival of cells capable of repopulating irradiated mice with B and T lymphocytes (Schrader et al., 1988).

Other evidence for the broad target specificity of IL-3 among hemopoietic cells came from Iscove and colleagues (1982), who characterized an activity from WEHI-3B conditioned medium that had the same biochemical properties as IL-3 and that synergized with erythropoietin to generate bursts of erythroid cells in semi-solid cultures (burst promoting activity—BPA) as well as stimulating the precursors of macrophages, granulocytes and megakaryocytes. Likewise, Metcalf's group purified an activity which stimulated the growth of colonies of multiple lineages of hemopoietic cells (multi-CSF) which also had the same biochemical properties as IL-3 (Cutler et al., 1985).

IL-3 was thus studied intensely by different laboratories under different guises—the variety of these assays and their cellular targets testify to the broad range of targets of IL-3 among hemopoietic stem cells and their derivatives. These multiple activities resulted in IL-3 being initially studied under a variety of names, which include persisting-cell-stimulating factor, histamine-producing cell-stimulating factor, multi-CSF, multilineage hemopoietic growth factor, Thy-1 inducing factor, CFU stimulating activity, CSF-2αa, CSF-2β, hemopoietic cell growth factor, mast-cell growth factor, eosinophil-CSF, megakaryocyte-CSF, erythroid-CSF, burst-promoting activity, neutrophil-granulocyte-CSF, hemopoietin-2, and synergistic activity. The structure of human IL-3 was determined following expression cloning of the cDNA (Yang et al., 1986). There were only minimal similarities in the amino-acid sequence of human and mouse IL-3, and an absolute and reciprocal lack of cross-species bioactivity. Human IL-3, however, shared with murine IL-3 the same broad spectrum of biological activities.

III. IL-3 AND PLURIPOTENTIAL HEMOPOIETIC STEM CELLS

There has been significant progress in understanding the heterogeneity of pluripotential hemopoietic stem cells and in particular in identifying the most primitive cells with greatest self-renewal potential. Since IL-3 was first recognized, the family of cytokines to which it belongs has grown dramatically and other members that affect pluripotential stem cells, such as SLF and Flt-3 ligand (Flt-3L), have been identified. IL-3, however, remains the most effective single cytokine for stimulating the division of primitive hemopoietic stem cells, although in general, the optimal promotion of their *in vitro* self-renewal or differentiation requires combinations of cytokines. IL-3, particularly in combination with other factors acts on stem cells with significant self-renewal potential in the mouse (Iscove and Yan, 1990) and in the human (Brugger et al., 1993; Haylock et al., 1992). IL-3 is, thus, used in most combinations of cytokines that have been used for the *ex vivo* expansion of human hemopoietic stem cells (McAlister et al., 1992). Because of the technical difficulties in rigorously assessing the self-renewal capacity of stem cells, particularly in the human, it is not clear what mixtures of cytokines best expand stem cells with long-term repopulating activity, and whether any combination of cytokines can support true self-renewal without loss of primitive characteristics.

IV. ACTIONS OF IL-3 ON MATURE, WELL-DIFFERENTIATED CELLS

In addition to these actions on hemopoietic stem or progenitor cells, homogenous IL-3 is also active on mature, well-differentiated cells of a number of hemopoietic lineages. Mature macrophages reacted to IL-3 by increasing their spreading on plastic and their phagocytosis of yeast (Crapper et al., 1985). IL-3 also induces macrophages to express increased levels of major histocompatibility complex antigens and LFA-1 (Frendl and Beller, 1990) and increased levels of mRNA for IL-1, IL-6, and TNFα (Frendl et al., 1990). Well-differentiated bone marrow derived mast cells ("P-cells") required IL-3 to survive *in vitro* (Schrader and Nossal, 1980). Moreover, IL-3 regulates their expression of Class II Major Histocompatibility Complex antigens, inhibiting the stimulatory effects on Class II MHC expression of interferon-γ (Wong et al., 1984).

Treatment of eosinophils with IL-3 enhanced their *in vitro* survival (Rothenberg et al., 1988), antibody-dependent cell-mediated cytotoxicity, phagocytosis, and generation of super-oxide ions in response to f-met-leu-phe (Lopez et al., 1987). IL-3 primes human basophils for the release of histamine (Hirai et al., 1988; Kurimoto et al., 1989; Mayer et al., 1989; Valent et al., 1989). Well-differentiated megakaryocytes also respond to IL-3 (Ishibashi and Bernstein, 1986)

V. IL-3 AND LYMPHOID CELLS

As discussed above an early report that IL-3 stimulated the generation of clones of helper T lymphocytes (Hapel et al., 1981) was erroneous. Palacios reported that he could reproduce this phenomenon with human T lymphocytes but this has not been seen by others. Palacios and colleagues also reported the generation of IL-3 responsive cell lines with the characteristics of prothymocytes (Palacios and Pelkunen, 1988) but these results have not been reproduced. There are suggestions that a subset of mature T lymphocytes in humans (Londei et al., 1989) respond to IL-3, but the physiological significance of these observations is not yet clear.

There is better evidence for effects of IL-3 on cells of the B lymphocyte lineage. A report from Palacios and colleagues (1984) that IL-3 stimulated the generation from fetal liver at high frequency of lines of IL-3-dependent pre B lymphocytes has not been reproduced. Data in the original paper that demonstrated rearrangement of immunoglobulin genes in these cells was apparently due to plasmid contamination (McKearn, personal communication). IL-3 has also been used to generate factor-dependent cell lines that are capable of generating T and B lymphocytes in irradiated mice (Palacios and Steinmetz, 1985) but these results have not been readily reproduced. Such an IL-3 dependent cell line has been shown to undergo differentiation along the B-cell lineage, exhibiting rearrangements of immunoglobulin genes (Kinashi et al., 1988). However, it is not clear the degree to which such cell lines, which appear to occur infrequently, represent counterparts of physiologically important cells.

Recently, however, data from a number of laboratories studying primary cultures have clarified the effects of IL-3 on stem cells capable of generating B and T lymphocytes in irradiated mice, and also on cells committed to the B lineage. Investigation of the potential effects of IL-3 on B-lymphopoiesis appears to have been complicated by the presence in the experimental systems used of myeloid cells, which inhibit B lymphopoiesis. Moreover, it has been suggested that alone, IL-3 directly affects pluripotential cells, inhibiting their subsequent generation of cells committed to the B lineage.

Much of this evidence comes from Ogawa and colleagues who have demonstrated the *in vitro* generation of murine pre B cells from single pluripotential hemopoietic stem cells (Hirayama et al., 1992; Ball et al., 1995). In the original two-stage culture system (Hirayama et al., 1992), small colonies of stem cells were identified in a primary culture and transferred to a second culture. When cytokine combinations, such as SLF and IL-6, were present in the first culture, and SLF and IL-7 in the second, $B220^+$ cells were generated. Moreover, cells from the primary culture were capable of forming B lymphocytes upon transfer to scid mice were generated. However, if IL-3 were substituted for SLF in the first stage, the subsequent generation of B lineage cells was suppressed. This effect of IL-3 could be due to an instructive effect that altered the differentiative potential of pluripotential stem cells, disfavoring generation of B-lineage cells. Ogawa has recently reported that exposure of pluripotential stem cells to IL-3 also blocks the subsequent development of T lymphocytes from these cells in scid mice (personal communication).

These observations, or the effect of IL-3 on the differentiative potential of stem cells are important, since IL-3 is usually a component of culture systems used in protocols aiming to transfer genes into pluripotential hemopoietic stem cells with the aim or reconstituting the lymphoid system. There is data that IL-3, at least in combination with other factors, supports the *in vitro* propagation of pluripotential stem cells that can generate B lineage cells *in vitro* (Galy et al., 1994). It is significant that Ogawa's original report indicates that supernatants of activated T lymphocytes, that would have contained not only IL-3, but also other cytokines, did support the growth of stem cells that could give rise to B lymphopoiesis (Hirayama et al., 1992). IL-4 appears to overcome the inhibitory effect of IL-3 on commitment of stem cells to the B lineage (Ogawa, personal communication).

IL-3 also appears to exert an indirect inhibitory effect on B lymphopoiesis by favoring the proliferation and differentiation of progenitors of myeloid cells that in turn inhibit the generation of B cells. Thus, the depletion of c-kit positive cells from bone marrow results in a reduction in myelopoiesis, but an increase in B lymphopoiesis (Rico-

Vargas et al., 1994). Likewise, the *in vitro* generation of B lineage cells from embryo stem cells is inhibited by the presence of macrophages (Nakano et al., 1994). Winkler et al. (1995) have reported convincing evidence that, provided myeloid progenitors have been removed, precursors of B lymphocytes (c-kit+, B220+) can grow on an IL-7 deficient stromal line in the presence of either IL-7 or IL-3. IL-4 could also conceivably abrogate these inhibitory effects of myeloid cells on the growth of B cell precursors through the well-known ability of IL-4 to inhibit many macrophage functions.

The physiological significance of IL-3 in B lymphopoiesis remains to be demonstrated. Because IL-3 is probably not present in normal animals and is only produced following immunological activation of T lymphocytes or mast cells, any role of IL-3 would likely be restricted to pathological situations. Moreover, the inhibitory effects of myeloid cells on committed B lineage cells discussed above, might suggest that stimulation of the bone marrow by IL-3 would result in a net inhibition of B lymphopoiesis. However, it should be noted that the release of IL-3 from T-lymphocytes or mast cells will always be accompanied by the release of other cytokines, such as IL-4, that may overcome these inhibitory effects as discussed previously.

VI. SOURCES OF IL-3

The only well established physiological sources of IL-3 are lymphohemopoietic cells. There are two general rules: IL-3 is only released after activation of its cellular source, and IL-3 is always released in concert with other cytokines. The major physiological source of IL-3 is the activated T lymphocyte (Schrader, 1981), of both the TH1 subset, which secretes IL-3 together with IL-2, GM-CSF, interferon-γ and TNF, and the TH_2 subset, which secretes IL-3 together with IL-4, IL-5, IL-6, IL-10, IL-13, GM-CSF, and TGFβ. IL-3 is also secreted by activated mast cells. Activation of mast cells by cross-linking by antigen of cytophilic antibodies bound to Fc receptors, results in the rapid accumulation of mRNA encoding IL-3 and other cytokines such as IL-4, IL-5, TNF, and IL-8 (Wodnar-Filipowicz et al., 1989; Plaut et al., 1994; Burd et al., 1989). IL-3 release following mast cell activation by cross-linking of FcR is rapid and transient. It may account for the rapid rise in histamine producing cell-stimulating activity detected in the serum of parasitized mice six hours after challenge with antigen (Abbud-Filho et al., 1983).

There is also evidence that contact of connective tissue mast cells with fibroblasts up-regulates levels of IL-3 mRNA (Razin et al., 1991).

Eosinophils, like mast cells are both a target of IL-3 and also are a source of IL-3 and other cytokines (Kita et al., 1991; Moqbel et al., 1994). Stimulation of eosinophils or by crosslinking of FcR has been reported to induce the release of IL-5 (Dubucquoi et al., 1994) and it is likely also IL-3. Adherence of eosinophils to fibronectin enhances their survival and stimulates the autocrine production of IL-3 and GM-CSF (Anwar et al., 1993).

In summary, stimulation of mast cells or eosinophils via FcR and integrins results in increased levels of IL-3 mRNA and IL-3 production. The potential importance of these autocrine loops in the pathogenesis of allergic diseases is still under investigation.

VII. PHYSIOLOGICAL ROLE OF IL-3

There is no evidence that IL-3 is involved in the normal, steady-state production of any cells of hemopoietic origin. Biologically significant levels of IL-3 are not detectable in normal serum, and early experiments demonstrated that the skin of normal mice could not support the survival of transplanted IL-3-dependent bone marrow derived mast cells (Crapper et al., 1984a). In contrast, transplanted IL-3-dependent mast cells survived in mice that were bearing a tumor that secreted IL-3 and thus had detectable levels of IL-3 in their serum. Nor is there good evidence to support the hypothesis that there might be special sites in the bone marrow for the production of very small but critical amounts of IL-3 that regulated normal hemopoiesis. Mice with a genetic defect in the α-chain of the IL-3 receptor that results in hyporesponsiveness to IL-3 (Hara et al., 1995; Leslie et al., 1996) do not have obvious abnormalities of hemopoiesis. A definitive answer should be provided by analysis of "knockout" mice homozygous for non-functional IL-3 genes. It would be very surprising, given the evidence accumulated over the past 15 years, if IL-3 were to prove a critical component of the mechanisms regulating normal hemopoiesis.

In contrast, early studies on IL-3 pointed strongly to its primary function as a component of the response of the immune system to invasion. For example, the local activation of lymphocytes in lymph nodes by specific antigens results in local secretion of IL-3 and increases in numbers of mast cells and their progenitors in that lymph node, but not other nodes involved in the response (Crapper and Schrader, 1983).

In graft-versus-host disease, with systemic activation of T lymphocytes, effects of IL-3 were seen throughout the body, and IL-3 was detected in the serum (Crapper and Schrader, 1986). IL-3 can thus be seen as one of the links between the lymphocytes, which sense invasion by foreign substances, and the various effector cells that directly mediate the defense and repair response.

Because the range of potential targets for IL-3 varies in different anatomical sites, the local effects of IL-3 vary depending upon the site in which it is produced. For example, at mucosal surfaces such as that of the gut, the only targets for IL-3 are committed but undifferentiated mast cell progenitors (Crapper and Schrader, 1983). Activation of T cells in the gut wall thus results in the local generation of mast cells and a local mastocytoses. In contrast, in the spleen IL-3 finds more targets and thus stimulates increases in numbers of neutrophils, hemopoietic progenitor cells, and megakaryocytes, as well as mast cells (Crapper and Schrader, 1986; Schrader et al., 1988).

VIII. STRUCTURE OF IL-3

Analysis of the cDNA encoding murine IL-3 (Yokota et al., 1984; Fung et al., 1984) indicated that it was polypeptide of 140 amino acids, with no obvious homologies to other cytokines. One exception was the still unexplained observation that IL-3, like many cytokines, has the amino acid alanine followed by a proline at the amino-terminus (Schrader et al., 1986b). The amino acid sequences of human IL-3 and murine IL-3 were only 30% homologous (Yang et al., 1986), reflecting a rapid evolutionary divergence of murine and human IL-3. The gene encoding IL-3 was very closely linked to that encoding GM-CSF, occurring on chromosome 11 in the mouse, and chromosome 5 at 5q23-q31 in the human, in a cluster together with IL-4 and IL-5. In the human, the GM-CSF and IL-3 genes are less than a Kb apart. The clustering of these cytokine genes probably relates to their coordinated expression. Indeed, there is evidence in the case of the IL-3 and GM-CSF genes that there is an enhancer located in the area between them that controls their expression (Osborne et al., 1995).

Murine IL-3 is heavily glycosylated, mainly by N-linked carbohydrate moieties (Ziltener et al., 1988). Whereas non-glycosylated synthetic or recombinant IL-3 has an apparent Mr of 14 K, T cell derived IL-3 existed in a variety of glycoforms of Mr 23-32k (Ziltener, 1993).

Careful analysis of the biological activity of these different glycoforms, however, has failed to detect any functional differences.

IL-3 was the first hemopoietic growth factor to be chemically synthesized in biologically active form (Clark-Lewis et al., 1984). Automated chemical synthesis provided a new strategy for analysis of structure-function relationships. Analysis of structural variants of IL-3 with the use of monoclonal antibodies identified regions of the IL-3 molecule critical for interactions with the IL-3 receptor. Analyses of structure-function relationships have been facilitated by the recognition that IL-3 is a close structural and functional relative of GM-CSF and IL-5, and that structure-function data on these molecules is relevant to analysis of IL-3. Moreover, recently direct data on the three dimensional structure of IL-3 has become available (Feng et al., 1995). These show that the IL-3 molecule has the 4 helix-bundle fold that characterizes members of the cytokine family—including growth hormone, prolactin, GM-CSF, IL-2, IL-4, IL-5, and CSF-1. Residues in the first α-helix of IL-3 are critical for its binding to the receptor (Barry et al., 1994), and mutation of residues 21 and 22 result in greatly diminished activity. Glu^{22} corresponds to Glu^{21} in GM-CSF, mutation of which interferes with binding to the β chain of the receptor while preserving the interaction with the α chain (Lopez et al., 1992). As discussed below, the β_c chain of the IL-3 receptor is shared by the receptors for GM-CSF and IL-5, and the areas of these three cytokines that interact with β_c are likely to exhibit corresponding structural features. Experiments with a hybrid cytokine in which the first α-helix of mIL-5 was substituted for that of hGM-CSF demonstrated that this α-helix conferred upon the hGM-CSF the ability to interact with the hGM-CSF receptor α-chain and the murine β_c chain (Shanafelt et al., 1991a).

A second, carboxy-terminal region was also shown to be important for receptor binding (Shanafelt et al., 1991b). This region appears to be critical for interactions with the α chain of the IL-3 receptor. Mutations at residues 101 and 111 have been shown to increase the affinity of IL-3 for both its low and high affinity receptors (Lopez et al., 1992). More recently McKearn and colleagues have also generated IL-3 analogues with increased potency by mutating multiple residues (Thomas et al., 1995).

IX. THE IL-3 RECEPTOR

The high affinity binding site for IL-3 identified in cross-linking studies is made up of two subunits (Duronio et al., 1992b). A smaller, 70k, α chain binds IL-3 with low affinity (Kitamura et al., 1991) and a larger 125k β chain that alone has no affinity for IL-3 but with the a chain forms a high affinity binding site (Kitamura and Miyajima, 1993). The β chain is also a component of the receptors for GM-CSF and IL-5—each of which has a "private," specific α chain (Miyajima et al., 1992). Thus, in the case of cells which express all three α chains and a limited number of β_c chains, for example, the human eosinophil, these cytokines cross-compete for binding (Lopez et al., 1990; Park et al., 1989). The functional significance of this cross-competition is unclear in that, as discussed below, no differences in the intracellular signals triggered by these three cytokines have been described.

In the mouse, the situation is complicated by the presence of a duplication of the β_c gene termed AIC-2B in the mouse, which has given rise to an additional βc chain termed AIC-2A (Itoh et al., 1990). The AIC-2A protein participates only in IL-3 binding, and itself has a low affinity for IL-3. The mouse, thus, has two IL-3 receptors, one in which the IL-3Rα chain is complexed with the IL-3-specific AIC-2A β chain, and one in which the IL-3-Rα interacts with AIC-2B, the β_c also shared by the receptors for GM-CSF and IL-5 (Gorman et al., 1990, 1992; Miyajima et al., 1993). The functional significance of this situation in the mouse is unclear; mice homozygous for non-functional AIC-2A genes show no phenotype. In contrast mice with non-functional AIC-2B β_c genes show a phenotype resembling that seen in mice unable to produce GM-CSF (Nishikamamura et al., 1995).

X. THE CYTOKINE-RECEPTOR SUPERFAMILY

Both the α and the β chains of the IL-3 receptor are members of new receptor superfamily—the hemopoietin- or cytokine-receptor superfamily (Bazan, 1990). All cytokines in the four-helix bundle family bind to this class of receptors— with three known exceptions: CSF-1, SLF, and Flt3-L, all of which bind to members of a structurally distinct family of homodimeric receptors, that are characterized by their intrinsic enzymatic activity as protein tyrosine-kinases.

Receptors of the cytokine-receptor superfamily are oligomeric structures made up of homodimers or hetero-dimers or -trimers. With the exception of IL-2 and IL-15, all of these subunits are members of the cytokine-receptor superfamily, and are characterized by extracellular Fibronectin type III domains with characteristically spaced cysteine residues, and a motif of the form, WSXWS. In the case of the human growth hormone receptor, the three dimensional structure of the extracellular domains of the receptor bound to growth hormone has been described (DeVos et al., 1992). The high-affinity complex is made up of a single GH molecule and a homodimer of the receptor subunit. Ligand-induced dimerization of the subunits depends upon the presence on GH of two distinct binding sites. The side chains of the WSXWS motif characteristic of receptors of this family form a patch on the surface of the ligand-receptor complex, raising the intriguing possibility that it interacts with another molecule.

The intracellular portions of receptors of the cytokine receptor family are notable in lacking any conserved domains indicative of enzymatic activity. Instead they contain structural motifs such as tyrosine-residues in the context of specific peptide motifs, which mediate interaction of these cytoplasmic regions with a range of cytoplasmic molecules. These include kinases, phosphatases, and molecules that link the receptor complex to components of multiple signal transduction paths. A good deal is now known about the intracellular signals activated by IL-3 and about the role of different regions of cytoplasmic domains of the receptor activating these paths.

XI. SUBUNIT STRUCTURE OF THE FUNCTIONAL IL-3 RECEPTOR

The first step in IL-3 stimulated signalling appear to be the low-affinity interaction of IL-3 with the α chain of the receptor (IL-3Rα). The IL-3-IL-3Rα complex then interacts with a β chain forming a high-affinity receptor. By analogy with experiments on GM-CSF (Shanafelt and Kastelein, 1992), it seems likely that the high-affinity binding of IL-3 to this IL-3Rα: IL-3Rβ complex is not essential for triggering signal transduction, but the approximation of the α and β chains is. What is not clear is whether the simple generation of a dimer of one IL-3Rα and one IL-3Rβ is sufficient for the induction of signaling.

In the case of the IL-6 receptor, the functional receptor appears to be a hexamer that is made up of a pair of complexes of IL-6 and the α and β chains of the IL-6 receptor (Paonessa et al., 1995; Ward et al., 1994). In this instance the cytoplasmic domain of the α chain of the IL-6 receptor is completely dispensable for signaling; the critical part of the α chain of the IL-6 receptor is the extracellular domain, which, after it is complexed with IL-6, undergoes a conformational change and interacts with the β chain. Two of these trimers interact to form a hexamer, a key event because it results in approximation of the cytoplasmic domains of the two β chains (gp130). It is dimerization of the cytoplasmic domains of two β chains that triggers the intracellular signaling process.

In one respect, the IL-3 receptor is clearly different. Thus, there is good evidence that the cytoplasmic domain of the IL-3Rα chain, while not needed for the generation of a high-affinity binding site, is essential for the IL-3-induced mitogenic signals (Orban and Schrader, unpublished). Thus, the cytoplasmic domain of the IL-3R α chain either plays a direct role in signal transduction, or is essential for generating some other critical structure, such as a homodimer of β chain cytoplasmic domains.

In favor of a possible role for the formation of dimers of the cytoplasmic domains of β chains in IL-3 signaling, is evidence that β-β dimers can induce mitogenesis, at least in certain situations. Two groups have reported experiments in which they expressed in factor-dependent cell lines, chimeric receptors, in which the extracellular domains of IL-5R-α or GM-CSFR-α were fused to the cytoplasmic domain of the βc chain. Cells expressing these chimeric chains together with bc grew in response to IL-5 (Takatsu et al., 1995) or GM-CSF (Muto et al., 1995), respectively, suggesting that the ligand-induced generation of homodimers of the cytoplasmic domains of βc, in the absence of the cytoplasmic domains of the α-chains, was a sufficient signal for mitogenesis.

Similar conclusions have been drawn from experiments with a chimeric receptor in which the extracellular domain of the erythropoietin receptor was fused to the cytoplasmic domain of the IL-3 specific β chain, AIC-2A. Cells expressing this chimera grew when erythropoietin was used to crosslink the extracellular domains, suggesting once again that dimerization of the cytoplasmic domains of the β chain could result in a mitogenic signal (Sakamaki et al., 1993).

However, neither of these sets of experiments establish that β-β dimers are sufficient for mitogenesis. Thus, it is conceivable that the ligand-induced dimerization of the extracellular domains of the receptor chains used, may have resulted in the recruitment of an additional chain, perhaps interacting with the WSXWS patch on the ligand-receptor complex referred to above.

To determine whether there is a requirement for the recruitment of additional molecules by the extracellular domains of cytokine receptor-family receptors, experiments have been performed with chimeric molecules with the cytoplasmic domains of IL-3Rα and/or IL-3Rβ, but with extracellular domains that are unrelated to the cytokine-receptor super family. One such set of chimeras consisted of the extracellular domain of the CD8 molecule fused to either the cytoplasmic domain of the IL-3R α chain or to that of the IL-3R β chain AIC-2A. Because the extracellular domains of human CD8 form disulphide linked dimers, these constructs should result in the expression of constitutive dimers. In contrast to the results of the experiments described above, the expression of CD8-β chimeras, which are predicted to generate constitutive dimers of the cytoplasmic domains of the β chain, did not support growth of the factor-dependent cells. Nor did expression of the CD8-a chimeras (Orban, Levings, Jalbert, & Schrader, unpublished data). However, coexpression of the CD8-a and CD8-b chimeras resulted in constitutive growth. These experiments suggest that, whereas homo-dimers of neither the cytoplasmic domains of the IL-3R α nor the β chains are alone sufficient for mitogenesis, simple heterodimers of the cytoplasmic domains of the α and β chain are.

Similar experiments have utilized chimeras consisting of the extracellular domain of CD16 and the transmembrane region of CD7, with the intracytoplasmic domains of the IL-3R α or IL-3R β chains. In cells expressing one or the other chimera, cross-linking of the chimeric molecules using antibodies to CD-16 failed to support growth. However, in cells expressing both the CD16-CD7-IL-3Rα and the CD16-CD7-IL-3Rβ chains, cross-linking of CD-16 with monoclonal antibodies chain domain did result in growth (Orban, Levings, Jalbert, & Schrader, unpublished data).

These experiments with chimeric molecules suggest that heterodimerization of the cytoplasmic domains of the α and the β chains is necessary for signaling. There remains the formal possibility that a single α-β dimer is insufficient, as it is conceivable that the cytoplasmic domains of IL-3Rα and β form complexes of the forms αββ, αββα, and so forth. Nevertheless, these experiments suggest that the cytoplasmic domain of the IL-3R α chain has an active role in signal transduction. This is supported by the fact that the cytoplasmic domain of IL-3R α contains functionally critical peptide motifs that are conserved in the IL-5Rα and GM-CSFRα chains (Takaki et al., 1994). These are likely to be targets for signalling molecules, such as cytoplasmic kinases.

XII. INTRACELLULAR SIGNALS

A great deal is know known about the variety of intracellular signals that are activated when the IL-3 receptor is engaged and about how they are initiated. Not one of these signals, however, is entirely specific for signaling mediated by IL-3 (or IL-3/GM-CSF/IL-5). There is, thus, currently a great deal of evidence concerning the similarities between signals triggered by IL-3 and by other growth factors, hormones, and cytokines but very little insight into how IL-3 (or indeed any other cytokines) exert its distinctive effects.

XIII. TYROSINE KINASES

It was recognized early that IL-3 induced the tyrosine-phosphorylation of intracellular proteins (Koyasu et al., 1987; Morla et al., 1988; Sorensen et al., 1989; Isfort and Ihle 1990; Kanakura et al., 1990). However, because, like other receptors of the cytokine superfamily, the cytoplasmic portions of the IL-3 receptor lacked sequences characteristic of the catalytic domains of protein kinases, the molecular mechanisms underlying these phosphorylation events were unclear. There was evidence that a cytoplasmic kinase of the src family, Lyn was activated following IL-3 stimulation (Torigoe et al., 1992). More recently another family of cytoplasmic kinases, the JAK kinases, has been recognized as functioning in the response to interferons and shown to be a critical component in signals activated by cytokines (reviewed in Schindler and Darnell, 1995). Jak-2 kinase was shown to be activated by IL-3 (Silvennoinen et al., 1993). Studies on deletion mutants indicate that the membrane proximal region of the β_c chain (between residues 455-517) is sufficient for activation of JAK-2 kinase by GM-CSF (Quelle et al., 1994), although a more distal part of the β chain is required for maximum activation of JAK-2, consistent with the data of Rao and Mufsan (1995). These studies indicated that JAK-2 associates with β_c (Quelle et al., 1944; Rao and Mufson, 1995). It is not clear whether the α chain of the IL-3 receptor associates with JAK-2, although the α chain of the GM-CSF receptor has been reported not to do so (Quelle et al., 1994). Certainly, truncation of the α chain of the IL-3 receptor (Orban et al., unpublished), or of the GM-CSF receptor (Quelle et al., 1994) abolishes IL-3- or GM-CSF-induced activation of JAK-2. JAK- and Src-family kinases may not be the only cytoplasmic kinases involved IL-3 signaling;

the btk kinase has been reported to be involved in IL-5 signaling (Sato et al., 1994).

XIV. STATs

Analysis of interferon-mediated signals identified a new family of transcription factors, termed signal transducers, and activators of transcription, or STATs. These are activated by tyrosine-phosphorylation and are likely to be direct substrate of JAK-kinases (reviewed in Schindler and Darnell, 1995). Proteins of the STAT family, which now includes STAT-1 through STAT-6, are characterized by a DNA-binding domain, SH-2 SH-3-like domains, and a conserved tyrosine residue. STATs preexist in the cytoplasm as inactive monomers, which, upon activation by tyrosine phosphorylation, form homo- or heterodimers with other family members. These dimers rapidly translocate to the nucleus where they bind to palindromic sequences in the promoters of a variety of cytokine-regulated genes. One current working model is that ligand-induced oligomerization of cytokine receptors results in approximation and autophosphorylation and activation of JAK-kinases which then phosphorylate tyrosines in the receptor complex. STAT proteins dock onto phosphorylated tyrosines on the receptor complex via their SH-2 domain, the fine specificity of this SH-2 domain, determining which STAT family members dock with which receptor complexes. The STAT is phosphorylated on the conserved tyrosine, which forms a site for the docking of the SH-2 domain of another STAT monomer. This results in the formation of an SH-2 mediated dimer, which rapidly translocates into the nucleus, where it binds to specific motifs in the promoters of cytokines-regulated genes. Although different cytokine receptors exhibit different patterns of utilization of JAK-kinases and STATs, there is a great deal of overlap. Moreover the same cytokine can activate different JAK-kinases or STATs in different cell types. IL-3 induces tyrosine phosphorylation of STAT-5 (Mui et al., 1995; Azam et al., 1995) which has two genes and multiple isoforms and is also tyrosine-phosphorylated in response to IL-2 and erythropoietin (Quelle et al., 1995). However, IL-3 also induces tyrosine phosphorylation of STAT-6, which is also activated by IL-4 (Quelle et al., 1995) and in some cases STAT-1 (van der Bruggen et al., 1995), which is phosphorylated on tyrosine in response to many cytokines, growth factors and hormones (see Schindler and Darnell, 1995).

XV. THE RAS/MAP KINASE PATHWAY

It was evident from consideration of their biological actions that hemopoietic growth factors must activate not only signals specific for different cytokines, but also common intracellular signals that converge on processes fundamental to cellular function. In the first category are the signals that mediate actions that are specifically associated with the action of a particular cytokine, for example, activation of a specific gene or differentiation program. In the second category would fall the signals that regulate cellular growth and mitosis, as well as those promoting cell-survival and suppressing apoptosis. Thus, it would be expected that even cytokines that utilize receptors that differ markedly in their structure, for example, IL-3 that interacts with a receptor from the cytokine-receptor superfamily, and SLF that interacts with a tyrosine-kinase receptor, would ultimately engage a series of common signal transduction paths. One such path, activated by almost all growth factors, cytokines, and hormones is initiated by activation of members of the Ras family of small G-proteins and leads to a cascade of signaling events involving kinases such as the MAP kinases and ultimately transcription factors. It was, thus, not surprising that IL-3 and other hemopoietic growth factors such as IL-5, GM-CSF, and SLF activate Ras (Satoh et al., 1991; Duronio et al., 1992), through a mechanism that requires tyrosine phosphorylation (Duronio et al., 1992). There was one interesting exception, IL-4 (Satoh et al., 1991; Duronio et al., 1992). The failure of IL-4 to activate the Ras pathway, also shared by a close relative of IL-4, IL-13, suggested an approach to identifying the specific tyrosine phosphorylation events that were involved in the activation of Ras by cytokines like IL-3 or SLF, and were lacking in cells stimulated with IL-4.

Murine-bone-marrow-derived mast cells (BMMC), generated by culturing bone marrow cells with IL-3 (Schrader, 1981), provide a convenient source of a homogenous population of primary cells which responded to a variety of cytokines. These include not only IL-3, but SLF, IL-4, IL-13, IL-9, IL-10, and IFN-γ, and thus permit analysis of the similarities and differences in signals induced by these different cytokines. Comparison of the proteins phosphorylated on tyrosine in response to IL-3, with those phosphorylated in response to SLF revealed a common series of proteins of Mr 55k, 44k, 42k, and 38k (Welham and Schrader, 1992). Other proteins were tyrosine-phosphorylated only in response to one or the other of the two cytokines. Thus, the receptor for

SLF, c-Kit, and an unidentified 77k protein were only tyrosine phosphorylated in response to SLF, whereas the β chains of the IL-3 receptor (Duronio et al., 1992b) and a 70k protein were tyrosine phosphorylated only in response to IL-3.

The 55k, 44k, and 42k phosphorylated proteins all proved to be components of the Ras signaling path. The 44k and 42k proteins were shown to correspond to the *erk*-1 and *erk*-2 MAP-kinases, which were also enzymatically activated in response to IL-3 and SLF (Welham et al., 1992). IL-4 failed to activate MAP-kinases (Welham et al., 1994a). The 55k protein tyrosine phosphorylated in response to both IL-3 and SLF was shown to be Shc (Welham et al., 1994b). Shc is characterized by two types of domain which recognize phosphorylated tyrosines in the context of specific peptide motifs, an SH-2 domain and a PTB domain (reviewed in vander Greer and Pawson, 1995). IL-3 and SLF not only induced tyrosine phosphorylation of Shc but also its association with Grb-2. The latter is an adapter protein consisting of an SH-2 domain and two SH-3 domains. The Grb-2 molecule is constitutively associated with mSos-1 (Bowtell et al., 1992), a cytosolic protein which can interact with the Ras family of proteins and promote the exchange of Ras-bound GDP for GTP. The binding of GTP results in a conformational change in Ras that switches it to its active configuration. In contrast, IL-4 failed to induce tyrosine-phosphorylation of Shc or its association with Grb-2 (Welham et al., 1994b). Because IL-4, unlike IL-3 or SLF, failed to activate Ras, these observations correlated the ability of cytokines to induce tyrosine-phosphorylation of Shc with subsequent activation of Ras.

Evidence from experiments in Rat-1 fibroblasts that were stimulated with EGF (Buday and Downard, 1993) had suggested that the first step in the activation of Ras was the association of linker molecules such as Grb-2 with phospho-tyrosines on the activated receptor. Because mSos-1 was constitutively bound to Grb-2, this resulted in translocation of mSos-1 from the cytosol to the membrane and its approximation to the membrane-bound Ras. Interaction between mSos-1 and Ras then resulted in the exchange of Ras-bound GDP for GTP. A variant of this mechanism involved Shc docking to a phosphotyrosine in the receptor complex, and itself becoming phosphorylated on a tyrosine that was the target of the SH-2 domain of Grb-2. Shc would thus form a link between Grb-2 and its associated mSos-1 and the receptor complex.

Stimulation of BMMC with either IL-3 and SLF induced a retardation in electrophoretic mobility of mSos-1 (Welham et al., 1994b) due to ser/thr phosphorylation. However, kinetic analysis suggested that mSos-

1 was modified subsequent to activation of Ras. As expected from other data, neither IL-4 (Welham et al., 1994b), nor its close relative IL-13 (Welham et al., 1995), induced this modification of mSos-1. However, surprisingly in the light of the model of activation of Ras suggested by Buday and Downard's observations on Rat-1 fibroblasts and EGF, IL-3 did not induce the mass translocation of mSos-1, or of tyrosine phosphorylated Shc to the membrane. Tyrosine-phosphorylated Shc did become associated with Grb-2 following IL-3 stimulation, but these complexes remained cytosolic (Welham et al., 1994b). It seems likely, therefore, that the relevant interactions of these proteins with the membrane must be transient and that the mass translocation of mSos-1 observed in response to EGF by Buday and Downard is not typical.

The effector loop of activated Ras forms a docking site at the membrane for cytoplasmic ser/thr kinases such as Raf-1. In the case of Raf-1, there is evidence that the resultant translocation from the cytoplasm to the membrane results in activation of its enzymatic activity. Activated Raf-1 and Mek-kinase phosphorylate and activate Mek 1 and Mek 2, dual specificity protein kinases which in turn activate the *erk*-1 and *erk*-2 MAP kinases.

XVI. IL-3 ACTIVATES CSBP/P38/HOG-1 KINASE

IL-3 also, however, activates other kinase cascades. Recently, we have identified the 38K protein phosphorylated that was noted to be phosphorylated on tyrosine in response to IL-3 and SLF (Welham and Schrader, 1992) as a proline-directed kinase distantly related to *erk*-1 and *erk*-2 MAP kinases (Foltz, Young, Lee, and Schrader, unpublished data). This protein is a member of a well conserved evolutionary family represented in yeast by the HOG-1 stress-activated kinase. The mammalian enzyme has been termed CSBP (Lee et al., 1994), p38 (Hans et al., 1994), or Mpk2 (Rouse et al., 1994) and is also activated by IL-1, TNFα, lipopolysaccharide, osmotic stress, and UV light. CSBP can be activated by phosphorylation by a kinase termed MKK3 (Derijard et al., 1995; Raingeaud et al., 1996) but another MKK6 is probably the physiological activator. CSBP itself activates MAP Kap-kinase that has as one of its substrates, the small heat shock protein, Hsp25/27 (Rouse et al., 1994). IL-3 has been previously reported to activate MAPKAP-kinase (Ahlers et al., 1994).

XVII. IL-3 ACTIVATES SHPTP-2

IL-3 induces tyrosine phosphorylation and activation of a tyrosine phosphatase. The 70k protein that was prominently phosphorylated on tyrosine in mast cells in response to IL-3 but not SLF (Welham and Schrader, 1992) was also tyrosine phosphorylated in other cells in response to GM-CSF. The 70K protein seemed an attractive candidate for a component of a path specific to the three cytokines IL-3/GM-CSF/IL-5 that shared the βc subunit of the IL-3 receptor. However, we identified the 70K protein as the tyrosine-phosphatase, SH-PTP2 or Syp (Welham et al., 1994c). SH-PTP2 is known to be phosphorylated in response to other growth factors, such as PDGF and to IL-6. IL-3 also induced activation of the enzymatic activity of SH-PTP2 (Welham et al., 1994b). Moreover, IL-3 induced the association of SH-PTP2 with both Grb-2 and with the βc of the activated IL-3 receptor. Because Grb-2 is known to be constitutively associated with mSos-1, these observations suggest another potential mechanism for bringing mSos-1 to the cell membrane and into proximity with Ras. IL-3 also induced association of SH-PTP2 with the p85 subunit of PI-3 kinase via the SH-2 domains of the latter (Welham et al., 1994b). It is thus possible that SH-PTP2, a very prominent substrate of the tyrosine kinases activated by IL-3/GM-CSF/IL-5, has multiple roles, serving both as an activator or inhibitor through its action as a tyrosine-phosphatase, and as a linker molecule that may be involved in complexing the active IL-3 receptor with PI-3 kinase and Grb-2 and mSos-1.

XVIII. PI-3 KINASE AND PROTEIN KINASE C

Like other growth factors, IL-3 also stimulates the production of products of the action of PI-3 kinase (Gold et al., 1994). There is no direct evidence for activation of the enzyme itself, and the enhanced generation of products of its action following stimulation of cells by IL-3 may reflect IL-3-induced translocation of PI-3 kinase from the cytoplasm into association with the plasma membrane and its lipid substrates. This would occur via binding of the p85 subunit of the enzyme via its SH-2 domain to phosphotyrosines in the activated receptor complex, or via association of the p110 catalytic domain of the enzyme with activated Ras. The functions of the products of PI3 kinase action are still under investigation. They may be involved in activation of the zeta isoform of protein-kinase-C (PKC). The S6-kinase $p70^{rsk}$ may also be downstream of PI-3 kinase. This enzyme is activated by IL-3 and, rapamycin, which blocks activation of $p70^{rsk}$, also blocks the ability of IL-3 to support IL-3-dependent cells (unpublished observations). IL-3 also activates PKC, possibly through a mechanism involving stimulation of phosphatidyl choline turnover (Duronio et al., 1989).

XIX. ACTIVATION OF IMMEDIATE-EARLY GENES

IL-3 shares with other growth factors the ability to induce the transcription of immediate early genes such as c-*myc*, c-*fos*, and c-*jun*. Studies with deletion mutants of the β_C chain indicate that the induction of c-*myc* involves a path distinct from that involved in induction of c-*fos* and c-*jun*.

One membrane proximal region, from residues 455 to 517, appears sufficient for the induction of c-*myc* and pim-1 genes (Sato et al., 1993). This occurs through a tyrosine-kinase dependent mechanism, but the kinase(s) involved have not been identified. A more distal region, from 626 to 763, is also required for activation of the Ras Raf-1/MAP-kinase pathway. Activation of $p70^{rsk}$ also requires this region. Miyajima and colleagues have more recently shown that this distal region of β_C beyond residue 544 is required for the apoptosis-sparing action of GM-CSF (Kinoshita et al., 1995a). This deficit could be complemented by expression of an activated Ras protein, suggesting that Ras was upstream of a mechanism blocking apoptosis. The ability of IL-3 to induce the *bcl-2* gene showed the same insensitivity to the tyrosine-kinase inhibitor genestein but sensitivity to staurosporine, as did activation of the Ras/Raf-1/MAP-K pathway and the anti-apoptotic effect of IL-3. Moreover, expression of activated Ras induced increased expression of bcl-2 and bcl-x_L, suggesting that Ras regulates apoptosis through regulation of levels of bcl-2 (Kinoshita et al., 1995b) However, there is also evidence that protein kinase-C is required for induction of *bcl-2* expression by IL-3 (Rinaudo et al., 1995), and the details of how IL-3 regulates *bcl-2* expression are still unclear.

In contrast, activation of c-*fos* and c-*jun* requires more distal regions of β_C and correlates with activation of the Ras pathway and phosphorylation of tyrosine 750 (Sato et al., 1994). However, the precise mechanism of IL-3-mediated induction of c-*fos* and c-*jun* is unknown. Analysis of mutant cell lines indicates that different mechanisms are involved in IL-3 induced induction of c-*fos* and c-*jun*, and that the regulation of c-*fos* by IL-3 and SLF involves different pathways (Wieler and Schrader, unpublished data).

XX. INTERACTIONS OF EFFECTS OF IL-3 AND OTHER CYTOKINES

IL-3 synergizes with many other cytokines, particularly in stimulating more primitive stem and progenitor cells. The qualitative similarities in the signal transduction paths stimulated by IL-3 and other cytokines of the 4-helix bundle family, irrespective of whether they utilize receptors of the cytokine-receptor or tyrosine-kinase super-families, suggests that many of these synergistic effects of cytokines may reflect the fact that numbers of receptors for single cytokines may be limiting. For example, the amount of a particular convergent signal that can be generated by stimulation of all available receptors for a single cytokine could be below a critical threshold required for mitogenesis. The additive effect of engagement of the receptors for a second cytokine could overcome this limitation resulting in synergy. Other interactions may have a more specific basis. For example, IL-3 is known to reduce the levels of mRNA encoding c-kit the receptor for SLF (Welham and Schrader, 1991).

XXI. IL-3 AND PATHOLOGY

The pivotal role of IL-3 in generating and supporting the survival of mucosal mast cells points to its likely importance in the pathogenesis of allergic disease. The development of antagonists, particularly small, orally active molecules that would block these actions of IL-3, could provide a new and specific approach to these diseases. As noted above it seems unlikely that IL-3 is necessary for the normal generation of the lympho-hemopoietic cells and a specific IL-3-antagonists seems unlikely to do harm, at least in normal circumstances. The actions of IL-3 in promoting the generation, survival and activation of basophils and particularly eosinophils may also be an important factor in the pathogenesis of chronic changes seen in bronchia asthma.

IL-3 shares these actions within GM-CSF and IL-5, but in the light of the fact that the receptors for all three of these cytokines share a common β_c subunit, it is conceivable that a common inhibitor of the actions of all three cytokines could be developed. Because the IL-3, IL-5, and GM-CSF that affect mast cells, basophils and eosinophils in bronchial asthma may be produced not only by T lymphocytes, but by the granulocytic cells themselves, inhibitors must be effective antagonists for cytokines produced from both autogenous and exogenous sources. Attempts to block the stimulation of leukemic cells by autogenous cytokines have revealed that this can be achieved by extracellular antagonists such as antibodies, but only with much higher concentrations of antibody relative to that of the cytokine than is required to neutralize exogenous cytokines (Orban and Schrader, 1996). These experiments suggest that autostimulatory signals require that the receptor-ligand complex is on the cell surface, but that receptor-ligand complexes may be formed within the cell. Antagonists that freely enter the cell and inhibit ligand-receptor interactions or downstream effects are likely to be required to effectively block the actions of autogenous cytokines.

XXII. IL-3 AND ONCOGENESIS

In mice the aberrant expression of IL-3 in cells that are normally targets of its action results in autostimulation of growth. If such aberrant expression of IL-3 occurs in a clone that already has genetic defect(s) resulting in immortality, it is sufficient to generate a transplantable leukemia (Schrader and Crapper, 1983). In mice aberrant activation of IL-3 genes is a frequent and important mechanism in myeloid leukemogenesis (e.g., Leslie and Schrader, 1989; Metcalf and Rasko, 1993). Interestingly, the myelomonocytic tumor WEHI-3B, which was used as a source of IL-3 by many of the pioneering groups, represents one such autostimulatory leukemia. One copy of the IL-3 gene in WEHI-3B has been activated by insertion of an IAP-element (Ymer et al., 1985). Similar mechanisms have been discovered in other murine myeloid leukemias at a surprisingly high frequency (Leslie et al., 1991). In contrast, autostimulatory mechanisms involving aberrant production of IL-3 do not appear to be important in the pathogenesis of human myeloid leukemia. There are only isolated reports of leukemic cells exhibiting rearrangements of growth-factor genes, including IL-3 (e.g., Uckun et al., 1989).

The absence of evidence for a role of IL-3 in the pathogenesis of human AML contrasts with indications of a role for GM-CSF, IL-1, and TNFα. Thus, purified populations of leukemic blast cells are frequently observed to release GM-CSF, IL-1, and TNF (Young and Griffin, 1986; Oster et al., 1989), although these have not been shown to be of primary significance for the pathogenesis of the disease. One factor to be taken into account is recent evidence that normal hemopoietic stem and progenitor cells express genes encoding IL-1β and the IL-1β converting enzyme (ICE; Watari et al., 1994). These data raise the possibility that the secretion of IL-1 by leukemic blasts might reflect physiological characteristics of their normal blast cell counterparts, and not a primarily pathogenic event. Our unpublished data (Watari, Lee, and Schrader) also indicate that, whereas normal hemopoietic progenitor cells can under some circumstances express GM-CSF genes, this is not the case for IL-3 genes. The production of GM-CSF, but not IL-3, by normal hemopoietic progenitors may be another indication that the expression of cytokine genes in the leukemic blasts of AML mirrors the properties of normal myeloid progenitors.

XXIII. CLINICAL USE OF IL-3 AND AGONISTS OR ANTAGONISTS

Administration of chemically synthesized or recombinant IL-3 to mice generates the predicted increases in mast cells in the gut and lymphoid tissues, and increases in myeloid cells and progenitors including megakaryocytes in the spleen (Kindler et al., 1986; Schrader et al., 1986a; Metcalf, 1988). Clinical interest has focused on the potential for IL-3 to accelerate recovery of hemopoiesis, particularly of platelet levels, after myelotoxic therapies. The possibility that IL-3, through its ability to stimulate early hemopoietic stem cells, would synergize with other, later-acting cytokines in accelerating the recovery of transplanted or damaged marrow is also under evaluation (Donahue et al., 1988). A novel hybrid molecule P1XY321 (Williams et al., 1987), in which IL-3 and GM-CSF are linked together is also in trials (Vadhan-Raj et al., 1994). No clear indications for the clinical use of IL-3 have yet been demonstrated, although it appears to have a favorable effect on the rate of increase in platelets following marrow damage.

As noted, analogues of IL-3 with increased potency and possibly differential effects on different targets have been reported and may be more efficacious, although the question of their immunogenicity will have to be resolved. As has been the case with IL-4 and GM-CSF, analogues of IL-3 with significant antagonist activity are likely to be developed. However, for the management of allergic diseases or asthma, not only IL-3 but IL-5 and GM-CSF would need to be antagonized. The most effective antagonists are thus likely to be small molecules that interfere with the ligand-binding or oligomerization of β_c, or with its ability to initiate critical downstream events.

ACKNOWLEDGMENTS

Unpublished work from the author's laboratory was supported by the Protein Engineering Network of Centres of Excellence of Canada, and The Medical

Research Council of Canada. I thank Cathy Davidson-Hall for expert assistance in preparation of the manuscript.

REFERENCES

Abbud-Filho, M., Dy, M., Lebel, B., Luffau, G., & Hamburger, J. (1983). *In vitro* and *in vivo* histamine-producing cell-stimulating factor (or IL 3) production during Nippostrongylus brasiliensis infection: Coincidence with self-cure phenomenon. Eur. J. Immunol. 13, 841.

Ahlers, A., Engel, K., Sott, C., Gaestel, M., Herrmann, F., & Brach, M.A. (1994). Interleukin-3 and granulocyte-macrophage colony-stimulating factor induce activation of the MAPKAP kinase 2 resulting in *in vitro* serine phosphorylation of the small heat shock protein (Hsp 27). Blood 83, 1791.

Anwar, A.R.E., Moqbel, R., Walsh, G.M., Kay, A.B., & Wardlaw, A.J. (1993) Adhesion to fibronectin prolongs eosinophil survival. J. Exp. Med. 177, 839.

Azam, M., Erdjument-Bromage, H., Kreider, B., Zia, M., Quelle, F., Basu, R., Saris, C., Tempst, P., Ihle, J.N., & Schindler, C. (1995). Interleukin-3 signals through multiple isoforms of Stat5. EMBO J. 14, 1402.

Ball, T.C., Hirayama, F., & Ogawa, M. (1995). Lymphohemoatopoietic progenitors of normal mice. Blood 85, 3086.

Bazan, J.F. (1990). Structural design and molecular evolution of a cytokine receptor superfamily. Proc. Natl. Acad. Sci. USA 87, 6934.

Bradley, T.R., & Metcalf, D. (1966). The growth of mouse bone-marrow cells in vitro. Australian J. of Exp. & Biol. Sciences 44, 287.

Brugger, W., Mocklin, W., Heimfeld, S., Berenson, R.J., Mertelsmann, R., & Kanz, L. (1993). *Ex vivo* expansion of enriched peripheral blood CD34+ progenitor cells by stem cell factor, interleukin-1β, IL-6, IL-3, interferon-γ and erythropoietin. Blood 81, 2579.

Buday, L., & Downard, J. (1993). Epidermal growth factor regulates p21ras through the formation of a complex of receptor, Grb2 adapter protein, and Sos nucleotide exchange factor. Cell 73, 611.

Burd, P.R., Rogers, H.W., Gordon, J.R., Martin, C.A., Jayaraman, S., Wilson, S.D., Dvorak, A.M., Galli, S.J., & Dorf, M. (1989). Interleukin 3-dependent and -independent mast cells stimulated with IgE and antigen express multiple cytokines. J. Exp. Med. 170, 245.

Clark-Lewis, I., Kent, S.B.H, & Schrader, J.W. (1984). Purification to apparent homogeneity of a factor stimulating the growth of multiple lineages of haemopoietic cells. J. Biol. Chem. 259, 7488.

Crapper, R.M., & Schrader, J.W. (1983). Frequency of mast-cell precursors in normal tissues determined by an in vitro assay: Antigen induces parallel increases in the frequency of P-cell precursors and mast cells. J. Immunol. 131, 923.

Crapper, R.M., & Schrader, J.W. (1984.) In vivo transfer of persisting (P) cells: Further evidence for their identity with T-dependent mast cells. J. Immunol. 133, 2174.

Crapper, R.M., & Schrader, J.W. (1986). Role of the T-cell lymphokine, persisting-cell stimulating factor (PSF), in graft-versus-host reactions. Immunology 57, 553.

Crapper, R.M., Thomas, W.R., & Schrader, J.W. (1984). *In vivo* transfer of persisting (P) cells: Further evidence for their identity with T-dependent mast cells. J. Immunol. 133, 2174.

Crapper, R.M., Vairo, G., Hamilton, J., Clark-Lewis, I., & Schrader, J.W. (1985). P cell stimulating factor stimulates peritoneal and bone marrow derived macrophages. Blood 66, 859.

Cutler, R.L., Metcalf, D., Nicola, N.A. & Johnson, G.R. (1985). Purification of a multipotential colony-stimulating factor from pokeweed mitogen-stimulated ouse spleen cell conditioned medium. J. Biol. Chem. 260, 6579.

De Vos, A.M., Ultsch, M., & Kossiakoff, A.A. (1992). Human growth hormone and extracellular domain of its receptor: Crytal structure of the complex. Science 255, 306.

Derijard, B., Raingeand, J., Barrett, T., Wu, I., Han, J., Ulevitch, R., & Davis, R. (1995). Independent human MAP kinase signal transduction pathways defined by MEK and MKK isoforms. Science 267, 682.

Dexter, T.M., Garland, J., Scott, D., Scolnick, E., & Metcalf, D. (1980). Growth of factor-dependent hemopoietic precursor cell lines. J. Exp. Med. 152, 1036.

Donahue, R.E., Seehra, J., Metzger, M., Lefebvre, D., Rock, B., Carbone, S., Nathan, D.G., Garnick, M., Sehgal, P.K., Laston, D., La Vallie, E., McCoy, J., Schendel, P.F., Norton, C., Turner, K., Yang, Y.C., & Clark, S.C. (1988). Human IL-3 and GM-CSF act synergistically in stimulating hematopoiesis in primates. Science 241, 1820.

Dubucquoi, S., Desreumaux, P. Janin, A., Klein, O., Goldman, M., Tavernier, J., Capron, A., & Capron, M. (1994). Interleukin 5 synthesis by eosinophils: Association with granules and immunolglobulin-dependent secretion. J. Exp. Med. 179, 703.

Duronio, V., Nip, L., & Pelech, S.L. (1989). Interleukin 3 stimulates phosphatidylcholine turnover in a mast/megakaryocyte cell line. Biochem. Biophys. Res. Commun. 164, 804.

Duronio, V., Clark-Lewis, I., Federsppiel, B., Wieler, J.S., & Schrader, J.W. (1992a). Tyrosine phosphorylation of receptor beta subunits and common sustrates in response to interleukin-3 and granulocyte-macrophage colony-stimulating factor. J. Biol. Chem. 267, 21856.

Duronio, V., Clark-Lewis, I., & Schrader, J.W. (1992b). Two polypeptides identified by interleukin-3 crosslinking represent distinct components of the interleukin-3 receptor. Exp. Hematol. 20, 505.

Duronio, V., Welham, M., Abraham, S., Dryden, P., & Schrader, J.W. (1992c). p21ras activation via haemopoietin receptors and c-kit requires tyrosine kinase activity but not tyrosine phosphorylation of GAP. Proc. Natl. Acad. Sci. USA 89, 1587.

Feng, Y., Klein, B.K., Vu, L., Aykent, S., & McWherter, C.A. (1995). 1H, 13C and 15N NMR resonance assignments, secondary structure, and backbone topology of a variant of human interleukin-3. Biochemistry 34, 6540.

Frendl, G., & Beller, D.I. (1990). Regulation of macrophage activation by IL-3: I. IL-3 functions as a macrophage-activating factor with unique properties, inducing Ia and lymphocyte function-associated antigen-1 but not cytotoxicity. J. Immunol. 144, 3392.

Frendl, G., Fenton, M.J., & Beller, D.I. (1990). Regulation of macrophage activation by IL-3: II. IL-3 and lipopolysaccharide act synergistically in the regulation of IL-1 expression. J. Immunol. 144, 3400.

Galy, A., Cen., D., Travis, M., Chen, S., & Chen., B., (1994). Evidence for a lymphoid-restricted progenitor cell within the CD34+ compartment of adult bone marrrow. Blood 84(10), 436a.

Gorman, D.M., Itoh, N., Kitamura, T., Schreurs, J., Yonehara, S., Yahara, I., Arai, K., & Miyajima, A. (1990.) Cloning and expression of a gene encoding an interleukin 3 receptor-like protein: Identification of another member of the cytokine receptor gene family. Proc. Natl. Acad. Sci. USA 87, 5459.

Gorman, D.M., Itoh, N., Jenkins, N.A., Gilbert, D.J., Copeland, N.G., & Miyajima, A. (1992). Chromosomal localization and organization of the murine genes encoding the β subunits (AIC2A and AIC2B) of the interleukin 3, granulocyte/macrophage colony-stimulating factor and interleukin 5 receptors. J. Biol. Chem. 267, 15842.

Greenberger, J.S. (1980). Self-renewal of factor-dependent hemopoietic progenitor cell lines derived from long-term bone marrow cultures demonstrates significant mouse strain genotypic variation. J. Supramolec. Struc. 13, 501.

Hans, J., Lee, J., Bibbs, L., & Ulevitch, R. (1994). A MAP kinase targeted by endotoxin and hyperomolarity in mammalian cells. Science 265, 808.

Hapel, A.J., Lee, J.C. Farrar, W.L., & Ihle, J.N. (1981). Establishment of continuous cultures of Thy-1.2$^+$, Lyt-1$^+$, 2-T cells with purified interleukin-3. Cell 25, 179.

Hapel, A.J., Osborne, J.M., Fung, M.C., Young, I.G., Allan, W., & Hume, D.A. (1985). Expression of 20-alpha-hydroxysteroid dehydrogenase in mouse macrophages, hemopoietic cells, and cell lines and its induction by colony-stimulating factors. J. Immunol. 134, 2492.

Hara, T., Ichihara., M., Takagi, M., & Miyajima, A. (1995). Interleukin-3 (IL-3) poor-responsive inbred mouse strains carry the identical deletion of a branch point in the IL-3 receptor alpha subunit gene. Blood 85, 2331.

Hasthorpe, S. (1980). A hemopoietic cell line dependent on a factor in pokeweed mitogen-stimulated spleen-cell conditioned medium. J. Cell. Physiol. 105, 379.

Haylock, D.N., To, L.B., Dowse, T.L., Juttner, C.A., & Simmons, P.J. (1992). Ex vivo expansion and maturation of peripheral blood CD34$^+$ cells into the myeloid lineage. Blood 80, 1405.

Hirai, K., Morita, Y., Misaki, Y., Ohta, K., Takashi, T., Suzuki, S., Motoyoshi, K., & Miyamoto, T. (1988). Modulation of human basophil histamine release by hemopoietic growth factors. J. Immunol. 141, 3958.

Hirayama, F., Shih, J.P., Awgulewitsch, A., Warr, G.W., Clark, S.C., & Ogawa, M. (1992). Clonal proliferation of murine lymphohemopoietic progenitors in culture. Proc. Natl. Acad. Sci. USA 89, 5907.

Ihle, J.N., Keller, J., Oroszlan, S., Henderson, L.E., Copeland, T.D., Fitch, F., Prystowsky, M.B., Goldwasser, E., Schrader, J.W., Palaszynski, E., Dy, M., & Lebel, B. (1983). Biologic properties of homogeneous interleukin-3. I. Demonstration of WEHI-3B growth factor activity, mast cell growth factor activity, P cell-stimulating factor activity, colony-stimulating factor activity, and histamine-producing cell-stimulating factor activity. J. Immunol. 131, 282.

Iscove, N.N., Roitsch, C.A., Williams, N., & Guilbert, L.J. (1982). Molecules stimulating early red cells, granulocyte, macrophage and megakaryocyte precursors in culture: Similarity in size, hydrophobicity and charge. J. Cell. Physiol. 1, 65 (Suppl.).

Iscove, N.N., & Yan, X.Q. (1990). Precursors (pre-CFCmulti) of multilineage hemopoietic colony-forming cells quantitated *in vitro*. Uniqueness of IL-1 reuqirement, partial separation from pluripotential colony-forming cells and correlation with long term reconstituting cells *in vivo*. J. Immunol. 145, 190.

Isfort, R.J., & Ihle, J.N. (1990). Multiple hematopoietic growth factors signal through tyrosine phosphorylation. Growth Factors 2, 213.

Isfort, R., Stevens, D., May, W.S., & Ihle, J.N. (1988). Interleukin 3 binds to a 140-κDa phosphotyrosine-containing cell surface protein. Proc. Natl. Acad. Sci. USA. 85, 7982.

Ishibashi, T., & Burstein, S.A. (1986). Interleukin-3 promotes the differentiation of isolated single megakaryocytes. Blood 67, 1512.

Itoh, N., Yonehara, S., Schreurs, J., Gorman, D.M., Maruyama, K., Ishii, A., Yahara, I., Arai, K., & Miyajima, A., (1990). Cloning of an interleukin-3 receptor: A member of a distinct receptor gene family. Science 247, 324.

Kanakura, Y., Druker, B., Cannistra, S.A., Furukawa, Y., Torimoto, Y., & Griffin, J.D. (1990). Signal transduction of the human granulocyte-macrophage colony-stimulating factor and interleukin-3 receptors involves tyrosine phosphorylation of a common set of cytoplasmic proteins. Blood 76, 706.

Kinashi, T., Inaba, K., Tsubata, T., Tash, K., Palacios, R., & Honjo, T. (1988). Differentiation of an interleukin-3-dependent precursor B-cell clone into immunoglobulin-producing cells *in vitro*. Proc. Natl. Acad. Sci. USA 85, 4473.

Kinoshita, T., Yokota, T., Arai, K.-I., & Miyajima, A. (1995a). Suppression of apoptotic death in hematopoietic cells by signalling through the IL-3/GM-CSF receptors. EMBO J. 14, 266.

Kinoshita, T., Yokota, T., Arai, K.-I., & Miyajima, A. (1995b). Regulation of bcl-2 expression by oncogenic Ras protein in hematopoietic cells. Oncogene 10, 2207.

Kita, H., Ohnishi, T., Okubo, Y., Weiler, D., Abrams, J.S., & Gleich, G.J. (1991). GM-CSF and interleukin-3 release from human peripheral blood eosinophils and neutrophils. J. Exp. Med. 174, 743.

Kitamura, T., & Miyajima, A. (1992). Functional reconstitutionof the human interleukin 3 receptor. Blood 80, 84.

Kitamura, T., Sato, N., Arai, K., & Miyajima, A. (1991). Expression cloning of the human IL-3 receptor cDNA reveals a shared β subunit for human IL-3 and GM-CSF receptors. Cell 66, 1165.

Koyasu, S., Tojo, A., Miyajima, A., Akiyama, T., Kasuga, M., Urabe, A., Schreurs, J., Arai, K., Takaku, F., & Yahara, I. (1987). Interleukin 3-specific tyrosine phosphorylation of a membrane glycoprotein of Mr 150,000 in muti-factor-dependent myeloid cell lines. EMBO J. 6, 3979.

Kurimoto, Y., de Weck, A.L., & Dahinden, C.A. (1989). Interleukin-3 dependent mediator release in basophils triggered by C5a. J. Exp. Med. 170, 467.

Lee, J.C., Hapel, A.J., & Ihle, J.N. (1982). Constitutive production of a unique lymphokine (IL-3) by the WEHI-3 cell line. J. Immunol. 128, 2393.

Lee, J., Laydon, J., McDonnell, P., Gallagher, T., Kumar, S., Green, D., McNulty, D., Blumenthal, M., Heys, J., Landvatter, S., Strickler, J., McLaughlin, M., Siemens, I., Fisher, S., Liui, G., White, J., Adams, J., & Young, P. (1994). A protein kinase involved in the regulation of inflammatory cytokine biosynthesis. Nature 372, 739.

Leslie, K.B., & Schrader, J.W. (1989). Growth factor gene activation and clonal heterogeniety in an autostimulatory myeloid leukemia. Mol. Cell Biol. 6, 2414.

Leslie, K.B., Lee, F., & Schrader, J.W. (1991). IAP-Mediated activations of cytokine genes in a murine myelomonocytic leukemia: Novel generation of functional cytokine mRNAs by retroviral splicing events. Mol. Cell. Biol. 11, 5562.

Lopez, A.F., Eglinton, J.M., Lyons, A.B., Tapley, P.M., To, L.B., Park, L.S., Clarke, S.C., & Vadas, M.A. (1990). Human interleukin-3 inhibits the binding of granulocyte-macrophage colony-stimulating factor and interleukin-5 to basophils and strongly enhances their functional activity. J. Cell Physiol. 145, 69.

Lopez, A.F., Vadas, M.V., Woodcock, J.M., Milton, S.E., Lewis, A., Elliott, M.J., Gillis, D., Ireland, R., Olwell, E., & Park, L.S. (1991). Interleukin-5, interleukin-3, and granulocyte-macrophage colony-stimulating factor cross-compete for binding to cell surface receptors on human eosinophils. J. Biol. Chem. 266, 24741.

Lopez, A.F., Shannon, M.F., Barry, S., Phillips, J.A., Cambareri, B., Dottore, M., Simmons, P., & Vadas, M.A. (1992). A human interleukin 3 analog with increased biological and binding activities. Proc. Natl. Acad. Sci. USA 89, 11842.

Mayer, P., Valent, P., Schmidt, G., Liehl, E., & Bettelheim, P. (1989). The *in vivo* effects of recombinant human interleukin-3: Demonstration of basophil differentiation factor, histamine-producing activity, and priming of GM-CSF-responsive progenitors in non-human primates. Blood, 74, 613.

McAlister, I.B., Teepe, M., Gillis, S., & Williams, D.E. (1992). *Ex vivo* expansion of peripheral blood progenitor cells with recombinant cytokines. Exp. Hematol. 20, 6262.

Metcalf, D. (1988). The multipotential colony-stimulating factor, multi-CSF (IL-3). In: *Lymphockines 15; Interleukin 3: The Panscpecific Hemopoietin* (Schrader, J.W., Ed.). Academic Press, San Diego, pp. 183-215.

Metcalf, D., & Rasko, J.E.J. (1993). Leukemic transformation of immortalized FDC-P1 cells engrafted in GM-CSF transgenic mice. Leukemia 7, 878.

Morla, A.O., Schreurs, J., Miyajima, A., & Wang, J.Y.J. (1988). Hematopoietic growth factors activate the tyrosine phosphorylation of distinct sets of proteins in interleukin-3-dependent murine cell lines. Mol. Cell Biol. 8, 2214.

Mui, A.L., Wakao, H., Harada, N., O'Farrell, A.M., & Miyajima, A. (1995). Interleukin-3, granlocyte-macrophage colony-stimulating factor, and interleukin-5 transduce signals through two forms fo STAT5. J. Leukoc. Biol. 57, 799.

Mui, A.L., Wakao, H., O'Farrell, A.M., Harada, N., & Miyajima, A. (1995). Interleukin-3 granulocyte-macrophage colony stimulating factor and interleukin-5 transduce signals through two STAT5 homologs. EMBO J. 15, 1166.

Muto, A., Watanabe, S., Miyajima, A., Yokota, T., & Arai, K. (1995). High affinity chimeric human granulocyte-macrophage colony-stimulating factor receptor carrying the cytoplasmic domain of the beta subunit but not the alpha subunit transduces growth promoting signals in Ba/F3 cells. Biochem. Biophys. Res. Commun. 208, 368.

Nabel., G., Galli, S.J., Dvorak, A.M., Dvorak, H.F., & Cantor, H. (1981). Inducer T lymphocytes synthesize a factor that stimulates proliferation of cloned mast cells. Nature 291, 333.

Nagao, K., Yokova, K., & Aarsonson, S.A. (1981). Continuous lines of basophil/mast cells derived from normal bone marrow. Science 212, 333.

Nishinakamura, R., Nakayama, N., Hirabayashi, Y., Inoue, T., Aud, D., McNeil, T., Azuma, S., Yoshida, S., Toyoda, Y., Arai, K., et al. (1995). Mice deficient for the IL-3/GM-CSF/IL-5 beta c receptor exhibit lung pathology and impaired immune response, while beta IL3 receptor-deficient mice are normal. Immunity 2, 211.

Orban, P.C., & Schrader, J.W. (1996). Antibodies to an autostimulatory growth factor (IL-2) or its receptor induce cell death. J. Immunol. 156(9), 3334-3341.

Osborne, C.S., Vadas, M.A., & Cockerill, P.N. (1995). Transcriptional regulation of mouse granulocyte-macrophage colony-stimulating factor/IL-3 locus. J. Immunol. 155, 226.

Oster, W., Cicco, N.A., Klein, H., Hirano, T., Kishimoto, T., Lindemann, A., Mertelsmann, R.H., & Hermmann, F. (1989). Participation of the cytokines, interleukin-6, tumour necorsis factor-alpha, and interleukin-1 beta secreted by acute myelogenous leukemia blasts in autocrine and paracrine leukemia growth control. J. Clin. Invest. 84, 451.

Palacios, R., Henson, G., Steinmetz, M., & McKearn, J.P. (1984). Interleukin-3 supports growth of mouse pre-B-cell clones *in vitro*. Nature (London) 309, 126.

Palacios, R., & Pelkonen, J. (1988). Prethymic and intrathymic mouse T-cell progenitors. Growth requirements and analysis of the expression of genes encoding TCR/T3 components and other T-cell-specific molecules. Immunol. Rev. 104, 158.

Paonessa, G., Graziani, R., De Serio, A., Savino, R., Ciapponi, L., Lahm, A., Salvati, A.L., Toniatti, C., & Ciliberto, G. (1995). Two distinct and independent sites on IL-6 trigger gp130 dimer formation and signalling. EMBO J. 14, 1942.

Park, L.S., Friend, D., Price, V., Anderson, D., Singer, J., Prickett, K.S., & Urdal, D.L. (1989). Heterogeneity in human interleukin-3 receptors. J. Biol. Chem. 264, 5420.

Plaut, M., Pierce, J.H., Watson, C.J., Hanley-Hyde, J., Nordan, R.P., & Paul, W.E. (1989). Mast cell lines produce lymphokines in response to crosslinkage of FcERI and calcium ionophores. Nature 339, 64.

Pluznik, D.H., & Sachs, S.L. (1965). The cloning of normal "mast" cells in tissue culture. J. of Cell. & Comp. Phys. 66, 319.

Quelle, F.W., Sato, N., Witthuhn, B.A., Inhorn, R.C., Eder, M., Miyajima, A., Griffin, J.D., & Ihle, J.N. (1994). JAK2 associates with the β_c chain of the receptor for granulocyte-macrophage colony-stimulating factor, and its activation requires the membrane-proximal region. Mol. Cell. Biol. 14, 4335.

Quelle, F.W., Shimoda, K., Thierfelder, W., Fischer, C., Kim, A., Ruben, S.M., Cleveland, J.L., Pierce, J.H., Keegan, A.D., Nelms, K., et al. (1995). Cloning of murine Stat6 and human Stat6, Stat proteins that are tyrosine phosphorylated in responses to IL-4 and IL-3 but are not required for mitogenesis. Mol. Cell. Biol. 15, 3336.

Raingeand, J., Gupta, S. Rogers, J., Dickens, M., Hans, J., Ulevitch, R., & Davis, R. (1995). Proinflammatory cytokines and environmental stress cause p38 mitogen-activated protein kinase activation by dual phosphorylation on tyrosine and threonine. J. Biol. Chem. 270, 7420.

Rao, P., & Mufson, R.A. (1995). A membrane proximal domain of the human interleukin-3 receptor beta c subunit that signals DNA synthesis in NIH 3T3 cells specifically binds a complex of Src and Janus family tyrosine kinases and phosphatidylinositol 3-kinase. J. Biol. Chem. 270, 6886.

Razin, E., Cordon-Cardo, C., & Good, R.A. (1981). Growth of a pure population of mast cells in vitro with conditioned medium derived from concanavalin-A stimulated splenocytes. Proc. Natl. Acad. Sci. USA 78, 2559.

Razin, E., Leslie, K.B., & Schrader, J.W. (1991) Connective tissue mast cells in contact with fibroblasts express IL-3 mRNA: Analysis of single cells by polymerase chain reaction. J. Immunol. 146, 981.

Rico-Vargas, S.A., Weiskopf, B., Nishikawa, S.-I., & Osmond, D.G. (1994). c-kit expression of B cell genesis by *in vivo* treatment with anti-c-kit antibody. J. Immunol. 152, 2845.

Rinaudo, M.S., Su, K., Falk, L.A., Haldar, S., & Mufson, R.A. (1995). Human interleukin-3 receptor modulates bcl-2 mRNA and protein levels thorugh protein kinase C in TF-1 cells. Blood 86, 80.

Rouse, J., Cohen, P., Trigon, S., Morange, M., Alonso-Llamazares, A., Zamanillo, D., Hunt, T., & Nebreda, A. (1994). A novel kinase cascade triggered by stress and heat shock that stimulates MAPKAP kinase-2 and phosphorylation of the small heat shock proteins. Cell 78, 1027.

Sato, N., Sakamaki, K., Terada, N., Arai, K.-I., & Miyajima, A. (1993). Signal transduction by the high affinity GM-CSF receptor: Two distinct cytoplasmic regins of the common b subunit responsible for different signalling. EMBO J. 12, 4181.

Sato, S., Katagiri, T., Takaki, S., Kikuchi, Y., Hitoshi, Y., Yonehara, S., Tsukada, S., Kitamura, D., Watanabe, T., Witte, O., & Takatsu, K. (1994). IL-5 receptor-mediated tyrosine phosphorylation of SH2/SH3-containing proteins and activation of Bruton's tyrosine and Janus 2 kinases. J. Exp. Med. 180, 2101.

Sato, T., Nakafuku, M., Miyajima, A., & Kaziro, Y. (1991). Involvement of ras p21 protein in signal-transduction pathways from interleukin 2, interleukin 3, and granulocyte/macrophage colony-stimulating factor, but not from interleukin 4. Proc. Natl. Acad. Sci. USA 88, 3314.

Schindler, C., & Darnell, J.E., Jr. (1995). Transcriptional responses to polypeptide ligands: The JAK-STAT pathway. Annu. Rev. Biochem. 64, 621.

Schrader, J.W., & Crapper, R.M. (1983). Autogenous production of a hemopoietic growth factor, persisting cell stimulating factor, as a mechanism for transformation of bone-marrow-derived cells. Proc. Soc. Natl. Acad. Sci. USA 80, 6892.

Schrader, J.W., Clark-Lewis, I., Ziltener, H.J., Hood, L.E., & Kent, S.B.H. (1986a). In vivo activity of a chemically synthesized hemopoietin growth factor pan-specific hemopoietin (IL-3). In: *Immune Regulation by Characterized Polypeptides* (Goldstein, G., Bach, J.F., & Wigzell, H., Eds.). Liss, New York, pp. 475-484.

Schrader, J.W., Ziltener, H.J., & Leslie, K.B. (1986). Structural homologies among the hemopoietins. Proc. Natl. Acad. Sci. USA 83, 2458.

Shanafelt, A.B., Johnson, K.E., & Kastelein, R.A. (1991a). Identification of critical amino acid residues in human and mouse granulocyte-macrophage colony-stimulating factor and their involvement in species specificity. J. Biol. Chem. 266, 13804.

Shanafelt, A.B., Miyajima, A., Kitamura, T., & Kastelein, R.A. (1991b). The amino-terminal helix of GM-CSF and IL-5 governs high affinity binding to their receptors. EMBO J. 10, 4105.

Shanafelt, A.B., & Kastelein, R.A. (1992). High affinity ligand binding is not essential for granulocyte-macrophage colony-stimulating factor receptor activation. J. Biol. Chem. 267, 25466.

Silvennoinen, O., Witthuhn, B.A., Quelle, F.W., Cleaveland, J.L., Yi, T., & Ihle, J.N. (1993). Structure of the JAK2 protein tyrosine kinase and its role in IL-3 signal transduction. Proc. Natl. Acad. Sci. USA. 90, 8429.

Sorenson, P., Mui, A.L-F. & Krystal, G. (1989). Interleukin-3 stimulates the tyrosine phosphorylation of the 140-kilodalton interleukin-3 receptor. J. Biol. Chem. 264, 19253.

Takaki, S., Kanazawa, H., Shiiba, M., & Takatsu, K. (1994). Critical cytoplasmic domain of the interleukin-5 (IL-5) receptor α chain and its function in IL-5-mediated growth signal transduction. Mol. Cell. Biol. 14, 7404.

Takatsu, K., Kanazawa, H., Kouro, T., Shiiba, M., Kikuchi, Y., & Takaki, S. (1995). Proline-rich regions in the cytoplasmic domain of the IL-5R α-chain play a critical role in the IL-5-mediated activation of JAK and BTK kinases. Proceedings of the International Congress of Immunology.

Tertian, G., Yung, T.-P., Guy-Grand, D., & Moore, M.A.S. (1981) Long-term in vitro culture of murine-mast cells. I. Description of a growth factor dependent culture technique. J. Immunol. 127, 788.

Thomas, J.W., Baum, C.M., Hood, W.F., Klein, B., Monhan, J.B., Paik, K., Staten, N., Abrams, M., & McKearn, J.P. (1995). Potent interleukin 3 receptor agonist with selectively enhanced hematopoietic activity relative to recombinant human interleukin 3. Proc. Natl. Acad. Sci. 92, 3779.

Torigoe, T., O'Connor, R., Santoli, D., & Reed, J.C. (1992). Interleukin-3 regulates the activity of the Lyn protein-tyrosine kinase in myeloid-committed leukemic cell lines. Blood 80, 617.

Uckun, F.M., Gesner, T.G., Song, C.W., Myers, D.E., & Mufson, A. (1989). Leukemic B-cell precursors express functional receptors for human interleukin-3. Blood 73(2) 533.

Vadhan-Raj, S., Papadopoulous, N.E., Burgess, M.A., Linke, K.A., Patel, S.R., Hays, C., Arcenas, A., Plager, C., Kudelka, A.P., Hittleman, W.N., Broxmeyer, H.E., Williams, D.E., Garrison, L., & Benjamin, R.S. (1994). Effects of PIXY321, a granulocyte-macrophage colony-stimulating factor/interleukin-3 fusion protein, on chemotherapy-induced multilineage myelo-suppression in patients with sarcoma. J. Clin. Oncol. 12, 715.

Valent, P., Schmidt, G., Besemer, J., Mayer, P., Zenke, G., Liehl, E., Hinterberger, W., Lechner, K., Maurer, D., & Bettelheim, P. (1989). Interleukin-3 is a differentiation factor for human basophils. Blood 73, 1763.

van der Bruggen, T., Caldenhoven, E., Kanters, D., Coffer, P., Raaijmakers, J.A.M., Lammers, J.-W.J., & Koenderman, L. (1995). Interleukin-5 signaling in human eosinophils involves JAK2 tyrosine kinase and Stat1a. Blood, 85, 1442.

van der Geer, P., & Pawson, T. (1995). The PTB domain: A new protein module implicated in signal transduction. TIBS 20, 277.

Ward, L.D., Howlett, G.J., Discolo, G., Yasukawa, K., Hammacher, A., Moritz, R., & Simpson, R.J. (1994). High affinity interleukin-6 receptor is a hexameric complex consisting of two molecules each of interleukin-6, interleukin-6 receptor and gp-130. J. Biol. Chem. 269, 23286.

Watari, K., Lansdorp, P.M., Dragowska, W., & Schrader, J.W. (1994). Detection of interleukin-1β mRNA in candidate human hematopoietic stem cells. Blood 84, 36.

Welham, M.J., & Schrader, J.W. (1991). Modulation of c-kit mRNA and protein by hemopoietic growth factors. Mol. Cell. Biol. 11, 2901.

Welham, M.J., & Schrader, J.W. (1992). Steel factor induced tyrosine phosphorylation in murine mast cells: Common elements with IL-3-induced signal transduction pathways. J. Immunol. 149, 2772.

Welham, M.J., Duronio, V., Sanghera, J.S., Pelech, S.L., & Schrader, J.W. (1992). Multiple hematopoietic growth factors stimulate activation of mitogen-activated protein kinase family members. J. Immunol. 149, 1683.

Welham, M.J., Dechert, U., Leslie, K.B., Jirik, F., & Schrader, J.W. (1994a). IL-3 and GM-CSF, but not IL-4, induce tyrosine phosphorylation, activation and association of SH-PTP2 with grb2 and PI3 kinase. J. Biol. Chem. 269, 23764.

Welham, M.J., Duronio, V., Leslie, K.B., Bowtell, D., & Schrader, J.W. (1994b). Multiple hemopoietins, with the exception of interleukin-4, induce modification of shc and mSos1, but not their translocation. J. Biol. Chem. 269, 21165.

Welham, M.J., Duronio, V., & Schrader, J.W. (1994c). Interleukin-4-dependent proliferation dissociates $p44^{erk\text{-}1}$, $p42^{erk\text{-}2}$, and $p21^{ras}$ activation from cell growth. J. Biol. Chem. 269, 5865.

Williams, D.E., Hangoc, G., Cooper, S., Boswell, H.S., Shadduck, R.K., Gillis,S., Waheed, A., Urdal, D.U., & Broxmeyer, H.E. (1987). The *in vivo* effects of purified recombinant IL-3 and purified natural CSF-1 on murine high proliferative potential colony forming cells: Demonstration of *in vivo* synergism. Blood 70, 401.

Wodnar-Filipowicz, A., Heusser, C.H., & Moroni, C. (1989). Production of the hemopoietic growth factors GM-CSF and interleukin-3 by mast cells in response to IgE receptor-mediated activation. Nature 339, 150.

Wong, G.H.W., Clark-Lewis, I., Hamilton, J.A., & Schrader, J.W. (1984). P cell stimulating factor and glucocorticoids oppose interferon-g in inducing Ia antigens on T-dependent mast cells (P cells). J. Immunol. 133, 2043.

Yang, Y.-C., Ciarletta, A.B., Temple, P.A., Chung, M.P., Kovacic, S., Witek-Giannotti, J.S., Leary, A.C., Kriz, R., Donahue, R.E., Wong, G.G., & Clark, S.C. (1986). Human IL-3 (multi-CSF): Identification by expression cloning of a novel hematopoietic growth factor related to murine IL-3. Cell 47, 3.

Young, D.C., & Griffin, J.D. (1986). Autocrine secretion of GM-CSF in acute myeloblastic leukemia. Blood 68, 1178.

Ziltener, H.J. (1993). Glycosylation does not affect *in vitro* biological activity of interleukin-3. Cytokine 5, 291.

Ziltener, H.J., Clark-Lewis, I., Fazekas, B., Hood, L.E., Kent, S.B.H., & Schrader, J.W. (1987). Antipeptide antibodies define the NH2-terminal structure of the pan-specific hemopoietin Interleukin-3. J. Immunol. 138, 1099.

Ziltener, H.J., Fazekas de St. Groth, B., Leslie, K.B., & Schrader, J.W. (1988). Multiple glycosylated forms of T cell-derived Interleukin-3 (IL-3) heterogeneity of IL-3 from physiological and nonphysiological sources. J. Biol. Chem. 263, 14511.

INTERLEUKIN-4

Jacques Banchereau and Sem Saeland

Growth Factors and Cytokines in Health and Disease
Volume 2A, pages 85-142.

ISBN: 0-7623-0117-1

I. INTRODUCTION

Interleukin-4 (IL-4) was identified in 1982 for its ability to induce the proliferation of murine B lymphocytes costimulated with anti-IgM (Howard et al., 1982) and for its capacity to induce lipopolysaccharide (LPS) activated B cells to produce IgG1 (Isacksson et al., 1982). In 1986, a cDNA encoding human IL-4 was isolated (Yokota et al., 1986). IL-4 is a member of a cytokine family including IL-3, IL-5, IL-13, and granulocyte-macrophage colony stimulating factor (GM-CSF).

Although first described as a stimulatory factor for B cells, IL-4 was thereafter found to promote the growth of mast cells and T cells (Grabstein et al., 1986; Lee et al., 1986; Mosmann et al., 1986; Smith and Rennick, 1986), indicating that its cellular targets encompass several hemopoietic lineages. Subsequently, IL-4 was recognized to modulate the functions of various cell types of non-hemopoietic origin. Herein, we will mainly refer to studies performed with human IL-4. Additional references on IL-4 can be found in several excellent reviews by Paul (1991), Rennick and colleagues (1992), and O'Garra and Spits (1993). Clinical trials are presently being performed aimed at determining the possible therapeutic use of IL-4 as an anti-tumor agent.

II. MOLECULAR ASPECTS OF IL-4

The principal characteristics of IL-4 are summarized in Table 1. Molecular cloning of human IL-4 cDNA revealed a single open reading frame of 153 AA yielding a secreted glycoprotein of 129 AA (Yokota et al., 1986). IL-4 displays 20% homology at the amino acid level, with the more recently identified human IL-13 (McKenzie et al., 1993; Minty et al., 1993). Expression of the recombinant IL-4 protein in mammalian cells demonstrates three variants with apparent Mr values of 15, 18, and 19 kDa. The microheterogeneity of rhu IL-4 appears to be related to the nature of the N-linked oligosaccharides in the 18 and 19 kDa variants. Human IL-4 contains three disulfide bridges between C3-C127, C4-C65, and C46-C99 (Windsor et al., 1990). X-ray diffraction of IL-4 crystals (Walter et al., 1992), as well as magnetic resonance spectroscopy of IL-4 in solution (Powers et al., 1992) indicate that IL-4 is a left-handed four α-helices bundle with short stretches of β-sheets. The four α-helices are situated between the residues 9-21, 45-64, 74-96, and 113-129, and the mini antiparallel β-sheets are between residues 32-34 and 110-112 (Garret et al., 1992). The structure of IL-4 bears close ressemblance to GM-CSF, M-CSF, and growth hormone. The human IL-4 gene, composed of four exons and three introns, is localized on the long arm of chromosome 5 on bands q23-31 together with genes of other related cytokines including IL-3, IL-5, IL-9, IL-13, and GM-CSF (Asano et al., 1987; Morgan et al., 1992; Figure 1). The IL-4 and IL-13 genes display considerable similarity in their structural organization and share many potential binding sites for transcription factors. The IL-13 gene is located 12 kb 5' to IL-4 and linked in a "tail-to-head" fashion (Smirnov et al., 1995). The promoter of the human IL-4 gene in the 5'-flanking region contains a unique repetitive motif (P-sequence) (Abe et al., 1992) that binds the T cell-specific activation inducible factor NF(P) (Szabo et al., 1993; Todd et al., 1993). Another site of the IL-4 promoter binds the ubiquitous and constitutive nuclear factor NF-Y. Furthermore, the IL-4 promoter contains two OAP-40 sites that bind factors which may consist of dimers of *jun-C* or *jun-D*. The human IL-4 promoter displays a negative regulatory element containing two protein binding sites (NRE-I and NRE-II), and which interacts with the T cell-specific NEG-1 and the ubiquitous NEG-2 nuclear factors (Li-Weber et al., 1992, 1993). The silencer suppresses the activity of an enhancer element in the human IL-4 promoter (Li-Weber et al., 1993). IL-4 production is also controlled post-transcriptionally through stabilization of IL-4 mRNA (Dokter et al., 1993b).

Recently, the existence of IL-4 mRNA which lacks 48 bp coding for AA residues 22-37 has been reported, but the existence and the role of such a deleted IL-4 protein remain to be established (Sorg et al., 1993). An IL-4 molecule in which the tyrosine residue 124 was substituted by an aspartic acid residue was found to act as an antagonist of IL-4 (Aversa et al., 1993; Kruse et al., 1992).

Table 1. Properties of Human IL-4 and IL-4 Receptor (CDW124)

	IL-4	*IL-4 Receptor*
Protein		
Precursor protein: amino acids (aa)	153	825
Mature protein: aa	129	800
Extracellular domain: aa		207
Transmembrane domain: aa		24
Cytoplasmic domain: aa		569
Molecular weight (Kda)	15-19	130
Glycosylation sites	2N	6N
Disulphide bonds	3 (6Cys)	
Intracytoplasmic Tyr		5
Gene		
Chromosome	5q23q31	16q12.1-p11.2
Exons	4	
Introns	3	
Cell sources		
	T cells (TH2)	Ubiquitous
	Basophils/mast cells	
	Stromal cells (?)	

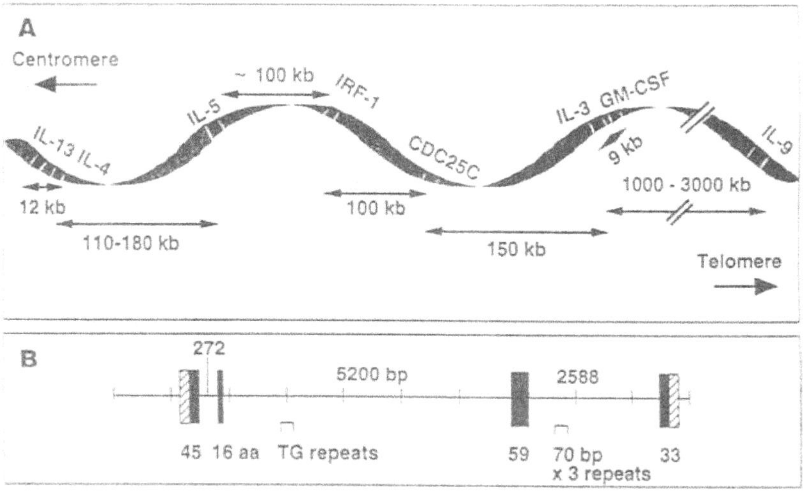

Figure 1. IL-4 gene. (A) A map showing genes around human chromosome 5q31-1. IRF-1, which encodes a transcription activator of IFN-α/β and other IFN-inducible genes. CDC 25C: cell division cycle 25. (B) Schematic representation of human IL-4 gene. Coding regions of exons are indicated by open boxes; untranslated regions are indicated by hatched boxes.

III. IL-4 RECEPTOR

IL-4 binds to high-affinity (Kd = 100 pM) receptors which are expressed in low number (a few hundred to a few thousand) on virtually every cell type tested, including T and B lymphocytes, monocytes, granulocytes, fibroblasts, epithelial, and endothelial cells (Cabrillat et al., 1987; Park et al., 1987). Human and murine IL-4 are species-specific with respect to receptor binding (Park et al., 1987). Studies with radiolabeled IL-4 showed that the IL-4 receptor includes molecular species of MW130-140, 70-75, and 65 kDa (Foxwell et al., 1989; Galizzi et al., 1989, 1990a; Park et al., 1987), with the 70kDa polypeptide representing a breakdown product of p130 (Keegan et al., 1991). The 130kDa glycoprotein has been purified and binds IL-4 with a high affinity close to that of the receptor expressed on intact cells (Garrone et al., 1991). A cDNA coding for the human 130-140 kDa IL-4 binding protein (p130) (IL-4 Rd) (CDw124) has been isolated, and encodes an open reading frame of 825aa which includes a 25aa signal sequence (Galizzi et al., 1990b; Idzerda et al., 1990; Table 1). p130 displays a 207aa extracellular domain which contains the characteristic features (four conserved cysteines and a WSXWS box) of the class-1 cytokine receptor superfamily (Bazan, 1990; Idzerda et al., 1990; Miyajima et al., 1992), a 24aa transmembrane domain, and a long 569aa intracellular portion. The intracellular domain of p130 (Figure 2) contains five tyrosine residues (at positions 497, 603, 631, 713, and 821), a membrane-proximal Box-1 (Murakami et al., 1991) sequence (residues 262-269), a conserved acidic portion (EAPV-X3-EEEEE), and an I4R motif (Keegan et al., 1994b), surrounding tyrosine-497 and shared with the insulin and IGF-1 receptors. Recently, the IL-2 receptor γc chain (MW 64kDa) has been demonstrated to represent a common component to the receptors for IL-2, IL-4, IL-7, IL-9, and IL-15 (Kondo et al., 1993; Noguchi et al., 1993a; Russel et al., 1993). The γc chain increases the affinity of IL-4 for p130 by 2-3 fold. The wide use of this common gc chain, whose gene is located on the X chromosome, explains the severe X-SCID immunodeficiency characterized by a genetic alteration of the γc gene (Noguchi et al., 1993b). The ability of a non-functional IL-4 mutant (Y124D) to block both IL-4 and IL-13 actions (Aversa et al., 1993; Kruse et al., 1992; Zurawski et al., 1993) suggested that the IL-4 and IL-13 receptors could share a common component. The blocking by antibodies to the 130 kDa IL-4Rα protein of the activities of IL-4 and

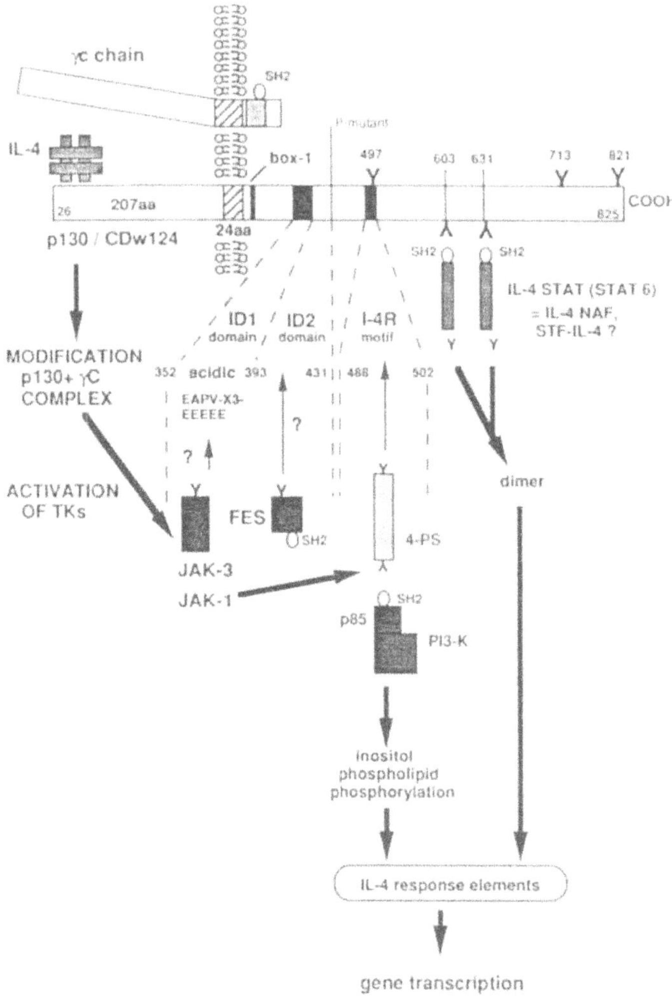

Figure 2. IL-4 signaling through the p130/γc IL-4 receptor complex. Binding of IL-4 activates tyrosine kinase activity (JAK-1, JAK-3, FES), likely as a result of conformational change in the p130/γc complex. The 4-PS substrate becomes tyrosine (Y)-phosphorylated, allowing activation of PI3-kinase. In addition, phosphorylation occurs on Y-residues of the p130 receptor, allowing recruitment of IL-4 STAT transcription factor and activation of IL-4 responsive genes. Association between phosphotyrosine groups and SH2-domains play a key role in IL-4 signal transduction. The ID1/ID2 domains and the I-4R motif of p130 have been described (Harada, 1992; Keegan, 1994). TK: tyrosine kinase. Numbers correspond to amino-acids starting from residue 26 (aa 1-25 = signal peptide). Signal transduction through the p130/IL-13 receptor remains to be determined.

IL-13 suggest (Lin et al., 1995; Zurawski et al., 1995) that the known 130 kDa IL-4Rα protein is the shared component rather than the c chain. Currently, it is proposed that IL-4 may bind to two distinct receptors, one composed of a 130 kDa component and a gc to which IL-13 does not bind, and one composed of the 130 kDa component and a 65-70 kDa component which represents the IL-13 receptor (He and Malek, 1995; Obiri et al., 1995; Smerz-Bertling and Duschl, 1995; Vita et al., 1995). Such a structure would explain why B cells from X-SCID patients with characterized γc chain gene mutations would respond to both IL-13 and IL-4 (Matthews et al., 1995). Finally, a distinct 65-75kDa low-affinity (Kd = 600 nM) IL-4 binding protein has been described. This low-affinity receptor was detected both at the cell surface and as a soluble protein whose partial N-terminal amino acid sequence is unrelated to p130 (Fanslow et al., 1993).

A40kDa soluble form of IL-4 binding protein has been demonstrated in mouse serum, and represents a truncated form of the murine p130 homolog, as a result of alternative splicing (Fernandez-Botran and Vitetta, 1990; 1991; Keegan et al., 1991; Mosley et al., 1989). Soluble IL-4 receptor may act as a transport protein which prevents enzymatic degradation of IL-4 (Fernandez-Botran and Vitetta, 1991). Because of a lack of alternative splicing, existence of an equivalent soluble human IL-4 binding protein is unlikely. Soluble murine IL-4 receptor can inhibit the lymphoproliferative response to a localized injection of allogeneic cells and prolongs the survival of cardiac allografts (Fanslow et al., 1990; Maliszewski et al., 1992). In contrast, however, soluble IL-4 receptor may potentiate induction of IgE synthesis (Sato et al., 1993), presumably due to its role as an IL-4 transport protein. Thus, soluble IL-4 receptor may have antagonistic or agonistic properties on IL-4-mediated biologic effects.

IL-4 receptor expression is regulated both through transcription rate and stabilization of p130 mRNA (de Wit et al., 1994; Dokter et al., 1992). The IL-4 receptor appears to be differentially regulated according to cell type, as its expression is affected by IL-4 and Ca^{++}-dependent pathways in T cells (Dokter et al., 1992; Galizzi et al., 1989) but not in monocytic cells (de Wit et al., 1994). Following internalization IL-4 is degraded, most likely in lysosomes. Preliminary studies suggest a model of receptor-ligand interaction involving the formation of a ternary complex consisting of two molecules of the extracellular portion of the receptor and one molecule of IL-4 (Ramanathan et al., 1993).

IV. BIOCHEMICAL PATHWAYS OF IL-4 ACTION

The signaling mechanisms through the IL-4 receptor has recently been the subject of excellent reviews (Keegan et al., 1994a; Keegan and Pierce, 1994). Crosslinking studies have shown that the p130 chain of the IL-4 receptor associates with the γc chain following binding of IL-4, suggesting that dimerization is important for biologic activity (Russel et al., 1993). The cytoplasmic region of the human γc chain has been shown to be essential for IL-4 induced growth signal transduction in murine BAF-B03 cells transfected with the human p130 IL-4 binding chain (Kawahara et al., 1994). Upon binding to its receptor, IL-4 induces tyrosine phosphorylation of multiple proteins, including species of 170, 140, 130, 110-120, 100, and 92 kDa (Izuhara and Harada, 1993; Wang et al., 1992; Yin et al., 1994). As neither the cytoplasmic domain of p130 nor that of γc contain consensus sequences established for catalytic activity, it appeared likely that binding of IL-4 would recruit non-receptor tyrosine kinases. Indeed, analogous to signaling through several other cytokine receptors, JAK family kinases are mobilized by IL-4 (Figure 2). Thus, JAK-1 (130kDa) and JAK-3 (120kDa) are tyrosine phosphorylated and activated following IL-4 binding (Barber and D'Andrea, 1994; Witthuhn et al., 1994; Yin et al., 1994). A recent study indicates that JAK-1 may play a subordinate role in IL-4 receptor signaling, and that in certain cells exclusive JAK-3 activation may mediate IL-4 induced cell growth (Brunn et al., 1995; Malabarba et al., 1995; Musso et al., 1995). JAK-1-3 complex with 4PS (170kDa) (Yin et al., 1994), which rapidly becomes tyrosine-phoshorylated in response to IL-4 (Wang et al., 1992). IL-13 shares with IL-4 the capacity to phosphorylate 4PS (Keegan et al., 1994a), thus explaining shared biologic activities of the two cytokines. The 4PS substrate is closely related to insulin receptor substrate-1 (IRS-1) (Wang et al., 1993), which, through tyrosine-phosphorylated sites, serves as a docking protein for signaling molecules that contain a src-homology 2 (SH2) domain (White and Kahn, 1994). The I4R motif surrounding tyrosine-497 in the cytoplasmic region of the p130 IL-4 binding chain bears homology to sequences in the insulin and IGF-1 receptors, suggesting that it has been conserved among receptors that link to the IRS-1 / 4PS pathway (Keegan et al., 1994b; Koettnitz and Kalthoff, 1993). Of interest, expression of the γc chain is required for IL-4 induced phosphorylation of 4PS (Russel et al., 1993). Phosphorylated 4PS associates with the SH2-containing p85 subunit of phosphatidylinositol 3-kinase (Wang et al., 1992), thus activating a pathway likely associated with cell proliferation. However, 4PS-independent signals may also lead to proliferation, as suggested from studies with an IL-4 receptor mutant lacking tyrosine-497 (Keegan et al., 1994a) and with the P-mutant lacking the entire carboxyl-terminal portion from residue 432 (Harada et al., 1992).

Recently, JAK-1 was shown to complex with the IL-4 receptor following ligand binding (Yin et al., 1994), thus likely playing a role in the previously observed tyrosine phosphorylation of the 130kDa IL-4 receptor (Izuhara and Harada, 1993; Wang et al., 1992). In addition to the JAK family kinases, IL-4 binding induces tyrosine phosphorylation of a 92kDa protein, identical or closely related to the FES tyrosine kinase (Izuhara et al., 1994). p92 FES associates with the p130 IL-4 receptor, within a domain between residues 352-431 of the receptor (Figure 2), and association is enhanced upon ligand binding (Izuhara et al., 1994). It is noteworthy that the stretch of 41aa between residues 352-392 is required to signal growth through the human IL-4 receptor in murine Ba/F3 transfectants, probably through an acidic EAPV-X3-EEEEE domain (Harada et al., 1992). These results would be consistent with the existence of binding sites for p92 FES and possibly also for the JAK-family kinases in this portion of the receptor.

Recently, it has been recognized that tyrosine-phosphorylation initiated by engagement of cytokine receptors can recruit factors, termed signal transducers and activators of transcription (STATs), that directly translocate to the nucleus, leading to activation of cytokine-responsive genes (Ihle et al., 1995; Taniguchi, 1995). In this context, IL-4 activates IL-4 STAT or STAT 6, a 100kDa DNA-binding protein, containing contiguous SH2 and SH3 domains, and recognizing an IL-4 response element upstream of the human FcγRI gene (Hou et al., 1994; Quelle et al., 1995). STAT6 binds to either or both of the peptides surrounding phosphorylated tyrosines at positions 603 and 631 in the p130 IL-4 receptor chain (Figure 2). Following binding of IL-4 to its receptor, STAT6 becomes tyrosine-phosphorylated by a yet undefined tyrosine kinase, allowing dimerization and subsequent release from the receptor. STAT6 is rapidly tyrosine phosphorylated following stimulation of appropriate cell lines with IL-4 or IL-3 but is not detectably phosphorylated following stimulation with IL-2, IL-12, or erythropoietin. Inducible tyrosine phosphorylation of STAT6 requires the membrane-distal region of the p130 IL-4Rα chain. This region of the receptor is not required for cell growth, demonstrating that tyrosine phosphorylation of STAT6 does not contribute to mitogenesis (Quelle et al., 1995). Other tyrosine-phosphorylated DNA-binding factors induced by IL-4 have been described, which may turn out to be identical to STAT6. Thus, IL-4 NAF binds to the IFN-γ response region of the human FcγRI gene, to the human FcεRIIb promoter, and to the murine Cγ1 promoter, all of which contain an inverted GAA repeat structure (Kotanides and Reich, 1993). The factor STF-IL-4 binds to an IFN-γ activation site (GAS element) of the IRF-1 promoter, and to a sequence at the initiation site of Ig germline ε transcription (Schindler et al., 1994). Furthermore, a factor has been reported which binds to an IL-4 responsive element upstream of the Iε exon of the human Cε gene (Ichiki et al., 1993). Of interest, the fact that IL-4-induced DNA-binding factors recognize IFN-γ response sites, which allows the two cytokines to modulate transcription of the same genes, provides an explanation for the fact that IL-4 and IFN-γ may have either antagonistic or analogous biologic effects.

IL-4, thus, mediates its actions through at least two distinct pathways (PI3-kinase and STAT) linked to the p130 IL-4 binding chain, in line with earlier speculations that more than one pathway is involved in IL-4 signaling (Rigley et al., 1991). The involvement of other pathways is likely, but less clearly defined. IL-4 inhibits transcription of the c-fos and c-jun genes in activated human monocytes and U937 cells (Dokter et al., 1993a; Ouaaz et al., 1994), and the effect on c-fos expression is partly controlled by the lipoxygenase directed pathway of arachidonic acid metabolism (Dokter et al., 1994). The contribution of phosphoinositide metabolism to IL-4 signaling is controversial. Thus, IL-4 has been shown to increase inositol 1,4,5-triphosphate (IP3) and intracellular Ca++ levels in human (Finney et al., 1990), but not murine (Justement et al., 1986; Mizuguchi et al., 1986) B cells. In support of an involvement of the phosphoinositide pathway, IL-4 has been shown to induce a redistribution of protein kinase C (PKC) from the cytosol to the nucleus in human monocytes (Arruda and Ho, 1992). In addition, IL-4 may recruit the adenylate cyclase pathway, as it has been shown to increase the levels of cAMP in human B lymphocytes (Finney et al., 1990) and NK cells (Blay et al., 1990). Of interest, IL-4 can increase the levels of cGMP in IFN-γ activated human monocytes, likely as a result of enhanced nitric oxide (NO)-synthase activity and NO release (Kolb et al., 1994). Two groups have reported the phosphorylation of a membrane-bound 42-44kDa protein in response to IL-4 in B cells (Justement et al., 1986; McGarvie and Cushley, 1989). Finally, IL-4-induced proliferation of human myeloid cell lines involves dephosphorylation of an 80kDa protein, suggesting involvement of a tyrosine phosphatase (Mire-Sluis and Thorpe, 1991).

The collective information thus indicates that IL-4 can signal through multiple pathways, and suggests that the signaling utilized is dependent upon cell type and cell activation status.

V. CELLULAR SOURCES OF IL-4

Unlike IL-1, IL-6 and IL-10, which are produced by many different cell types, IL-4, like IL-2 and IFN-γ, is secreted by restricted cell types.

A. T Lymphocytes

T cells constitute a major source of IL-4. Thus, Th2 cells secrete IL-4 and IL-5, and lead to a preferential stimulation of humoral immunity. In contrast, Th1 cells which produce IL-2 and IFN-γ, lead to a preferential stimulation of cellular immunity (Figure 3).

Recently, studies have been undertaken to attempt to identify T cell surface markers whose expression may correlate with either a Th1 or a Th2 phenotype. In particular CD30, a member of the Tumor Necrosis Factor Receptor family (Gruss and Dower, 1995), has been the object of

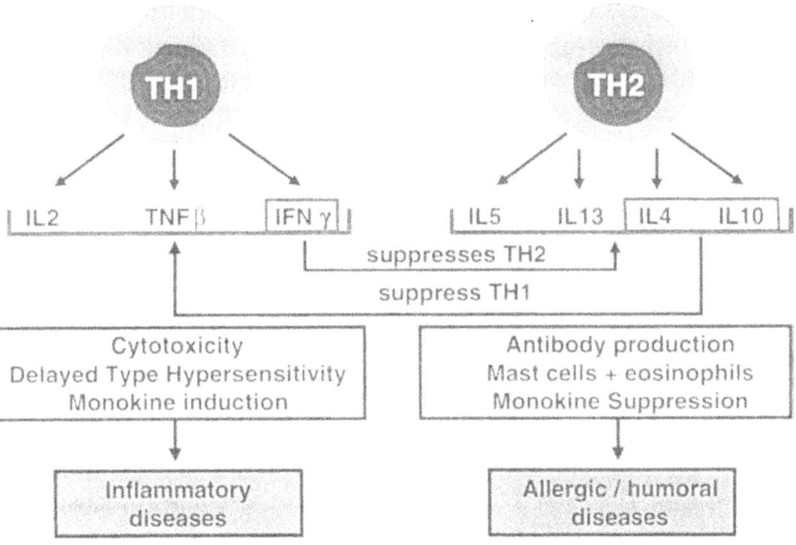

Figure 3. Functional roles and cross-regulation of Th1 and Th2 cells. Protective immunity results from the balanced activation of both subsets. Hyperactivation of either subset leads to immune-mediated disease.

much attention by Romagnani's group (Del Prete et al., 1995b): (1) Activated Th2, but not Th1, $CD4^+$ T cell clones express membrane CD30 and release sCD30; Th0 T cell clones exhibit an intermediate pattern of CD30 expression and release (Del Prete et al., 1995a); (2) Peripheral blood T cells activated *in vitro* with Th2-inducing antigens express CD30 while T cells activated with Th1-inducing antigens do not; (3) Allergen-specific $CD4^+$ $CD30^+$ T cells, that can be induced to produce Th2-type cytokines, are present in the circulation of atopic donors following allergen exposure; (4) $CD30^+$ T cells are present in the lymph nodes of patients with Omenn's syndrome, a rare congenital immunodeficiency syndrome in which abnormal Th2-like cells appear to play a pathogenic role (Schandene et al., 1993); (5) Elevated levels of sCD30 are detectable in the serum of patients suffering from diseases in which a prevalent Th2 response has been suggested including atopy, lupus, Omenn's syndrome, measles virus infection, and HIV (Pizzolo et al., 1994). In contrast, other studies have demonstrated that CD30 defines a subset of activated human memory (CD45 Ro^+) T cells that produce IFNγ and IL-5, but no IL-4 (Alzona et al., 1994, 1995). This subset possesses enhanced B cell helper activity. Note that IL-2 and IL-10 are found in $CD30^+$ and $CD30^-$ T cells.

CD27, another member of the TNFR family, appears useful to delineate CD4 T cell subpopulations inasmuch as T helper 1, T helper 2, and T helper 0 cells are identified within the small population of $CD4^+$ $CD27^-$ T cells which thus appear as defining functionally differentiated memory T cells (Elson et al., 1994, 1995; Sunder-Plassmann et al., 1995). Finally, a small proportion (≈10%) of blood $CD4^+$ T cells fails to express CD7 (Autran et al., 1995). This population produces much more IL-4 and IL-10 than $CD4^+$ $CD7^+$ T cells which produce most of the IL-2. IFNγ production is similar in both subsets.

An important feature of Th1 and Th2 cells is the ability of one subset to regulate the activities of the other. It occurs at the levels of the effector cells triggered by these subsets, as indicated by the inhibitory effects of IFNγ on IL-4 induced B cell activation or those of IL-4 on IL-2 induced T and B lymphocyte proliferation. It also occurs directly at the level of these subsets as the products of one subset can antagonize the activation of the other: IFNγ inhibits proliferation of Th2 cells (Gajewski et al., 1989), whereas IL-4 and IL-10 inhibit cytokine production by Th1 cells (Fiorentino et al., 1989; Peleman et al., 1989; Vieira et al., 1991).

The mechanisms controlling the development of Th1 and Th2 cells are becoming elucidated (Figure 4). The different functional subsets of

Th cells arise post-thymically from naive precursors (Thp) as a consequence of antigen activation. Thp are short-lived $CD4^+$ cells which secrete IL-2 but not other cytokines. After activation by antigen, Thp develop into effector cells (Th0) that can be induced to secrete very high titers of IL-3, IL-4, IL-5, GM-CSF, and IFNγ. In contact with IL-4, activated T cells mature into Th2 cells, whereas IFNγ and IL-12 allow the generation of Th1 cells (Coffman et al., 1991; Maggi et al., 1992; Manetti et al., 1994b; Seder et al., 1992; Swain et al., 1991). Interestingly, the B7-2 / CD86 molecule, one of the CD28-CTLA-4 counterreceptors, appears to preferentially costimulate the initial production of IL-4 by naive T cells (Freeman et al., 1995). While the above model of Th1Th2 maturation has been developped from studies performed with mouse T cells, recent studies with human naive T helper cells suggest that generation of IL-4-producing helper T cells may be independent of exogenous IL-4 (Kalinski et al., 1995).

More recent evidence suggests that $CD8^+$ T cells may also be subdivided into Th1 and Th2 subpopulations (Seder and Le Gros, 1995). In particular, murine $CD8^+$ T cells, activated in the presence of IL-4, develop into a $CD8^-$ $CD4^-$ population that is not cytolytic, lacks perforin and does not produce IFNγ. Furthermore, these cells produce large amounts of IL-4, IL-5, and IL-10 and can induce the growth and differentiation of activated B cells (Erard et al., 1993). IL-4 also appears to play an important role in the development of Th2 $CD8^+$ T cells (Sad et al., 1995). In man, $IL-4^+$ producing $CD8^+$ T cells have been observed in diseases such as leprosy (Salgame et al., 1991) and more recently AIDS (Maggi et al., 1994). As described above for $CD4^+$ T cells, these Th2 $CD8^+$ T cells have been found to express membrane CD30 and release sCD30 while Th1 $CD8^+$ T cells do not (Manetti et al., 1994a). The emergence of such cells may contribute to the disease possibly because of reduced overall cytolytic activity against microbe-infected cells.

B. NK cells.

A minor population of mouse $CD4^+$, NK 1.1^+ T cells has been shown to produce large amounts of IL-4 in response to *in vivo* challenge with anti-CD3 (Yoshimoto and Paul, 1994). This T cell population, which represents an important lymphokine-producing cell of the thymus and bone marrow (Arase et al., 1993), does not require exogenous IL-4 priming to secrete IL-4.

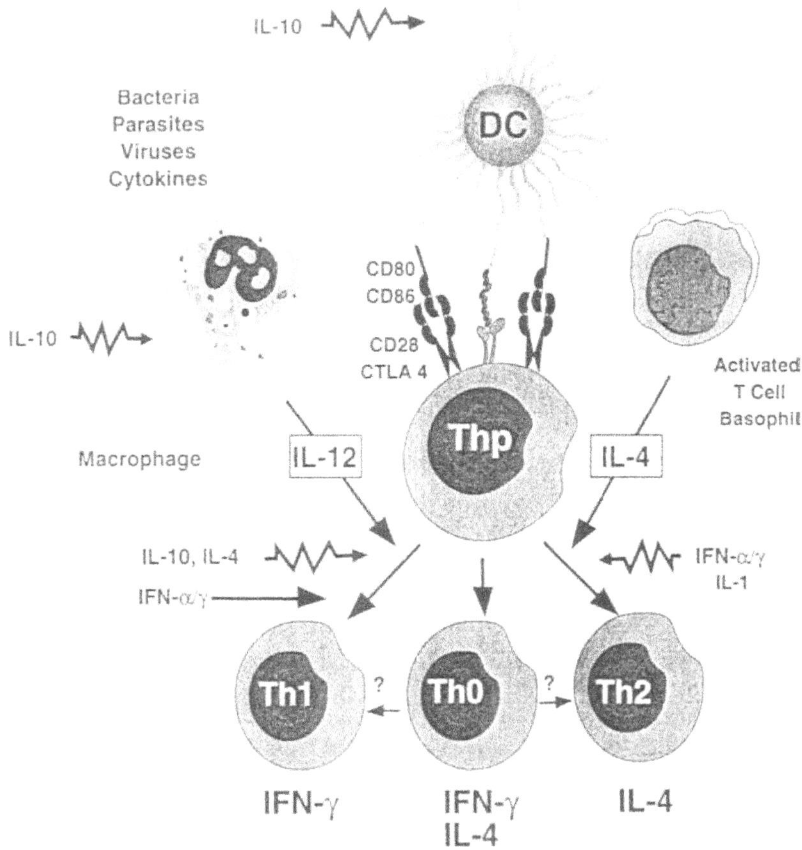

Figure 4. Cytokine control of Th1/Th2 development.

C. Mast Cells/Basophils

IL-4 is also produced by basophils/mast cells activated by cross-linkage of FcεRI and FcγRII (Brown et al., 1987; Brunner et al., 1993; Mueller et al., 1994; Piccinni et al., 1991). Nasal biopsy specimens from patients with allergic rhinites and bronchial biopsy specimens from patients with allergic asthma display mast cells with intracellular IL-4 (Bradding et al., 1992). Interestingly, anti-IgE induces maximum release of IL-4 within one hour suggesting that it is present within the cells in a preformed state similar to that reported for TNFα.

Recent data, however, implies that basophil/mast cell derived IL-4 is not required to prime Th2 cells to express IL-4 (Schmitz et al., 1994).

D. Stromal Cells

Murine fibroblastic bone marrow stromal cells can produce IL-4 constitutively, and this function is upregulated by exposure to IL-1 (King et al., 1988, 1989a). Heterogeneous primary stromal cell populations, as well as cell lines of mouse origin, have been demonstrated to produce IL-4 (Billips et al., 1991; Kinashi et al., 1988; King et al., 1988). In contrast, however, no IL-4 transcripts were detected in primary cultured bone marrow stromal cells that produced a number of cytokines including IL-7, M-CSF, c-kit ligand, IGF-1, and LIF (Witte et al., 1993). These findings suggest that IL-4-producing stromal cell lines reflect a relatively rare subset of the stromal cell population in the bone marrow.

VI. EFFECTS OF IL-4 ON IMMUNE EFFECTOR CELLS

A. B Lymphocytes

IL-4 increases the volume of resting B cells (Vallé et al., 1989) and induces hyperexpression of MHC class II and CD23 (the low affinity receptor for IgE-FcεRII and a ligand for CD21) on normal and leukemic B lymphocytes (Defrance et al., 1987a). Two CD23 cDNAs, FcεRIIα and FcεRIIβ, have been isolated which differ in the first few amino acids of the cytoplasmic tail (Yokota et al., 1988). In normal B cells, FcεRIIα but not FcεRIIβ, is expressed spontaneously at low levels and IL-4 enhances expression of FcεRIIα and induces expression of FcεRIIβ. Both IFN-α and IFN-γ block the IL-4-dependent increase of CD23 on B cells. IL-4 up-regulates expression of CD40, whose ligand is expressed on activated T cells and plays a key role in B cell proliferation and differentiation (Banchereau et al., 1994; Van Kooten and Banchereau, 1995). IL-4 also up-regulates expression of surface IgM (Shields et al., 1989), B7-1/CD80 (Ranheim and Kipps, 1993; Vallé et al., 1991), and B7-2/CD86, both being counterstructures of T cell CD28 and CTLA4 (Linsley and Ledbetter, 1993; Thompson, 1995). All these effects of IL-4 on B cells are suggestive of a role of IL-4 in the enhancement of antigen-presenting capacity of B cells toward T cells. Furthermore, IL-4 induces activated B cells to produce IL-6 and TNF (Smeland et al., 1989), cytokines which play an important role in the activation and clonal expansion of activated T cells. Thus, IL-4 favors T-B cell interactions, concurring to enhanced antibody production.

IL-4 allows the proliferation of B cells activated through their antigen receptor (Defrance et al., 1987b; Llorente et al., 1990). Addition of IL-4 to anti-CD40 activated B cells results in a strong B cell proliferation allowing the generation of factor-dependent long-term normal B cell lines (Banchereau et al., 1991). Paradoxically, IL-4 inhibits the IL-2 induced proliferation of B cells activated through their antigen receptor (Defrance et al., 1988; Jelinek and Lipsky, 1988). This effect may be due to the preferential sequestering by IL-4 of the γc chain which is also part of the IL-2 receptor (Karray et al., 1990).

Early studies have shown that murine IL-4 can induce murine B cells to produce IgG1 (Isacksson et al., 1982) and IgE (Coffman et al., 1986). Addition of $CD4^+$ T cells and monocytes (Pène et al., 1988), anti-CD40 monoclonal antibody (Jabara et al., 1990; Rousset et al., 1991a), or hydrocortisone to human B lymphocytes allow them to secrete IgE in response to IL-4. The IL-4 dependent production of IgE is the consequence of IL-4 induced isotype switching (Jabara et al., 1993; Mills et al., 1992; Shapira et al., 1991, 1992; Zhang et al., 1994). Human IL-4 also appears to direct switching toward the production of IgG4 (Gascan et al., 1991). *In vivo* experiments in mice have confirmed the crucial role of IL-4 and interferons in the regulation of IgE production (Finkelman et al., 1990, 1991). Inactivation of the IL-4 gene in mice through gene targeting is associated with normal T and B cell development but with a strong reduction of seric IgG1 and disappearance of IgE (Kühn et al., 1991). Conversely, IL-4 transgenic mice have increased IgE and an allergic-like disease with ocular lesions infiltrated with mast cells and eosinophils (Tepper et al., 1990).

IL-13 has IL-4-like effects on B cells (Zurawski and de Vries, 1994).

B. T Lymphocytes and NK Cells

The T cell growth promoting effects of IL-4 were initially discovered on continuous T cell lines (Mosmann et al., 1986). IL-4 can act on activated normal $CD4^+$ and $CD8^+$ T cells (Hu-Li et al., 1987; Spits et al., 1987) in an IL-2 independent fashion. Studies with antisense oligonucleotides showed that IL-4 and IL-2 are autocrine growth factors of T_{H2} and T_{H1} T cell clones, respectively (Harel-Bellan et al., 1988). Combinations of IL-2 and IL-4 result in a proliferation of activated T cells which is greater than that obtained with each cytokine alone. However, IL-4 blocks the IL-2 induced proliferation of naive $CD4^+/CD45RA^+$ T cells (Gaya et al., 1990).

IL-4 is also able to inhibit the production of IFNγ by activated T cells (Peleman et al., 1989) and, in doing so, favors the generation of T cells of the Th2 type (see preceding).

IL-4 turns on the expression of CD8 on $CD4^+$ T cell clones and neonate T cells (Paliard et al., 1988; Reason et al., 1990). The induced $CD4^+$ $CD8^+$ neonatal cells subsequently differentiate to express only CD8.

Analogous to its effect of B cells, IL-4 can induce the expression of CD23 on activated T lymphocytes (Armitage et al., 1989; Prinz et al., 1990).

When added to mixed leukocyte cultures, IL-4 increases antigen-specific cytotoxic activity against allogeneic stimulator cells (Spits et al., 1988; Widmer et al., 1987; Widmer and Grabstein, 1987). It also enhances the development of virus-specific cytotoxic T cells (Horohov et al., 1988). The endogenous production of IL-4 in response to viral challenge is likely to play an important role in the generation of antigen-specific cytotoxic cells. T cell clones, cultured in the presence of IL-4, have been shown to display higher cytolytic ability than when cultured in IL-2. Administration of soluble IL-4 receptor to mice inhibits an allogenic response *in vivo* and enhances cardiac allograft survival, thus suggesting *in vivo* relevance for IL-4 in the development of cytotoxic T cells (Fanslow et al., 1991). Finally, IL-4 increases the adhesiveness of T cells for endothelial cells (Thornhill et al., 1990).

$CD3^-$ NK cells proliferate in response to IL-4 provided that the cells have received a mitogenic stimulus (Blay et al., 1990; Kawakami et al., 1989; Nagler et al., 1988; Spits et al., 1988). However, IL-4 inhibits IL-2 dependent proliferation of these cells and accordingly blocks IL-2 dependent generation of human LAK cells. Recently, activated human NK cells have been shown to secrete IL-5 (Warren et al., 1995) and IL-4 was found to enhance that secretion while it inhibited IFNγ production.

It is interesting to point out that IL-13 does not act on either T cells or NK cells, and these cells do not bind IL-13 even after activation.

C. Myelomonocytic Cells

Mononuclear Phagocytes

IL-4 enhances the antigen-presenting capacity of monocytes / macrophages through an increased expression of MHC class II (Te Velde et al.,

1988; Zlotnik et al., 1987a), and down-regulates expression of the differentiation antigen CD14, a receptor for LPS (Vercelli et al., 1988). In addition, IL-4 up-regulates expression of the low-affinity IgE receptor/CD23 (Vercelli et al., 1988), But down-regulates expression of the three types of IgG receptors (Te Velde et al., 1990) resulting in inhibition of antibody-dependent cytotoxicity. Adherent blood mononuclear cells cultured in the presence of both GM-CSF and IL-4 mature into dendritic cells that can efficiently present soluble antigen to specific T cell clones (Romani et al., 1994; Sallusto and Lanzavecchia, 1994). IL-4 activates tumoricidal activity of macrophages (Crawford et al., 1987) and induces the formation of giant multinucleated cells (McInnes and Rennick, 1988). An important property of IL-4 is its ability to block spontaneous and induced production of the proinflammatory cytokines IL-1, IL-6, IL-8, and TNFα (Hart et al., 1989; Te Velde et al., 1988; Vercelli et al., 1988) while stimulating the production of IL-1 receptor antagonist (IL-1RA) (Vannier et al., 1992). IL-4 also inhibits the secretion of interstitial collagenase and 92 kDa type IV collagenase (Corcoran et al., 1992; Lacraz et al., 1992) thus reducing the ability of macrophages to degrade extracellular matrix. IL-4 has a profound inhibitory effect on the release of superoxide (Abramson and Gallin, 1990; Ho et al., 1992) and the secretion of PGE_2 (Hart et al., 1989). In addition, IL-4 inhibits IFN-γ-induced expression of the proinflammatory α-chemokine IP-10 in murine macrophages (Deng et al., 1994; Larner et al., 1993). Taken together these properties suggest a powerful anti-inflammatory effect of IL-4. Accordingly, in vivo administration of IL-4 was found to inhibit the development of an antigen-specific T cell mediated inflammatory response in a hapten-induced model of contact sensitivity (Gautam et al., 1992). The anti-inflammatory effects of IL-4 have recently been postulated to be due to up-regulation of expression and activity of cell membrane aminopeptidase-N (CD13) (van Hal et al., 1994), which inactivates proinflammatory peptides (Shipp and Look, 1993).

IL-4 inhibits the killing by macrophages of various parasites, such as *Leishmania* (Ho et al., 1992; Lehn et al., 1989) and asexual forms of *Plasmodium falciparum* (Kumaratilake and Ferrante, 1992).

Finally, complete resistance to HIV-1 infection was observed in monocytes that had been treated with IL-4 or IL-13 (Denis and Ghadirian, 1994; Montaner et al., 1993; Schuitemaker et al., 1992).

Granulocytes

Eosinophils express reduced levels of FcγR upon culturing with IL-4 (Baskar et al., 1990). The down-regulation of FcγR results in a decreased secretion of glucuronidase and arylsulphatase in response to IgG-coated beads, and in reduced antibody-dependent killing of schistosomulas (Baskar et al., 1990). Of importance, IL-4 enhances the transendothelial migration of eosinophils, an effect related to upregulation of adhesion molecules such as VCAM-1 on endothelial cells (Moser et al., 1992; Schleimer et al., 1992; Thornhill and Kaskard, 1990). Infiltration of eosinophils contributes to the anti-tumor effects of IL-4 *in vivo* (Tepper et al., 1992), and is a feature of IL-4 transgenic mice (Tepper et al., 1990). IL-4 also displays an anti-inflammatory effect on neutrophils by enhancing respiratory burst and phagocytic properties (Bober et al., 1995; Boey et al., 1989), and inhibiting their secretion of IL-8 (Wertheim et al., 1993). IL-4 enhances expression and secretion of membrane and soluble type II IL-1 receptor by neutrophils (Colotta et al., 1993). This may also contribute to IL-4 anti-inflammatory effects by blocking IL-1 effects.

Finally, IL-4 induces the adherence of basophils, but not of neutrophils, to endothelial cells (Schleimer et al., 1992).

VII. IL-4 AND HEMOPOIESIS

A. Myelopoiesis

Normal Myeloid Development

Several studies indicate that IL-4 alone has no effect on normal myeloid progenitors, but acts as a cofactor in modulating myelopoiesis (Rennick et al., 1992; Vellenga et al., 1993).

No effect of IL-4 was observed on mixed-erythroid colony formation from T cell and monocyte-depleted human bone marrow in the presence of Epo (Sonoda et al., 1990). However, a recent study indicates that IL-4 does enhance the output of human CFU-GM in stromal-cell dependent long-lerm cultures, a likely result of synergy between IL-4 and stem cell factor (SCF) / c-kit ligand (Keller et al., 1994).

IL-4 acts as a cofactor for the generation of neutrophilic granulocytes in the presence of G-CSF, but not of IL-3 or GM-CSF. This cofactor effect of IL-4 was noted in early murine studies (Peschel et al., 1987; Rennick et al., 1987) and subsequently extended to human granulocytic precursors (Broxmeyer et al., 1988; Sonoda et al., 1990; Vellenga et al., 1990). Of interest, this potentiating effect of IL-4 is antagonized by IFN-γ (Snoeck et al., 1993b), thereby extending the well-established counteracting properties of IL-4 / IFN-γ to hemopoietic progenitor cells. Early claims that the effect of IL-4 is directly exerted on the granulocytic progenitors rather than indirectly mediated by accessory cells have been confirmed by single-cell cultures of human bone marrow $CD34^+$ cells (Snoeck et al., 1993b). The synergy between IL-4 and G-CSF is borne by $CD34^+$ $HLA\text{-}DR^{++}$ cells, through recruitment by IL-4 of a higher number of G-CSF-responsive progenitors and their subsequent enhanced proliferation in response to IL-4 + G-CSF as compared to G-CSF alone, resulting both in more colonies and an increase in the size of colonies (Snoeck et al., 1993a).

Basophils and Mast Cells

Historically, the capacity of IL-4 to enhance the proliferation of murine factor-dependent mast cell lines (Grabstein et al., 1986; Lee et al., 1986; Mosmann et al., 1986; Schmitt et al., 1987; Smith and Rennick, 1986) was the first indication that IL-4 acted on myeloid cells. Subsequently, IL-4 was shown to enhance the size of mast cell colonies from murine bone marrow Thy 1-low progenitors in the presence of IL-3, although this effect was only noted after a considerable (21 days) culture period (Rennick et al., 1987). A further synergy resulting in increased numbers of mucosal-like mast cell colonies was noted when mesenteric lymph node cells from *Nippostrongylus brasiliensis* infected mice were cultured in a combination of IL-4, IL-3, and IL-10 (Rennick et al., 1992; Thompson-Snipes et al., 1991). It is likely that IL-4 is an important component in the process leading to mastocytosis in parasitic infections. IL-4 also supports the formation of colonies from murine peritoneal connective tissue-type mast cells (Hamaguchi et al., 1987; Tsuji et al., 1990). Bone marrow-derived mast cells cultured in the presence of IL-4 can respond to Endothelin-1 by secreting histamine, serotonin, and leukotriene C_4 to the extent observed with IgE cross-linking (Egger et al., 1995). Thus, some of the potent broncho- and vasoconstricting activities of Endothelin-1 may not be exclusively explained by its direct action on smooth muscle cells but in addition by a series of mast cell mediators whose release depend on Endothelin-1. In humans, IL-4 results in enhanced generation of basophils from T and monocyte-depleted cord blood cells cultured in the presence of IL-3 (Favre et al., 1990). The cells thus generated have a morphology and a cytochemical composition (alcian-blue positive, safranin-0 negative) compatible with blood basophils and not of mast cells. IL-4 down-regulates c-*kit* gene expression in human bone marrow progenitors and in the immature mast cell line HMC-1 (Sillaber et al., 1991). In line with the observed down-regulation of c-kit, IL-4 has been reported to inhibit SCF/c-kit ligand dependent human mast cell development (Sillaber et al., 1992). These results suggest that IL-4 has agonistic or antagonistic effects on the generation of basophils / mast cells depending on its association either with IL-3 or SCF.

We observed that human IL-4 enhances the generation of eosinophils from T and monocyte-depleted human cord blood cells in the presence of IL-3 (Favre et al., 1990). However, another study did not identify significant effects of IL-4 on eosinophilic colony formation in response to IL-3 (Vellenga et al., 1990). Murine IL-4 does clearly play a role in the eosinophilic pathway, as demonstrated by *in vivo* murine experiments in which both IL-4 transgenic animals and nude mice injected with IL-4-producing tumor cells exhibit tissue infiltration of eosinophils (Tepper et al., 1989, 1990). However, the exact contribution of IL-4 to the development of eosinophilic progenitors remains to be clarified.

Early experiments revealed suppressive effects of IL-4 on macrophage colony formation by murine bone marrow cells cultured in IL-3 (Rennick et al., 1987). This inhibitory effect of IL-4 was confirmed on human progenitors and is direct and independent of the stimulation signal provided (i.e., M-CSF or GM-CSF) (Jansen et al., 1989; Sonoda et al., 1990). Analogous to its action on neutrophil progenitors, IFN-γ also counteracts the effects of IL-4 on macrophage development. Thus, IFN-γ can overcome the inhibitory effect of IL-4 through direct action on macrophage progenitors (Snoeck et al., 1993b). Single-cell cultures indicate that IL-4 inhibits human macrophage development.

Combinations of erythropoietin and IL-4 induce partially purified progenitor cells to generate erythroid colonies (Broxmeyer et al., 1988; Peschel et al., 1987) but this may be due to an indirect effect of IL-4 on accessory cells. In contrast, IL-4 inhibits the formation of pure and mixed megakaryocyte colonies from enriched human hemopoietic progenitors (Sonoda et al., 1993).

Myeloid Leukemias

IL-4 inhibits *in vitro* the spontaneous DNA replication of acute myeloblastic leukemia cells (Akashi et al., 1991a; Ouaaz et al., 1994; Wagteveld et al., 1991), an effect likely due to interference with the production of autocrine growth factors such as GM-CSF, IL-6 and TNFα (Jansen et al., 1991; Wagteveld et al., 1992). However, IL-4 was shown to display highly divergent effects on AML cells cultured in the presence of exogenous cytokines (Akashi et al., 1991a; Imai et al., 1991; Miyauchi et al., 1991; Vellenga et al., 1990). An inhibitory effect of IL-4 on chronic myelomonocytic leukemias has also been reported (Akashi et al., 1991b; Estrov et al., 1993).

B. Lymphopoiesis

Normal B Cell Lymphopoiesis

Proliferation of early B cell progenitors is enhanced in long-term Whitlock-Witte type murine bone marrow cultures in the presence of IL-4 (Peschel et al., 1989). Thus, IL-4 favors the expansion of early B lineage cells that bear DJ_H rearrangements, but do not yet express B220 antigen. This effect likely results from IL-4-mediated induction of a potentiating factor by stromal cells. Other studies have demonstrated a direct growth-promoting effect of IL-4 on murine pro-B cell clones (germ-line Ig gene configuration) with the capacity to differentiate either into B cells or myeloid cells (Kinashi et al., 1989; Sideras and Palacios, 1987). In contrast to these stimulatory effects, IL-4 can also suppress B lymphopoiesis. In the mouse, IL-4 inhibits the development of $B220^+$ B cell precursors (Peschel et al., 1989; Rennick et al., 1987), and both the emergence (Billips et al., 1990) and proliferation (Rennick et al., 1987) of pre-B cells (expressing cytoplasmic μ-chain) in stroma-dependent cultures. Thus, murine studies would suggest that negative regulatory activities of IL-4 are mainly exerted at later developmental stages in B lymphopoiesis. Several groups have extended the growth-inhibitory properties of IL-4 to human bone marrow B cell progenitors (Hofman et al., 1988; Pandrau et al., 1992; Ryan et al., 1994). In apparent contrast to murine data, primitive human lymphoid progenitors ($CD34^+$ cells which form TdT^+ colonies that include early B lineage cells) are also inhibited by IL-4 (Ryan et al., 1994). It has been suggested that IL-4 is involved in a network with other cytokines such as IL-1α and TNF-α in orienting the bone marrow microenvironment towards supporting myeloid development at the expense of B lymphopoiesis (Ryan et al., 1994). Of interest, the growth-inhibitory effect of IL-4 on human B cell precursors is shared by IL-13 (Renard et al., 1994).

In contrast to mature B cells, IL-4 does not appear to induce CD23 expression in human B cell precursors (Law et al., 1991; our unpublished observations). This does not, however, represent an intrinsic failure to express CD23, as ligation of surface CD40 antigen results in induction of CD23 in B cell precursors (Saeland et al., 1993).

IL-4 plays a role in the final maturation step during bone marrow B lymphopoiesis, by enhancing the output of $sIgM^+$ cells in murine cultures (Kinashi et al., 1988; King et al., 1988). This effect likely results from enhanced cell survival rather than from induction of maturation (Simons and Zharhary, 1989). In humans, controversy exists as to whether IL-4 induces an increase in the numbers of $sIgM^+$ cells generated from precursors (Hofman et al., 1988), or only enhances sIgM expression on $sIgM^+$ cells that have emerged independently of IL-4 (Law et al., 1991; Pandrau et al., 1992). Finally, human bone marrow B cell precursors have been described to differentiate into Ig-secreting cells when cultured with IL-4 in association with activated T cells (Punnonen et al., 1993). Thus, although IL-4 clearly functions in positively regulating the final stages of B lymphopoiesis, the exact contribution of direct versus indirect pathways remains to be clarified.

Leukemic B Cell Precursors

B cell precursor acute lymphoblastic leukemia (BCP-ALL) cells express IL-4 receptors, although the proportion of freshly-isolated cells which bind IL-4 is heterogeneous as compared to leukemic cell lines (Law et al., 1991). Several studies indicate that IL-4 inhibits growth and survival of BCP-ALL cells. Thus, DNA-replication is suppressed both in Ph^1-positive [t(9;22)] (Okabe et al., 1991) and Ph^1-negative BCP-ALL (Pandrau et al., 1992; Renard et al., 1994; Greil et al., 1994), although not all cases are responsive. The effect of IL-4 on DNA-replication is directly exerted on the leukemic cells in IL-4 sensitive cell lines, and does not appear to be due to production of inhibitory factors such as IFN-α,-β,-γ, or TNF-α (Okabe et al., 1991). *In vitro* viability of freshly isolated ALL cells can be maintained when cultured on bone marrow stromal layers (Manabe et al., 1992). Under such conditions, IL-4 has a cytotoxic effect in the majority (16/21) of the BCP-ALL cases tested

(Manabe et al., 1994). Cytotoxicity has been shown to be due to induction of apoptosis in G0/G1 phases of cell cycle, and is believed to be a direct effect of IL-4 (Manabe et al., 1994). IL-4 can induce CD23 (Law et al., 1991) and CD20 (Pandrau et al., 1992) in a proportion of BCP-ALL cases. Given the documented inhibitory effects of IL-4 on leukemic cell growth and survival, this cytokine appears as a candidate for clinical trials in cases of high-risk BCP-ALL. In this context, it is noteworthy that IL-4 was cytotoxic for all the cases analyzed (n = 9) that correspond to poor prognosis ALL associated either with t(9;22) or t(4;11) translocations (Manabe et al., 1994). Recently, administration of high-dose (200 µg/kg/d) human IL-4 was reported to suppress human ALL xenografts in scid mice (Mitchell et al., 1994). Finally, IL-13 shares with IL-4 the capacity to inhibit BCP-ALL growth (Renard et al., 1994). Of interest, dichotomy was observed in the pattern of IL-13 versus IL-4 responsiveness in a number of cases (Renard et al., 1994).

T Cell Lymphopoiesis

Both murine (Zlotnik et al., 1987b) and human (Barcena et al., 1990; Spits et al., 1987) thymocytes proliferate in response to IL-4 and a costimulant such as phorbol esters (PMA). In the mouse, IL-4 induces the proliferation of the most immature $CD4^-$ $CD8^-$ subset which also has the ability to produce IL-4. The intermediate $CD8^+$ $CD4^+$ subset virtually fails to proliferate under these conditions, whereas the most mature $CD4^+$ $CD8^-$ and $CD4^-$ $CD8^+$ subsets also proliferate in response to IL-4 and PMA. More recently, IL-4 has been shown to protect murine thymocytes from dexamethasone-induced apoptosis (Migliorati et al., 1993). This effect was shown to be borne predominantly by $CD4^-CD8^+$ and $CD4^+CD8^-$ cells. Fetal thymocytes also respond to IL-4 and *in situ* hybridization studies (Sideras et al., 1988) have demonstrated transcription of the IL-4 gene in fetal thymus. In addition to its effects on cell proliferation and survival, IL-4 can induce the maturation of thymocytes through induction of mature T cell antigens (CD3, CD5, T cell receptor) and loss of CD1. Concomitantly, the $CD4^+$ $CD8^+$ subset disappears and $CD4^-$ $CD8^-$, $CD4^+$ $CD8^-$, $CD4^-$ $CD8^+$ cells are generated (Ueno et al., 1989). These data indicate that IL-4 plays an important role in T cell ontogeny. Paradoxically, IL-4 transgenic mice were found to have involuted thymuses (Tepper et al., 1990). In accordance with this latter finding, IL-4 inhibits early T cell development in fetal thymus organ culture, probably during the differentiation of CD4⁻ CD8⁻ cells into $CD4^+CD8^+$ thymocytes. The generation of TCRαβ thymocytes appears to be more impaired than that of TCRγδ thymocytes (Plum et al., 1990). Likewise, human IL-4 induces a preferential differentiation of TCR γ/δ pre-T cells (Barcena et al., 1990).

VIII. EFFECTS OF IL-4 ON OTHER CELL TYPES

A. IL-4 and Fibroblasts

Fibroblasts are chemoattracted by IL-4 (Postlethwaite and Seyer, 1991). IL-4 also induces dermal fibroblasts to secrete extracellular matrix proteins, such as type I and type III collagen and fibronectin (Fertin et al., 1991; Gillery et al., 1992; Postlethwaite et al., 1992). IL-4 stimulates fibroblasts to secrete glycosaminoglycans and more specifically dermatan/chondroitin sulphates while hyaluronan synthesis is not affected. Decorin has been identified as the main component whose synthesis is stimulated by IL-4 (Wegrowski et al., 1995). IL-4 blocks cytokine (IL-1, PDGF) induced proliferation of synoviocytes (Dechanet et al., 1993). Cell cycle studies have indicated a rapid and persistent blocking effect of IL-4 on the G1 phase which is accompanied by an increased cell volume, which may explain the increase in thymidine uptake reported earlier (Feghali et al., 1992; Monroe et al., 1988). However, viable cell counts clearly demonstrate that IL-4, in fact, inhibits IL-1 and PDGF induced fibroblast proliferation. Whereas IL-4 inhibits IL-6 production by monocytes, it enhances that of fibroblasts (Dechanet et al., 1994). The production of G-CSF and GM-CSF was also found increased by treatment of fibroblasts with IL-4 (Tushinski et al., 1991). However, IL-4 blocks spontaneous and cytokine-induced secretion of LIF by fibroblasts. Finally, IL-4 blocks the production of PGE_2 by fibroblasts. Treatment of dermal fibroblasts with IL-4 results in increased ICAM-1 expression which allows increased adhesion of LFA-1-bearing T lymphocytes, as well as binding of human rhinovirus (Piela-Smith et al., 1992).

B. IL-4 and Endothelial Cells

Capillary endothelial cells are induced to proliferate in response to IL-4 (Toi et al., 1992). Treatment of endothelial cells with IL-4 increases their adhesiveness for T cells, eosinophils, and basophils but not for neutrophils. This results from an increased VCAM 1 expression (Masinovsky et al., 1990; Schleimer et al., 1992; Thornhill et al., 1991). IL-4 induction of VCAM-1 involves activation of a protein tyrosine kinase which phosphorylates the 130 kDa IL-4R (Palmer-Crocker and Pober, 1995). As well as IL-4, IL-13 can induce expression of VCAM-1 on endothelial cells thus resulting in increased eosinophil adhesiveness (Bochner et al., 1995). IL-4 pretreatment of vascular constructs, composed of endothelial cells cultivated on extracellular matrix from human fibroblasts, induces adherence and layer penetration of eosinophils but not neutrophils. For layer penetration, blood eosinophils from non-allergic donors need priming with GM-CSF, IL-3 or IL-5, whereas those from allergic donors spontaneously transmigrate (Moser et al., 1992).

Cytokine induced up-regulation of ICAM-1 and ELAM-1 on endothelial cells is inhibited by IL-4. This contrasts with the IL-4 induced increase of ICAM-1 expression observed on dermal fibroblasts, macrophages and mast cells (Piela-Smith et al., 1992; Thornhill et al., 1991; Valent et al., 1991). IL-4 synergizes with TNF or IFNγ to cause a change of endothelial cells toward a more fibroblastic morphology (Thornhill et al., 1990). These morphological changes are accompanied by a reorganization of the intracellular vimentin matrix from a diffuse pattern to a perinuclear concentration (Klein et al., 1993). A synergy between IL-4 and TNFα for VCAM-1 expression by endothelial cells is observed *in vitro* (Iademarco et al., 1995) and *in vivo*, following injection in the skin of baboons (Briscoe et al., 1992). IL-4 increases IL-6 production by endothelial cells in synergy with IL-1 and IFNγ (Colotta et al., 1992; Howells et al., 1991). Both IL-4 and IL-13 can block the production of RANTES by endothelial cells treated with a combination of TNFα and IFNγ. Reducing the secretion of RANTES, a chemokine that attracts macrophages and $CD4^+$ T lymphocytes at sites of delayed-type hypersensitivity (DTH) reaction, may contribute to an inhibition of the formation of DTH granulomas (Marfaing-Koka et al., 1995). Finally, IL-4 counteracts the effect of LPS, IL-6, and TNFα on the expression of procoagulant activity on endothelial cells and down-regulates the thrombomodulin anticoagulation pathway (Kapiotis et al., 1991). Since prothrombotic vascular changes are associated with inflammatory reactions, this illustrates a novel aspect of the anti-inflammatory effect of IL-4 (Mantovani et al., 1992).

C. IL-4 and Epithelial Cells

IL-4 enhances the production of soluble CD23 by nasopharyngeal carcinoma, one of the human tumors associated with Epstein Barr virus (Rousselet et al., 1990). IL-4 also enhances expression of the polymeric Ig receptor by colon adenocarcinoma cell lines (Phillips et al., 1990). On human thymic epithelial cells, IL-4 increases IL-1 induced IL-6 production and inhibits that induced by GM-CSF (Galy and Spits, 1991).

D. IL-4 and Hepatocytes

IL-4 can also affect human hepatocytes in primary cultures. In particular, it decreases the spontaneous production of haptoglobin and to a lesser extent that of albumin and C-reactive protein while α1-antitrypsin and fibrinogen production remain unaffected (Loyer et al., 1993). Furthermore, IL-4 antagonizes the IL-6 enhanced secretion of haptoglobin therefore demonstrating another level of anti-inflammatory action. Indeed, IL-4 also inhibits the stimulation of hepatic lipogenesis by tumor necrosis factor, IL-1, and IL-6 without altering that induced by IFNα (Grunfeld et al., 1991). In addition, IL-4 enhances the expression of the cytochrome P-450, 2E1 in a specific manner, since levels of cytochromes P-450, 1A2, 2C, and 3A are not affected or weakly inhibited (Razzak et al., 1993).

IX. CLINICAL ASPECTS OF IL-4

A. IL-4 and Allergy

Atopic patients blood mononuclear cells display an increased capacity to produce IL-4 (Parronchi et al., 1992) and/or a decreased capacity to produce IFNγ in response to polyclonal activation when compared to non-allergic/atopic patients (Lester et al., 1995; Rousset et al., 1991b). Further support for the involvement of IL-4 in deregulated IgE synthesis in atopics was provided by the observation that spontaneous IgE synthesis *in vitro* by lymphocytes from patients with atopic dermatitis is partially inhibited by anti-IL-4 antibodies and IFNγ (Vollenweider et al., 1991).

Other cell compartments appear to be affected by IL-4 in atopic patients. In particular, increased values of total eosinophil counts, $CD20^+$

$CD23^+$ B cells and sCD23 levels, are found in these patients (Reddy et al., 1992). Higher proportions of IL-4 producing T cell clones and lower proportions of IFNγ producing T cell clones are found in the blood and skin of patients with severe atopic disorders (Kapsenberg et al., 1991, 1992; Parronchi et al., 1991). The *Dermatophagoides pteronyssinus* (Dp)-specific Th2 cells from atopic patients are efficient helper T cells for IgE production (Kapsenberg et al., 1992). Dominance of allergen-specific Th2 cells in atopic dermatitis patients, secreting high levels of IL-4 without substantial IFNγ can explain the elevated levels of allergen-specific IgE. $CD4^+$ T cells from allergic individuals produce IL-4 in response to allergens and allergen immunotherapy greatly reduces IL-4 production in an allergen-specific fashion (Jutel et al., 1995; Secrist et al., 1993; Varney et al., 1993). Interestingly, $CD4^+$ T cells from allergic donors produce high levels of IL-4 when cultured with low concentrations of allergen particularly when B cell enriched populations are used as antigen presenting cells. In contrast, the same $CD4^+$ populations produced little IL-4 when stimulated with high concentrations of allergen and when monocytes are used as antigen presenting cells (Secrist et al., 1995). Such observations may contribute to explaining and better applying immunotherapy in the treatment of allergic diseases.

Interestingly, preliminary therapeutic trials in humans have indicated that IFNγ and IFNα can induce a decrease of IgE circulating levels in hyper IgE patients (Boguniewicz et al., 1990; King et al., 1989b; Souillet et al., 1989). A switch of T cells from Th1 to Th2 has also been claimed to represent a critical step in the etiology of HIV infection, a clinical situation often associated with hypergammaglobulinemia. This is still a matter of controversy (Clerici and Shearer, 1993).

A recent linkage analysis between total serum IgE concentration and markers of chromosome 5q31-1 suggests that IL-4 or a nearby gene in 5q31-1 regulates IgE production in a non-antigen-specific (non cognate) fashion (Marsh et al., 1994). One or more functional polymorphisms may exist in the coding or regulatory region of the IL-4 gene. However, it may also affect other genes such as IL-13 and IL-5, whose products directly influence IgE production. Another possible candidate is IRF-1 (Interferon regulatory factor-1) whose gene product up-regulates production of interferon α which, in turn, can down-regulate IgE production.

IL-4 appears to play a central role in host-versus-graft disease as well as graft-versus-host disease that occur after bone marrow transplantation and are associated with high serum levels of IgE (Claman and Spiegelberg, 1990; Heyd et al., 1988). Accordingly, administration of anti-IL-4 to animals prior to grafting prevents the development of both diseases (Schurmans et al., 1990; Takai et al., 1993; Ushiyama et al., 1995).

B. IL-4 and Neoplasia

IL-4 possesses potent antitumor activity *in vivo* in mice. Tumorigenic plasmacytoma cell lines expressing IL-4 as a result of IL-4 gene transfection fail to grow in nude mice and IL-4 producing tumors inhibit the growth of other transplantable tumors injected at the same site (Tepper et al., 1989). The anti-tumor effect is clearly due to IL-4 as anti-IL-4 antibodies allow reexpression of the tumorigenic potential. A marked cellular infiltrate composed of eosinophils and activated macrophages can be seen at the site of tumor injection with a notable absence of lymphocytes. Eosinophil-dependent cytotoxicity appears to be an important mechanism of action for the anti-tumor activity of IL-4 (Tepper et al., 1992). This finding correlates with earlier clinical observations showing that gastric and colonic malignancies infiltrated with eosinophils have a better prognosis. Studies carried out in immunocompetent mice have proved that $CD8^+$ T cells also play a role in the antitumor effects of IL-4 (Bosco et al., 1990; Golumbek et al., 1991).

IL-4 may also display a direct anti-tumor effect as shown with human chronic lymphocytic leukemia, lymphoma, acute pre B cell leukemia, and myelomas (Pandrau et al., 1992), chronic myelogenous leukemias (Akashi et al., 1991a), colon and breast carcinomas and human renal cell carcinomas (Obiri et al., 1993). Finally, IL-4 can inhibit the proliferation of some human cancer cell lines (including non-small cell lung, colon, head, and neck carcinomas as well as glioblastomas) *in vitro* as well as *in vivo* in nude mice (Topp et al., 1993, 1995). Recent studies have established that, following establishment of liver micrometastases after intrasplenic injections, delivery of retrovirally infected cell lines producing IL-4 can direct tumoricidal inflammatory responses that inhibit metastasis formation and cause significant reduction of tumor burden (Hurford et al., 1995). Preliminary clinical trials in man indicate that administration of IL-4 may permit the stabilization of non-small cell lung carcinomas (M.E. Rybak, personal communication).

C. IL-4 and Inflammation

Inflammatory responses are characterized by a complex cascade of events that can lead to tissue destruction and remodeling.

The initial injury which can be an infectious agent (as in tuberculosis), a foreign body (as in asbestos fibrosis), or an unknown agent (as in rheumatoid arthritis) induces neighboring cells to release chemotactic factors that will trigger an influx of inflammatory cells, such as neutrophils, basophils, monocytes/macrophages, dendritic cells, and T cells that will all contribute to the secretion of mediators (histamine, prostaglandins, superoxyde), complement components, and proinflammatory cytokines (IL-1α/β, TNFα). These induce the secretion of proteolytic enzymes that are involved in tissue remodeling and whose uncontrolled production eventually leads to tissue destruction. The anti-inflammatory effects of IL-4 identified using highly purified populations of monocytes/macrophages, endothelial cells, and fibroblasts have been confirmed using whole inflammatory tissues, such as pieces of synovium from rheumatoid arthritis (Miossec et al., 1992) whose spontaneous production of IL-1, IL-6, TNF, LIF, and PGE_2 is inhibited by IL-4. Furthermore, IL-4 inhibits the production of IL-6, LIF, and collagen cross-links (collagen degradation product) by juxta-articular bone pieces. Finally, IL-4 inhibits bone resorption by blocking osteoclast activity and survival (Lacey et al., 1995; Miossec et al., 1994; Watanabe et al., 1990), and may contribute to bone healing by attracting osteoblasts (Lind et al., 1995).

The anti-inflammatory effect of IL-4 has now been demonstrated *in vivo* in various animal models. In particular, the treatment of Experimental Allergic Encephalomyelitis by retinoids correlates with strongly increased IL-4 production, a signature of a Th_2-like response (Racke et al., 1995). Furthermore, administration of Th_2 cells with a diabetogenic T cell receptor did not provoke diabetes while injection of Th_1 cells with such a T cell receptor actively promoted diabetes (Katz et al., 1995).

Administration of IL-4 to rats receiving an arthropathic dose of streptococcal cell wall fragments significantly suppressed the chronic destructive phase of arthritis, though it had minimal effects on the acute phase (Allen et al., 1993). IL-4 treated animals showed a substantial decrease in the influx of inflammatory cells and virtual elimination of pannus and bone erosions. Moreover, the remaining infiltrating mononuclear leukocytes had decreased proinflammatory functions and IL-4 induced increased IL-1ra expression. Furthermore, IL-4 administration is protective in a rat model of IgG immune complex induced lung injury, which is due to products released by neutrophils (toxic oxygen products and proteases) and by lung macrophages (NO, TNFα and IL-1; Mulligan et al., 1993). Finally, *in vivo* administration of IL-4 to prediabetic NOD (non-obese diabetic) mice protects them from diabetes (Rapoport et al., 1993). Therefore, numerous studies strongly argue for a potent anti-inflammatory effect of IL-4 that will hopefully be tested soon in humans suffering from severe inflammatory diseases.

D. Preliminary Clinical Trials with IL-4

The anti-tumor effects of IL-4 observed in mice have encouraged the first clinical trials. Patients with various malignancies including solid tumors (melanoma, renal cell cancer, etc.) have received IL-4. Toxic symptoms including nasal congestion, diarrhea, nausea, vomiting, fatigue, anorexia, headache, dyspnea, weight gain, oliguria, postural hypotension, and rare transaminase elevation have been reported (Gilleece et al., 1992). Such adverse effects have been observed with other cytokines such as IL-2. Fever, gastritis, and gastric ulcerations have been observed in some but not all studies. A nasal congestion due to edema and vascular engorgement from histamine stimulation and gastritis may be more specific to IL-4. Surprisingly, IL-4 treated patients display lymphopenia which may result from localized tissular lymphocytosis but may also come from an inhibition of thymocyte development as shown in transgenic mice (Tepper et al., 1990). All these adverse effects were resolved following discontinuation of treatment. Some clinical responses have been observed in these preliminary studies, in particular, rapid shrinkage of the lymph nodes and spleen in a patient with low grade lymphoma and complete clinical remission in a patient with Hodgkin's disease who had relapsed following autologous bone marrow transplantation. Also, anti-tumor responses were observed with combinations of low doses of IL-2 and IL-4 with complete resolution of pulmonary metastatic disease from renal carcinoma (Lotze, 1992).

X. CONCLUSIONS

In vitro as well as *in vivo* studies have demonstrated that IL-4 represents a typical cytokine with pleiotropic effects which are either direct (Figure 5)

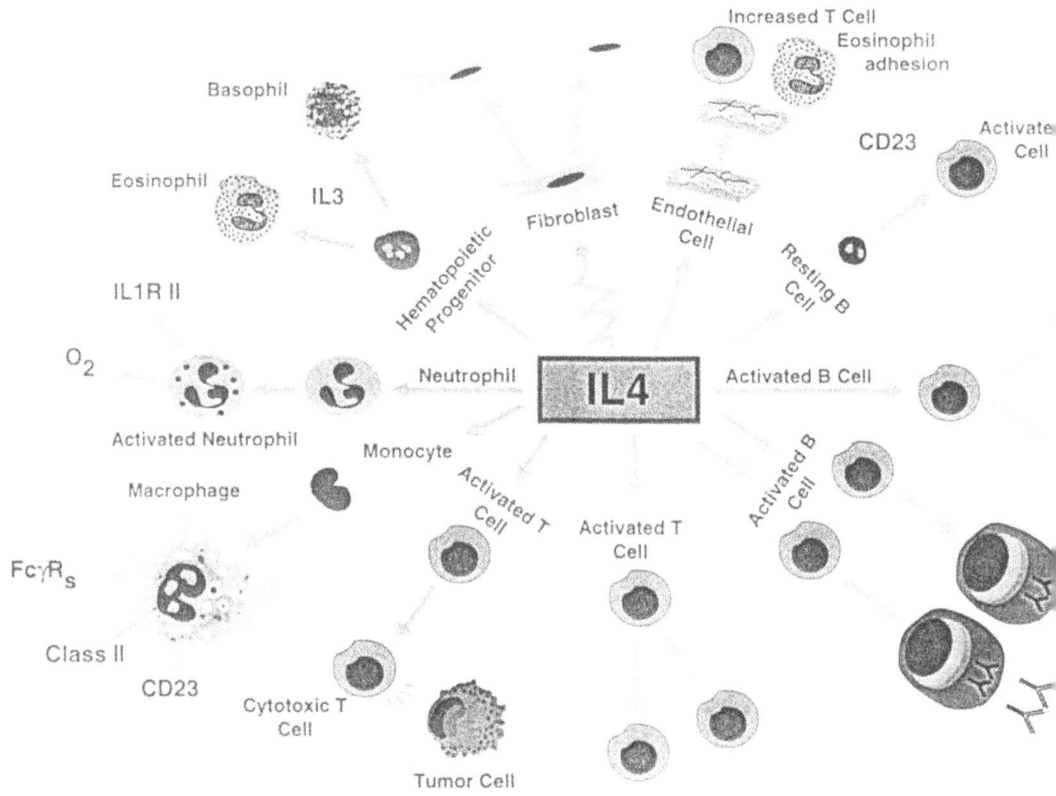

Pleiotropic effects of IL-4.

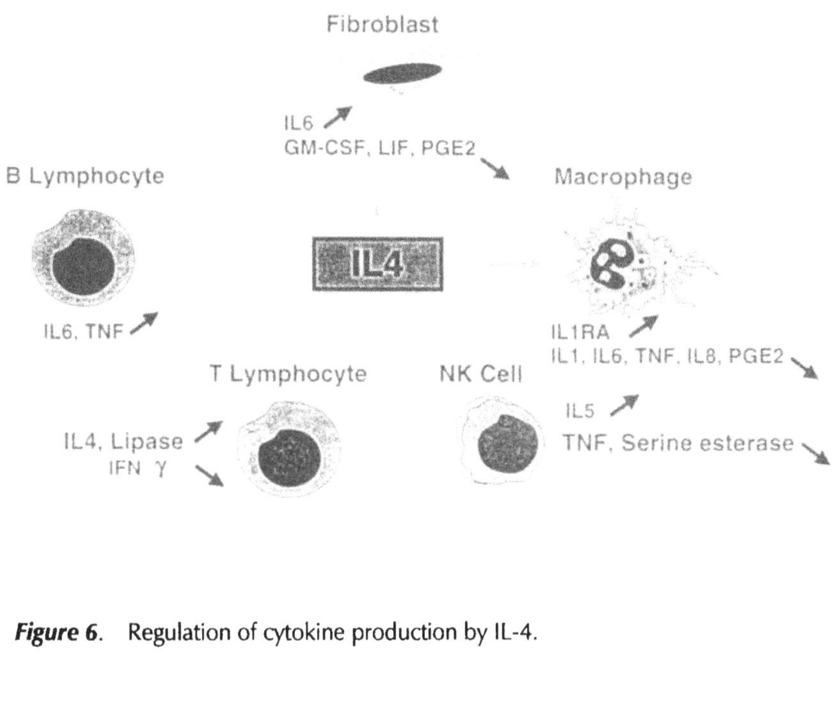

Figure 6. Regulation of cytokine production by IL-4.

or indirect through modulation of the secretion of other cytokines (Figure 6). In particular, IL-4 plays a key role in the development of the allergic reaction both at effector level, by inducing the switch toward IgE, and at a regulatory level by permitting the differentiation of naive Thp cells into Th2 cells. Administration of IL-4 to mice, and more recently humans, indicates that there may be a clinical application as an antitumor agent through activation of both T cell-dependent and cell independent effector mechanisms. IL-4 may also prove useful in the management of chronic inflammation diseases by blocking the production of proinflammatory cytokines by monocytes/macrophages and by blocking the development of the Th1 pathway, the exacerbation of which leads to inflammatory reactions. The intracellular mechanisms turned on by IL-4 are now being deciphered. This information will hopefully permit the design of chemical agents mimicking or blocking part or all of the biological activities of IL-4.

ACKNOWLEDGMENT

The authors wish to thank Sandrine Bonnet-Arnaud for her outstanding editorial assistance.

REFERENCES

Abe, E., De Waal Malefyt, R., Matsuda, I., Arai, K., & Arai, N. (1992). An 11-base-pair DNA sequence motif apparently unique to the human interleukin 4 gene confers responsiveness to T-cell activation signals. PNAS 89, 2864-2868.

Abramson, S. L., & Gallin, J. I. (1990). IL-4 inhibits superoxide production by human mononuclear phagocytes. J. Immunol. 144, 625-630.

Akashi, K., Harada, M., Shibuya, T., Eto, T., Takamatsu, Y., Teshima, T., & Niho, Y. (1991a). Effects of interleukin-4 and interleukin-6 on the proliferation of CD34+ and CD34- blasts from acute myelogenous leukemia. Blood 78, 197-204.

Akashi, K., Shibuya, T., Harada, M., Takamatsu, Y., Uike, N., Eto, T., & Niho, Y. (1991b). Interleukin 4 suppresses the spontaneous growth of chronic myelomonocytic leukemia cells. J. Clin. Invest. 88, 223-230.

Allen, J. B., Wong, H. L., Costa, G. L., Bienkowski, M. J., & Wahl, S. M. (1993). Suppression of monocyte function and differential regulation of IL-1 and IL-1ra by IL-4 contribute to resolution of experimental arthritis. J. Immunol. 151, 4344-4351.

Alzona, M., Jäck, H. M., Fisher, R. I., & Ellis, T. M. (1994). CD30 defines a subset of activated human T cells that produce IFN-γ and IL-5 and exhibit enhanced B cell helper activity. J. Immunol. 153, 2861-2867.

Alzona, M., Jäck, H. M., Fisher, R. I., & Ellis, T. M. (1995). IL-12 activates IFN-γ production through the preferential activation of CD30+ T cells. J. Immunol. 154, 9-16.

Arase, H., Arase, N., Nakagawa, K., Good, R. A., & Onoe, K. (1993). NK1.1+ CD4+ CD8+ thymocytes with specific lymphokine secretion. Eur. J. Immunol. 23, 307-310.

Armitage, R. J., Goff, L. K., & Beverley, P. C. L. (1989). Expression and functional role of CD23 on T cells. Eur. J. Immunol. 19, 31-35.

Arruda, S., & Ho, J. L. (1992). IL-4 receptor signal transduction in human monocytes is associated with protein kinase C translocation. J. Immunol. 149, 1258-1264.

Asano, Y., Shibuya, T., Okamura, S., Yamaga, S., Otsuka, T., & Niho, Y. (1987). Effect of human recombinant granulocyte/macrophage colony-stimulating factor and native granulocyte colony-stimulating factor on clonogenic leukemic blast cells. Cancer Res. 47, 5647-5648.

Autran, B., Legac, E., Blanc, C., & Debré, P. (1995). A ThO/Th2-like function of CD4+ CD7- T helper cells from normal donors and HIV-infected patients. J. Immunol. 154, 1408-1417.

Aversa, G., Punnonen, J., Cocks, B. J., de Waal Malefyt, R., Vega, F., Zurawski, S. M., Zurawski, G., & de Vries, J. E. (1993). An interleukin 4 (IL-4) mutant protein inhibits both IL-4 or IL-13-induced human immunoglobulin G4 (IgG4) and IgE synthesis and B cell proliferation: Support for a common component shared by IL-4 and IL-13 receptors. J. Exp. Med. 178, 2213-2218.

Banchereau, J., de Paoli, P., Vallé, A., Garcia, E., & Rousset, F. (1991). Long term human B cell lines dependent on interleukin 4 and antibody to CD40. Science 251, 70-72.

Banchereau, J., Bazan, F., Blanchard, D., Brière, F., Galizzi, J.-P., van Kooten, C., Liu, Y.-J., Rousset, F., & Saeland, S. (1994). The CD40 antigen and its ligand. Ann. Rev. Immunol. 12, 881-922.

Barber, D. L., & D'Andrea, A. D. (1994). Erytropoietin and interleukin-2 activate distinct JAK kinase family members. Mol. Cel. Biol. 14, 6506-6514.

Barcena, A., Toribio, M. L., Pezzi, L., & Martinez-A, C. (1990). A role for interleukin 4 in the differentiation of mature T cell receptor γ/δ+ cells from human intrathymic T cell precursors. J. Exp. Med. 172, 439-446.

Baskar, P., Silberstein, D. S., & Pincus, S. H. (1990). Inhibition of IgG-triggered human eosinophil function by IL-4. J. Immunol. 144, 2321-2326.

Bazan, J. F. (1990). Haemopoietic receptors and helical cytokines. Immunol. Today 11, 350-354.

Billips, L., Petitte, D., & Landreth, K. S. (1990). Bone marrow stromal cell regulation of B lymphopoiesis: Interleukin-1 (IL-1) and IL-4 regulate stromal cell support of pre-B cell production *in vitro*. Blood 75, 611-619.

Billips, L. G., Petitte, D., Hostutler, R., Tsai, P., & Landreth, K. S. (1991). Suppression of bone marrow stromal cell function. Ann. NY Acad. Sci. 628, 313-322.

Blay, J. Y., Branellec, D., Robinet, E., Dugas, B., Gay, F., & Chouaib, S. (1990). Involvment of cAMP in the IL-4 inhibitory effect on IL-2 induced LAK generation. J. Clin. Invest. 85, 1909-1913.

Bober, L. A., Waters, T. A., Pugliese-Sivo, C. C., Sullivan, L. M., Narula, S. K., & Grace, M. J. (1995). IL-4 induces neutrophilic maturation of HL-60 cells and activation of human peripheral blood neutrophils. Clin. Exp. Immunol. 99, 129-136.

Bochner, B. S., Klunk, D. A., Sterbinsky, S. A., Coffman, R. L., & Schleimer, R. P. (1995). IL-13 selectively induces vascular cell adhesion molecule-1 expression in human endothelial cells. J. Immunol. 154, 799-803.

Boey, H., Rosenbaum, R., Castracane, J., & Borish, L. (1989). Interleukin-4 is a neutrophil activator. J. Allergy Clin. Immunol. 83, 978-984.

Boguniewicz, M., Jaffe, H. S., Izu, A., Sullivan, M. J., York, D., Geha, R. S., & Leung, D. Y. M. (1990). Recombinant gamma interferon in treatment of patients with atopic dermatitis and elevated IgE levels. Am. J. Med. 88, 365-370.

Bosco, M. C., Giovarelli, M., Forni, M., Modesti, A., Scarpa, G., Masuelli, L., & Forni, G. (1990). Low doses of interleukin 4 injected perilymphatically in tumor bearing mice inhibit the growth of poorly and apparently nonimmunogenic tumors and induce a tumor-specific immune memory. J. Immunol. 145, 3136-3143.

Bradding, P., Feather, I. H., Howarth, P. H., Mueller, R., Roberts, J. A., Britten, K., Bews, J. P. A., Hunt, T. G., Okayama, Y., & Heusser, C. H. (1992). Interleukin 4 is localized to and released by human mast cells. J. Exp. Med. 176, 1381-1386.

Briscoe, D. M., Cotran, R. S., & Pober, J. S. (1992). Effects of tumor necrosis factor, lipopolysaccharide, and IL-4 on the expression of vascular cell adhesion molecule-1 *in vivo*. Correlation with CD3+ T cell infiltration. J. Immunol. 149, 2954-2960.

Brown, M. A., Pierce, J. H., Watson, C. J., Falco, J., Ihle, J. N., & Paul, W. E. (1987). B cell stimulatory factor-1/interleukin-4 mRNA is expressed by normal and transformed mast cells. Cell 50, 809-818.

Broxmeyer, H. E., Lu, L., Cooper, S., Tushinski, R., Mochizuki, D., Rubin, B. Y., Gillis, S., & William, D. E. (1988). Synergistic effects of purified recombinant human and murine B cell growth factor-1/IL-4 on colony formation *in vitro* by hematopoietic progenitor cells. J. Immunol. 141, 3852-3862.

Brunn, G. J., Falls, E. L., Nilson, A. E., & Abraham, R. T. (1995). Protein-tyrosine kinase-dependent activation of STAT transcription factors in interleukin-2- or interleukin-4-stimulated T lymphocytes. J. Biol. Chem. 270, 11628-11635.

Brunner, T., Heusser, C. H., & Dahinden, C. A. (1993). Human peripheral blood basophils primed by interleukin-3 (IL-3) produce IL-4 in response to immunoglobulin E receptor stimulation. J. Exp. Med. 177, 605-611.

Cabrillat, H., Galizzi, J. P., Djossou, O., Arai, N., Yokota, T., Arai, K., & Banchereau, J. (1987). High affinity binding of human interleukin 4 to cell lines. Biochem. Biophys. Res. Commun. 149, 995-1001.

Claman, H. N., & Spiegelberg, H. L. (1990). Immunoglobulin dysregulation in murine graft-versus-host disease: A hyper-IgE syndrome. Clin. Immunol. Immunopathol. 56, 46-53.

Clerici, M., & Shearer, G. M. (1993). A TH_1 / TH_2 switch is a critical step in the etiology of HIV infection. Immunol. Today 14, 107-111.

Coffman, R. L., Ohara, J., Bond, M. W., Carty, J., Zlotnick, A., & Paul, W. E. (1986). B cell stimulatory factor-1 enhances the IgE response of lipopolysaccharide-activated B cells. J. Immunol. 136, 4538-4541.

Coffman, R. L., Varkila, K., Scott, P., & Chatelain, R. (1991). Role of cytokines in the differentiation of CD4+ T cell subsets *in vivo*. Immunol. Rev. 123, 189-207.

Colotta, F., Sironi, M., Borrè, A., Luini, W., Maddalena, F., & Mantovani, A. (1992). Interleukin 4 amplifies monocyte chemotactic protein and interleukin 6 production by endothelial cells. Cytokine 4, 24-28.

Colotta, F., Re, F., Muzio, M., Bertini, R., Polentarutti, N., Sironi, M., Giri, J. G., Dower, S. K., Sims, J. E., & Mantovani, A. (1993). Interleukin-1 Type II receptor: as a decoy target for IL-1 that is regulated by IL-4. Science 261, 472-475.

Corcoran, M. L., Stetler-Stevenson, W. G., Brown, P. D., & Wahl, L. M. (1992). Interleukin 4 inhibition of prostaglandin E_2 synthesis blocks interstitial collagenase and 92-kDa type IV collagenase/gelatinase production by human monocytes. J. Biol. Chem. 267, 515-519.

Crawford, R. M., Finbloom, D. S., Ohara, J., Paul, W. E., & Meltzer, M. S. (1987). B cell stimulatory factor-1 (interleukin-4) activates macrophages for increased tumouricidal activity and expression of Ia antigens. J. Immunol. 139, 135-141.

de Wit, H., Hendriks, D. W., Halie, M. R., & Vellenga, E. (1994). Interleukin-4 receptor regulation in human monocytic cells. Blood 84, 608-615.

Dechanet, J., Briolay, J., Rissoan, M.-C., Chomarat, P., Galizzi, J.-P., Banchereau, J., & Miossec, P. (1993). Interleukin 4 inhibits growth factor-stimulated rheumatoid synoviocyte proliferation by blocking the early phases of the cell cycle. J. Immunol. 151, 4908-4917.

Dechanet, J., Taupin, J. L., Chomarat, P., Rissoan, M. C., Moreau, J. F., Banchereau, J., & Miossec, P. (1994). Interleukin-4 but not interleukin-10 inhibits the production of leukemia inhibitory factor by rheumatoid synovium and synoviocytes. Eur. J. Immunol. 24, 3222-3228.

Defrance, T., Aubry, J. P., Rousset, F., Vanbervliet, B., Bonnefoy, J. Y., Arai, N., Takebe, Y., Yokota, T., Lee, F., Arai, K., De Vries, J., & Banchereau, J. (1987a). Human recombinant interleukin 4 induces Fcε receptors (CD23) on normal human B lymphocytes. J. Exp. Med. 165, 1459-1467.

Defrance, T., Vanbervliet, B., Aubry, J. P., Takebe, Y., Arai, N., Miyajima, A., Yokota, T., Lee, T., Arai, K., de Vries, J. E., & Banchereau, J. (1987b). B cell growth-promoting activity of recombinant human interleukin-4. J. Immunol. 139, 1135-1141.

Defrance, T., Vanbervliet, B., Aubry, J. P., & Banchereau, J. (1988). Interleukin 4 inhibits the proliferation but not the differentiation of activated human B cells in response to interleukin 2. J. Exp. Med. 168, 1321-1337.

Del Prete, G., De Carli, M., Almerigogna, F., Daniel, C. K., D'Elios, M. M., Zancuochi, G., Vinante, F., Pizzolo, G., & Romagnani, S. (1995a). Preferential expression of CD30 by human CD4+ T cells producing Th2-type cytokines. FASEB J. 9, 81-86.

Del Prete, G., Maggi, E., Pizzolo, G., & Romagnani, S. (1995b). CD30, Th2 cytokines and HIV infection: A complex and fascinating link. Immunol. Today 16, 76-80.

Deng, W., Ohmori, Y., & Hamilton, T. A. (1994). Mechanisms of IL-4-mediated suppression of IP-gene expression in murine macrophages. J. Immunol. 153, 2130-2136.

Denis, M., & Ghadirian, E. (1994). Interleukin 13 and interleukin 4 protect bronchoalveolar macrophages from productive infection with human immunodeficiency virus type 1. AIDS Res. Hum. Retroviruses 10, 795-802.

Dokter, W. H. A., Borger, P., Hendriks, D., van der Horst, I., Halie, M. R., & Vellenga, E. (1992). Interleukin-4 (IL-4) receptor expression on human T cells is affected by different intracellular signaling pathways and by IL-4 at transcriptional and posttranscriptional level. Blood 80, 2721-2728.

Dokter, W. H. A., Esselink, M. T., Halie, M. R., & Vellenga, E. (1993a). Interleukin-4 inhibits the lipopolysaccharide-induced expression of c-*jun* and c-*fos* messenger RNA and activator protein-1 binding activity in human monocytes. Blood 81, 337-343.

Dokter, W. H. A., Esselink, M. T., Sierdsema, S. J., Halie, M. R., & Vellenga, E. (1993b). Transcriptional and posttranscriptional regulation of the interleukin-4 and interleukin-3 genes in human T cells. Blood 81, 35-40.

Dokter, W. H. A., Sierdsema, S. J., Esselink, M. T., Halie, M. R., & Vellenga, E. (1994). Interleukin-4-mediated inhibition of *c-fos* mRNA expression: Role of the lipoxygenase directed pathway. Leukemia 8, 1181-1184.

Egger, D., Geuenich, S., Denzlinger, C., Schmitt, E., Mailhammer, R., Ehrenreich, H., Dörmer, P., & Hültner, L. (1995). IL-4 renders mast cells functionally responsive to Endothelin-1. J. Immunol. 154, 1830-1837.

Elson, L. H., Shaw, S., Van Lier, R. A. W., & Nutman, T. B. (1994). T cell subpopulation phenotypes in filarial infections: CD27 negativity defines a population greatly enriched for Th2 cells. Int. Immunol. 6, 1003-1009.

Elson, L. H., Nutman, T. B., Metcalfe, D. D., & Prussin, C. (1995). Flow cytometric analysis for cytokine production identifies T helper 1, T helper 2, and T helper 0 cells within the human CD4+ CD27- lymphocyte subpopulation. J. Immunol. 154, 4294-4301.

Erard, F., Wild, M.-T., Garcia-Snaz, J. A., & Le Gros, G. (1993). Switch of CD8 T cells to noncytolytic CD8- CD4- cells that make Th2 cytokines and help B cells. Science 260, 1802-1805.

Estrov, Z., Markowitz, A. B., Kurzrock, R., Wetzler, M., Kantarjian, H. M., Ferrajoli, A., Gutterman, J. U., & Talpaz, M. (1993). Suppression of chronic myelogenous leukemia colony growth by interleukin-4. Leukemia 7, 214-220.

Fanslow, W. C., Sims, J. E., Sassenfeld, H., Morrissey, P. J., Gillis, S., Dower, S. K., & Widmer, M. (1990). Regulation of alloreactivity *in vivo* by a soluble form of the interleukin-1 receptor. Science 248, 739-742.

Fanslow, W. C., Clifford, K. N., Park, L. S., Rubin, A. S., Voice, R. F., Beckmann, M. P., & Widmer, M. B. (1991). Regulation of alloreactivity *in vivo* by IL-4 and the soluble IL-4 receptor. J. Immunol. 147, 535-540.

Fanslow, W. C., Spriggs, M. K., Rauch, C. T., Clifford, K. N., Macduff, B. M., Ziegler, S. F., Schooley, K. A., Mohler, K. M., March, C. J., & Armitage, R. J. (1993). Identification of a distinct low-affinity receptor for human interleukin-4 on pre-B cells. Blood 81, 2998-3005.

Favre, C., Saeland, S., Caux, C., Duvert, V., & De Vries, J. E. (1990). Interleukin 4 has basophilic and eosinophilic cell growth-promoting activity on cord blood cells. Blood 75, 67-73.

Feghali, C. A., Bost, K. L., Boulware, D. W., & Levy, L. S. (1992). Human recombinant interleukin-4 induces proliferation and interleukin-6 production by cultured human skin fibroblasts. Clin. Immunol. Immunopathol. 63, 182-187.

Fernandez-Botran, R., & Vitetta, E. S. (1990). A soluble, high-affinity, interleukin-4 binding protein is present in the biological fluids of mice. Proc. Natl. Acad. Sci. USA 87, 4202-4206.

Fernandez-Botran, R., & Vitetta, E. S. (1991). Evidence that natural murine soluble interleukin 4 receptors may act as transport proteins. J. Exp. Med. 174, 673-681.

Fertin, C., Nicolas, J. F., Gillery, P., Kalis, B., Banchereau, J., & Maquart, F. X. (1991). Interleukin-4 stimulates collagen synthesis by normal and scleroderma fibroblasts in dermal equivalents. Cell. Mol. Biol. 37, 823-829.

Finkelman, F. D., Holmes, J., Katona, I. M., Urban, J. F., Beckmann, M. P., Schooley, K. A., Coffman, R. L., Mosmann, T. R., & Paul, W. E. (1990). Lymphokine control of *in vivo* immunoglobulin isotype selection. Annu. Rev. Immunol. 8, 303-333.

Finkelman, F. D., Svetic, A., Gresser, I., Snapper, C., Holmes, J., Trotta, P. P., Katona, I. M., & Gause, W. C. (1991). Regulation by interferon α of immunoglobulin isotype selection and lymphokine production in mice. J. Exp. Med. 174, 1179-1188.

Finney, M., Guy, G. R., Michell, R. H., Gordon, J., Dugas, B., Rigley, K. P., & Callard, R. E. (1990). Interleukin 4 activates human B lymphocytes via transient inositol lipid hydrolysis and delayed cyclic adenosine monophosphate generation. Eur. J. Immunol. 20, 151-156.

Fiorentino, D. F., Bond, M. W., & Mosmann, T. R. (1989). Two types of mouse T helper cell. IV. Th2 clones secrete a factor that inhibits cytokine production by Th1 clones. J. Exp. Med. 170, 2081-2095.

Foxwell, B. M. J., Woerly, G., & Ryffel, B. (1989). Identification of interleukin-4 receptor-associated proteins and expression of both high and low affinity binding on human lymphoid cells. Eur. J. Immunol. 19, 1637-1641.

Freeman, G. J., Boussiotis, V. A., Anumanthan, A., Bernstein, G. M., Ke, X. Y., Rennert, P. D., Gray, G. S., Gribben, J. G., & Nadler, L. M. (1995). B7-1 and B7-2 do not deliver identical costimulatory signals, since B7-2 but not B7-1 preferentially costimulates the initial production of IL-4. Immunity 2, 523-532.

Gajewski, T. F., Joyce, J., & Fitch, F. W. (1989). Anti-proliferative effect of IFN-γ in immune regulation. III. Differential selection of Th1 and Th2 murine helper T lymphocyte clones using recombinant IL-2 recombinant IFN-γ. J.Immunol. 143, 15-22.

Galizzi, J.-P., Zuber, C. E., Cabrillat, H., Djossou, O., & Banchereau, J. (1989). Internalization of human interleukin 4 and transient down-regulation of its receptor in the CD23-inducible Jijoye cells. J. Biol. Chem. 264, 6984-6989.

Galizzi, J.-P., Castle, B., Djossou, O., Harada, N., Cabrillat, H., Ait Yahia, S., Barrett, R., Howard, M., & Banchereau, J. (1990a). Purification of a 130-kDa T cell glycoprotein that binds human interleukin 4 with high affinity. J. Biol. Chem. 265, 439-444.

Galizzi, J. P., Zuber, C. E., Harada, N., Gorman, D. M., Djossou, O., Kastelein, R., Banchereau, J., Howard, M., & Miyajima, A. (1990b). Molecular cloning of a cDNA encoding the human interleukin-4 receptor. Int. Immunol. 2, 669-675.

Galy, A. H., & Spits, H. (1991). IL-1, IL-4, and IFN-γ differntially regulate cytokine production and cell surface molecule expression in cultured human thymic epithelial cells. J. Immunol. 147, 3823-3830.

Garret, D. S., Powers, R., March, C. J., Frieden, E. A., Clore, G. M., & Gronenborn, A. M. (1992). Determination of the secondary structure and folding topology of human interleukin-4 using three-dimensional heteronuclear magnetic resonance spectroscopy. Biochemistry 31, 4347-4353.

Garrone, P., Djossou, O., Galizzi, J. P., & Banchereau, J. (1991). A recombinant extracellular domain of the human interleukin 4 receptor inhibits the biological effects of interleukin 4 on T and B lymphocytes. Eur. J. Immunol. 21, 1365-1369.

Gascan, H., Gauchat, J.-F., Roncarolo, M.-G., Yssel, H., Spits, H., & de Vries, J. E. (1991). Human B cell clones can be induced to proliferate and to switch to IgE and IgG4 synthesis by interleukin 4 and a signal provided by activated CD4+ T cell clones. J. Exp. Med. 173, 747-750.

Gautam, S. C., Chikkala, N. F., & Hamilton, T. A. (1992). Anti-inflammatory action of IL-4. J. Immunol. 148, 1411-1415.

Gaya, A., Alsinet, E., Martorell, J., Places, L., De La Calle, O., Yagüe, J., & Vives, J. (1990). Inhibitory effect of IL-4 on the sepharose-CD3-induced proliferation of the CD4CD45RO human T cell subset. Int. Immunol. 2, 685-689.

Gilleece, M. H., Scarffe, J. H., & Ghosh, A. (1992). Recombinant human interleukin-4 (IL-4) given as daily subcutaneous injections—a phase I dose toxicity trial. Br. J. Cancer 66, 204-210.

Gillery, P., Fertin, C., Nicolas, J. F., Chastang, F., Kalis, B., Banchereau, J., & Maquart, F. X. (1992). Interleukin-4 stimulates collagen gene expression in human fibroblast monolayer cultures. FEBS Lett. 302, 231-234.

Golumbek, P. T., Lazenby, A. J., Levitsky, H. I., Jaffee, L. M., Karasuyama, H., Baker, M., & Pardoll, D. M. (1991). Treatment of established renal cancer by tumor cells engineered to secrete interleukin-4. Science 254, 713-716.

Grabstein, K., Eisenman, J., Mochizuki, K., Shanebeck, K., Conlon, P., Hopp, T., March, C., & Gillis, S. (1986). Purification to homogeneity of B cell stimulating factor. A molecule that stimulates proliferation of multiple lymphokine-dependent cell lines. J. Exp. Med. 163, 1405-1414.

Greil, J., Gramatzki, M., Burger, R., Marschalek, R., Peltner, M., Trautmann, U., Hansen-Hagge, T. E., Bartram, C. R., Fey, G. H., Stehr, K., & Beck, J. (1994). The acute lymphoblastic leukaemia cell line SEM with t(4;11) chromosomal rearrangement is biphenotypic and responsive to interleukin-7. Br. J. Haematol. 86, 275-283.

Grunfeld, C., Soued, M., Adi, S., Moser, A. H., Fiers, W., Dinarello, C. A., & Feingold, K. R. (1991). Interleukin-4 inhibits stimulation of hepatic lipogenegis by tumor necrosis factor, interleukin-1 and interleukin-6 but not interferon-α. Cancer Res. 51, 2803-2807.

Gruss, H. J., & Dower, S. K. (1995). Tumor necrosis factor ligand superfamily: Involvement in the pathology of malignant lymphomas. Blood 85, 3378-3404.

Hamaguchi, Y., Kanakura, Y., Fusita, J., Takeoa, S. I., Nakano, T., Tarui, S., Honjo, T., & Kitamura, Y. (1987). Interleukin 4 as an essential factor for *in vitro* clonal growth of murine connective tissue-type mast cells. J. Exp. Med. 165, 268-273.

Harada, N., Yang, G., Miyajima, A., & Howard, M. (1992). Identification of an essential region for growth signal transduction in the cytoplasmic domain of the human interleukin-4 receptor. J. Biol. Chem. 267, 22752-22758.

Harel-Bellan, A., Durum, S., Muegge, K., Abbas, A. K., & Farrar, W. L. (1988). Specific inhibition of lymphokine biosynthesis and autocrine growth using antisense oligonucleotides in Th1 and Th2 helper T cell clones. J. Exp. Med. 168, 2309-2318.

Hart, P. H., Vitti, G. F., Burgess, D. R., Whitty, G. A., Piccoli, D. S., & Hamilton, J. A. (1989). Potential anti-inflammatory effects of interleukin 4: Suppression of human monocyte tumor necrosis factor α, interleukin 1, and prostaglandin E2. Proc. Natl. Acad. Sci. USA 86, 3803-3807.

He, Y. W., & Malek, T. R. (1995). The IL-2 receptor γc chain does not function as a subunit shared by the IL-4 and IL-13 receptors. Implication for the structure of the IL-4 receptor. J. Immunol. 155, 9-12.

Heyd, J., Donnenberg, A. D., Burns, W. H., Segal, R., & Santos, G. W. (1988). Immunoglobulin IgE levels following allogeneic, autologous and syngeneic bone marrow transplantation: An indirect association between hyperproduction and acute graft-versus-host disease in allogeneic BMT. Blood 72, 442-446.

Ho, J. L., He, S. H., Rios, M. J., & Wick, E. A. (1992). Interleukin-4 inhibits human macrophage activation by tumor necrosis factor, granulocyte-monocyte colony-stimulating factor, and interleukin-3 for antileishmanial activity and oxidative burst capacity. J. Inf. Diseases 165, 344-351.

Hofman, F. M., Brock, M., Taylor, C. R., & Lyons, B. (1988). IL-4 regulates differentiation and proliferation of human precursor B cells. J. Immunol. 141, 1185-1190.

Horohov, D. W., Crim, J. A., Smith, P. L., & Siegel, J. P. (1988). IL-4 (B cell-stimulatory factor 1) regulates multiple aspects of influenza virus-specific cell-mediated immunity. J. Immunol. 141, 4217-4223.

Hou, J., Schindler, U., Henzel, W. J., Ho, T. C., Brasseur, M., & McKnight, S. L. (1994). An interleukin-4-induced transcription factor: IL-4 stat. Science 265, 1701-1706.

Howard, M., Farrar, J., Hilfiker, M., Johnson, B., Takatsu, K., Hamaoka, T., & Paul, W. E. (1982). Identification of a T-cell derived B cell growth factor distinct from interleukin-2. J. Exp. Med. 155, 914-921.

Howells, G., Pham, P., Taylor, D., Foxwell, B., & Feldman, M. (1991). Interleukin 4 induces interleukin 6 production by endothelial cells: Synergy with interferon-γ. Eur. J. Immunol. 21, 97-101.

Hu-Li, J., Shevach, E. M., Mizuguchi, J., Ohara, J., Mosmann, T., & Paul, W. E. (1987). B cell stimulatory factor 1 (interleukin 4 is a potent costimulant for normal resting T lymphocytes. J. Exp. Med. 165, 157-172.

Hurford, R. K., Dranoff, G., Mulligan, R. C., & Tepper, R. I. (1995). Gene therapy of metastatic cancer by *in vivo* retroviral gene targeting. Nature Genetics 10, 430-435.

Iademarco, M. F., Barks, J. L., & Dean, D. C. (1995). Regulation of vascular cell adhesion molecule-1 expression by IL-4 and TNF-alpha in cultured endothelial cells. J. Clin. Invest. 95, 264-271.

Ichiki, T., Takahashi, W., & Watanabe, T. (1993). Regulation of the expression of human Cε germline transcript. J. Immunol. 150, 5408-5417.

Idzerda, R. L., March, C. J., Mosley, B., Lyman, S. D., Vanden Bos, T., Gimpel, S. D., Din, W. S., Grabstein, K. H., Widmer, M. B., Park, L. S., Cosman, D., & Beckmann, M. P. (1990). Human Interleukin 4 receptor confers biological responsiveness and defines a novel receptor superfamily. J. Exp. Med. 171, 861-873.

Ihle, J. N., Witthuhn, B. A., Quelle, F. W., Yamamoto, K., & Silvennoinen, O. (1995). Signaling through the hematopoietic cytokine receptors. Annu. Rev. Immunol. 13, 369-398.

Imai, Y., Tohda, S., Nagata, K., Suzuki, T., Nara, N., & Aoki, N. (1991). Effects of recombinant interleukin 4 on the growth and differentiation of blast progenitors stimulated with G-CSF, GM-CSF and IL-3 from acute myeloblastic leukaemia patients. Br. J. Haematol. 78, 173-179.

Isacksson, P. C., Pure, E., Vitetta, S., & Krammer, P. H. (1982). Effect on the isotype switch of murine B cells. J. Exp. Med. 155, 734-748.

Izuhara, K., Feldman, R. A., Greer, P., & Harada, N. (1994). Interaction of the c-fes proto-oncogene product with the interleukin-4 receptor. J. Biol. Chem. 269, 18623-18629.

Izuhara, K., & Harada, N. (1993). Interleukin-4 (IL-4) induces protein tyrosine phosphorylation of the IL-4 receptor and association of phosphatidylinositol 3-kinase to the IL-4 receptor in a mouse T cell line, HT2. J. Biol. Chem. 268, 13097-13102.

Jabara, H. H., Fu, S. M., Geha, R. S., & Vercelli, D. (1990). CD40 and IgE: Synergism between anti-CD40 monoclonal antibody and interleukin 4 in the induction of IgE synthesis by highly purified human B cells. J. Exp. Med. 172, 1861-1864.

Jabara, H. H., Loh, R., Ramesh, N., Vercelli, D., & Geha, R. S. (1993). Sequential switching from μ to epsilon via gamma4 in human B cells stimulated with IL-4 and hydrocortisone. J. Immunol. 151, 4528.

Jansen, J. H., Wientjens, G.-J. H. M., Fibbe, W. E., Willemze, R., & Kluin-Nelemans, H. C. (1989). Inhibition of human macrophage colony formation by interleukin 4. J. Exp. Med. 170, 577-582.

Jansen, J. H., Fibbe, W. E., Wientjens, G. J. H. M., Van Damme, J., Landegent, J. E., Willemze, R., & Kluin-Nelemans, J. C. (1991). Interleukin 4 down-regulates the expression of CD14 and the production of interleukin 6 in acute myeloid leukemia cells. Lymph. Cyt. Res. 10, 457-461.

Jelinek, D. F., & Lipsky, P. E. (1988). Inhibitory influence of IL-4 on human B cell responsiveness. J. Immunol. 141, 164-173.

Justement, L., Chen, Z. Z., Harris, L. K., Ransom, J. T., Sandoval, V. S., Smith, C., Rennick, D., Roehm, N., & Cambier, J. (1986). BSF-1 induces membrane protein phosphorylation but not phosphoinositide metablism, Ca^{2+} mobilization, protein kinase C translocation, or membrane depolarization in resting murine B lymphocytes. J. Immunol. 137, 3664-3670.

Jutel, M., Pichler, W. J., Skrbic, D., Urwyler, A., Dahinden, C., & Müller, U. R. (1995). Bee venom immunotherapy results in decrease of IL-4 and IL-5 and increase of IFN-γ secretion in specific allergen-stimulated T cell cultures. J. Immunol. 154, 4187-4194.

Kalinski, P., Hilkens, C. M. U., Wierenga, E. A., Van der Pouw-Kraan, T. C. T. M., Van Lier, R. A. W., Bos, J. D., Kapsenberg, M. L., & Snijdewint, F. G. M. (1995). Functional maturation of human naive T helper cells in the absence of accessory cells. J. Immunol. 154, 3753-3760.

Kapiotis, S., Besemer, J., Bevec, D., Valent, P., Bettelheim, P., Lechner, K., & Speiser, W. (1991). Interleukin-4 counteracts pyrogen-induced downregulation of thrombomodulin in cultured human vascular endothelial cells. Blood 78, 410-415.

Kapsenberg, M. L., Wierenga, E. A., Bos, J. D., & Jansen, H. M. (1991). Functional subsets of allergen-reactive human CD4+ T cells. Immunol. Today 12, 392-395.

Kapsenberg, M. L., Wierenga, E. A., Van Der Heijden, F. L., & Bos, J. D. (1992). Atopic dermatitis and CD4+ atopen-specific Th2 lymphocytes. Eur. J. Dermatol. 8, 601-607.

Karray, S., Dautry-Varsat, A., Tsudo, M., Merle-Beral, H., Debré, P., & Galanaud, P. (1990). IL-4 inhibits the expression of high affinity IL-2 receptors on monoclonal human B cells. J. Immunol. 145, 1152-1158.

Katz, J. D., Benoist, C., & Mathis, D. (1995). T helper cell subsets in insulin-dependent diabetes. Science 268, 1185-1188.

Kawahara, A., Minami, Y., & Taniguchi, T. (1994). Evidence for a critical role for the cytoplasmic region of the interleukin-2 (IL-2) receptor γ chain in IL-2, IL-4, and IL-7 signalling. Mol. Cel. Biol. 14, 5433-5440.

Kawakami, Y., Custer, M. C., Rosenberg, S. A., & Lotze, M. T. (1989). IL-4 regulates IL-2 induction of lymphokine-activated killer activity from human lymphocytes. J. Immunol. 142, 3452-3461.

Keegan, A. D., & Pierce, J. H. (1994). The interleukin-4 receptor: Signal transduction by a hematopoietin receptor. J. Leuk. Biol. 55, 272-279.

Keegan, A. D., Beckmann, M. P., Park, L. S., & Paul, W. E. (1991). The IL-4 receptor: Biochemical characterization of IL-4-binding molecules in a T cell line expressing large numbers of receptors. J. Immunol. 146, 2272-2279.

Keegan, A. D., Nelms, K., Wang, L. M., Pierce, J. H., & Paul, W. E. (1994a). Interleukin-4 receptor: Signaling mechanisms. Immunol. Today 15, 423-432.

Keegan, A. D., Nelms, K., White, M., Wang, L. M., Pierce, J. H., & Paul, W. E. (1994b). An IL-4 receptor region containing an insulin receptor motif is important for IL-4-mediated IRS-1 phosphorylation and cell growth. Cell 76, 811-820.

Keller, U., Aman, M. J., Derigs, G., Huber, C., & Peschel, C. (1994). Human interleukin-4 enhances stromal cell-dependent hematopoiesis: Costimulation with stem cell factor. Blood 84, 2189-2196.

Kinashi, T., Inaba, K., Tsubata, T., Tashiro, K., Palacios, R., & Honjo, T. (1988). Differentiation of an interleukin 3-dependent precursor B-cell clone into immunoglobulin-producing cells *in vitro*. Proc. Natl. Acad. Sci. USA 85, 4473-4477.

Kinashi, T., Tashiro, K., Inaba, K., Takeda, T., Palacios, R., & Honjo, T. (1989). An interleukin-4-dependent precursor clone is an intermediate of the differentiation pathway from an interleukin-3-dependent precursor clone into myeloid cells as well as B lymphocytes. Int. Immunol. 1, 11-19.

King, A. G., Wierda, D., & Landreth, K. S. (1988). Bone marrow stromal cell regulation of B- lymphopoiesis. I. The role of macrophages, IL-1, and IL-4 in pre-B cell maturation. J. Immunol. 141, 2016-2026.

King, A. G., Landreth, K. S., & Wierda, D. (1989a). Bone marrow stromal cell regulation of B-lymphopoiesis. II. Mechanisms of hydroquinone inhibition of pre-B cell maturation. J. Pharm. Exp. Ther. 250, 582-590.

King, C. L., Gallin, J. I., Malech, H. L., Abramson, S. T., & Nutman, T. B. (1989b). Regulation of immunoglobulin production in hyperimmuno-globulin E recurrent-infection syndrome by interferon γ. Proc. Natl. Acad. Sci. USA 86, 10085-10089.

Klein, N. J., Rigley, K. P., & Callard, R. E. (1993). IL-4 regulates the morphology, cytoskeleton, and proliferation of human umbilical vein endothelial cells: Relationship between vimentin and CD23. Intern. Immunol. 5, 293-301.

Koettnitz, K., & Kalthoff, F. S. (1993). Human interleukin-4 receptor signalling requires sequences contained within two cytoplasmic regions. Eur. J. Immunol. 23, 988-991.

Kolb, J. P., Paul-Eugène, N., Damais, C., Yamaoka, K., Drapier, J. C., & Dugas, B. (1994). Interleukin-4 stimulates cGMP production by IFN-γ-activated human monocytes. J. Biol. Chem. 269, 9811-9816.

Kondo, M., Takeshita, T., Ishii, N., Nakamura, M., Watanabe, S., Arai, K.-A., & Sugamura, K. (1993). Sharing of the interleukin-2 (IL-2) receptor γ chain between receptors for IL-2 and IL-4. Science 262, 1874-1877.

Kotanides, H., & Reich, N. C. (1993). Requirement of tyrosine phosphorylation for rapid activation of a DNA binding factor by IL-4. Science 262, 1265-1267.

Kruse, N., Tony, H. P., & Sebald, W. (1992). Conversion of human interleukin-4 into a high affinity antagonist by a single amino acid replacement. EMBO J. 11, 3237-3244.

Kühn, R., Rajewsky, K., & Müller, W. (1991). Generation and analysis of interleukin-4 deficient mice. Science 254, 707-710.

Kumaratilake, L. M., & Ferrante, A. (1992). IL-4 inhibits macrophage-mediated killing of *Plasmodium Falciparum in vitro*. J. Immunol. 149, 194-199.

Lacey, D. L., Erdmann, J. M., Teitelbaum, S. L., Tan, H. L., Ohara, J., & Shioi, A. (1995). Interleukin-4, interferon-gamma, and prostaglandin E impact the osteoclastic cell-forming potential of murine bone marrow macrophages. Endocrinology 136, 2367-2376.

Lacraz, S., Nicod, L., Galve-de-Rochemonteix, B., Baumberger, C., Dayer, J.-M., & Welgus, H. G. (1992). Suppression of metalloproteinase biosynthesis in human alveolar macrophages by interleukin-4. J. Clin. Invest. 90, 382-388.

Larner, A. C., Petricoin, E. F., Nakagawa, Y., & Finbloom, D. S. (1993). IL-4 attenuates the transcriptional activation of both IFN-α and IFN-γ-induced cellular gene expression in monocytes and monocytic cell lines. J. Immunol. 150, 1944-1950.

Law, C.-L., Armitage, R. J., Villablanca, J. G., & LeBien, T. W. (1991). Expression of interleukin-4 receptors on early human B-lineage cells. Blood 78, 703-710.

Lee, F., Yokota, T., Otsuka, T., Meyerson, P., Villaret, D., Coffman, R., Mosmann, T., Rennick, D., Roehm, N., Smith, C., Zlotnick, A., & Arai, K. (1986). Isolation and characterization of a mouse interleukin cDNA clone that expresses B-cell stimulatory factor 1 activities and T cell- and mast-cell-stimulating activities. Proc. Natl. Acad. Sci. USA 83, 2061-2065.

Lehn, M., Weiser, W. Y., Engelhorn, S., Gillis, S., & Remold, H. G. (1989). IL-4 inhibits H2O2 production and antileishmanial capacity of human cultured monocytes mediated by IFN-γ. J. Immunol. 143, 3020-3024.

Lester, M. R., Hofer, M. F., Gately, M., Trumble, A., & Leung, D. Y. M. (1995). Down-regulating effects of IL-4 and IL-10 on the IFN-γ response in atopic dermatitis. J. Immunol. 154, 6174-6181.

Li-Weber, M., Eder, A., Krafft Czepa, H., & Krammer, P. H. (1992). T cell-specific negative regulation of transcription of the human cytokine IL-4. J. Immunol. 148, 1913-1918.

Li-Weber, M., Krafft, H., & Krammer, P. H. (1993). A novel enhancer element in the human IL-4 promoter is suppressed by a position-independent silencer. J. Immunol. 151, 1371-1382.

Lin, J. X., Migone, T. S., Tsang, M., Friedmann, M., Weatherbee, J. A., Zhou, L., Yamauchi, A., Bloom, E. T., Mietz, J., John, S., & Leonard, W. J. (1995). The role of shared receptor motifs and common stat proteins in the generation of cytokine pleiotropy and redundancy by IL-2, IL-4, IL-7, IL-13 and IL-15. Immunity 2, 331-339.

Lind, M., Deleuran, B., Yssel, H., Fink-Eriksen, E., & Thestrup-Pedersen, K. (1995). IL-4 and IL-13, but not IL-10, are chemotactic factors for human osteoblasts. Cytokine 7, 78-82.

Linsley, P. S., & Ledbetter, J. A. (1993). The role of the CD28 receptor during T cell responses to antigen. Ann. Rev. Immunol. 11, 191-212.

Llorente, L., Mitjavila, F., Crevon, M. C., & Galanaud, P. (1990). Dual effects of interleukin 4 on antigen-activated human B cells: induction of proliferation and inhibition of interleukin 2-dependent differentiation. Eur. J. Immunol. 20, 1887-1892.

Lotze, M. T. (1992). Role of IL-4 in the antitumor response. In: *IL4: Structure and Function* (Spits,H., Ed.). CRC Press Inc., New York, pp. 238-258.

Loyer, P., Ilyin, G., Razzak, Z. A., Dézier, J.-F., Banchereau, J., Campion, J.-P., Guguen-Guillouzo, C., & Guillouzo, A. (1993). IL-4 modulates production of acute phase proteins in adult human hepatocytes. FEBS Lett. 336, 215-220.

Maggi, E., Giudizi, M. G., Biagiotti, R., Annunziato, F., Manetti, R., Piccinni, M.-P., Parronchi, P., Sampognaro, S., Giannarini, L., Zuccati, G., & Romagnani, S. (1994). Th2-like CD8+ T cells showing B cell helper function and reduced cytolytic activity in human immunodeficiency virus type 1 infection. J. Exp. Med. 180, 489-495.

Maggi, E., Parronchi, P., Manetti, R., Simonelli, C., Piccinni, M.-P., Rugiu, F. S., De Carli, M., Ricci, M., & Romagnani, S. (1992). Reciprocal regulatory effects of IFN-γ and IL-4 on the *in vitro* development of human Th1 and Th2 clones. J. Immunol. 148, 2142-2147.

Malabarba, M. G., Kirken, R. A., Rui, H., Koettnitz, K., Kawamura, M., O'Shea, J. J., Kalthoff, F. S., & Farrar, W. L. (1995). Activation of JAK3, but not JAK1, is critical to interleukin-4 (IL-4) stimulated proliferation and requires a membrane-proximal region of IL-4 receptor alpha. J. Biol. Chem. 270, 9630-9637.

Maliszewski, C. R., Morrissey, P. J., Fanslow, W. C., Sato, T. A., Willis, C., & Devison, B. (1992). Delayed allograft rejection in mice transgenic for a soluble form of the IL-4 receptor. Cell. Immunol. 143, 434-448.

Manabe, A., Coustan-Smith, E., Behm, F. G., Raimondi, S. C., & Campana, D. (1992). Bone marrow-derived stromal cells prevent apoptotic cell death in B-lineage acute lymphoblastic leukemia. Blood 79, 2370-2377.

Manabe, A., Coustan-Smith, E., Kumagai, M.-A., Behm, F. G., Raimondi, S. C., Pui, C.-H., & Campana, D. (1994). Interleukin-4 induces programmed cell death (apoptosis) in cases of high-risk acute lymphoblastic leukemia. Blood 83, 1731-1737.

Manetti, R., Annunziato, F., Biagiotti, R., Giudizi, M. G., Piccinni, M. P., Giannarini, L., Sampognaro, S., Parronchi, P., Vinante, F., Pizzolo, G., Maggi, E., & Romagnani, S. (1994a). CD30 expression by CD8+ T cells producing type 2 helper cytokines. Evidence for large numbers of CD8+ CD30+ T cell clones in human immunodeficiency virus infection. J. Exp. Med. 180, 2407-2411.

Manetti, R., Gerosa, F., Giudizi, M. G., Biagiotti, R., Parronchi, P., Piccini, M. P., Sampognaro, S., Maggi, E., Romagnani, S., & Trinchieri, G. (1994b). Interleukin 12 induces stable priming for interferon gamma (IFN-gamma) production during differentiation of human T helper (Th) cells and transient IFN-gamma production in established Th2 cell clones. J. Exp. Med. 179, 1273-1283.

Mantovani, A., Bussolino, F., & Dejana, E. (1992). Cytokine regulation of endothelial cell function. FASEB J. 6, 2591-2599.

Marfaing-Koka, A., Devergne, O., Gorgone, G., Portier, A., Schall, T. J., Galanaud, P., & Emilie, D. (1995). Regulation of the production of the RANTES chemokine by endothelial cells. Synergistic induction by IFN-gamma plus TNF-alpha and inhibition by IL-4 and IL-13. J. Immunol. 154, 1870-1878.

Marsh, D. G., Neely, J. D., Breazeale, D. R., Ghosh, B., Freidhoff, L. R., Ehrlich-Kautzky, E., Schou, C., Krishnaswamy, G., & Beaty, T. H. (1994).

Linkage analysis of IL4 and other chromosome 5q31-1 markers and total serum immunoglobulin E concentrations. Science 264, 1152-1156.

Masinovsky, B., Urdal, D., & Gallatin, M. W. (1990). IL-4 acts synergistically with IL-1β to promote lymphocyte adhesion to microvascular endothelium by induction of vascular cell adhesion. J. Immunol. 145, 2886-2895.

Matthews, D. J., Clark, P. A., Herbert, J., Morgan, G., Armitage, R. J., Kinnon, C., Minty, A., Grabstein, K. H., Caput, D., Ferrara, P., & Callard, R. (1995). Function of the interleukin-2 (IL-2) receptor γ-chain in biologic responses of X-linked severe combined immunodeficient B cells to IL-2, IL-4, IL-13 and IL-15. Blood 85, 38-42.

McGarvie, G. M., & Cushley, W. (1989). The effect of recombinant interleukin 4 upon protein kinase activities associated with murine and human B lymphocyte plasma membranes. Cell Signal 1, 447-460.

McInnes, A., & Rennick, D. M. (1988). Interleukin 4 induces cultured monocytes/macrophages to form giant multinucleated cells. J. Exp. Med 167, 598-611.

McKenzie, A. N. J., Culpepper, J. A., de Waal Malefyt, R., Brière, F., Punnonen, J., Aversa, G., Sato, A., Dang, W., Cocks, B. G., Menon, S., de Vries, J. E., Banchereau, J., & Zurawski, G. (1993). Interleukin-13, a novel T cell-derived cytokine that regulates human monocyte and B cell function. Proc. Natl. Acad. Sci. USA 90, 3735-3739.

Migliorati, G., Nicoletti, I., Pagliacci, M. C., D'Adamio, L., & Riccardi, C. (1993). Interleukin-4 protects double-negative and CD4 single-positive thymocytes from dexamethasone-induced apoptosis. Blood 81, 1352-1358.

Mills, F. C., Thyphronitis, G., Finkelman, F. D., & Max, E. E. (1992). Immunoglobulin μ-ε isotype switch in IL-4 treated human B lymphoblastoid cells: Evidence for a sequential switch. J. Immunol. 149, 1075-1085.

Minty, A., Chalon, P., Derocq, J.-M., Dumont, X., Guillemot, J.-C., Kaghad, M., Labit, C., Leplatois, P., Liauzun, P., Miloux, B., Minty, C., Casellas, P., Loison, G., Lupker, J., Shire, D., Ferrara, P., & Caput, D. (1993). Interleukin-13 is a new human lymphokine regulating inflammatory and immune responses. Nature 362, 248-250.

Miossec, P., Briolay, J., Dechanet, J., Wijdenes, J., Martinez-Valdez, H., & Banchereau, J. (1992). Inhibition of the production of proinflammatory cytokines and immunoglobulins by interleukin-4 in an *ex vivo* model of rheumatoid synovitis. Arthritis Rheum. 35, 874-883.

Miossec, P., Chomarat, P., Dechanet, J., Moreau, J.-F., Roux, J.-P., Delmas, P., & Banchereau, J. (1994). Interleukin 4 inhibits bone resorption through an effect on osteoclasts and proinflammatory cytokines in an *ex vivo* model of bone resorption in rheumatoid arthritis. Arthritis Rheum. 37, 1715-1722.

Mire-Sluis, A. R., & Thorpe, R. (1991). Interleukin-4 proliferative signal transduction involves the activation of a tyrosine-specific phosphatase and the dephosphorylation of an 80-kDa protein. J. Biol. Chem. 266, 18113-18118.

Mitchell, P., Clutterbuck, R., Powles, R., De Lord, C., Titley, J., & Millar, J. (1994). Interleukin-4 (IL-4) improves survival and suppresses human ALL xenografts in SCID mice, confirming *in vitro* growth inhibition. Blood 84 (suppl), 48a.

Miyajima, A., Kitamura, T., Harada, N., Yokota, T., & Arai, K. (1992). Cytokine receptors and signal transduction. Annu. Rev. Immunol. 10, 295-331.

Miyauchi, J., Clark, S. C., Tsunematsu, Y., Shimizu, K., Park, J.-W., Ogawa, T., & Toyama, K. (1991). Interleukin-4 as a growth regulator of clonogenic cells in acute myelogenous leukemia in suspension culture. Leukemia 5, 108-115.

Mizuguchi, J., Beaven, M. A., O'Hara, J., & Paul, W. E. (1986). BSF-1 action on resting B-cells does not require elevation of inositol phospholipid metabolism or increased (Ca^{2+}). J. Immunol. 137, 2215-2219.

Monroe, J. G., Haldar, S., Prystowsky, M. B., & Lammie, P. (1988). Lymphokine regulation of inflammatory processes: Interleukin-4 stimulates fibroblast proliferation. Clin. Immunol. Immunopathatol. 49, 292-298.

Montaner, L. J., Doyle, A. G., Collin, M., Herbein, G., Illei, P., James, W., Minty, A., Caput, D., Ferrara, P., & Gordon, S. (1993). Interleukin 13 inhibits human immunodeficiency virus type 1 production in primary blood-derived human macrophages *in vitro*. J. Exp. Med. 178, 743-747.

Morgan, J. G., Dolganov, G. M., Robbins, S. E., Hinton, L. M., & Lovett, M. (1992). The selective isolation of novel cDNAs encoded by the regions surrounding the human interleukin 4 and 5 genes. Nucleic Acids Res. 20, 5173-5179.

Moser, R., Fehr, J., & Bruijnzeel, P. L. B. (1992). IL-4 controls the selective endothelium-driven transmigration of eosinophils from allergic individuals. J. Immunol. 149, 1432-1438.

Mosley, B., Beckmann, M. P., March, C. J., Idzerda, R. L., Gimpel, S. D., Vanden Bos, T., Friend, D., Alpert, A., Anderson, D., Jackson, J., Wignall, J. M., Smith, C., Gallis, B., Sims, J. E., Urdal, D., Widmer, M. B., Cosman, D., & Park, L. S. (1989). The murine interleukin-4 receptor: Molecular cloning characterization of secreted and membrane bound forms. Cell 59, 335-348.

Mosmann, T. R., Bond, M. W., Coffman, R. L., Ohara, J., & Paul, W. E. (1986). T-cell and mast cell lines respond to B-cell stimulatory factor 1. Proc. Natl. Acad. Sci. USA 83, 5654-5658.

Mueller, R., Heusser, C. H., Rihs, S., Brunner, T., Bullock, G. R., & Dahinden, C. A. (1994). Immunolocalization of intracellular interleukin-4 in normal human peripheral blood basophils. Eur. J. Immunol. 24, 2935-2940.

Mulligan, M. S., Jones, M. L., Vaporciyan, A. A., Howard, M. C., & Ward, P. A. (1993). Protective effects of IL-4 and IL-10 against immune complex-induced lung injury. J. Immunol. 151, 5666-5674.

Murakami, M., Narazaki, M., Hibi, M., Yawata, H., Yasukawa, K., Hamaguchi, M., Taga, T., & Kishimoto, T. (1991). A critical cytoplasmic region of the interleukin 6 signal transducer gp 130 is conserved in the cytokine receptor family. Proc. Natl. Acad. Sci. USA 88, 11349-11353.

Musso, T., Johnston, J. A., Linnekin, D., Varesio, L., Rowe, T. K., O'Shea, J. J., & McVicar, D. W. (1995). Regulation of JAK3 expression in human monocytes: Phosphorylation in response to interleukins 2, 4, and 7. J. Exp. Med. 181, 1425-1431.

Nagler, A., Lanier, L. L., & Phillips, J. H. (1988). The effects of IL 4 on human natural killer cells. A potent regulator of IL-2 activation and proliferation. J. Immunol. 141, 2349-2351.

Noguchi, M., Nakamura, Y., Russel, S. M., Ziegler, S. F., Tsang, M., Cao, X., & Leonard, W. J. (1993a). Interleukin-2 receptor γ chain: A functional component of the interleukin-7 receptor. Science 262, 1877-1880.

Noguchi, M., Yi, H., Rosenblatt, H. M., Filipovich, A. H., Adelstein, S., Modi, W. S., McBride, O. W., & Leonard, W. J. (1993b). Interleukin-2 receptor γ chain mutation results in X-linked severe combined immunodeficiency in humans. Cell 73, 147-157.

O'Garra, A., & Spits, H. (1993). The immunobiology of interleukin 4. Res. Immunol. 144, 567-643.

Obiri, N. I., Debinski, W., Leonard, W. J., & Puri, R. K. (1995). Receptor for interleukin 13. Interaction with interleukin 4 by a mechanism that does not involve the common γ chain shared by receptors for interleukins 2, 4, 7, 9 and 15. J. Biol. Chem. 270, 8797-8804.

Obiri, N. I., Hillman, G. G., Haas, G. P., Sud, S., & Puri, R. K. (1993). Expression of high affinity interleukin-4 receptors on human renal cell carcinoma cells and inhibition of tumor cell growth *in vitro* by interleukin-4. J. Clin. Invest. 91, 88-93.

Okabe, M., Kuni-eda, Y., Sugiwura, T., Tanaka, M., Miyagishima, T., Saiki, I., Minagawa, T., Kurosawa, M., Itaya, T., & Miyazaki, T. (1991). Inhibitory effect of interleukin-4 on the *in vitro* growth of Ph^1-positive acute lymphoblastic leukemia cells. Blood 78, 1574-1580.

Ouaaz, F., Sola, B., Issaly, F., Kolb, J.-P., Davi, F., Mentz, F., Arock, M., Paul-Eugène, N., Körner, M., Dugas, B., Debré, P., & Mossalayi, M. D. (1994). Growth arrest and terminal differentiation of leukemic myelomonocytic cells induced through ligation of surface CD23 antigen. Blood 84, 3095-3104.

Paliard, X., de Waal Malefijt, R., de Vries, J. E., & Spits, H. (1988). Interleukin 4 can mediate the induction of CD8 on human CD4+ T cell clones. Nature 335, 642-644.

Palmer-Crocker, R. L., & Pober, J. S. (1995). IL-4 induction of VCAM-1 on endothelial cells involves activation of a protein tyrosine kinase. J. Immunol. 154, 2838-2845.

Pandrau, D., Saeland, S., Duvert, V., Durand, I., Manel, A. M., Zabot, M. T., Philippe, N., & Banchereau, J. (1992). Interleukin-4 inhibits *in vitro* proliferation of leukemic and normal human B cell precursors. J. Clin. Invest. 90, 1697-1706.

Park, L. S., Friend, D., Sassenfeld, H. M., & Urdal, D. L. (1987). Characterization of the human B cell stimulatory factor 1 receptor. J. Exp. Med. 166, 476-488.

Parronchi, P., De Carli, M., Manetti, R., Simonelli, C., Piccini, M. P., Macchia, D., Maggi, E., Del Prete, G., Ricci, M., & Romagnani, S. (1992). Aberrant interleukin (IL)4 and IL5 production *in vitro* by CD4+ helper T cells from atopic subjects. Eur. J. Immunol. 87, 58-69.

Parronchi, P., Macchia, D., Piccini, M. P., Biswas, P., Simonelli, C., Maggi, E., Ricci, M., Ansari, A. A., & Romagnani, S. (1991). Allergen- and bacterial antigen-specific T-cell clones established from atopic donors show a different profile of cytokine production. Proc. Natl. Acad. Sci. USA 88, 4538-4542.

Paul, W. E. (1991). Interleukin-4: A prototypic immunoregulatory lymphokine. Blood 77, 1859-1870.

Peleman, R., Wu, J., Fargeas, C., & Delespesse, G. (1989). Recombinant interleukin-4 suppresses the production of interferon γ by human mononuclear cells. J. Exp. Med. 170, 1751-1756.

Pène, J., Rousset, F., Brière, F., Chrétien, I., Bonnefoy, J. Y., Spits, H., Yokota, T., Arai, N., Arai, K. I., Banchereau, J., & de Vries, J. E. (1988). IgE production by normal human lymphocytes is induced by interleukin 4 and suppressed by interferons gamma and alpha and prostaglandin E2. Proc. Natl. Acad. Sci. USA 85, 6880-6884.

Peschel, C., Paul, W. E., Ohara, J., & Green, I. (1987). Effects of B cell stimulatory factor-1/interleukin 4 on hematopoietic progenitor cells. Blood 70, 254-263.

Peschel, C., Green, I., & Paul, W. E. (1989). Preferential proliferation of immature B lineage cells in long-term stromal cell-dependent cultures with IL-4. J. Immunol. 142, 1558-1568.

Phillips, J. O., Everson, M. P., Moldoveanu, Z., Lue, C., & Mestecky, J. (1990). Synergistic effect of IL-4 and IFN-γ on the expression of polymeric Ig receptor (secretory component) and IgA binding by human epithelial cells. J. Immunol. 145, 1740-1744.

Piccinni, M. P., Macchia, D., Parronchi, P., Giudizi, M. G., Bani, D., Alterini, R., Grossi, A., Ricci, M., Maggi, E., & Romagnani, S. (1991). Human bone marrow non-B, non-T cells produce IL-4 in response to cross-linkage of Fcε and Fcγ receptors. Proc. Natl. Acad. Sci. USA 88, 8656-8660.

Piela-Smith, T. H., Broketa, G., Hand, A., & Korn, J. H. (1992). Regulation of ICAM-1 expression and function in human dermal fibroblasts by IL-4. J. Immunol. 148, 1375-1381.

Pizzolo, G., Vinante, F., Morosato, L., Nadali, G., Chilosi, M., Gandini, G., Sinicco, A., Raiteri, R., Semenzato, G., Stein, H., & Perona, G. (1994). High serum level of the soluble form of CD30 molecule in the early phase of HIV-1 infection as an independent predictor of progression to AIDS. AIDS 8, 741-745.

Plum, J., De Smedt, M., Leclercq, G., & Tison, B. (1990). Inhibitory effect of murine recombinant IL-4 on thymocyte development in fetal thymus organ cultures. J. Immunol. 145, 1066-1073.

Postlethwaite, A. E., & Seyer, J. M. (1991). Fibroblast chemotaxis induction by human recombinant interleukin-4. Identification by synthetic peptide analysis of two chemotactic domains residing in amino acid sequences 70-88 and 89-122. J. Clin. Invest. 87, 2147-2152.

Postlethwaite, A. E., Holness, M. A., Katai, H., & Raghow, R. (1992). Human fibroblasts synthetize elevated levels of extracellular matrix proteins in response to interleukin 4. J. Clin. Invest. 90, 1479-1485.

Powers, R., Garrett, D. S., March, C. J., Frieden, E. A., Gronenborn, A. M., & Clore, G. M. (1992). Three-dimensional solution structure of human interleukin-4 by multidimensional heteronuclear magnetic resonance spectroscopy. Science 256, 1673-1677.

Prinz, J. C., Baur, X., Mazur, G., & Rieber, E. P. (1990). Allergen-directed expression of Fc receptors for IgE (CD23) on human T lymphocytes is modulated by interleukin-4 and interferon-γ. Eur. J. Immunol. 20, 1259-1264.

Punnonen, J., Aversa, G., & de Vries, J. E. (1993). Human pre-B cells differentiate into Ig-secreting plasma cells in the presence of interleukin-4 and activated CD4+ T cells or their membranes. Blood 82, 2781-2789.

Quelle, F. W., Shimoda, K., Thierfelder, W., Fischer, C., Kim, A., Ruben, S. M., Cleveland, J. L., Pierce, J. H., Keegan, A. D., Nelms, K., Paul, W. E., & Ihle, J.

N. (1995). Cloning of murine Stat6 and human Stat6, Stat proteins that are tyrosine phosphorylated in responses to IL-4 and IL-3 but are not required for mitogenesis. Mol. Cel. Biol. 15, 3336-3343.

Racke, M. K., Burnett, D., Pak, S. H., Albert, P. S., Cannella, B., Raine, C. S., McFarlin, D. E., & Scott, D. E. (1995). Retinoid treatment of experimental allergic encephalomyelitis. IL-4 production correlates with improved disease course. J. Immunol. 154, 450-458.

Ramanathan, L., Ingram, R., Sullivan, L., Greenberg, R., Reim, R., Trotta, P. P., & Le, H. V. (1993). Immunochemical mapping of domains in human interleukin 4 recognized by neutralizing monoclonal antibodies. Biochemistry 32, 3549-3556.

Ranheim, E. A., & Kipps, T. J. (1993). Activated T cells induce expression of B7/BB1 on normal or leukemic B cells through a CD40-dependent signal. J. Exp. Med. 177, 925-935.

Rapoport, M. J., Jaramillo, A., Zipris, D., Lazarus, A. H., Serreze, D. V., Leiter, E. H., Cyopick, P., Dnaska, J. S., & Deloviqtch, T. L. (1993). Interleukin 4 reverses T cell proliferative unresponsiveness and prevents the onset of diabetes in nonobese diabetic mice. J. Exp. Med. 178, 87-99.

Razzak, Z. A., Loyer, P., Fautrel, A., Gautier, J.-C., Corcos, L., Turlin, B., Beaune, P., & Guillouzo, A. (1993). Cytokines down-regulate expression of major cytochrome P-450 enzymes in adult human hepatocytes in primary culture. Molec. Pharm. 44, 707-715.

Reason, D. C., Ebisawa, M., Saito, H., Nagakura, T., & Likura, Y. (1990). Interleukin 4 induces CD4+/CD8- to CD8+/CD4- transformation of human neonatal T cells by way of a double positive intermediate. Bioch. Biophys. Research Communications. 168, 830-836.

Reddy, M. M., Weissman, A. M., Mazza, D. S., Meriney, D. K., Johns, M., Carrabis, S., & Grieco, M. H. (1992). Circulating elevated levels of soluble CD23, interleukin 4, and CD20+ CD23+ lymphocytes in atopic subjects with elevated serum IgE concentrations. Ann. Allergy 69, 131-134.

Renard, N., Duvert, V., Banchereau, J., & Saeland, S. (1994). Interleukin-13 inhibits the proliferation of normal and leukemic human B-cell precursors. Blood 84, 2253-2260.

Rennick, D. M., Moore, J. G., & Thompson-Snipes, L. (1992). IL-4 and hematopiesis. In: *IL-4: Structure and Function* (Spits, H., Ed.). CRC Press, London, pp. 151-168.

Rennick, D., Yang, G., Muller-Sieburg, C., Smith, C., Arai, N., Takabe, Y., & Gemmell, L. (1987). Interleukin 4 (B-cell stimulatory factor 1) can enhance or antagonize the factor-dependent growth of hemopoietic progenitor cells. Proc. Natl. Acad. Sci. USA 84, 6889-6893.

Rigley, K. P., Thurstan, S. M., & Callard, R. E. (1991). Independent regulation of interleukin 4 (IL-4)-induced expression of human B cell surface CD23 and IgM: Functional evidence for two IL-4 receptors. Int. Immunol. 3, 197-203.

Romani, N., Gruner, S., Brang, D., Kämpgen, E., Lenz, A., Trockenbacher, B., Konwalinka, G., Fritsch, P. O., Steinman, R. M., & Schuler, G. (1994). Proliferating dendritic cell progenitors in human blood. J. Exp. Med. 180, 83-93.

Rousselet, G., Busson, P., Billaud, M., Guillon, J. M., Scamps, C., Wakasugi, H., Lenoir, G., & Tursz, T. (1990). Structure and regulation of the Blast-2/CD23 antigen in epithelial cells from nasopharyngeal carcinoma. Int. Immunol. 2, 1159-1166.

Rousset, F., Garcia, E., & Banchereau, J. (1991a). Cytokine-induced proliferation and immunoglobulin production of human B lymphocytes triggered through their CD40 antigen. J. Exp. Med. 173, 705-710.

Rousset, F., Robert, J., Andary, M., Bonnin, J. P., Souillet, G., Chrétien, I., Brière, F., Pène, J., & de Vries, J. E. (1991b). Shifts in interleukin-4 and interferon-γ production by T cells of patients with elevated serum IgE levels and the modulatory effects of these lymphokines on spontaneous IgE synthesis. J. Allergy Clin. Immunol. 87, 58-69.

Russel, S. M., Keegan, A. D., Harada, N., Nakamura, Y., Noguchi, M., Leland, P., Friedmann, M. C., Miyajima, A., Puri, R. K., Paul, W. E., & Leonard, W. J. (1993). Interleukin-2 receptor γ chain: A functional component of the interleukin-4 receptor. Science 262, 1880-1883.

Ryan, D. H., Nuccie, B. L., Ritterman, I., Liesveld, J. L., & Abboud, C. N. (1994). Cytokine regulation of early human lymphopoiesis. J. Immunol. 152, 5250-5258.

Sad, S., Marcotte, R., & Mosmann, T. R. (1995). Cytokine-induced differentiation of precursor mouse CD8+ T cells into cytotoxic CD8+ T cells secreting Th1 or Th2 cytokines. Immunity 2, 271-279.

Saeland, S., Duvert, V., Moreau, I., & Banchereau, J. (1993). Human B cell precursors proliferate and express CD23 after CD40 ligation. J. Exp. Med. 178, 113-120.

Salgame, P., Abrams, J. S., Clayberger, C., Goldstein, H., Convit, J., Modlin, R. L., & Bloom, B. R. (1991). Differing lymphokine profiles of functional subsets of human CD4 and CD8 T cell clones. Science 254, 279-282.

Sallusto, F., & Lanzavecchia, A. (1994). Efficient presentation of soluble antigen by cultured human dendritic cells is maintained by granulocyte/macrophage colony-stimulating factor plus interleukin 4 and downregulated by tumor necrosis factor alpha. J. Exp. Med. 179, 1109-1118.

Sato, T. A., Widmer, M. B., Finkelman, F. D., Madani, H., Jacobs, C. A., Grabstein, K. H., & Maliszewski, C. R. (1993). Recombinant soluble murine IL-4 receptor can inhibit or enhance IgE responses *in vivo*. J. Immunol. 150, 2717-2723.

Schandene, L., Ferster, A., Mascart-Lemone, F., Crusiaux, A., Gerard, C., Marchant, A., Lybin, M., Velu, T., Sariban, E., & Goldman, M. (1993). T helper type 2-like cells and therapeutic effects of interferon-gamma in combined immunodeficiency with hypereosinophilia (Omenn's syndrome). Eur. J. Immunol. 23, 56-60.

Schindler, C., Kashleva, H., Pernis, A., Pine, R., & Rothman, P. (1994). STF-IL-4: a novel IL-4-induced signal transducing factor. EMBO J. 13, 1350-1356.

Schleimer, R. P., Sterbinsky, S. A., Kaiser, J., Bickel, C. A., Klunk, D. A., Tomioka, K., Newman, W., Luscinskas, F. W., Gimbrone, M. A., McIntyre, B. W., & Bochner, B. S. (1992). IL-4 induces adherence of human eosinophils and basophils but not neutrophils to endothelium. Association with expression of VCAM-1. J. Immunol. 148, 1086-1092.

Schmitt, E., Fassbender, B., Beyreuther, K., Spaeth, E., Schwarzkopf, R., & Rude, E. (1987). Characterization of a T-cell-derived lymphokine that acts synergistically with IL-3 on the growth of murine mast cells and is identical with IL-4. Immunobiology 174, 406-419.

Schmitz, J., Thiel, A., Kühn, R., Rajewsky, K., Müller, W., Assenmacher, M., & Radbruch, A. (1994). Induction of interleukin 4 (IL-4) expression in T helper (Th) cells is not dependent on IL-4 from non-Th cells. J. Exp. Med. 179, 1349-1353.

Schuitemaker, H., Kootstra, N. A., Koppelman, M. H. G. M., Bruisten, S. M., Huisman, H. G., Tersmette, M., & Miedema, F. (1992). Proliferation-dependent HIV-1 infection of monocytes occurs during differentiation into macrophages. J. Clin. Invest. 89, 1154-1160.

Schurmans, S., Heusser, C. H., Qin, H., Merino, J., Brighouse, G., & Lambert, P. (1990). *In vivo* effects of anti-IL-4 monoclonal antibody on neonatal induction of tolerance and on an associated autoimmune syndrome. J. Immunol. 145, 2465-2473.

Secrist, H., Chelen, C. J., Wen, Y., Marshall, J. D., & Umetsu, D. T. (1993). Allergen immunotherapy decreases interleukin 4 production in $CD4^+$ T cells from allergic individuals. J. Exp. Med. 178, 2123-2130.

Secrist, H., DeKruyff, R. H., & Umetsur, D. T. (1995). Interleukin 4 production by $CD4^+$ T cells from allergic individuals is modulated by antigen concentration and antigen-presenting cell type. J. Exp. Med. 181, 1081-1089.

Seder, R. A., & Le Gros, G. G. (1995). The functional role of $CD8^+$ T helper type 2 cells. J. Exp. Med. 181, 5-7.

Seder, R. A., Paul, W. E., Davis, M. M., & Fazekas de St. Groth, B. F. (1992). The presence of interleukin 4 during *in vitro* priming determines the lymphokine-producing potential of CD4+ T cells from T cell receptor transgenic mice. J. Exp. Med. 176, 1091-1098.

Shapira, S. K., Jabara, H. H., Thienes, C. P., Ahern, D. J., Vercelli, D., Gould, H. J., & Geha, R. S. (1991). Deletional switch recombination occurs in interleukin-4 induced isotype switching to IgE expression by human B cells. Proc. Natl. Acad. Sci. USA 88, 7528-7532.

Shapira, S. K., Vercelli, D., Jabara, H. H., Fu, S. M., & Geha, R. S. (1992). Molecular analysis of the induction of immunoglobulin E synthesis in human B cells by interleukin 4 and engagement of CD40 antigen. J. Exp. Med. 175, 289-292.

Shields, J. G., Armitage, R. J., Jamieson, B. N., Beverley, P. C. L., & Callard, R. E. (1989). Increased expression of surface IgM but not IgD or IgG on human B cells in response to IL-4. Immunol. 66, 224-227.

Shipp, M. A., & Look, A. T. (1993). Hematopoietic differentiation antigens that are membrane-associated enzymes: Cutting is the key! Blood 82, 1052-1070.

Sideras, P., & Palacios, R. (1987). Bone marrow pro-T and pro-B lymphocyte clones express functional receptors for interleukin (IL)3 and IL-4/BSF-1 and nonfunctional receptors for IL2. Eur. J. Immunol. 17, 217-221.

Sideras, P., Funa, K., Zalcberg-Quintana, I., Xanthopoulos, K. G., Kisielow, P., & Palacios, R. (1988). Analysis by *in situ* hybridization of cells expressing mRNA for interleukin 4 in the developing thymus and in peripheral lymphocytes from mice. Proc. Natl. Acad. Sci. USA 85, 218-221.

Sillaber, C., Strobl, H., Bevec, D., Ashman, L. K., Butterfield, J. H., Lechner, K., Maurer, D., Bettelheim, P., & Valent, P. (1991). IL-4 regulates c-*kit* proto-oncogene product expression in human mast and myeloid progenitor cells. J. Immunol. 147, 4224-4228.

Sillaber, C., Sperr, W. R., Agis, H., Geissler, K., Spanblöchl, E., Ashman, L. K., Bettelheim, P., & Lechner, K. (1992). IL-4 inhibits SCF-dependent growth of human mast cells and downregulates *c-kit* expression. Blood 80 (suppl), 92a

Simons, A., & Zharhary, D. (1989). The role of IL-4 in the generation of B lymphocytes in the bone marrow. J. Immunol. 143, 2540-2545.

Smeland, E. B., Blomhoff, H. K., Funderud, S., Shalaby, M. R., & Espevik, T. (1989). Interleukin 4 induces selective production of interleukin 6 from normal human B lymphocytes. J. Exp. Med. 170, 1463-1468.

Smerz-Bertling, C., & Duschl, A. (1995). Both interleukin 4 and interleukin 13 induce tyrosine phosphorylation of the 140-kDa subunit of the interleukin 4 receptor. J. Biol. Chem. 270, 966-970.

Smirnov, D. V., Smirnova, M. G., Korobko, V. G., & Frolova, E. I. (1995). Tandem arrangement of human genes for interleukin-4 and interleukin-13: Resemblance in their organization. Gene 155, 277-281.

Smith, C. A., & Rennick, D. M. (1986). Characterization of a murine lymphokine distinct from interleukin 2 and interleukin 3 (IL-3) possessing a T-cell growth factor activity and a mast-cell growth factor activity that synergizes with IL-3. Proc. Natl. Acad. Sci. USA 83, 1857-1861.

Snoeck, H.-W., Lardon, F., Lenjou, M., Nys, G., Van Bockstaele, D. R., & Peetermans, M. E. (1993a). Differential regulation of the expression of CD38 and human leukocyte antigen-DR on CD34+ hematopoietic progenitor cells by interleukin-4 and interferon-γ. Exp. Hematol. 21, 1480-1486.

Snoeck, H.-W., Lardon, F., Lenjou, M., Nys, G., Van Bockstaele, D. R., & Peetermans, M. E. (1993b). Interferon-γ and interleukin-4 reciprocally regulate the production of monocytes/macrophages and neutrophils through a direct effect on committed monopotential bone marrow progenitor cells. Eur. J. Immunol. 23, 1072-1077.

Sonoda, Y., Okuda, T., Yokota, S., Maekawa, T., Shizumi, Y., Nishigaki, H., Misawa, S., Fujii, H., & Abe, T. (1990). Actions of human interleukin-4/B-cell stimulatory factor-1 on proliferation and differentiation of enriched hematopoietic progenitor cells in culture. Blood 75, 1615-1621.

Sonoda, Y., Kuzuyama, Y., Tanaka, S., Yokota, S., Maekawa, T., Clark, S. C., & Abe, T. (1993). Human interleukin-4 inhibits proliferation of megakaryocyte progenitor cells in culture. Blood 81, 624-630.

Sorg, R. V., Enczmann, J., Sorg, U. R., Schneider, E. M., & Wernet, P. (1993). Identification of an alternatively spliced transcript of human interleukin-4 lacking the sequence encoded by exon 2. Exp. Hematol. 21, 560-563.

Souillet, G., Rousset, F., & de Vries, J. E. (1989). Alpha-interferon treatment of patient with hyper IgE syndrome. Lancet 1, 1384-1384.

Spits, H., Yssel, H., Takebe, Y., Arai, N., Yokota, T., Lee, F., Arai, K., Banchereau, J., & de Vries, J. E. (1987). Recombinant interleukin-4 promotes the growth of human T cells. J. Immunol. 135, 1142-1147.

Spits, H., Yssel, H., Paliard, X., Kastelein, R., Figdor, D., & de Vries, J. E. (1988). Interleukin-4 inhibits interleukin-2 mediated induction of human lymphokine activated killer cells, but not the generation of antigen specific cytotoxic T lymphocytes in mixed leucocyte cultures. J. Immunol. 141, 29-36.

Sunder-Plassmann, R., Majdiç, O., Knapp, W., & Holter, W. (1995). High IL-4 production is a stable phenotype of CD8negCD45RAnegCD27neg T cells. Cell. Immunol. 160, 309-313.

Swain, S. L., Bradley, L. M., Croft, M., Tonkonogy, S., Atkins, G., Weinberg, A. D., Duncan, D. D., Hedrick, S. M., Dutton, R. W., & Huston, G. (1991). Helper T-cell subsets: Phenotype, function and the role of lymphokines in regulating their development. Immunol. Rev. 123, 115-144.

Szabo, S. J., Gold, J. S., Murphy, T. L., & Murphy, K. M. (1993). Identification of cis-acting regulatory elements controlling interleukin-4 gene expression in T cells. Mol. Cel. Biol. 13, 4793-4805.

Takai, S., Tateno, M., Hirano, T., Kondo, N., Hirose, S., & Yoshiki, T. (1993). Increased IgE level as a marker of host-versus-graft disease: Inhibition of this HVGD with a monoclonal antibody to IL-4. Cell. Immunol. 149, 1-10.

Taniguchi, T. (1995). Cytokine signaling through nonreceptor protein tyrosine kinase. Science 268, 251-255.

Te Velde, A. A., Rousset, F., Péronne, C., de Vries, J. E., & Figdor, C. (1990). IL-4 induced expression of FcεRIIb and soluble FcεRIIb release from human monocytes is downregulated by IFNα and IFNγ. J. Immunol. 144, 3052-3059.

Te Velde, A. A., Klomp, J. P. G., Yard, B. A., de Vries, J. E., & Figdor, C. G. (1988). Modulation of phenotypic and functional properties of human peripheral blood monocytes by IL-4. J. Immunol. 140, 1548-1554.

Tepper, R. I., Pattengale, P. K., & Leder, P. (1989). Murine interleukin-4 displays potent anti-tumor activity *in vivo.* Cell 57, 503-512.

Tepper, R. I., Levinson, D. A., Stanger, B. Z., Campos-Torres, J., Abbas, A. K., & Leder, P. (1990). IL-4 induces allergic-like inflammatory disease and alters T cell development in transgenic mice. Cell 62, 457-467.

Tepper, R. I., Coffman, R. L., & Leder, P. (1992). An eosinophil-dependent mechanism for the antitumor effect of interleukin-4. Science 257, 548-551.

Thompson, C. B. (1995). Distinct roles for the costimulatory ligands B7-1 and B7-2 in T helper cell differentiation? Cell 81, 979-982.

Thompson-Snipes, L., Dhar, V., Bond, M. W., Mosmann, T. R., Moore, K. W., & Rennick, D. (1991). Interleukin-10: A novel stimulatory factor for mast cells and their progenitors. J. Exp. Med. 173, 507.

Thornhill, M. H., & Kaskard, D. O. (1990). IL-4 regulates endothelial cell activation by IL-1, tumor necrosis factor, or IFN-γ. J. Immunol. 145, 865-872.

Thornhill, M. H., Kyan-Aung, U., & Haskard, D. O. (1990). IL-4 increases human endothelial cell adhesiveness for T cells but not for neutrophils. J. Immunol. 144, 3060-3065.

Thornhill, M. H., Wellicome, S. M., Mahiouz, D. L., Lanchbury, J. S. S., Kyan-Aung, U., & Haskard, D. O. (1991). Tumor necrosis factor combines with IL-4 or IFN-γ to selectively enhance endothelial cell adhesiveness for T cells. The contribution of vascular cell adhesion molecule-1-dependent and-independent binding mechanisms. J. Immunol. 146, 592-598.

Todd, M. D., Grusby, M. J., Lederer, J. A., Lacy, E., Lichtman, A. H., & Glimcher, L. H. (1993). Transcription of the interleukin 4 gene is regulated by multiple promoter elements. J. Exp. Med. 177, 1663-1674.

Toi, M., Bicknell, R., & Harris, A. L. (1992). Inhibition of colon and breast carcinoma cell growth by interleukin-4. Cancer Res. 52, 275-279.

Topp, M. S., Koenigsmann, M., Mire-Sluis, A., Oberberg, D., Eitelbach, F., Von Marschall, Z., Notter, M., Reufi, B., Stein, H., Thiel, E., & Berdel, W. E. (1993).

Recombinant human interleukin-4 inhibits growth of some human lung tumor cell lines *in vitro* and *in vivo*. Blood 82, 2837-2844.

Topp, M. S., Papadimitriou, C. A., Eitelbach, F., Koenigsmann, M., Oelmann, E., Koehler, B., Oberberg, D., Reufi, B., Stein, H., Thiel, E., & Berdel, W. E. (1995). Recombinant human interleukin-4 has antiproliferative activity on human tumor cell lines derived from epithelial and nonepithelial histologies. Cancer Res. 55, 2173-2176.

Tsuji, K., Nakahata, T., Takagi, M., Kobayashi, T., Ishiguro, A., Kikuchi, T., Naganuma, K., Koike, K., Miyajima, A., Arai, K., & Akabane, T. (1990). Effects of interleukin 3 and interleukin 4 on the development of connective tissue type mast cells: Interleukin 3 supports their survival and interleukin 4 triggers and supports their proliferation synergistically with interleukin 3. Blood 75, 421-427.

Tushinski, R. J., Larsen, A., Park, L. S., & Spoor, E. (1991). Interleukin 4 alone or in combination with interleukin 1 stimulates 3T3 fibroblasts to produce colony-stimulating factors. Exp. Hematol. 19, 238-244.

Ueno, Y., Boone, T., & Uittenbogaart, C. H. (1989). Selective stimulation of human thymocyte subpopulations by recombinant IL-4 and IL-3. Cell. Immunol. 118, 382-393.

Ushiyama, C., Hirano, T., Miyajima, H., Okumura, K., Ovary, Z., & Hashimoto, H. (1995). Anti-IL-4 antibody prevents graft-versus-host disease in mice after bone marrow transplantation. J. Immunol. 154, 2687-2696.

Valent, P., Bevec, D., Maurer, D., Besemer, J., Di Padova, F., Butterfield, J. H., Speiser, W., Majdic, O., Lechner, K., & Bettelheim, P. (1991). Interleukin 4 promotes expression of mast cell ICAM-1 antigen. Immunol. 88, 3339-3342.

Vallé, A., Zuber, C. E., Defrance, T., Djossou, O., De Rie, M., & Banchereau, J. (1989). Activation of human B lymphocytes through CD40 and interleukin 4. Eur. J. Immunol. 19, 1463-1467.

Vallé, A., Aubry, J.-P., Durand, I., & Banchereau, J. (1991). IL-4 and IL-2 upregulate the expression of antigen B7, the B cell counterstructure to T cell CD28: An amplification mechanism for T-B cell interactions. Int. Immunol. 3, 229-235.

Van Hal, P. T. W., Hopstaken-Broos, J. P. M., Prins, A., Favaloro, E. J., Huijbens, R. J. F., Hilvering, C., Figdor, C. G., & Hoogsteden, H. C. (1994). Potential indirect anti-inflammatory effects of IL-4. Stimulation of human monocytes, macrophages, and endothelial cells by IL-4 increases aminopeptidase-N activity (CD13; EC 3.4.11.2). J. Immunol. 153, 2718-2728.

Van Kooten, C., & Banchereau, J. (1995). CD40-CD40 ligand, a multifunctional receptor-ligand pair. Adv. Immunol. In press.

Vannier, E., Miller, L. C., & Dinarello, C. A. (1992). Coordinated anti-inflammatory effects of interleukin 4: IL-4 suppresses IL-1 production but up-regulates gene expression and synthesis of interleukin 1 receptor antagonist. Proc. Natl. Acad. Sci. USA 89, 4076-4080.

Varney, V. A., Hamid, Q. A., Gaga, M., Ying, S., Jacobson, M., Frew, A. J., Kay, A. B., & Durham, S. R. (1993). Influence of grass pollen immunotherapy on cellular infiltration and cytokine mRNA expression during allergen induced late phase cutaneous responses. J. Clin. Invest. 92, 644-651.

Vellenga, E., De Wolf, J. T., Beentjes, J. A. M., Esselink, M. T., Smit, J. W., & Haee, M. R. (1990). Divergent effects of interleukin 4 (IL-4) on the granulocyte colony stimulating factor and IL-3 supported myeloid colony formation from normal and leukemic bone marrow cells. Blood 75, 633-637.

Vellenga, E., Dokter, W., & Halie, R. M. (1993). Interleukin-4 and its receptor: Modulating effects on immature and mature hematopoietic cells. Leukemia 7, 1131-1141.

Vercelli, D., Jabara, H. H., Lee, W., Woodland, N., Geha, R. S., & Leung, D. Y. M. (1988). Human recombinant interleukin 4 induces Fcε R2/CD23 on normal human monocytes. J. Exp. Med. 167, 1406-1416.

Vieira, P., de Waal-Malefyt, R., Dang, M. N., Johnson, K. E., Kastelein, R., Fiorentino, D. F., de Vries, J. E., Roncarolo, M. G., Mosmann, T. R., & Moore, K. W. (1991). Isolation and expression of human cytokine synthesis inhibitory factor (CSIF/IL-10) cDNA clones: Homology to Epstein-Barr Virus open reading frame BCRF1. Proc. Natl. Acad. Sci. USA 88, 1172-1176.

Vita, N., Lefort, S., Laurent, P., Caput, D., & Ferrara, P. (1995). Characterization and comparison of the interleukin 13 receptor with the interleukin 4 receptor on several cell types. J. Biol. Chem. 270, 3512-3517.

Vollenweider, S., Saurat, J.-H., Röcken, M., & Hauser, C. (1991). Evidence suggesting involvement of interleukin-4 (IL-4) production in spontaneous *in vitro* IgE synthesis in patients with atopic dermatitis. J. Allergy Clin. Immunol. 87, 1088-1095.

Wagteveld, A. J., van Zanten, A. K., Esselink, M. T., Halie, M. R., & Vellenga, E. (1991). Expression and regulation of IL-4 receptors on human monocytes and acute myeloblastic leukemic cells. Leukemia 5, 782-788.

Wagteveld, A. J., Esselink, M. T., Limburg, P., Halie, M. R., & Vellenga, E. (1992). The effects of IL-1ß and IL-4 on the proliferation and endogenous secretion of growth factors by acute myeloblastic leukemic cells. Leukemia 6, 1020-1024.

Walter, M. R., Cook, W. J., Zhao, B. G., Cameron, R. P., Ealick, S. E., Walter, R. L., Reichert, P., Nagabhushan, T. L., Trotta, P. P., & Bugg, C. E. (1992). Crystal structure of recombinant human interleukin-4. J. Biol. Chem. 267, 20371-20376.

Wang, L.-M., Keegan, A. D., Paul, W. E., Heidaran, M. A., Gutkind, J. S., & Pierce, J. H. (1992). IL-4 activates a distinct signal transduction cascade from IL-3 in factor-dependent myeloid cells. EMBO J. 11, 4899-4908.

Wang, L.-M., Keegan, A. D., Li, W., Lienhard, G. E., Pacini, S., Gutkind, J. S., Myers, M. G., Sun, X.-J., White, M. F., Aaronson, S. A., Paul, W. E., & Pierce, J. H. (1993). Common elements in interleukin-4 and insulin signaling pathways in factor-dependent hematopoietic cells. Proc. Natl. Acad. Sci. USA 90, 4032-4036.

Warren, H. S., Kinnear, B. F., Phillips, J. H., & Lanier, L. L. (1995). Production of IL-5 by human NK cells and regulation of IL-5 secretion by IL-4, IL-10, and IL-12. J. Immunol. 154, 5144-5152.

Watanabe, K., Tanaka, Y., Morimoto, I., Yahata, K., Zeki, K., Fujihira, T., Yamashita, U., & Eto, S. (1990). Interleukin-4 as a potent inhibitor of bone resorption. Bioch. Bioph. Res. Comm. 172, 1035-1041.

Wegrowski, Y., Paltot, V., Gillery, P., Kalis, B., Randoux, A., & Maquart, F. X. (1995). Stimulation of sulphated glycosaminoglycan and decorin production in adult dermal fibroblasts by recombinant human interleukin-4. Biochem. J. 307, 673-678.

Wertheim, W. A., Kunkel, S. L., Standiford, T. J., Burdick, M. D., Becker, F. S., Wilke, C. A., Gilbert, A. R., & Strieter, R. M. (1993). Regulation of neutrophil-derived

IL-8: The role of prostaglandin E2, dexamethasone, and IL-4. J. Immunol. 151, 2166-2175.
White, M. F., & Kahn, C. R. (1994). The insulin signaling system. J. Biol. Chem. 269, 1-4.
Widmer, M. B., Acres, R. B., Sassenfeld, H. M., & Grabstein, K. H. (1987). Regulation of cytolytic cell populations from human peripheral blood by B cell stimulatory factor 1 (Interleukin 4). J. Exp. Med. 166, 1447-1455.
Widmer, M. B., & Grabstein, K. H. (1987). Regulation of cytolytic T lymphocyte generation by B-cell stimulatory factor. Nature 326, 795-798.
Windsor, W. T., Syto, R., Durkin, J., Das, P., Reichert, P., Pramanik, B., Tindall, S., Le, H. V., Labdon, J., Nagabhushan, T. L., & Trotta, P. P. (1990). Disulfide bond assignment of mammalian cell-derived recombinant human interleukin 4. Biophys. J. 57, 423a.
Witte, P. L., Frantsve, L. M., Hergott, M., & Rahbe, S. M. (1993). Cytokine production and heterogeneity of primary stromal cells that support B lymphopoiesis. Eur. J. Immunol. 23, 1809-1817.
Witthuhn, B. A., Silvennoinen, O., Miura, O., Lai, K. S., Cwik, C., Liu, E. T., & Ihle, J. N. (1994). Involvement of the Jak-3 Janus kinase in signalling by interleukin 2 and 4 in lymphoid and myeloid cells. Nature 370, 153-157.
Yin, T., Tsang, M. L.-S., & Yang, Y.-C. (1994). JAK1 kinase forms complexes with interleukin-4 receptor and 4PS/insulin receptor substrate-1-like protein and is activated by interleukin-4 and interleukin-9 in T lymphocytes. J. Mol. Biol. 269, 26614-26617.
Yokota, T., Otsuka, T., Mosmann, T., Banchereau, J., Defrance, T., Blanchard, D., de Vries, J. E., Lee, F., & Arai, K. (1986). Isolation and characterization of a human interleukin cDNA clone, homologous to mouse B-cell stimulatory factor 1, that expresses B-cell-stimulatory activities. Proc. Natl. Acad. Sci. USA 83, 5894-5898.
Yokota, A., Kikutani, H., Tanaka, T., Sato, R., Barsumian, E. L., Suemura, M., & Kishimoto, T. (1988). Two species of human Fc epsilon receptor II (Fc epsilon RII/CD23): Tissue-specific and IL-4-specific regulation of gene expression. Cell 55, 611-618.
Yoshimoto, T., & Paul, W. E. (1994). $CD4_{pos}$, $NK1.1_{pos}$ T cells promptly produce interleukin 4 in response to in vivo challenge with anti-CD3. J. Exp. Med. 179, 1285-1295.
Zhang, K., Mills, F. C., & Saxon, A. (1994). Switch circles from IL-4-directed epsilon class switching from human B lymphocytes. Evidence for direct, sequential, and multiple step sequential switch from μ to ε Ig heavy chain gene. J. Immunol. 152, 3427-3435.
Zlotnik, A., Fischer, M., Roehm, N., & Zipori, D. (1987a). Evidence for effects of interleukin 4 (B cell stimulatory factor 1) on macrophages: Enhancement of antigen presenting ability of bone marrow-derived macrophages. J. Immunol. 138, 4275-4279.
Zlotnik, A., Ramsom, J., Franck, G., Fischer, M., & Howard, M. (1987b). Interleukin 4 is a growth factor for activated thymocytes: Possible role in T cell ontogeny. Proc. Natl. Acad. Sci. USA 84, 3856-3860.
Zurawski, G., & de Vries, J. E. (1994). Interleukin 13, an interleukin 4-like cytokine that acts on monocytes and B cells, but not on T cells. Immunol. Today 15, 19-26.

Zurawski, S. M., Vega Jr., F., Huyghe, B., & Zurawski, G. (1993). Receptors for interleukin-13 and interleukin-4 are complex and share a novel component that functions in signal transduction. EMBO J. 12, 2663-2670.

Zurawski, S. M., Chomarat, P., Djossou, O., Bidaud, C., McKenzie, A. N. J., Miossec, P., Banchereau, J., & Zurawski, G. (1995). The primary binding subunit of the human interleukin-4 receptor is also a component of the Interleukin-13 receptor. J. Biol. Chem. 270, 13869-13878.

INTERLEUKIN-5

Kiyoshi Takatsu, Richard R. Dickason, and
David P. Huston

Growth Factors and Cytokines in Health and Disease
Volume 2A, pages 143-200.

ISBN: 0-7623-0117-1

I. INTRODUCTION

Interleukin-5 (IL-5) is a cytokine with important biological activity in inflammation (Takatsu, 1995). The gene for IL-5 clusters with the genes for IL-3, IL-4, and granulocyte macrophage colony stimulating factor (GM-CSF), and their similar exon-intron organization suggests an evolutionary relationship (Tanabe et al., 1987; Le Beau et al., 1989; Takahashi et al., 1989; Rozwarski et al., 1994). The three-dimensional structure of IL-5 demonstrates a four alpha helical bundle motif that is conserved among several hematopoietic cytokines (Kaushansky and Karplus, 1993; Milburn et al., 1993; Sprang and Bazan, 1993; Rozwarski et al., 1994). However, IL-5 is unusual in that two monomers interdigitate to form a pair of these helical bundle motifs (Milburn et al., 1993). The major cellular sources of IL-5 are mast cells, eosinophils, and lymphocytes (Azuma et al., 1986; Kinashi et al., 1986; Plaut et al., 1989; Bohjanen et al., 1990; Bradding et al., 1993, 1995; Dubucquoi et al., 1994; Bressler et al., forthcoming). Although IL-5 has pleiotrophic effects on a variety of leukocytes (Koike and Takatsu, 1994; Takatsu, 1995), it is unique among cytokines in its ability to specifically promote the terminal maturation of eosinophils (Clutterbuck et al., 1989; Coffman et al., 1989).

The IL-5 receptor (IL-5R) is a heterodimer (Mita et al., 1989; Tavernier et al., 1991; Migita et al., 1991). The IL-5R alpha chain (IL-5Rα) specifically binds IL-5 (Takaki et al., 1990; Tavernier et al., 1991). The IL-5R beta chain is shared with the cytokine-specific alpha chain subunit of the receptor for IL-3 and GM-CSF and, therefore, is called the common beta chain (βc; Takaki et al., 1991; Tavernier et al., 1991). βc alone cannot bind IL-5 but is necessary for signal transduction (Tavernier et al., 1991; Takaki et al., 1993). The IL-5Rα and βc subunits each have the conserved motifs of the hematopoietic receptor superfamily (Bazan, 1990a, 1990b).

The cytoplasmic domains of both IL-5Rα and βc lack regions of sequence homology with known signaling molecules (Kitamura et al., 1991a; Senda et al., 1992). Nevertheless, receptor-associated factors such as tyrosine kinases directly mediate IL-5 signal transduction upon receptor engagement by phosphorylating downstream secondary messengers (Sakamaki et al., 1992). Several domains on each subunit have been linked to discrete cytoplasmic signaling cascades that result in gene activation (Murata et al., 1990; Sato, N. et al., 1993; Sato, S., et al., 1994; Takaki et al., 1994).

In humans, the biologic effects of IL-5 are best characterized for eosinophils. In addition to inducing terminal maturation of eosinophils (Clutterbuck et al., 1989), IL-5 prolongs eosinophil survival by delaying apoptotic death (Yamaguchi et al., 1988, 1991), possesses eosinophil chemotactic activity (Sehmi et al., 1992), increases eosinophil adhesion to endothelial cells (Walsh et al., 1990, 1991) and enhances eosinophil effector function (Lopez et al., 1988; Yamaguchi et al., 1988). Physiologically, IL-5 and its associated eosinophil activation serve a protective role against helminth infections and facilitate the immunologic destruction of tumors (Urban et al., 1991; Wu et al., 1992; Nakashima et al., 1993). An increase in serum and tissue levels of IL-5 and eosinophil numbers have also been described in several pathophysiological conditions. Of clinical relevance is a role for IL-5 in atopic diseases (Bradding et al., 1993; Ying et al., 1993; Hamid et al., 1994), eosinophilic bronchopulmonary inflammation of asthma (Bousquet et al., 1990; Hamid et al., 1991; Kay et al., 1993), hypereosinophilic syndromes (Owen et al., 1989, 1990), and solid organ allograft rejection (Martinez et al., 1992, 1993a, 1993b). In addition to its effects on eosinophils, IL-5 enhances the differentiation of cytotoxic T lymphocytes (CTL; Takatsu et al., 1987; Nagasawa et al., 1991) and activated B cells (Bertolini et al., 1993; Morikawa et al., 1993), and enhances the activation state of lymphokine-activated killer cells (Aoki et al., 1989) and basophils (Bischoff et al., 1990; Dent et al., 1990). In mice, an IL-5 sensitive stromal cell-dependent pathway has been defined for the development of a subpopulation of B cells (Tominaga et al., 1989; Takatsu, 1995). In aggregate, these diverse biological consequences of IL-5 provide the impetus for elucidating the functional structure of both IL-5 and its receptor complex, as well as the mechanisms of IL-5 signal transduction.

Herein, the current understanding of the functional structure of IL-5, its receptor complex, and signal transducing pathways will be reviewed. Emerging concepts in understanding the role of IL-5 in health and disease will be presented through the integration of results from animal studies and analyses of clinical disorders. The paradigms developed should enable a rational approach for developing modulators of IL-5-mediated immune responses.

II. STRUCTURE OF IL-5

The gene for IL-5 has been mapped and many of its upstream regulatory elements have been identified. Transcriptional regulation of IL-5 expression has also been elucidated. The crystal structure of recombinantly expressed IL-5 has revealed that IL-5 is a short chain helical bundle cytokine. The physical properties of human IL-5 are shown in Table 1.

A. Gene Map and Transcriptional Regulation of IL-5

A single four exon gene found on human chromosome 5 codes for IL-5 (Figure 1) and is clustered near the genes for IL-3, IL-4, and GM-CSF, which all share a common exon-intron structure (Miyatake et al., 1985; Tanabe et al., 1987; Arai et al., 1989; Le Beau et al., 1989; Rozwarski et al., 1994). Mature human IL-5 mRNA encodes a 134 amino acid protein (Figure 1), which includes a 19 amino acid leader sequence and the 115 amino acids of mature IL-5 (Tanabe et al., 1987). The murine IL-5 gene is on chromosome 11 and also lies near the genes for IL-3,

Table 1. Physical Properties of Human IL-5

Gene locus	Chromosome 5q 23.3-31.1
Gene structure	4 exons
Primary structure	19 amino acid leader sequence 115 amino acid mature protein 2 N-glycosylation sites
Secondary structure	4 alpha helices 2 beta strands
Tertiary structure	Interdigitating homodimer 2 helical bundle motifs 2-fold axis of symmetry
Molecular weight	30 kD nonreduced 15 kD reduced

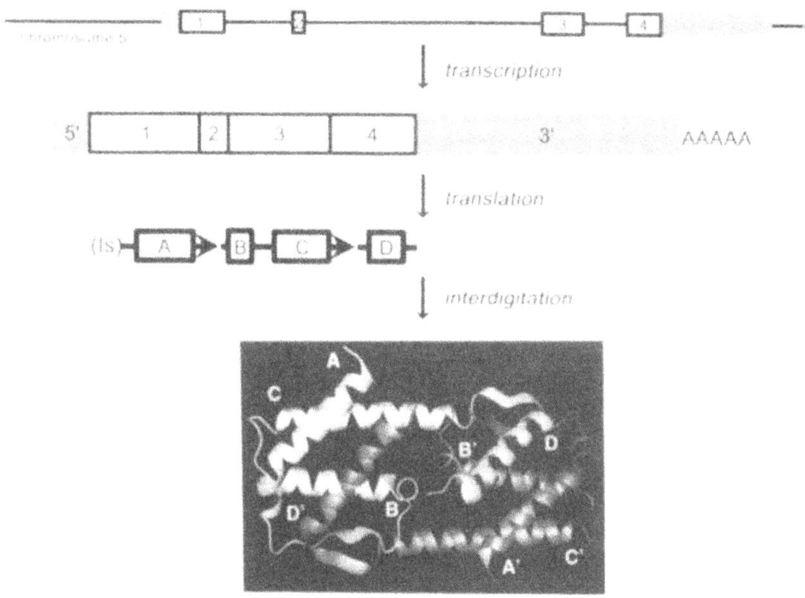

Figure 1. Expression of IL-5. The IL-5 gene is located on human chromosome 5 and consists of four exons (1-4). Spliced IL-5 mRNA contains, in addition to its coding sequence, 5' and 3' untranslated tails. The block diagram below the spliced transcript represents the secondary structures of IL-5 aligned with the appropriate mRNA coding regions. Rectangles represent helices (A-D), arrowheads represent beta-strands, lines represent random loop structures, and the (ls) stands for the leader sequence. Two IL-5 monomers then interdigitate to form native IL-5, which is illustrated by the IL-5 ribbon structure.

IL-4, and GM-CSF (Wilson et al., 1990). The murine gene encodes a 113 amino acid mature protein that is greater than 70% homologous to human IL-5 (McKenzie et al., 1991a; Figure 2).

The initial genetic map of human IL-5 revealed several areas in the 5' untranslated sequence that have homology with similar regions of the genes for IL-3, IL-4, and GM-CSF (Tanabe et al., 1987). Since these cytokines also have an exon-intron structure similar to IL-5, an evolutionary relationship is suggested (Rozwarski et al., 1994). Just upstream of the coding sequence is a TATA-like and a CAAT-like transcriptional initiation site (Tanabe et al., 1987). Other positive and negative regulatory elements identified upstream of the IL-5 transcription initiation site include an octomer binding site, a GATA binding site, a CLEO binding

```
      ______________________10____________________________20
IleProThrGluIleProThrSerAlaLeuValLysGluThrLeuAlaLeuLeuSerThr
 *  * Met         Met   ThrVal               ThrGln      Ala

____________________________30___............             40
HisArgThrLeuLeuIleAlaAsnGluThrLeuArgIleProValProValHisLysAsn
      Ala      ThrSer         Met   Leu         Thr

            ________________50________________________    60
HisGlnLeuCysThrGluGluIlePheGlnGlyIleGlyThrLeuGluSerGlnThrVal
            IleGly               LeuAspIle   LysAsn

            ________________70____________________________80
GlnGlyGlyThrValGluArgLeuPheLysAsnLeuSerLeuIleLysLysTyrIleAsp
Arg               Met      Gln

____________            ....90......_____________________100
GlyGlnLysLysLysCysGlyGluGluArgArgArgValAsnGlnPheLeuAspTyrLeu
Arg      Glu                        ThrArg

________________________   110
GlnGluPheLeuGlyValMetAsnThrGluTrpIleIleGluSer
                     Ser         AlaMet   Gly
```

Figure 2. Alignment of the primary sequences for human and murine IL-5. The upper sequence is human and the lower is murine (McKenzie et al., 1991a). Differences in the murine IL-5 sequence are indicated below the mature human IL-5 sequence of 115 amino acids. Murine IL-5 is only 113 amino acids and lacks the first two residues of the human sequence (*). The four helices of IL-5 are indicated by solid lines and its beta-strands by dotted lines.

site, a CTF binding site, and two GRE binding sites (Le Beau et al., 1989; Wang et al., 1993; Bourke et al., 1995; Gruart-Gouilleux et al., 1995; Yamagata et al., 1995).

Expression of IL-5 is primarily regulated at the level of transcription (Naora and Young, 1995). This regulation is dependent on *de novo* protein synthesis since cyclohexamide, a potent inhibitor of protein translation, blocks transcription of IL-5 mRNA (Van Straaten et al., 1994; Naora and Young, 1995). The induction of IL-5 transcription by phorbol myristate acetate (PMA) indicates a regulatory role for protein kinase C (Van Straaten et al., 1994). This response can be further enhanced with the calcium ionophore, A23187, or the cAMP analogue, N^6,O^2-dibutryl cAMP (Bt_2cAMP), demonstrating that IL-5 transcriptional regulation is also dependent on calcium influx and activation of protein kinase A, respectively (Tominaga et al., 1988; Miyatake et al., 1991; Lee et al., 1993). Furthermore, the enhancement of IL-5 promoter activity induced by protein kinase C and calcium influx requires the nuclear binding protein GATA-4 (Yamagata et al., 1995). Corticosteroids suppress IL-5 transcription indicating that the GRE binding sites serve as negative regulatory elements (Bentley et al., 1993; Wang et al., 1993; Okayama et al., 1994).

B. Protein Structure of IL-5

Solution of the IL-5 crystal structure (Figure 1) derived from both prokaryote and deglycosylated eukaryote expressed protein revealed that IL-5 is a member of the helical bundle family of proteins (Milburn et al., 1993; Johanson et al., 1995). The structure of each family member is based on a helical bundle motif composed of four helices (A-D) arranged in an up-up-down-down configuration. This conserved topology is attained in most family members by unimolecular folding and requires that helices A and B and helices C and D be connected by two long overhand loops, loops 1 and 3, while helices B and C be connected by a short turn, loop 2 (Milburn et al., 1993; Rozwarski et al., 1994). Additional structural features have been used to define subfamilies of these helical bundle cytokines (Rozwarski et al., 1994). IL-5 is in the short chain subfamily (Rozwarski et al., 1994), whose other crystal confirmed members are IL-2, IL-4, GM-CSF, and M-CSF (Diederichs et al., 1991; Bazan and McKay, 1992; Pandit et al., 1992; Walter et al., 1992). These helical bundle proteins are distinguished by helices of approximately 15 amino acids in length, the existence of two interfacing anti-parallel β-strands (β1 and β2 within loops 1 and 3, respectively), and a loop 1 that passes behind helix D (Rozwarski et al., 1994; Figure 3).

IL-5 is unique within this subfamily in that it cannot fold unimolecularly into a single canonical helical bundle (Figure 3). Rather, IL-5 exists as a homodimer formed by the interdigitation of two identical monomers that complement each other by contributing a D helix to the others A-C helices, resulting in a pair of helical bundles that are related by a two-fold axis of symmetry (Rozwarski et al., 1994). Thus, a given subunit is a unique analogue to the overhand loop 3 that is necessary for completion of the other subunit's helical bundle motif. This homodimeric configuration is anchored by a pair of identical intermolecular disulfide bonds between Cys44 and Cys86. Reduction and alkylation of these cysteines, as well as mutation of these cysteines to serines, demonstrated that a

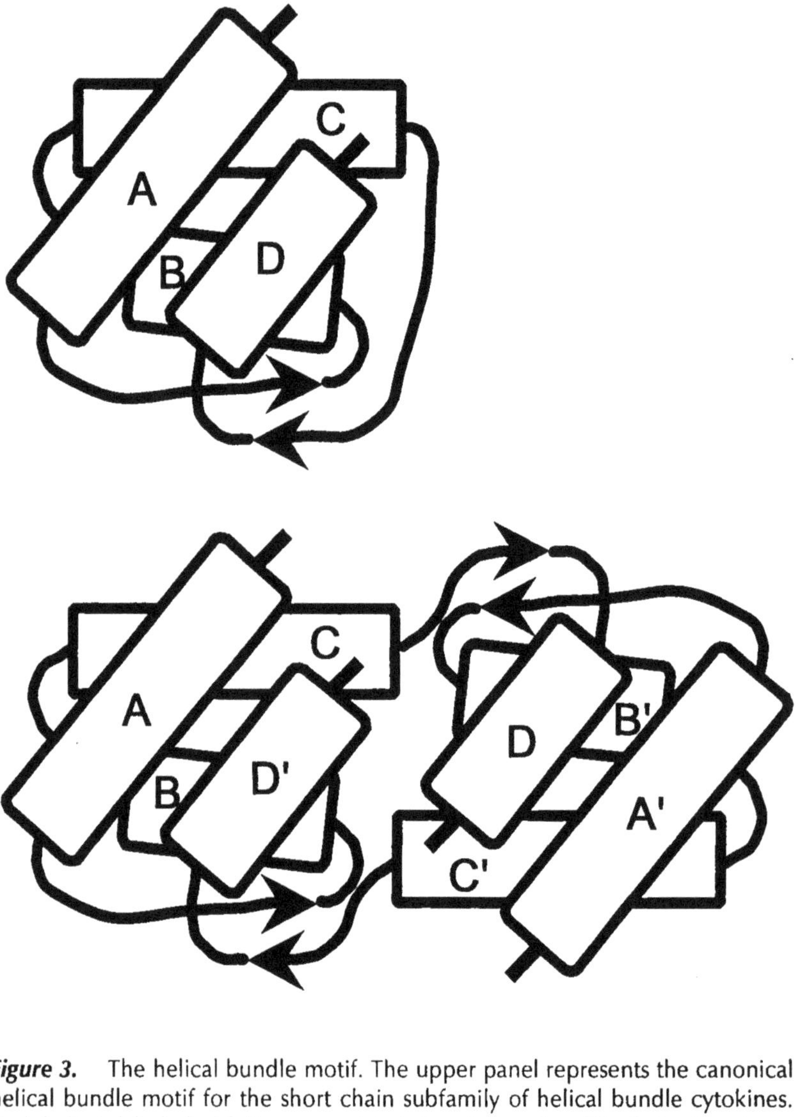

Figure 3. The helical bundle motif. The upper panel represents the canonical helical bundle motif for the short chain subfamily of helical bundle cytokines. Each helix is labeled with its appropriate letter designation. The beta-strands are indicated by arrowheads. Lines represent random loop sequence. The lower panel is the homodimeric structure of IL-5. Note that each helix of the second subunit is designated by a letter prime. The two helical bundles of IL-5 are related by a twofold axis of symmetry that is perpendicular to the page between the two helical bundles.

native IL-5 monomer lacks all function (Takatsu et al., 1988; Tsuruoka et al., 1990; McKenzie et al., 1991b). Although IL-5 is post-translationally modified with both N-linked and O-linked carbohydrates, neither is necessary for function (Tominaga et al., 1990; Kodama et al., 1993). However, these modifications may enhance the half-life of IL-5 *in vivo* and impart added thermostability (Tominaga et al., 1990; Kodama et al., 1993).

III. STRUCTURE OF THE IL-5 RECEPTOR COMPLEX

The heterodimeric IL-5 receptor complex is comprised of an IL-5 specific alpha subunit (IL-5Rα) and a common beta subunit (βc) shared by IL-3, IL-5, and GM-CSF (Miyajima et al., 1993). Both subunits have been cloned and expressed. Identification of conserved features in the extracellular domains of both subunits reveals that they belong to the superfamily of cytokine receptors (Miyajima, 1992).

A. Structure of IL-5Rα

The gene for IL-5Rα maps to the long arm of human chromosome 3 (Isobe et al., 1992). Although the human gene map is not complete, the murine gene is known to consist of 11 exons (Imamura et al., 1994). The 5' regulatory elements of the murine gene include binding sites for AP-1, Sp1, NF-kB, NF-IL-6, and GATA-1 (Imamura et al., 1994). The human gene codes for a 420 amino acid protein (Figure 4) consisting of a 20 amino acid leader sequence, a 324 amino acid extracellular domain, a 20 amino acid transmembrane domain, and a 55 amino acid cytoplasmic domain (Tavernier et al., 1991; Murata et al., 1992). In addition, a soluble form of IL-5Rα (sIL-5Rα) is expressed (Figure 4). The human sIL-5Rα is derived from alternatively spliced mRNA which excludes the last three exons that code for the transmembrane and cytoplasmic domains and either includes a special mini-exon or simply terminates after the exon which encodes the membrane proximal extracellular domain (Tavernier et al., 1992; Imamura et al., 1994).

IL-5Rα belongs to the superfamily of cytokine receptors (Miyajima et al., 1993; Figure 4). The extracellular portion of IL-5Rα is composed of three fibronectin-type-III-like modules, each approximately 100 amino acids in length, which are connected by a proline hinge (Bazan, 1990b; Cornelis et al., 1995). Each module is predicted to contain seven

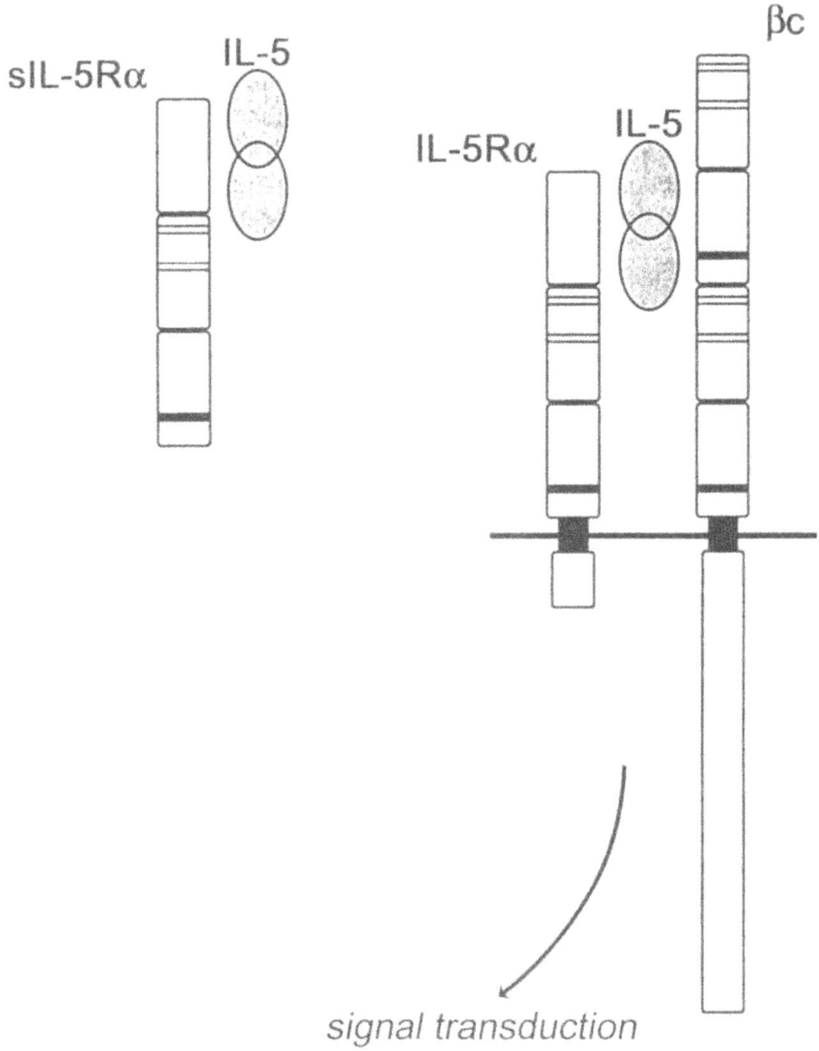

Figure 4. Structure of the IL-5 receptor. To the right is the high affinity signal transducing heterodimeric IL-5 receptor that is composed of IL-5Rα and βc. Each fibronectin-type-III-like module is represented in the extracellular domain by a rectangle. The conserved cysteines are represented by thin horizontal black lines, and the WSXWS motifs are represented by thick horizontal gray lines. The narrower solid rectangles and open rectangles represent the transmembrane and cytoplasmic domains, respectively. To the left is the soluble form of the alpha subunit, sIL-5Rα, which is capable of binding IL-5 in solution, but does not interact with surface bound βc.

antiparallel β-strands arranged in two sheets (Bazan, 1990b). The two membrane proximal modules contain the conserved superfamily structures, while the N-terminal module is an added nonconserved feature of IL-5Rα (Cornelis et al., 1995). Four conserved cysteines in the middle domain (module 2) form two intramolecular disulfide bonds (Bazan, 1990b). In addition, the membrane proximal domain (module 3) contains the conserved tryptophane-serine-X-tryptophane-serine or WSXWS (where X is any amino acid) motif (Bazan, 1990b).

Alone, human IL-5Rα can bind either human or murine IL-5 with a dissociation constant of approximately 1 nM (Morton et al., 1994; Johanson et al., 1995). This is in contrast to murine IL-5Rα which binds murine IL-5 with much greater affinity than it binds human IL-5, despite the 70% homology for IL-5 between these two species (Tsuruoka et al., 1993; Cornelis et al., 1995). Although deglycosylated IL-5 is functional, only glycosylated IL-5Rα is capable of binding IL-5 (Murata et al., 1992; Johanson et al., 1995). Alone IL-5Rα is not capable of transducing a signal (Takaki et al., 1993).

The sIL-5Rα binds IL-5 with an affinity similar to membrane bound IL-5Rα suggesting that the transmembrane and cytoplasmic domains of IL-5Rα contribute little energetically to IL-5 binding (Tavernier et al., 1991; Murata et al., 1992; Johanson et al., 1995). The biological role of sIL-5Rα is unknown but is hypothesized to act as an IL-5 carrier *in vivo* and thereby prolong the cytokine's half-life. Conversely, the ability of sIL-5Rα to neutralize IL-5 (IC_{50} = 12 nM) suggests a role in down-regulating IL-5 mediated immune responses (Johanson et al., 1995).

The dimeric configuration of IL-5 suggests the possibility of receptor bivalency. However, several complementary techniques, including analytical ultracentrifugation, titration microcalorimetry, analysis of IL-5:sIL-5Rα co-crystals, gel filtration, and plasmon resonance have exhaustively demonstrated that the native IL-5 dimer binds IL-5Rα in a 1:1 stoichiometry (Devos et al., 1993; Johanson et al., 1995). Based on titration microcalorimetry and plasmon resonance results, it is also speculated that a possible conformation change occurs upon IL-5 engagement of IL-5Rα although the mechanism is entirely unknown (Johanson et al., 1995).

B. Structure of βc

The βc subunit of the IL-5 receptor complex is shared by IL-3, IL-5, and GM-CSF (Takaki et al., 1991; Tavernier et al., 1991). The gene for βc is on human chromosome 22, but details of its structure are limited. The murine gene is on chromosome 15 and contains 14 exons spanning 28 kilobases (Gorman et al., 1992). In the 5' flanking sequence, potential recognition sequences for the hematopoietic transcription factors GATA-1 and PU.1 are present, in addition to a TATA-like transcriptional initiation sequence (Gorman et al., 1992). Expression of βc is induced with LPS or IL-1 stimulation (Mita et al., 1988, 1989; Kitamura et al., 1991b; Watanabe et al., 1992). The translated protein consists of a 16 amino acid leader sequence, a 422 amino acid extracellular domain, a 28 amino acid transmembrane domain, and a 432 amino acid cytoplasmic domain (Figure 4). The extracellular domains consist of a double repeat of the classic two module superfamily domain (Gorman et al., 1992). Each domain includes two fibronectin-type-III-like modules with the more N-terminal containing the four conserved cysteines and the more C-terminal including the WSXWS motif (Figure 4).

The βc subunit does not bind IL-5 alone, but in the human system βc associates with IL-5Rα to enhance the affinity for IL-5 by a few fold, resulting in a high affinity dissociation constant on the order of 300 pM (Banchereau and Galizzi, 1990; Plaetinck et al., 1990; Migita et al., 1991; Tavernier et al., 1991, 1992). This enhancement of affinity is accounted for by a reduction in the off rate of IL-5, as the on rate is unchanged from that of IL-5Rα alone (Mita et al., 1993). The βc subunit also enhances the efficiency of IL-5 internalization which explains the reduction in surface expression of high affinity IL-5R following IL-5 stimulation (Mita et al., 1989, 1993). The βc subunit is absolutely necessary for signal transduction (Takaki et al., 1993). There is no significant association between surface βc and sIL-5Rα (Kikuchi et al., 1994), which explains the ability of sIL-5Rα to neutralize IL-5 activity. This is in contrast to the related IL-6 receptor system in which soluble IL-6R is capable of binding to gp130 and IL-6 to transduce a signal (Taga et al., 1989; Yasukawa et al., 1990). The binding stoichiometry between IL-5 and βc in the high affinity complex also is suggested by indirect cross-linking studies to be 1:1 like that for IL-5Rα (Devos et al., 1991; Tavernier et al., 1992).

IV. INTERACTION OF IL-5 WITH ITS RECEPTOR COMPLEX

A variety of approaches are being utilized to target IL-5 functional domains and individual residues important for receptor engagement. As a complement to these studies, the functional domains of IL-5 receptor subunits also are being defined. This understanding of IL-5 and IL-5R functional structure is providing the foundation for the development of IL-5 agonists and antagonists.

A. Functional Domains of IL-5

Early experiments established that IL-5 must exist as a homodimer covalently anchored by a pair of intermolecular disulfide bonds in order to be functional (Milburn et al., 1993; Johanson et al., 1995). Solution of the IL-5 crystal structure further established that the monomers interdigitate to form a pair of helical bundles and that the completion of these helical bundle motifs could only be satisfied by this unique configuration (Takatsu et al., 1988; Tsuruoka et al., 1990; McKenzie et al., 1991b). Though the three-dimensional structure of IL-5 is known, its relationship to function remains incomplete. Initially, the C-terminal region of IL-5 was found to be important for bioactivity, since either digestion with cyanogen bromide C-terminal to Met107 or oxidation of Met107 by chloramine T resulted in a loss of function (Kodama et al., 1991). Analysis of human-murine chimeras later targeted the C-terminal 36 amino acids of human IL-5 as the region responsible for human IL-5 species specific behavior (McKenzie et al., 1991a). As the species specific activity of human IL-5 is attributed to a difference in IL-5Rα binding, these studies began to define a broad putative IL-5Rα binding domain. Likewise, a chimera of IL-5 and GM-CSF, in which helix A of GM-CSF was replaced with helix A of IL-5, revealed that this N-terminal helix was involved in βc engagement (Shanafelt et al., 1991).

The understanding of IL-5 functional structure was next advanced by the generation of a panel of neutralizing anti-IL-5 monoclonal antibodies (Schumacher et al., 1988; Dickason et al., 1994) whose epitopes were mapped utilizing a library of human-murine IL-5 chimeras (Dickason et al., 1996a). Two clusters of neutralizing epitopes were localized. One cluster included loop 2 and a proximal portion of helix A, while the other cluster localized to the N-terminal end of helix A, the two beta strands of loops 1 and 3, the C-terminal end of helix C, and the N-terminal end of helix D (Dickason et al., 1996a). The human-murine IL-5 chimeras were also utilized to better define the region of human IL-5 responsible for its species specific activity. The species specific residues were limited to Lys84 at the C-terminal end of helix C and Val93 and Asn94 of the N-terminal end of helix D (Dickason et al., 1996a). Site-directed approaches which replaced each putative human residue with that of the murine analogue confirmed the involvement of Lys84 in determining species specificity, and added Asn108, though this latter residue is controversial (Cornelis et al., 1995; Dickason et al., 1996a; Tavernier et al., 1995).

Recently, alanine replacement mutants have been utilized to dissect regions of IL-5 which interact with IL-5Rα and βc. Two independent reports found Glu13 to be very important in binding to βc (Graber et al., 1995; Tavernier et al., 1995). This residue is analogous to Glu21 of GM-CSF and Glu22 of IL-3 which are also important in engaging βc (Shanafelt et al., 1991). Residues implicated in IL-5Rα engagement clustered on both beta strands, the N-terminal end of helix B, the C-terminal end of helix C, and the N-terminal end of helix D (Graber et al., 1995; Tavernier et al., 1995). Together these results associate IL-5Rα binding to a region near the junction of loop 3 with helix D and the N-terminal end of helix D, while βc binding localizes to a portion of helix A centered around Glu13 (Figure 5).

Although additional mutational analysis will eventually assign all residues essential for the engagement of IL-5 and its receptor complex, a central question remains. Why is IL-5 a homodimer? The pair of helical

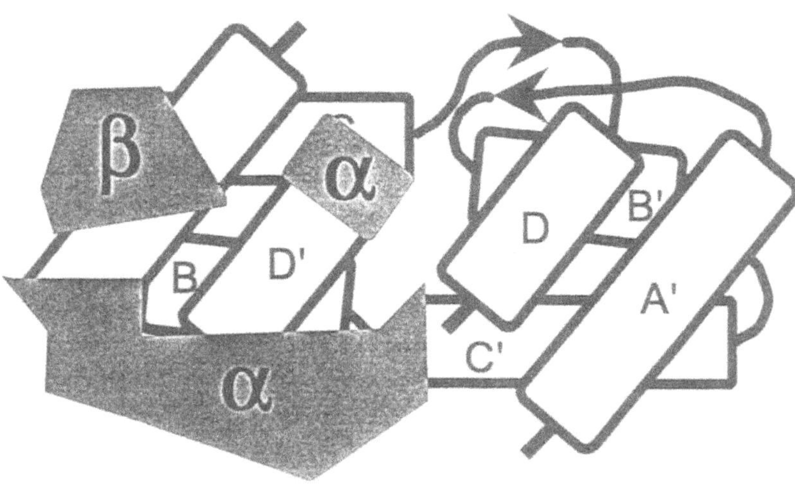

Figure 5. IL-5 functional domains. Regions of IL-5 involved in receptor engagement are shaded in gray over the IL-5 structure. Each region predicted to be involved in IL-5Rα binding is represented by an α, and the region predicted to be involved in βc engagement is represented by a β. Note that each domain is represented twice on the IL-5 homodimer, but for simplicity only one helical bundle motif has been labeled.

bundles on IL-5 suggests that IL-5 could crosslink IL-5R by virtue of its bivalency. However, comprehensive investigation has established that a single IL-5 homodimer binds only one IL-5Rα, while less direct analysis argues that the same is true for βc. By examining the crystal structure, and in particular the region involved in binding to IL-5Rα, a steric argument against the capacity of two IL-5Rα subunits to simultaneously bind IL-5 also can be made. The twofold symmetry of IL-5 places these two domains, which are centered on helix D and the junction of loop 3 and helix D, adjacent to each other. This arrangement suggested that two IL-5Rα simply cannot fit and raised the possibility that IL-5 binds to IL-5Rα via a single broad domain that spans both helical bundle motifs (Morton et al., 1995; Li et al., 1996). However, a novel IL-5 monomer, designated mono5, was recently engineered by insertion of eight codons in the loop 3 of the native IL-5 gene (Dickason and Huston, 1996). Structural analysis of mono5 established that no intermolecular association occurred and functional analysis demonstrated full biological activity compared to native IL-5 at saturating concentrations. Furthermore, signal transduction pathways utilized by mono5 and IL-5 were similar as determined by Western Blot analysis of early tyrosine phosphorylation events, JAK2 activation, and mitigen-activated protein kinase activation (Dickason et al., forthcoming). Although the formation of interchain disulfide bonds between residues 44 and 86 in the native IL-5 monomers is necessary for biological activity, mono5 is functional in the absence of a disulfide bond. Therefore, the disulfide bonds formed in native IL-5 likely serve to stabilize a conformational motif which mono5 can assume in the absence of a disulfide bond.

Thus, all structural features necessary for IL-5 biological activity are contained within a single helical bundle motif (Dickason and Huston, 1996). The functional requirement for two native IL-5 monomers to interdigitate appears necessary to overcome a folding constraint imparted by loop 3 which is approximately eight amino acids shorter than loop 3 in the other short chain helical bundle cytokines which are able to fold unimolecularly to form a single helical bundle motif. The loop 3 hypothesis which led to the creation of mono5 potentially provides insight globally to all helical bundle cytokines. The structural hypothesis, which states that the length of a helical bundle cytokine's loop 3 dictates whether this protein is actively expressed as a monomer or an interdigitating homodimers, may apply to both the conversion of native interdigitating homodimers into monomers (as in the case with IL-5 and mono5) by lengthening loop 3, and the conversion of native monomers into interdigitating homodimers by shortening loop 3 (Dickason et al., forthcoming). In addition to providing the structural framework for designing novel IL-5 agonists and antagonists, the knowledge gained from the development of mono5 3 through loop 3 hypothesis should enable other helical bundle proteins to be redesigned with therapeutic potential.

B. Functional Domains of IL-5Rα

To complement the investigation of IL-5 functional domains, the functional structure of IL-5Rα has also been examined. Utilizing recombinant sources of sIL-5Rα several solid-phase assays have been established to help identify key residues involved in binding IL-5 (Devos et al., 1993; Bennett et al., 1995; Johanson et al., 1995). The N-terminal module of IL-5Rα was first targeted as a region involved in IL-5 binding by covalent modification using isothiazolones (Devos et al., 1994). Derivatization of Cys66 on its free sulfhydryl group disrupted IL-5 binding (Figure 6). However, it should be noted that a cysteine to serine mutant at this position, which was resistant to isothiazolones treatment, still bound IL-5 (Devos et al., 1994). Therefore, it is unlikely that Cys66 is directly involved in IL-5 binding, but perhaps lies near other key residues. By assessing the affinity of IL-5 to a panel of human-murine IL-5Rα chimeras another region within this N-terminal module was identified (Cornelis et al., 1995). Alanine scanning mutagenesis was then able to identify Asp55, Asp56, Tyr57, and Glu58 within this region as crucial for binding IL-5 (Cornelis et al., 1995). These residues are predicted to lie in the loop that connects β-strands C and D on the proposed tertiary structure of IL-5Rα (Figure 6). This cluster of negatively charged amino acids may interact electrostatically with the cluster of positively charged amino acids in the β-strand region of IL-5 that has been identified as important for IL-5Rα binding. Key residues within the second module of IL-5Rα have not yet been identified, however, others in the third module have. By aligning the amino acid sequence of IL-5Rα with the known structure of the related growth hormone receptor (Devos et al., 1992), five putative binding determinants were targeted for site-directed mutagenesis, which found Arg188 of the third module to be important for binding IL-5 (Cornelis et al., 1995). Additionally, serendipitous mutations of IL-5Ra revealed that Cys249 and Cys296 of the third module (Figure 6) are also important residues for binding to il-5 (Devos et al., 1994).

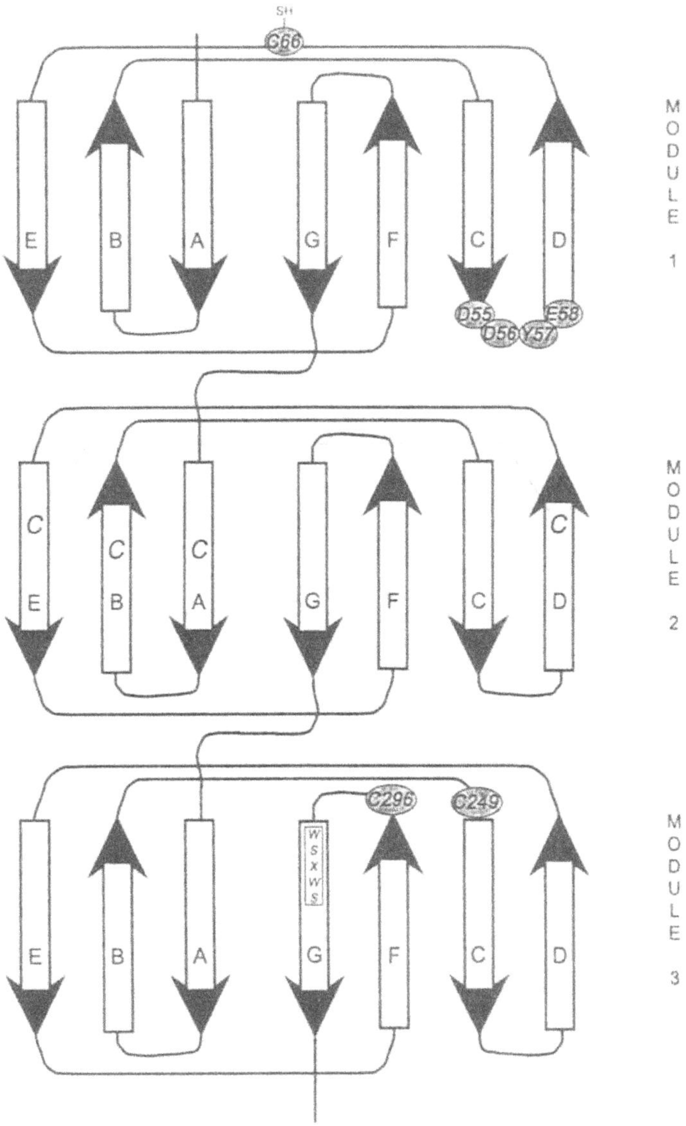

Figure 6. IL-5Rα functional residues. Each of the three fibronectin-type-III-like modules is represented by seven arrows (A-G). Each of the two conserved disulfide bonds in module 2 is represented by a dashed line. The WSXWS motif is boxed in module 3. Residues that have been associated with function are represented by labeled gray ovals. The free sulfhydryl group of Cys66 that is sensitive to isothiazolone derivatization is represented by **SH**.

C. Functional Domains of βc

The inability of βc to bind IL-5 alone and the lack of a soluble analogue have hindered investigation of βc functional structure. Only limited mutagenesis studies have been conducted. Again, the known structure of the growth hormone receptor (Devos et al., 1992) was utilized to target an analogous functional domain on βc. This alignment highlighted the region between residues 358 and 373. Alanine substitutions revealed that within this region Tyr365, His367, and Ile368 are important in binding IL-5 (Woodcock et al., 1994). These residues are also important for binding to GM-CSF, but not to IL-3 (Woodcock et al., 1994). These data provide the first demonstration that high affinity binding of IL-5 can be dissociated from that of the other cytokines which share the βc subunit and suggest a mechanism of differential signaling by related cytokines.

V. SIGNAL TRANSDUCTION THROUGH THE IL-5 RECEPTOR COMPLEX

The molecular mechanism by which IL-5 binding to its receptor elicits diverse responses by subsets of leukocytes remains an enigma. Accumulating evidence suggests that phosphatidyl inositol turnover, calcium mobilization, and activation of protein kinase A or C are not essential for IL-5-induced differentiation signals (Yamaguchi et al., 1989). Although tyrosine phosphorylation is usually one of the earliest biochemical events in the signal transduction cascade induced by growth factors whose receptors possess endogenous enzymatic activity (Ullrich and Schlessinger, 1990), the cytoplasmic domains of IL-5Rα and βc, like other cytokine receptors, have no homology with signaling molecules such as kinases, phosphatases, nucleotide binding proteins, and src homology domains. Nevertheless, IL-5 does induce rapid tyrosine phosphorylation of cellular proteins (Morla et al., 1988; Murata et al., 1990) and enhances the transcription of nuclear proto-oncogenes (Conscience et al., 1986).

A. Tyrosine Phosphorylation of Cellular Proteins

Tyrosine phosphorylation of several cellular proteins, including Tyr750 of βc, occurs within minutes of exposure to IL-5 (Murata et al.,

1990; Sakamaki et al., 1992; Sato et al., 1994; Takaki et al., 1994). The suppression of IL-5 induced proliferation of a cloned early B cell line with herbimycin A, which inhibits tyrosine kinases (Sato et al., 1994), indicates that activation of tyrosine kinases and tyrosine phosphorylation of cellular proteins are involved in IL-5 signal transduction. Several tyrosine phosphorylated effector molecules controlling growth have been identified on growth factor receptors carrying intrinsic protein tyrosine kinase activity. These effector molecules include phospholipase C-γ (PLC-γ), phosphatidylinositol-3 kinase (PI-3K), p95vav (Vav), p75HS1 (HS1), Shc, and Ras-associated GTPase-activating protein (GAP; Koch et al., 1991); all contain Src-homology 2 (SH2) or Src-homology 3 (SH3) domains, the former of which binds to specific phosphorylated tyrosines (Matsuda et al., 1991) while the latter binds to proline-rich domains (Bar-Sagi et al., 1993). To evaluate the IL-5 induced tyrosine phosphorylation profile, tyrosine phosphorylation of known signaling molecules including PLC-γ, GAP, PI-3K, Shc, p95Vav, and p75HS1 has been investigated. IL-5 induced tyrosine phosphorylation of Shc, Vav, PI-3K, and HSI (Sato et al., 1994). However, no tyrosine phosphorylation of any of these proteins was observed in transfectants expressing an IL-5Rα with a deleted cytoplasmic domain (Sato et al., 1994; Takaki et al., 1994). This indicates that the cytoplasmic domain of IL-5Rα is essential for tyrosine phosphorylation. Deletional mutants of βc revealed that a region between Leu626 and Ser763 is also important for tyrosine phosphorylation of Shc and PI-3K (Figure 7), while the region between Val578 and Leu626 is associated with the tyrosine phosphorylation of other cellular proteins.

B. Activation of JAK Kinases and STAT Proteins

There is a body of evidence that induction of protein tyrosine phosphorylation is critical to cytokine function (Shuai et al., 1993; Kishimoto et al., 1994). The fact that tyrosine phosphorylation of cellular proteins, including βc, is rapidly induced after IL-5 stimulation suggests that protein tyrosine kinases associate directly with IL-5R. A new family of cytoplasmic protein tyrosine kinases called Janus kinases (alternatively referred to as "just another kinase family") has been described and includes JAK1, JAK2, JAK3, and TYK2 (Velazquez et al., 1992; Argetsinger et al., 1993; Shuai et al., 1993; Silvennoinen et al., 1993; Witthuhn et al., 1993, 1994; Johnston et al., 1994; Lutticken et al., 1994; Stahl et al., 1994). These proteins contain a kinase domain and a

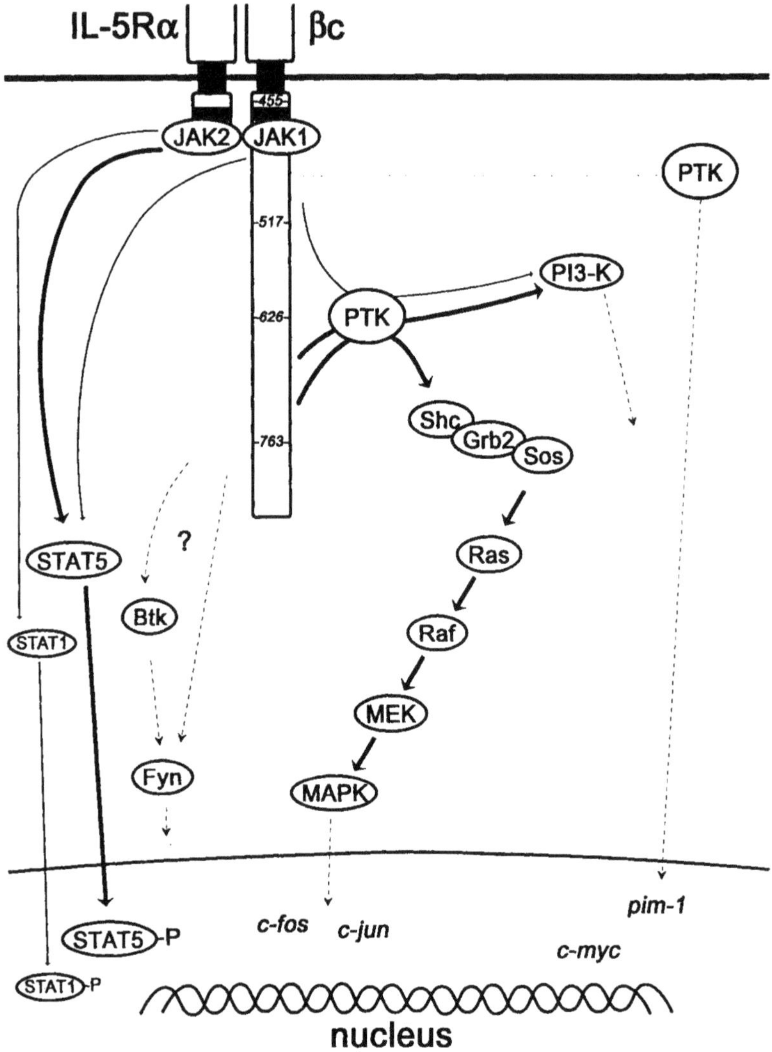

Figure 7. Signal transduction through IL-5R. The cytoplasmic domains of IL-5Rα and βc are drawn below the cell membrane (thick horizontal black line). Both box 1 regions of IL-5Rα and βc are represented proximal to the membrane. Residues which demarcate functional domains on βc are also labeled. All known IL-5 signaling pathways are represented. Direct connections between signaling molecules are indicated by thick arrows. Weaker responses are represented by thin arrows. Unknown connections are indicated by dashed lines. The question mark (?) represents the uncertainty of the connection between Btk and Fyn. The cytoplasm is separated from the nucleus by the lower horizontal line.

kinase-like domain, but lack SH2, SH3, and membrane-spanning domains. Recent studies showed that JAK1, JAK2, JAK3, and TYK2 couple ligand binding to cytokine receptors with tyrosine phosphorylation of cellular proteins (Firmbach-Kraft et al., 1990; Silvennoinen et al., 1993; Kishimoto et al., 1994). JAK kinases associate with a proline-rich segment of a cytokine receptor's cytoplasmic domain called the box 1 region. Upon ligand engagement of the receptor, activated JAK kinases phosphorylate tyrosines on cytoplasmic proteins which function as signal transducers and activators of transcription (STAT). These phosphorylated STAT proteins translocate to the nucleus and bind to DNA response elements (Akira et al., 1994; Darnell, 1994; Kishimoto et al., 1994). IL-5 induced tyrosine phosphorylation of JAK1, JAK2, and JAK3 has been examined in IL-5R expressing Y16 cells (Sato et al., 1994; Kouro et al., 1996). JAK2 is rapidly tyrosine phosphorylated in response to IL-5 (Sato et al., 1994; Kouro et al., 1996). A significant but lesser extent of tyrosine phosphorylation of JAK1 is also observed in response to IL-5 (Figure 7). However, JAK3 is not tyrosine phosphorylated. The membrane proximal proline-rich sequence of the cytoplasmic domain of IL-5Rα, which is conserved among the alpha chains of the IL-3R, IL-5R, and GM-CSFR, is essential for tyrosine phosphorylation of JAK2 (Takaki et al., 1994). It is also clear that sequence just C-terminal to the proline-rich region is important, though not essential (Takaki et al., 1994). A role for the box 1 region of βc is also suggested (Takaki et al., 1994).

Treatment of human peripheral blood monocytes or basophils with IL-3, IL-5, IL-10, or GM-CSF activates DNA binding proteins which are tyrosine phosphorylated (Larner et al., 1993). An IL-3 and GM-CSF responsive STAT protein has been identified as STAT5 (Mui et al., 1992), which is one of the substrates for JAK2 kinase. IL-5 also induces STAT5 activity (Kouro et al., 1996), which is mediated by both JAK2 and to a lesser extent JAK1 (Figure 7). Recently, activation of JAK2 and STAT1 (Figure 7) has been shown in eosinophils in response to IL-5 (Pazdrak et al., 1995; Van der Brugger et al., 1995). These results suggest different JAK-STAT signaling pathways for B cells and eosinophils.

The activation of STAT5 in response to IL-3, IL-5, and GM-CSF suggests a common signaling pathway through IL-3R, IL-5R, and GM-CSFR, all of which share βc. As IL-3, GM-CSF, and IL-5 exert different functions in some cell types, more subtle modulation of STAT5 or other signaling molecules may result in a cytokine specific response. Although the specificity of IL-5 is mainly due to the restricted expression of IL-5Rα (Takagi et al., 1995), efforts are continuing to determine whether

IL-5Rα-specific cytoplasmic residues also contribute to IL-5 specific signaling pathways.

C. Activation of the Ras Pathway and Proto-oncogenes

One of the earliest events which occurs after IL-5 binds to its receptor is the rapid tyrosine phosphorylation of βc and several cellular substrates including Shc and Vav. The tyrosine phosphorylated Shc can initiate the Ras pathway (Figure 7) by binding to an adapter protein called Grb2 through its SH2 domain (Pelicci et al., 1992; Egan et al., 1993). Grb2 also binds through its SH3 domain to a proline-rich motif of Sos, a nucleotide exchange protein. Sos then activates Ras, which is a GTP-binding protein that works downstream of tyrosine kinases in the growth signaling pathway (Smith et al., 1986; Satoh et al., 1991; Duronio et al., 1992). Downstream of Ras is a serine/threonine kinase called Raf which activates the unique bifunctional kinase MEK that can phosphorylate both tyrosine and serine/threonine. The only known substrate for MEK is the serine/threonine kinase MAPK. Once activated, MAPK triggers nuclear transcription factors that lead to the induction of several gene products including proto-oncogenes. Activation of the Ras cascade is dependent on both the cytoplasmic region of IL-5Rα and the region between Leu626 and Ser763 of βc (Sato et al., 1993).

It is well established that induction of tyrosine phosphorylation and nuclear proto-oncogene expression is critical for cytokine signal transduction (Conscience et al., 1986). Expression of *c-fos, c-jun,* and *c-myc* proto-oncogene mRNA is rapidly induced upon stimulation with IL-5 (Takaki et al., 1994). This proto-oncogene expression is not induced in cells transfected with IL-5Rα mutants that lack either the proline-rich motif or a sequence just downstream of the proline-rich region in the cytoplasmic domain (Sato et al., 1994; Takaki et al., 1994). The region between Leu626 and Ser763 of βc (Figure 7) is also necessary for induction of *c-fos* and *c-jun* expression, while the region between Arg455 and Glu517 is responsible for induction of *c-myc* expression and *pim-1*, a proto-oncogene which encodes a serine/threonine kinase that cooperates with *c-myc* and plays a role in cell proliferation (Sato et al., 1993).

D. The Activation of Bruton and Fyn Tyrosine Kinases

Nonreceptor protein tyrosine kinases are key regulators of the development and function of lymphocytes. The structures of these kinases are characterized by the src homology domains which serve as protein-protein interaction sites. The Bruton tyrosine kinase (Btk) belongs to a new subfamily of nonreceptor tyrosine kinases, which includes TecI, Itk, and DSrc28C (Gregory et al., 1987; Mano et al., 1990; Siliciano et al., 1992). Btk has been cloned and is a B cell specific cytoplasmic tyrosine kinase. Btk is the defective gene in human X-linked agammaglobulinemia (XLA; Tsukada et al., 1993, 1994; Vetrie et al., 1993) and murine X-linked immunodeficiency (XID) in which a single conserved residue within the amino terminal unique region of Btk is mutated (Rawlings et al., 1993; Thomas et al., 1993). XLA patients have less than 1% of the normal levels of mature B cells in the peripheral blood and dramatically reduced immunoglobulin levels. In XID mice, B cell numbers are reduced to around 50% of normal and levels of IgM and IgG_3 are greatly decreased. B cells from XID mice are unresponsiveness to IL-10, CD38, and CD40 stimulation (Go et al., 1990; Hasbold and Klaus, 1994; Kikuchi et al., 1995). This suggests that Btk may be critical for multiple pathways important in B cell function.

The number of IL-5 responsive B cells and their capacity to respond to IL-5 is greatly reduced in XID mice (Hitoshi et al., 1993). However, XID mice show IL-5 induced eosinophilia in peripheral blood to a similar extent as normal mice. Eosinophils from IL-5 injected XID mice express both α and b chains of the IL-5R complex and respond normally to IL-5 with prolonged *in vitro* survival (Hitoshi et al., 1993). These results indicate that low responsiveness to IL-5 in XID mice is lineage restricted and that IL-5 mediated signaling cascades in eosinophils are somehow different from those in B cells.

The kinase activity of Btk in IL-5-stimulated Y16 cells was examined by an immune complex-kinase assay. Significantly enhanced Btk kinase activity was detected in anti-Btk immunoprecipitates following 5-20 minutes of IL-5 stimulation (Sato et al., 1994). No enhanced Btk activity was detected in IL-5 stimulated cells which lacked the cytoplasmic domain of IL-5Rα. Btk contains SH2 and SH3 domains, which are important for signal transduction in many nonreceptor tyrosine kinases. Recently, it has been reported that signaling through the IL-5R, which ordinarily provides a co-stimulus with other mitogenic factors, was completely blocked in fyn^{Tnull} B cells, while an IL-5 inducible increase in Btk kinase activity was observed from both wild type and fyn^{Tnull} animals (Appleby et al., 1995). Btk and Fyn interact with each other (Cheng et al., 1994), and if Btk and Fyn are both part of the same signaling cascade, these observations suggest that Btk is the more IL-5R proximal element (Figure 7).

Using a random mutagenesis scheme, a gain-of-function Btk mutant, Btk*, which drives growth of NIH3T3 cells in soft agar, has been isolated (Li et al., 1995). Btk* results from a single point mutation in the pleckstrin homology domain, where a glutamic acid is mutated to a lysine at position 41. Btk* shows an increase in phosphorylation of tyrosine residues and an increase in membrane targeting. Expression of Btk* also relieves IL-5 dependence of a lineage cell line. This transforming activity required kinase activity, a putative autophosphorylation site, and a functional pleckstrin homology domain. These results show that Btk may play a critical role in B cell development and activation by stimuli such as IL-5 and CD38.

E. Functional Significance of the IL-5Rα Proline-rich Domain

Deletion of the membrane proximal cytoplasmic domain of IL-5Rα containing the sequence Pro352-Pro353-Val354-Pro355 (ppvp) disrupted all IL-5 mediated signal transduction including tyrosine phosphorylation of cellular proteins, activation of JAK-STAT pathways, and induction of nuclear proto-oncogene expression (Takaki et al., 1994). Deletion of the adjacent C-terminal region of this proline-rich motif also had a profound effect on IL-5 responsiveness, suggesting an important but not essential role for this region. To better characterize the functional residues within the proline-rich domain, a panel of IL-5Rα alanine replacement mutants was generated (Takaki et al., 1994). JAK1 and JAK2 were phosphorylated following IL-5 stimulation of cells expressing the mIL-5Rα mutant pAvA, AAvp, or to a much lesser extent ApvA (where A represents an alanine substitution), while cells expressing the AAvA mutant were not. Furthermore, the pAvA, AAvp, and ApvA mutants significantly activated STAT5 (Kouro et al., 1996). Similar effects on proliferation and proto-oncogene induction were also demonstrated (Takaki et al., 1994). These results suggest that the existence of at least one of three proline residues in the Pro-Pro-x-Pro motif is indispensable for signal transduction and that Pro352 and Pro355 are more critical than Pro353 (Figure 8).

The proline-rich region (Pro352-Pro353-Val354-Pro355-Ala356-Pro357) in the cytoplasmic domain of mIL-5Rα appears to belong to the Pro1-X-Al-Pro2-X-Pro3 (where Al is any aliphatic residue) motif of the PRM1 family (O'Neal and Yu-Lee, 1993). Recently, another proline-rich motif, Pro-X-Pro, important for signaling of other cytokines has been elucidated (Murakami et al., 1991; O'Neal and Yu-Lee, 1993). Results

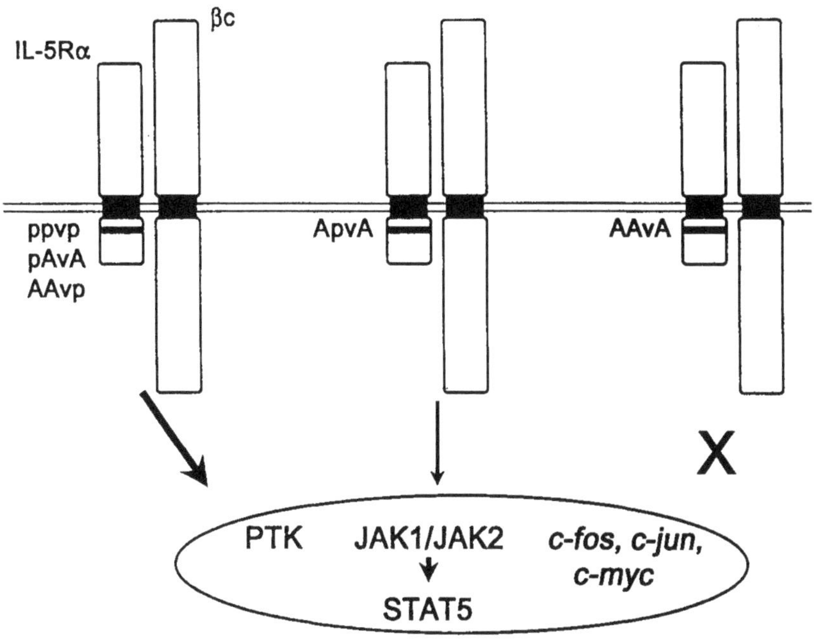

Figure 8. The role of the IL-5Rα proline-rich domain in signal transduction. The proline-rich cytoplasmic domain of IL-5Rα has a native sequence of Pro352-Pro353-Val354-Pro355, which is represented on the left model of the IL-5R by **ppvp**. Each alanine substitution to this sequence is represented by an **A**. The thick arrow represents full activity, while the thinner arrow represents reduced activity. The AAvA mutant demonstrated no activity which is indicated by an **X**.

with mIL-5Rα mutants showed that neither of the proline residues of the Pro-X-Pro sequence in the Pro-Pro-X-Pro motif is critical for IL-5R induced JAK2-STAT5 activation or for cell growth. However, Pro357 of mIL-5Rα, which is not conserved in IL-3Rα or GM-CSFRα, may play a role in IL-5 signaling. It is possible that Pro357 of IL-5Rα plus one of the N-terminal prolines in the Pro-Pro-X-Pro motif is sufficient for IL-5 signal transduction (O'Neal and Shearer, 1995).

Other cytokine receptors undergo ligand induced dimerization or multimerization. To test whether multimerization of βc is involved in IL-5 mediated signal transduction, two IL-5Rα-βc chimeras were constructed (Takaki et al., 1994). The ααβ chimera possessed the IL-5Rα extracellular and transmembrane domains and the βc cytoplasmic domain. The αββ chimera possessed the extracellular domain of IL-5Rα and the transmembrane and cytoplasmic domains of βc. Transfectants expressing either chimeric receptor together with a native βc showed activation of the JAK1-STAT5 pathway and proliferated in response to IL-5 (Figure 9), whereas JAK2 tyrosine phosphorylation was hardly detectable (Kouro et al., 1996). These results imply that dimerization of the cytoplasmic domain of βc confers the ability, in part, to transduce a growth signal through JAK1-STAT5 in the absence of the cytoplasmic domain of IL-5Rα (Figure 9). Therefore, it is speculated that an IL-5 signal is transduced through a IL-5Rα-βc heterodimer via JAK2-STAT5 and through cytoplasmic dimerization of βc via JAK1-STAT5 (Figure 9). Thus, the proline-rich domain of IL-5Rα may function to dimerize or multimerize βc.

VI. BIOLOGICAL ROLE OF IL-5

Biologic activities originally attributed to B cell growth factor II (Dutton et al., 1984), IgA enhancing factor (Coffman et al., 1987), T cell replacing

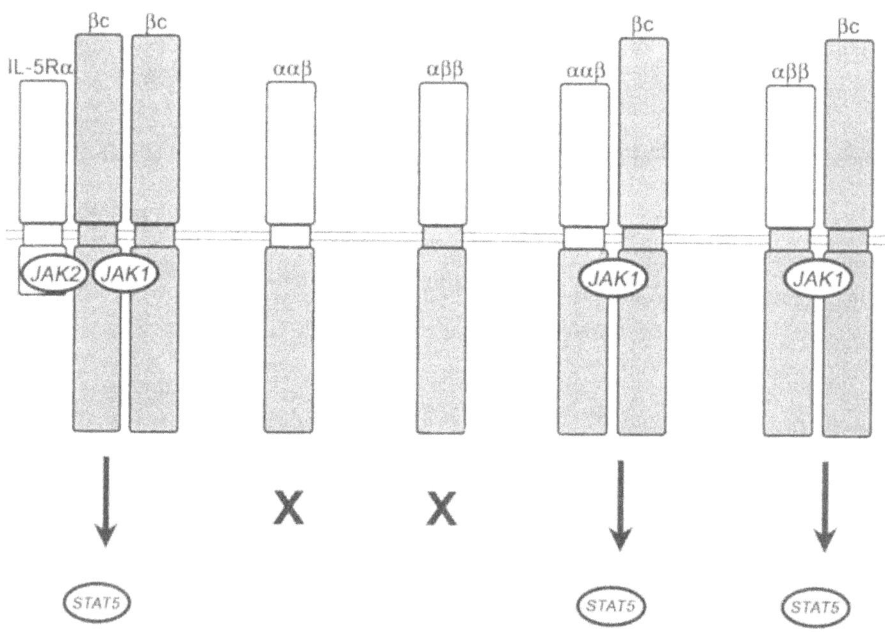

Figure 9. Dimerization of βc. A series of receptor combinations is illustrated. Each receptor subunit has three domains (extracellular, transmembrane, and cytoplasmic). Each chimera is designated by three Greek letters depending on from which IL-5R subunit (α orβ) a respective domain was derived. Sequence from IL-5Rα is white and sequence from βc is gray. Alone neither chimera is functional (**X**), however together with β_c, IL-5Rα ααβ, and αββ, it can signal through the JAK1-STAT5 pathway. Only the native receptor complex can active the JAK2-STAT5 pathway.

factor (Takatsu et al., 1980), eosinophil differentiation factor (Campbell et al., 1987), and eosinophil colony-stimulating factor (Lopez et al., 1986) are now known to be properties of IL-5 (Kinashi et al., 1986). The physiologic and pathophysiologic consequences of these pleiotrophic effects have more recently been facilitated by the availability of recombinant IL-5 and IL-5R subunits, monoclonal antibodies to these proteins, IL-5 transgenic animals, and IL-5 or IL-5R knock-out animals. Many effects of IL-5 demonstrated in animal models have now been confirmed in humans studies.

A. Cellular Sources of IL-5

The capacity to produce IL-5 is restricted to leukocytes (Figure 10). Within lymphocyte lineages, the major source of IL-5 is a subset of T helper cells. Among the myeloid lineages, mast cells, and eosinophils are the major producers of IL-5.

T Helper Cells

Naive T helper (TH) cells differentiate into two subsets, TH1 and TH2, which are distinguished by their patterns of cytokine production (Mosmann and Coffman, 1989; Wierenga et al., 1990). Differentiation into TH1 cells is promoted by IL-12 (Hsieh et al., 1993b) and differentiation into TH2 cells is promoted by IL-4 (Le Gros et al., 1990). TH1 cells produce large quantities of IL-2, interferon gamma, and tumor necrosis factor beta, but little IL-4, IL-5, and IL-10 (Mosmann and Coffman, 1989). Conversely, TH2 cells produce large quantities of IL-4, IL-5, and IL-10, but little IL-2, interferon gamma, or tumor necrosis factor beta (Mosmann and Coffman, 1989). These distinct cytokine profiles are associated with susceptibility or resistance to specific pathogens and with development of cellular or humoral immune responses (Le Gros et al., 1990; Urban et al., 1992; Hsieh et al., 1993a, 1993b). Differentiation into TH1 versus TH2 subsets may be dependent on T cell (CD28) engagement of the B7-1 (CD80) or B7-2 (CD86) molecule, respectively, on antigen-presenting cells (Thompson, 1995). Also, TH1 and TH2 subsets may be $CD4^+ CD8^-$, $CD4^- CD8^+$, or $CD4^- CD8^-$ (Erard et al., 1993; Till et al., 1995). Following antigen activation of TH2 cells, IL-5 transcription is upregulated and secreted IL-5 is detected within hours and peaks by 48 hours (Figure 11; Dickason and Huston, 1995).

Natural Killer Cells

Natural killer cells (NK cells) were originally described for their ability to lyse certain target cells in a non-MHC restricted manner or by

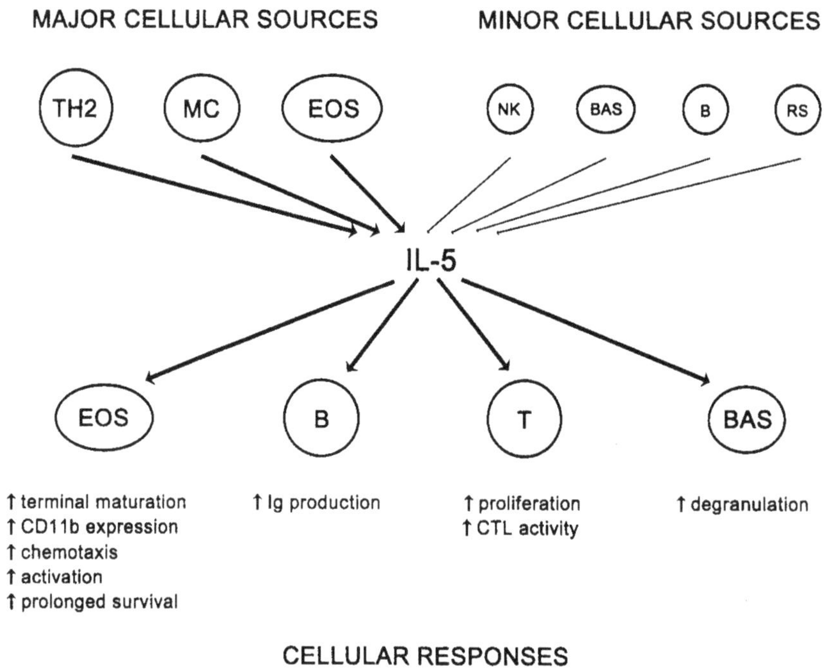

Figure 10. Cells producing and responding to IL-5. TH2 = type 2 T helper lymphocytes, MC = mast cells, EOS = eosinophils, NK = natural killer cells, BAS = basophils, RS = Reed Sternberg cells, B = B lymphocytes, T = T lymphocytes.

antibody-dependent cell-mediated cytotoxicity (Trinchieri, 1989). Subsequent studies also demonstrate a role for NK cells in the regulation of both cellular and humoral immunity through the elaboration of various cytokines (Anegon et al., 1988). As with T cells, IL-5 production by NK cells is dependent on their state of activation and requires both antigen and IL-2 stimulation (Warren et al., 1995). Also analogous to T cells, IL-4 enhances while IL-12 inhibits production of IL-5 by NK cells (Warren et al., 1995).

B Cells

The potential for B cells to produce IL-5 is limited to studies with Epstein-Barr virus (EBV) transformed B lymphocytes (Paul et al., 1990; Baumann and Paul, 1992). Constitutive production of IL-5 message was demonstrated by in situ hybridization, and immunoreactive IL-5 was detected in culture supernatants. However, IL-5 production by nontransformed B cells

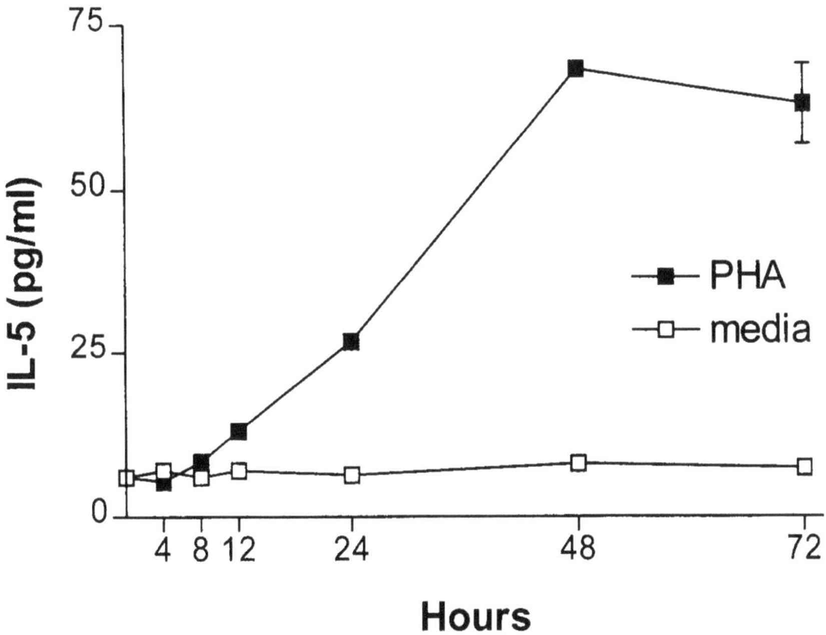

Figure 11. Kinetics of IL-5 production by T lymphocytes. Human peripheral blood T cells were cultured in media containing phytohemagglutinin (PHA) or media alone (Dickason and Huston, 1995). IL-5 in culture supernatants at the indicated times was quantified by enzyme-linked immunosorbent assay (Dickason et al., 1994).

has not been demonstrated. Likewise, murine B cells have not been shown to produce IL-5. Interestingly, IL-5 message has been detected in Reed-Sternberg cells of Hodgkin's lymphoma (Samoszuk and Nansen, 1990).

Mast Cells and Basophils

In situ hybridization can detect IL-5 message and immunocytochemistry can detect IL-5 protein in mast cells and basophils present in cutaneous and mucosal biopsies (Bradding et al., 1993, 1994, 1995). Kinetic analyses of IL-5 production by mast cells have been facilitated by the *in vitro* differentiation of naive mast cells from stem cell precursors (Bressler et al., Forthcoming). These naive mast cells do not constitutively produce IL-5 or express IL-5 message. However, upon ligation of the FcεR1, IL-5 message is rapidly upregulated and followed by protein synthesis with kinetics similar to that of antigen activated T cells (Figure 12). Thus, the *in vivo* presence of IL-5 message or protein in mast cells indicates their prior activation.

Eosinophils

Production of IL-5 by eosinophils is evidenced by the presence of IL-5 message and protein which localizes to their cytoplasmic granules (Kay et al., 1993; Dubucquoi et al., 1994). Rapid secretion of IL-5 occurs upon activation of eosinophils with immune complexes (Dubucquoi et al., 1994). IgA immune complexes provide the most potent stimulus for both the synthesis of IL-5 by eosinophils and the secretion of preformed IL-5 stored in their secretory granules.

B. Cellular Responses to IL-5

Biologic responsiveness to IL-5 is limited to subpopulations of leukocytes (Figure 10). Although interspecies comparisons are not exact at

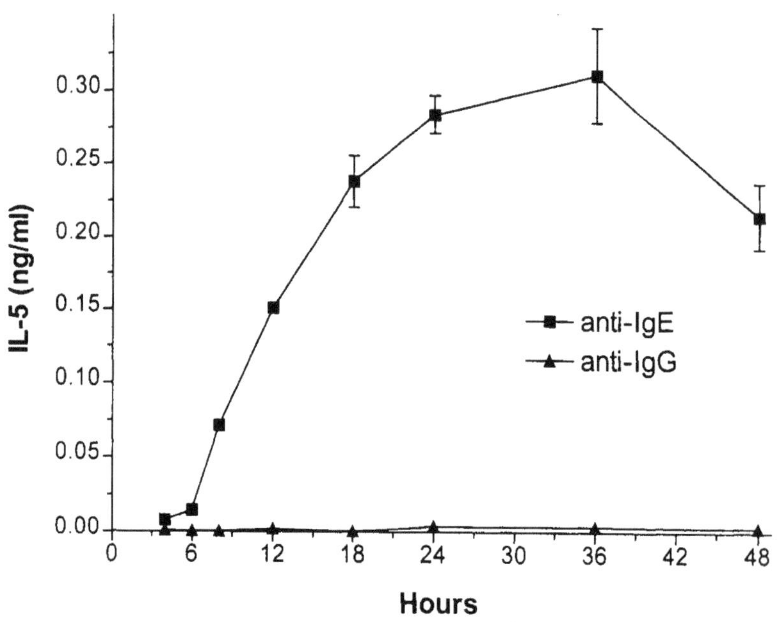

Figure 12. Kinetics of IL-5 production by mast cells. Naive human mast cells were derived *in vitro* from stem cells, sensitized with IgE, and activated with anti-IgE antibody or not activated with anti-IgG antibody (Bressler et al., forthcoming). IL-5 in culture supernatants at the indicated times was quantified by enzyme-linked immunosorbent assay (Dickason et al., 1994).

this time, the bulk of studies suggests a universality of IL-5 effects for enhancement of immune responses.

B Cells

IL-5 was initially described as a growth factor for the $CD5^+$ murine B cell lymphoma BCL1 (Dutton et al., 1984). In normal mice, B cell expression of the IL-5R complex is readily demonstrated on $CD5^+$ B cells which belong to the B1 lineage that resides largely in the peritoneum (Hitoshi et al., 1990; Rolink et al., 1990; Takatsu et al., 1992). The B1 progenitors develop along a stromal cell-dependent IL-5-sensitive pathway (Ogawa et al., 1988; Tominaga et al., 1989; Hayashi et al., 1990), which can bifurcate to $CD5^+$ macrophages under the influence of GM-CSF (Katoh et al., 1990). The alternative B1 pathway leads to $CD5^+$ B cells that produce natural antibodies which include a spectrum of autoantibodies (Shirai et al., 1995). By contrast, the B2 lineage of B cells is $CD5^-$ and constitutes most of the bone marrow and peripheral B cells (Hayakawa et al., 1985). In mice expressing the IL-5 transgene, the $CD5^+$ B1 population is markedly increased with concomitant hypergammaglobulinemia and autoantibody production (Dent et al., 1990; Tominaga et al., 1991). Likewise, IL-5 responsive $CD5^+$ B1 cells are increased in the spontaneously autoimmune NZB and (NZBxNZW) F1 mice (Herron et al., 1988; Umland et al., 1989). The potential for $CD5^-$ B2 cells to express the IL-5R complex and respond to IL-5 can also be demonstrated by stimulation with anti-Ig-dextran, a type II T-independent antigen (Allison et al., 1991). Thus, evidence has accumulated in murine systems for both an IL-5-sensitive developmental pathway for $CD5^+$ B1 cells and a role for IL-5 in the terminal differentiation of mature $CD5^+$ B1 cells or appropriately activated $CD5^-$ B2 cells. Evidence that IL-5 increases $C_H\mu$ secretory mRNA by influencing polymerase termination demonstrates that IL-5 also acts on appropriately activated B cells to increase the production of secretory immunoglobulin mRNA for isotypes determined by prior antigenic and cytokine stimulation (Webb et al., 1989; Yuan et al., 1990). In addition, a role for IL-5 in promoting C_H gene recombination has been demonstrated by synergy with IL-4 which induces $S\mu$-$S\gamma 1$ recombination in murine B cells activated with anti-IgD dextran (Mandler et al., 1993). The ability of IL-5 to prevent apoptosis of B cells, whose growth is arrested by cross-linking of surface IgM (Kamesaki et al., 1994; Takatsu, 1995), also suggests a potential role for IL-5 in the survival of germinal center B cells.

In contrast to murine B cells, evidence of a role for IL-5 in human B cell growth and differentiation has been controversial. A large collaborative study found no effects on proliferation and Ig production when IL-5 was tested for functional activity on human B cells purified from peripheral blood, tonsils, or spleen, or on human B cell lymphoma lines using anti-μ antibodies, phorbolesters, calcium ionophores, *Staphylococcus aureus* Cowan strain I (SAC), or EBV (Clutterbuck et al., 1987). Conversely, several studies subsequently confirmed the potential of human B cells to respond to IL-5 (Bertolini et al., 1993; Morikawa et al., 1993; Huston et al., 1996). As with mice, IL-5 responsiveness of human B cells is dependent on the activation signal employed. IL-5 responsiveness of SAC stimulated B cells is affected by the duration of B cell activation and dependent on the presence of IL-2 (Bertolini et al., 1993). IL-5 responsiveness of pokeweed mitogen stimulated B cells was also IL-2 dependent (Morikawa et al., 1993). Detection of message for IL-5Rα and βc in resting B cells as well as IL-5 enhancement of immunoglobulin production upon the stimulation of B cells with the novel bacteria B cell mitogen *Moraxella catarrhalis* (Huston et al., 1996) have provided direct evidence that human B cells can respond to IL-5. In addition, IL-5R have been detected on the surface of some EBV transformed B cell lines that constitutively produce IL-5 (Paul et al., 1990). Decreased survival of these EBV transformed B cells in the presence of neutralizing anti-IL-5 mAb argues for an autocrine role for IL-5 in the growth of these B cells (Baumann and Paul, 1992). Thus, data is accruing for the potential of human B cells to respond to IL-5. Whether IL-5 responsiveness is limited to a subset or discrete lineage of human B cells or is truly dependent on the nature of the activation signal remains to be determined.

T Cells

Evidence that T cells respond to IL-5 was initially demonstrated by the capacity of IL-5 to support the IL-2-dependent generation of murine CTL from their thymocyte precursors (Takatsu et al., 1987). In addition, IL-5 up-regulated the expression of IL-2Rα on these CTL precursors (Kikuchi et al., 1994). In humans, IL-5 enhances alloantigen-induced T cell proliferation (Martinez and Krams, 1993), the generation of alloantigen-induced CTL (Nagasawa et al., 1991), and the generation of IL-2-induced lymphokine activated killer cells (Aoki et al., 1989). As with murine T cells, IL-5 increases IL-2Rα expression on human peripheral blood T cells (Noma et al., 1987). Detection of high levels of IL-5 in the bronchoalveolar lavage fluid from lung transplant recipients may also be an in vivo correlate to the in vitro studies demonstrating a role for IL-5 in the differentiation of alloreactive CTL (Dickason et al., 1994).

Eosinophils

The cloning of IL-5 cDNA and availability of recombinant IL-5 was crucial to resolving the functional association of murine eosinophil differentiation factor and B cell growth factor II in supernatants of activated T cells (Campbell et al., 1987). Both murine and human studies now demonstrate that IL-5 uniquely and specifically directs the terminal maturation of eosinophils from precursors which have differentiated from stem cells under the influence of IL-3, GM-CSF, or G-CSF (Campbell et al., 1987; Clutterbuck et al., 1989; Clutterbuck and Sanderson, 1990). IL-5 also is chemotactic for eosinophils (Yamaguchi et al., 1988) and increases expression of the β2 integrin CD11b (Mac-1) on eosinophils for adhesion to CD54 (ICAM-1) on endothelial cells (Walsh et al., 1990, 1991). In addition, eosinophils cultured in the presence of IL-5 become activated and their survival is prolonged by inhibition of apoptosis (Yamaguchi et al., 1988, 1991). IL-5 enhanced eosinophil effector functions include an increased antibody-dependent phagocytic activity and respiratory burst, as well as a decreased threshold for degranulation (Lopez et al., 1988; Yamaguchi et al., 1988; Fujisawa et al., 1990). These effects serve to rapidly generate and mobilize eosinophils for participation in inflammatory immune responses (Kay et al., 1993).

Basophils

IL-5 enhances basophil histamine release induced by ligation of FcεRI, formyl-methionyl-leucyl-phenylolamine, or calcium ionophores (Bischoff et al., 1990; Hirai et al., 1990). Basophilic differentiation of the erythroleukemic cell line HL60 in the presence of IL-5 also suggests a basophilopoietin activity (Denburg, 1992; Metwali et al., 1993).

C. IL-5 in Health and Disease

Efforts to delineate the physiologic and pathophysiologic role of IL-5 have utilized animal models of disease, animals modified by genetic engineering, and analysis of human diseases in which IL-5 producing and responding cells are involved. From such studies, diverse paradigms have been developed which are consistent with the cellular origins and pleiotrophic effects of IL-5.

Allergy and Asthma

Immediate hypersensitivity allergic reactions are dependent on degranulation of activated mast cells and peak within 30 minutes (McNeil and Austen, 1995). Over the subsequent several hours, the late phase allergic reaction (LPAR) evolves, peaks at 6-12, and may persist for several days (Lemanske and Kaliner, 1994). This LPAR is dependent on mast cell activation. In contrast to the immediate hypersensitivity reaction in which histologic findings are predominantly edema and vasodilatation, the LPAR is characterized by a prominent leukocytic infiltrate comprised of eosinophils, TH2 cells, and neutrophils. The eosinophils are activated and the size of LPAR correlates with the number of activated eosinophils (Schwartz and Austen, 1988; Diaz et al., 1989; Gaga et al., 1991; Kay et al., 1991; Lemanske and Kaliner, 1994). A strong correlation also exists between the presence of eosinophils (Frew and Kay, 1988) and the presence of $CD4^+$ T cells which exhibit a TH2 cytokine secretotype (Frew and Kay, 1988; Kay et al., 1991). However, IL-5 message and protein are also found in mast cells located within allergen-challenged tissues (Bradding et al., 1993, 1994, 1995). Based on these observations, an attractive paradigm (Figure 13) for eosinophil involvement in the LPAR is (1) the up-regulation of IL-5 synthesis by mast cells activated in an immediate hypersensitivity allergic reaction, resulting in (2) eosinophil recruitment and activation. Concomitantly, (3) T cells are recruited by other inflammatory mediators of the allergic reaction and undergo antigen-specific activation and acquisition of the TH2 phenotype which (4) further enhances eosinophil recruitment and activation.

The histopathologic features of allergic rhinitis and chronic asthma are consistent with the LPAR (Diaz et al., 1989; Bousquet et al., 1990; Gaga et al., 1991; Rossi et al., 1991; Kay et al., 1993). Mucosal biopsies contain activated mast cells expressing IL-5 message, increased numbers of IL-5 producing TH2 cells, and increased numbers of activated eosinophils (Bousquet et al., 1990; Hamid et al., 1991; Bradding et al., 1994). Bronchoalveolar lavage fluid also contains IL-5, TH2 cells, and eosinophils (DeMonchy et al., 1985; Robinson et al., 1992). These findings strongly implicate IL-5 in the pathogenesis of inflammation in allergies

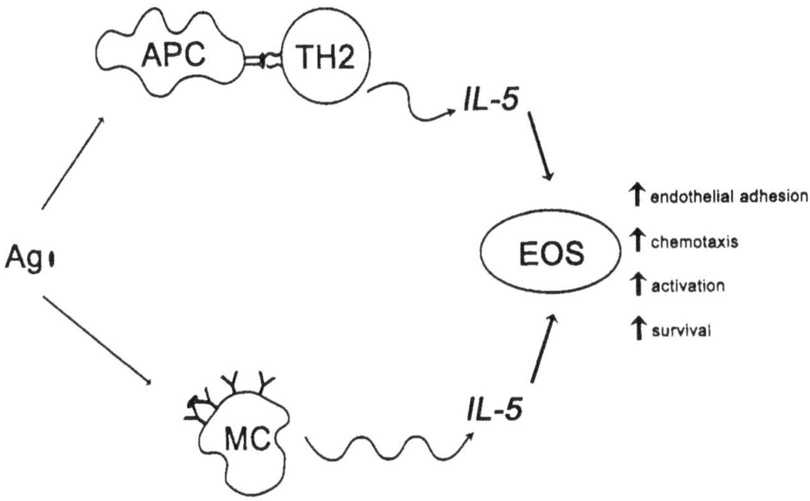

Figure 13. Paradigm for antigen induced eosinophil accumulation and activation in allergic diseases and asthma. Antigen (Ag) is presented to TH2 cells by antigen presenting cells (APC) or is bound to IgE which crosslinks FcεRI on mast cells (MC). Both events result in IL-5 secretion which has several effects on eosinophils (EOS).

and asthma. Animal models further support such a role for IL-5 (Nakajima et al., 1992; Renz et al., 1992; Kung et al., 1995a, 1995b; Van Oosterhout et al., 1995; Watanabe et al., 1995). Most studies utilize animals which are immunized with a T cell-dependent antigen (e.g., ovalbumin) and subsequently rechallenged with the antigen by inhalation. Eosinophil infiltration, accumulation, and activation typically occur within several hours and peak over 24-48 in conjunction with an increase in TH2 cells. A delay in the eosinophil influx during the first 12 after antigen rechallenge in mast cell deficient WBB6F1 (W-Wv) mice suggests a role for mast cells in the initial recruitment of eosinophils (Kung et al., 1995b). The demonstration that activated mast cells produce IL-5 fully supports this observation (Bradding et al., 1994; Bressler et al., Forthcoming). However, accumulation of bronchial eosinophils in antigen-primed W-Wv mice following several days of antigen respiratory rechallenge indicates that mast cells are not essential for the eosinophil mediated inflammatory response, although they may initially contribute to early eosinophil recruitment and activation (Brusselle et al., 1994). The delayed eosinophil influx in this model is most consistent with recruitment by TH2 cells. Although most studies utilize antigens which induce a $CD4^+$ TH2 response, the potential for $CD8^+$ T cells to exhibit the TH2 phenotype and to produce IL-5 has also been demonstrated (Bradding et al., 1994). The $CD8^+$ TH2 response has been induced by virus sensitization of mice (Coyle et al., 1995). A virus respiratory rechallenge is associated with pulmonary eosinophilia, thus serving as a model for the pathophysiology of virus induced exacerbations of asthma. The capacity for both $CD4^+$ and $CD8^+$ T cells from bronchoalveolar lavage cells in asthmatic patients to produce IL-5 has also been demonstrated (Till et al., 1995).

A pivotal role for IL-5 in the chronic asthmatic response has been confirmed by the capacity for neutralizing anti-IL-5 mAb to inhibit antigen or virus-induced airway hyperresponsiveness and eosinophil infiltration in the airways of mice, guinea pigs, and primates (Akutsu et al., 1995; Chand et al., 1992; Mauser et al., 1993, 1995; Nagai et al., 1993; Kung et al., 1995a; Van Oosterhout et al., 1995). Importantly, administration of neutralizing anti-IL-5 mAb reduced pulmonary inflammation when administered within hours preceding respiratory antigen challenge or when administered up to five days after antigen challenge (Kung et al., 1995a). Thus, inhibitors of IL-5 may be both prophylactic and therapeutic for eosinophilic inflammation of the lung. Although a variety of other cytokines (e.g., IL-3, IL-4, GM-CSF) and chemokines (monocyte and chemotactic proteins, macrophage inhibitory proteins, and Rantes) can be detected in asthmatic pulmonary inflammation, the beneficial effect from neutralizing IL-5 indicates its central importance in the pathogenesis of asthma (Foster et al., 1996). The lack of these bronchial hyperreactivity and eosinophils in the lungs of these antigen-sensitized and aeroallergen-challenged IL-5 knock-out mice strengthens the linkage of IL-5 to the pathogenesis of asthma. This linkage is further strengthened by the occurrence of bronchial hyperreactivity and eosinophilic inflammation in the lungs of these IL-5 knock-out mice when IL-5 is transiently expressed by vector directed IL-5 gene delivery during antigen sensitization and aeroallergen rechallenge (Foster et al., 1996).

Helminth Infection

Helminth infections are typically accompanied by an elevation of serum IgE and peripheral blood eosinophilia. Isotype switch to IgE production is mediated by IL-4, while eosinophilia is dependent on IL-3 and GM-CSF for generation of eosinophil progenitors (Sonoda et al.,

1989) and on IL-5 for the terminal maturation of the eosinophil, as well as the accumulation and activation of eosinophils in tissues (Lopez et al., 1986, 1988). Accordingly, patients with strongyloidiasis, filariasis, onchochereciasis, or loiasis have increased numbers of $CD4^+$ TH2 cells which produce high levels of IL-5 (Mahanty et al., 1992). The association between peripheral blood eosinophilia and resistance to schistosomiasis supports the argument that IL-5 plays an important role in protective immunity against parasitic infections (Hagan et al., 1985).

Murine studies support the paradigm of a helminth-induced $CD4^+$ TH2 immune response resulting in IL-5-induced eosinophilia in conjunction with the IL-4-induced switch to IgE production. Direct evidence that IL-5 is required for the blood and tissue eosinophilia induced by helminth infections was demonstrated by administration of neutralizing antibody to IL-5 (Coffman et al., 1989) which inhibited the eosinophil response to *Schistosoma mansoni.* A similar depletion of eosinophils by anti-IL-5 mAb was demonstrated in mice infected with *Trichinella spirasis* (Herndon and Kayes, 1992). However, the necessity of IL-5 and eosinophils to provide immunity against parasites remains controversial. The failure for anti-IL-5 or anti-IL-4 antibodies to affect immunity against *Schistosoma mansoni,* despite ablation of eosinophilia and IgE responses, respectively, argues against an essential role for IL-5 in parasite immunity (Sher et al., 1990). Conversely, ablation of eosinophilia with anti-IL-5 mAb does decrease the survival of mice infected with *Angiostrongylus cantonensis* (Sasaki et al., 1993), which is consistent with in vitro studies demonstrating enhanced destruction of parasites by IL-5 activated eosinophils. The ability of IL-5 knock-out mice (Kopf et al., 1995) or IL-5 receptor subunit knock-out mice (Yoshida et al., 1996) to recover from at least some parasitic infections further confounds efforts to determine the necessity of IL-5-mediated eosinophilia for immunologic protection against helminths.

Hypereosinophilic Syndromes

Increased levels of IL-5 are detected in the serum of patients with idiopathic hypereosinophilia or with L-tryptophane-induced eosinophilic myalgia syndrome (Owen et al., 1990). Likewise, patients with episodic angioedema and eosinophilia have transient elevation of serum IL-5 levels which precede the peripheral blood eosinophilia by several days (Inoue et al., 1990). This is consistent with the kinetics of IL-5-induced eosinophilia and suggests an important role for IL-5 in the pathogenesis of these syndromes. Elevated levels of IL-5 in the pleural fluid and serum of patients with acute or chronic pulmonary eosinophilia provide further evidence for the association of IL-5 up-regulation and eosinophilic inflammation (Matsumoto et al., 1992; Okubo et al., 1995; Yamaguchi et al., 1995).

Tumor Rejection

Increased survival of patients with carcinomas of the stomach, lung, colon, or uterine cervix positively correlates with histologic evidence of eosinophil infiltration (Iwasaki et al., 1981; Pretlow et al., 1983; Pasternark and Jansa, 1984). A role for IL-5 in tumor rejection is further supported in murine studies wherein administration of recombinant IL-5 enhanced rejection of lymphoma and sarcoma cells (Wu et al., 1992; Nakashima et al., 1993). Administration of anti-IL-5 mAb also partially inhibited eosinophil infiltration and rejection of tumor cells transfected with cDNA for IL-4 or IL-2, suggesting intercytokine networks for upregulating IL-5 production (Tepper et al., 1989, 1992). In humans this concept is supported by the observation that IL-2 infusion results in increased IL-5 production by IL-2 activated T cells and correlates with eosinophilia and tumor regression (MacDonald et al., 1990; Rivoltini et al., 1993). IL-5 mediated anti-tumor effects may also result from enhanced generation of CTL (Nagasawa et al., 1991).

Allograft Rejection

Rejection of hepatic allografts correlates with eosinophil infiltration and increased message for IL-5 by allograft infiltrating cells (Foster et al., 1989; Martinez et al., 1992, 1993b, 1993c). Furthermore, high concentrations of IL-5 have been detected in the bronchoalveolar lavage fluid of patients undergoing chronic rejection of transplanted lungs (Dickason et al., 1994). IL-5 may participate in allograft rejection by both eosinophil-mediated inflammation and enhancement of CTL generation.

VII. FUTURE PERSPECTIVES

Elucidation of the functional residues on IL-5 and its receptor subunits will expedite the design and development of therapeutic antagonists and agonists of IL-5-mediated immune responses. Likewise, identification of intracellular signaling pathways unique to IL-5 may enable the development of alternative therapeutic agents. However, a more precise definition of the physiologic and pathophysiologic role of IL-5 will be necessary to accurately pursue such clinical applications. Rapidly accruing information on the characteristics of mice in whom the gene for IL-5, IL-5Rα, or βc is disrupted is providing a major advance toward this goal. The successful generation and breeding of mice with each of these genetic deficiencies indicates that none of these mutations are lethal. Disruption of the bc receptor subunit results in a reduction in basal levels of eosinophils, although a few eosinophils are still present and are morphologically normal (Nishinakamura et al., 1995). Immunoglobulin levels are also normal for all isotypes as are B cell numbers, including $CD5^+$ B cells. Infection with the nematode *Nippostrongylus brasiliensis* does not induce an eosinophilic response and yet the animals recover from the infection. Similar results are observed in IL-5 knock-out mice (Kopf et al., 1996). However, although immunoglobulin levels and isotypes and B cells in both the B1 and B2 lineages are normal in the adult mice, the neonatal IL-5 knock-out mice have up to an 80% reduction in B1 cells. The delayed establishment of normal members of B1 cells in these mice indicates that IL-5 may be important but is not essential for the development of this lineage. As with the βc knock-out mice, IL-5 knock-out mice unexpectedly have morphologically normal eosinophils which are present in small numbers. However, an eosinophilic response is not mounted to either helminth infection (Kopf et al., 1996) or aeroallergen rechallenge (Foster et al., 1996), confirming a major role for IL-5 in the rapid expansion of the eosinophil population from progenitor cells and in the migration, accumulation, and activation of eosinophils in inflamed tissues.

In aggregate, the structural, functional, and clinical studies described herein provide insight into the role of IL-5 in health and disease, and a strong impetus for investigating the means of modulating IL-5 effects. In concert with advances in understanding the biology of other cytokines and growth factors, entirely new approaches to patient care should emerge in the near future.

REFERENCES

Akira, S., Nishio, Y., Inoue, M., Wang, X. J., Wei, S., Matsusaka, T., Yoshida, K., Sudo, T., Naruto, M., & Kishimoto, T. (1994). Molecular cloning of APRF, a novel IFN-stimulated gene factor 3 p91-related transcription factor involved in the gp130-mediated signaling pathway. Cell 77, 63-71.

Akutsu, I., Kojima, K., Kariyone, A., et al. (1995). Passive administration of anti-IL-5 antibody into chronically ovalbumin-sensitized guinea pigs induces suppression of eosinophilia and bronchial hyperreactivity. Immunol. Lett. 45, 109-116.

Allison, K. C., Strober, W., & Harriman, G. R. (1991). Induction of IL-5 receptors on normal B cells by cross-linking surface Ig with anti-Ig-dextran. J. Immunol. 146,4197-4203.

Anegon, I., Cuturi, M. C., Trinchieri, G., & Perussia, B. (1988). Interaction of Fc receptor (CD16) ligands induces transcription of interleukin genes and expression of their products in human natural killer cells. J. Exp. Med. 167,452-472.

Aoki, T., Kikuchi, H., Miyatake, S., Oda, Y., Iwasaki, K., Yamasaki, T., Kinashi, T., & Honjo, T. (1989). Interleukin 5 enhances interleukin 2-mediated lymphokine-activated killer activity. J. Exp. Med. 170,583-588.

Appleby, M. W., Kerner, J. D., Chien, S., Maliszewski, C. R., Bondadaa, S., & Perlmutter, R. M. (1995). Involvement of p59fynT in IL-5 receptor signaling. J. Exp. Med. 182,811-820.

Arai, N., Nomura, D., Villaret, D., DeWaal Malefijt, R., Seik, M., Yoshida, M., Minoshima, S., Fukuyama, R., Maekawa, M., Kudoh, J., Shimizu, K., Yokota, K., Abe, E., Yokota, T., & Takebe, Y. (1989). Complete nucleotide sequence of the chromosomal gene for human IL-4 and its expression. J. Immunol. 142,274-282.

Argetsinger, L. S., Campbell, G. S., Yangs, X., Witthuhn, B. A., Silvennoinen, O., Ihle, J. N., & Carter-Su, C. (1993). Identification of JAK2 as a growth hormone receptor-associated tyrosine kinase. Cell 74,237-244.

Azuma, C., Tanabe, T., Konishi, M., Kinashi, T., Noma, T., Matsuda, F., Yaoita, Y., Takatsu, K., Hammarstrom, L., Smith, C. I., Severinson, E., & Honjo, T. (1986). Cloning of cDNA for human T-cell replacing factor (interleukin-5) and comparison with the murine homologue. Nuc. Acids Res. 14,9149-9158.

Banchereau, J. & Galizzi, J. P. (1990). Receptors for interleukin 4, interleukin 5 and interleukin 6. Ann. Ist. Super. Sanita. 26,453-467.

Bar-Sagi, D., Rotin, D., Batzer, A., Mandiyam, V., & Schlessinger, J. (1993). SH3 domains direct cellular localization of signaling molecules. Cell 74,83-91.

Baumann, M. A. & Paul, C. C. (1992). Interleukin-5 is an autocrine growth factor for Epstein-Barr virus-transformed B lymphocytes. Blood 79,1763-1767.

Bazan, J. F. (1990a). Haemopoietic receptor and helical cytokines. Immunol. Today. 11,350-354.

Bazan, J. F. (1990b). Structural design and molecular evolution of a cytokine receptor superfamily. Proc. Natl. Acad. Sci. USA 87,6934-6938.

Bazan, J. F., & McKay, D. B. (1992). Unraveling the structure of IL-2. Science 257,410-413.

Bennett, D., Morton, T., Breen, A., Hertzberg, R., Cusimano, D., Appelbaum, E., McDonnell, P., Young, P., Matico, R., & Chaiken, I. (1995). Kinetic characterization of the interaction of biotinylated human interleukin 5 with an Fc chimera of its receptor alpha subunit and development of an ELISA screening assay using real-time interaction biosensor analysis. J. Mol. Recognit. 8,52-58.

Bentley, A. M., Meng, Q., Robinson, D. S., Hamid, Q., Kay, A. B., & Durham, S. R. (1993). Increases in activated T lymphocytes, eosinophils, and cytokine mRNA expression for interleukin-5 and granulocyte/macrophage colony-stimulating factor in bronchial biopsies after allergen inhalation challenge in atopic asthmatics. Am. J. Respir. Cell. Mol. Biol. 8,35-42.

Bertolini, J. N., Sanderson, C. J., & Benson, E. M. (1993). Human interleukin-5 induces staphylococcal A Cowan 1 strain-activated human B cells to secrete IgM. Eur. J. Immunol. 23,398-402.

Bischoff, S. C., Brunner, T., De-Weck, A. L., & Dahinden, C. A. (1990). Interleukin 5 modifies histamine release and leukotriene generation by human basophils in response to diverse agonists. J. Exp. Med. 172,1577-1582.

Bohjanen, P. R., Okajima, M., & Hodes, R. J. (1990). Differential regulation of interleukin 4 and interleukin 5 gene expression: a comparison of T-cell gene induction by anti-CD3 antibody or by exogenous lymphokines. Proc. Natl. Acad. Sci. USA 87,5283-5287.

Bourke, P. F., van Leeuwen, B. H., Campbell, H. G., & Young, I. G. (1995). Localization of the inducible enhancer in the mouse interleukin-5 gene that is responsive to T-cell receptor stimulation. Blood 85,2069-2077.

Bousquet, J., Chanez, P., Lacoste, J. Y., Barneon, G., Ghavanian, N., Enander, I., Venge, P., Ahlstedt, S., Simony-Lafontaine, J., Godard, P., & Michel, F. B. (1990). Eosinophilic inflammation in asthma. N. Engl. J. Med. 323,1033-1039.

Bradding, P., Feather, I. H., Wilson, S., Bardin, P. G., Heusser, C. H., Holgate, S. T., & Howarth, P. H. (1993). Immunolocalization of cytokines in the nasal mucosa of normal and perennial rhinitic subjects. The mast cell as a source of IL-4, IL-5, and IL-6 in human allergic mucosal inflammation. J. Immunol. 151,3853-3865.

Bradding, P., Roberts, J. A., Britten, K. M., Montefort, S., Djukanovic, R., Mueller, R., Huesser, C. H., Howarth, P. H., & Holgate, S. T. (1994). Interleukin-4, -5, -6 and tumour necrosis factor-alpha in normal and asthmatic airway: Evidence for the human mast cell as a source of these cytokines. Am. J. Respir. Cell. Mol. Biol. 10,471-480.

Bradding, P. , Okayama, Y. , Howarth, P. H ., Church, M. K., & Holgate, S. T. (1995). Heterogeneity of human mast cells based on cytokine content. J. Immunol. 155,297-307

Bressler, R. B., Lesko, J., Jones, M. L., Wasserman, M., Dickason, R. R., Huston, M. M., Cook, S. W., & Huston, D. P. (Forthcoming). Production of interleukin-5 and granulocyte-macrophage colony stimulating factor by naive human mast cells activated by high affinity IgE receptor ligation. J. Allergy Clin. Immunol.

Brusselle, G. G., Kips, J. C., Tavernier, J. H., Van der Heyden, J. G., Cavelier, C. A., Pauwels, R. A., & Bleuthmann, H. (1994). Attenuation of allergic airway inflammation in IL-4 deficient mice. Clin. Exp. Allergy. 24,73-80.

Campbell, H. D., Tucker, W. Q. J., Hort, Y., Martinson, M. E., Mayo, G., Clutterbuck, E. J., Sanderson, C. J., & Young, I. G. (1987). Molecular cloning, nucleotide sequence, and expression of the gene encoding human eosinophil differentiation factor (interleukin-5). Proc. Natl. Acad. Sci. USA 84,6629-6633.

Chand, N., Harrison, J. E., Rooney, S., Pillar, J., Jakubicki, R., Nolan, K., Diamantis, W., & Sofia, R. D. (1992). Anti-IL-5 monoclonal antibody inhibits allergic late phase bronchial eosinophilia in guinea pigs: A therapeutic approach. Eur. J. Pharmacol. 211,121-123.

Cheng, G., Ye, Z. S., & Baltimore, D. (1994). Binding of Bruton's tyrosine kinase to Fyn, Lyn, or Hck through a Src homology domain-mediated interaction. Proc Natl Acad Sci. USA. 91,8152-8155.

Chihara, J., & Nakajima, S. (1992). Characterization of a receptor for interleukin-5 on pulmonary eosinophils with eosinophilic pneumonia: Nippon Kyobu Shikkan Gakkai Zasshi. 30 Suppl,70-76.

Clutterbuck, E., Shields, J. G., Gordon, J., Smith, S. H., Boyd, A., Callard, R. E., Campbell, H. D., Young, I. G., & Sanderson, C. J. (1987). Recombinant human interleukin 5 is an eosinophil differentiation factor but has no activity in standard human B cell growth factor assays. Eur. J. Immunol. 17,1743-1750.

Clutterbuck, E. J., Hirst, E. M., & Sanderson, C. J. (1989). Human interleukin-5 (IL-5) regulates the production of eosinophils in human bone marrow cultures: Comparison and interaction with IL-1, IL-3, IL-6, and GMCSF. Blood 73,1504-1512.

Clutterbuck, E. J., & Sanderson, C. J. (1990). Regulation of human eosinophil precursor production by cytokines: A comparison of recombinant human interleukin-1 (rhIL-1), rhIL-3, rhIL-5, rhIL-6, and rh granulocyte-macrophage colony-stimulating factor. Blood 75,1774-1779.

Coffman, R. L., Shrader, B., Carty, J., Mosmann, T. R., Bond, M. W. (1987). A mouse T cell product that preferentially enhances IgA production. I. Biologic characterization. J. Immunol. 139,3685-3690.

Coffman, R. L., Seymour, B. W., Hudak, S., Jackson, J., & Rennick, D. (1989). Antibody to interleukin-5 inhibits helminth-induced eosinophilia in mice. Science 245,308-310.

Conscience, J. F., Verrier, B., & Martin, G. (1986). Interleukin-3-dependent expression of the c-*myc* and c-*fos* proto-oncogenes in hematopoietic cell lines. EMBO J. 5,317-323.

Cornelis, S., Plaetinck, G., de Vos, R., Van der Heyden, J., Tavernier, J., Sanderson, C. J., Guisez, Y., & Fiers, W. (1995). Detailed analysis of the IL-5-IL-5R alpha interaction characterization of crucial residues on the ligand and the receptor. EMBO J. 14,3395-3402.

Coyle, A. J., Erard, F., Bertrand, C., Walti, S., Pircher, H., & Le Gros, G. (1995). Virus-specific CD8+ cells can switch to interleukin 5 production and induce airway eosinophilia. J. Exp. Med. 181,1229-1233.

Darnell, J. E. (1994). Jr, Kerr IM, Stark GR. Jak-STAT pathways and transcriptional activation in response to IFNs and other extracellular signaling transduction. Science 264,1415-1418.

DeMonchy, J. G., Kauffman, H. F., Venge, P., Koeter, G. H., Jansen, H. M., Sluiter, H. J., & DeVries, K. (1985). Bronchoalveolar eosinophils during allergen-induced late asthmatic reactions. Am. Rev. Resp. Dis. 131,373-376.

Denburg, J. A. (1992). Basophil and mast cell lineages *in vitro* and *in vivo*. Blood 79,846-860.

Dent, L. A., Strath, M., Mellor, A. L., & Sanderson, C. J. (1990). Eosinophilia in transgenic mice expressing interleukin 5. J. Exp. Med. 172,1425-1431.

Devos, R., Plaetinck, G., Van-der-Heyden, J., Cornelis, S., Vandekerckhove, J., Fiers, W., & Tavernier, J. (1991). Molecular basis of a high affinity murine interleukin-5 receptor. EMBO J. 10,2133-2137.

Devos, A. M., Ultsch, M., & Kossiakoff, A. A. (1992). Human growth hormone and extracellular domain of its receptor: Crystal structure of the complex. Science 255,306-312.

Devos, R., Guisez, Y., Cornelis, S., Verhee, A., Van-der-Heyden, J., Manneberg, M., Lahm, H. W., Fiers, W., Tavernier, J., & Plaetinck, G. (1993). Recombinant soluble human interleukin-5 (hIL-5) receptor molecules. Cross-linking and stoichiometry of binding to IL-5. J. Biol. Chem. 268,6581-6587.

Devos, R., Guisez, Y., Plaetinck, G., Cornelis, S., Tavernier, J., van der Heyden, J., Foley, L. H., & Scheffler, J. E. (1994). Covalent modification of the interleukin-5 receptor by isothiazolones leads to inhibition of the binding of interleukin-5. Eur. J. Biochem. 225,635-640.

Diaz, P., Gonzalez, M. C., Galleguillos, F. R., Ancic, P., & Cromwell, O. (1989). Leukocytes and mediators in bronchoalveolar lavage during allergen-induced late-phase asthmatic reactions. Am. Rev. Respir. Dis. 139,1383-1389.

Dickason, R. R., English, J. D., & Huston, D. P. (Forthcoming). Engineering a functional IIL-5 monomer: A paradigm for redesigning helical bundle cytokines with therapeutic potential in allergy and asthma. J. Mol. Med.

Dickason, R. R., Huston, M. M., & Huston, D. P. (1996a). Delineation of IL-5 domains predicted to engage the IL-5 receptor complex. J. Immunol. 156, 1030-1037.

Dickason, R. R., & Huston, D. P. (1996b). Creation of biologically active interleukin-5 monomer. Nature 379-652-655.

Dickason, R. R., & Huston, D. H. (1995). Immunobiology of interleukin-5. Allergy & Immunol. (Life Sci. Adv.) 13,87-91.

Dickason, R. R., Huston, M. M., & Huston, D. P. (1994). Enhanced detection of human IL-5 in biological fluids utilizing murine monoclonal antibodies which delineate distinct neutralizing epitopes. Cytokine 6,647-656.

Diederichs, K., Boone, T., & Karplus, A. (1991). Novel fold and putative receptor binding site of granulocyte-macrophage colony-stimulating factor. Science 254,1779-1782.

Dubucquoi, S., Desreaumaux, P., Janin, A., Klein, O., Goldman, M., Tavernier, J., Capron, A., & Capron, M. (1994). Interleukin 5 synthesis by eosinophils: Association with granules and immunoglobin-dependent secretion. J. Exp. Med. 179,703-708.

Duronio, V., Welham, M. J., Abraham, S., Dryden, P., & Schrader, J. W. (1992). Welham M J, Abraham S et al. p21ras activation via hemopoietic receptors and c-kit requires tyrosine kinase activity but not tyrosine phosphorylation of p21ras GTPase-activating protein. Proc. Natl. Acad. Sci. USA 89,1587-1591.

Dutton, R. W., Wetzel, G. D., & Swain, S. L. (1984). Partial purification and characterization of a BCGFII from EL4 culture supernatants. J. Immunol. 132,2451-2456.

Egan, S. E., Giddings, B. W., Brooks, M. W., Buday, L., Sizeland, A. M., & Weinberg, R. A. (1993). Association of Sos Ras exchange protein with Grb2 is implicated in tyrosine kinase signal transduction and transformation. Nature 363,45-51.

Erard, F., Wild, M.-T., Garcia-Sanz, J. A., & Le Gros, G. (1993). CD8+ T lymphocytes can develop into non-cytolytic CD8- CD4- cells which produce IL-4, IL-5, IL-10, and help B cells. Science 260,1802-1805.

Firmbach-Kraft, I., Byers, M., Shows, T., Dalla-Favera, R., & Krolewski, J. J. (1990). Tyk2, prototype of a novel class of non-receptor tyrosine kinase genes. Oncogene 5,1329-1336.

Foster, P. F., Sankary, H. N., Hart, M., Ashmann, M., & Williams, J. W. (1989). Blood and graft eosinophilia as predictors of rejection in human liver transplantation. Transplant. 47,72-74.

Foster, P., Hogan, S., Ramsay, A., Young, I., & Matthaei, K. (1996). The role of IL-5 in aeroallergen induced airway eosinophilia and lung damage. J. Exp. Med. 183, 195-201.

Frew, A. J. & Kay, A. B. (1988). The relationship between infiltrating CD4+ lymphocytes, activated eosinophils, and the magnitude of the allergen-induced late phase cutaneous reaction in man. J. Immunol. 141,4158-4164.

Fujisawa, T., Abu-Ghazaleh, R., Kita, H., Sanderson, C. J., & Gleich, G. J. (1990). Regulatory effect of cytokines on eosinophil degranulation. J. Immunol. 144,642-646.

Gaga, M., Frew, A. J., Varney, V. A., & Kay, A. B. (1991). Eosinophil activation and T lymphocyte infiltration in allergen-induced late phase skin reactions and classical delayed-type hypersensitivity. J. Immunol. 147,816-822.

Go, N. F., Castle, B. E., Barrett, R., Kastelein, R., Dang, W., Mosmann, T. R., Moore, K. W., & Howard, M. (1990). Interleukin 10, novel B cell stimulatory factor: Unresponsiveness of X chromosome-linked immunodeficiency B cells. J. Exp. Med. 172,1625-1631.

Gorman, D. M., Itoh, N., Jenkins, N. A., Gilbert, D. J., Copeland, N. G., & Miyajima, A. (1992). Chromosomal localization and organization of the murine genes encoding the beta subunits (AIC2A and AIC2B) of the interleukin 3, granulocyte/macrophage colony-stimulating factor, and interleukin 5 receptors. J. Biol. Chem. 267,15842-15848.

Graber, P., Proudfoot, A. E. I., Talabot, R., Bernard, A., McKinnon, M., Banks, M., Fattah, D., Solari, R., Peitsch, M. C., & Wells, T. N. C. (1995). Identification of key charged residues of human interleukin-5 in receptor binding and cellular activitation. J. Biol. Chem. 270,15762-15769.

Gregory, R. J., Kammermeyer, K. L., Vincent, W. S., & Wadsworth, S. G. (1987). Primary sequence and developmental expression of a novel *Drosophila* melanogaster srk gene. Mol. Cell. Biol. 7,2119-2127.

Gruart-Gouilleux, V., Engels, P., & Sullivan, M. (1995). Characterization of the human interleukin-5 gene promoter: Involvement of octamer binding sites in the gene promoter activity. Eur. J. Immunol. 25,1431-1435.

Hagan, O. P., Wilkins, H. A., Blumental, U. J., Hayes, R. J., & Greenwood, B. M. (1985). Eosinophilia and resistance of Schistoma hematobium in man. Parasite. Immunol. 7,625-632.

Hamid, Q., Azzawi, M., Ying, S., Moqbel, R., Wardlaw, A. J., Corrigan, C. J., Bradley, B., Durham, S. R., Collins, J. V., Jeffery, P. K., et al. (1991). Expression of mRNA for interleukin-5 in mucosal bronchial biopsies from asthma. J. Clin. Invest. 87,1541-1546.

Hamid, Q., Boguniewicz, M., & Leung, D. Y. (1994). Differential *in situ* cytokine gene expression in acute versus chronic atopic dermatitis. J. Clin. Invest. 94,870-876.

Hasbold, J. & Klaus, G. G. B. (1994). B cells from CBA/N mice do not proliferate following ligation of CD40. Eur. J. Immunol. 24,152-157.

Hayakawa, K., Hardy, R. R., Herzenberg, L. A., & Herzenberg, L. A. (1985). Progenitors for Ly-1 B cells are distinct from progenitors for other B cells. J. Exp. Med. 161,1554-1568.

Hayashi, S. I., Kunisada, T., Ogawa, M., Sudo, T., Kodama, H., Suda, T., Nishikawa, S., & Nishikawa, S. (1990). Sepwise progression of B-lineage differentiation supported by interleukin-7 and other stromal cell moleculars. J. Exp. Med. 171,1683-1695.

Herndon, F. J. & Kayes, S. G. (1992). Depletion of eosinophils by anti-IL-5 monoclonal antibody treatment of mice infected with *Trichinella spiralis* does not alter parasite burden or immunologic resistance to reinfection. J. Immunol. 149,3642-3647.

Herron, L. R., Coffman, R. L., Bond, M. W., & Kotzin, B. L. (1988). Increased autoantibody production by NZB/NZW B cells in response to IL-5. J. Immunol. 141,842-848.

Hirai, K., Yamaguchi, M., Misaki, Y., Takaishi, T., Ohta, K., Morita, Y., Ito, K., & Miyamoto, T. (1990). Enhancement of human basophil histamine release by interleukin 5. J. Exp. Med. 172,1525-1528.

Hitoshi, Y., Yamaguchi, N., Mita, S., Sonoda, E., Takaki, S., Tominaga, A., & Takatsu, K. (1990). Distribution of IL-5 receptor-positive B cells. Expression of IL-5 receptor on Ly-1(CD5)+ B cells. J. Immunol. 144,4218-4225.

Hitoshi, Y., Sonoda, E., Kikuchi, Y., Yonehara, S., Nakauchi, H., & Takatsu, K. (1993). Sonoda E, Kikuchi Y et al. Interleukin-5 receptor positive B cells, but not eosinophils are functionally and numerically influenced in the mice carried with X-linked immunodeficiency. Int. Immunol. 9,1183-1190.

Hsieh, C. S., Macatonia, S. E., O'Garra, A., & Murphy, K. M. (1993a). Pathogen induced Th1 phenotype development in CD4+ alpha-beta-TCR transgenic T cells is macrophage dependent. Int. Immunol. 5,371-382.

Hsieh, C. S., Macatonia, S. E., Tripp, C. S., O'Garra, A., & Murphy, K. M. (1993b). Development of Th1 CD4+ T cells through IL-12 produced by listeria-induced macrophages. Science 260,547-549.

Huston, M. M., Moore, J. P., Mettes, H., Tavana, G., & Huston, D. P. (1996). Human B cells express IL-5 receptor mRNA and respond to IL-5 with enhanced IgM production after mitogenic stimulation with Moraxella catarrhalis. J. Immunol. 156, 1392-1401.

Imamura, F., Takaki, S., Akagi, K., Ando, M., Yamamura, K., Takatsu, K., & Tominaga, A. (1994). The murine interleukin-5 receptor alpha-subunit gene: Characterization of the gene structure and chromosome mapping. DNA. Cell. Biol. 13,283-292.

Inoue, C., Ichikawa, A., Hotta, T., & Saito, H. (1990). Constitutive gene expression of interleukin-5 in Kimura's disease. Br. J. Haematol. 76,554-555.

Isobe, M., Kumura, Y., Murata, Y., Takaki, S., Tominaga, A., Takatsu, K., & Ogita, Z. (1992). Localization of the gene encoding the alpha subunit of human interleukin-5 receptor (IL5RA) to chromosome region 3p24-3p26. Genomics 14,755-758.

Iwasaki, K., Torisu, M., & Fujimura, T. (1981). Tumor-associated eosinophilia: A review. J. Clin. Pathol 34,1343-1348.

Johanson, K., Appelbaum, E., Doyle, M., Hensley, P., Zhao, B., Abdel-Meguid, S. S., Young, P., Cook, R., Carr, S., Matico, R., Cusimano, D., Dul, E., Angelichio, M., Brooks, I., Winborne, E., McDonnell, P., Morton, T., Bennett, D., Sokoloski, T., McNulty, D., Rosenberg, M., & Chaiken, I. (1995). Binding interactions of human interleukin 5 with its receptor a subunit. J. Biol. Chem. 270,9459-9471.

Johnston, J. A., Kawamura, M., Kirken, R. A., Chen, Y. Q., Blake, T. B., Shibuya, K., Ortaldo, J. R., McVicar, D. W., & O'Shea, J. J. (1994). Phosphorylation and activation of the Jak-3 Janus kinase in response to interleukin-2. Nature 70,151-153.

Kamesaki, H., Zwiebel, J. A., Reed, J. C., & Cossman, J. (1994). Role of bcl-2 and IL-5 in the regulation of anti-IgM-induced growth arrest and apoptosis in immature B cell lines. A cooperative regulation model for B cell clonal deletion. J. Immunol. 152,3294-3305.

Katoh, S., Tominaga, A., Migita, M., Kudo, A., & Takatsu, K. (1990). Conversion of normal LY-1-positive B lineage cells into Ly-1-positive macrophages in long-term bone marrow cultures. Dev. Immunol. 1,113-125.

Kaushansky, K. & Karplus, P. A. (1993). Hematopoietic growth factors: understanding functional diversity in structural terms. Blood 82,3229-3240.

Kay, A. B., Ying, S., Varney, V., Gaga, M., Durham, S. R., Moqbel, R., Wardlaw, A. J., & Hamid, Q. (1991). Messenger RNA expression of the cytokine gene cluster, interleukin 3 (IL-3), IL-4, IL-5, and granulocyte/macrophage colony-stimulating factor, in allergen-induced late-phase cutaneous reactions in atopic subjects. J. Exp. Med. 173,775-778.

Kay, A. B., Hamid, Q., Robinson, D. S., Bentley, A. M., Tsicopoulos, A., Ying, S., Moqbel, R., Corrigan, C. J., & Durham, S. R. (1993). Asthma, eosinophils, and interleukin 5. In: *Eosinophils in Allergy and Inflammation* (Gleich, G. J., & Kay, A. B., Eds.). Marcel Dekker, Inc., New York, pp. 395-403.

Kikuchi, Y., Migita, M., Takaki, S., Tominaga, A., & Takatsu, K. (1994). Biochemical and functional characterization of soluble form of IL-5 receptor alpha (sIL-5R alpha). Development of ELISA system for detection of sIL-5R alpha. J. Immunol. Methods. 167,289-298.

Kikuchi, Y., Yasue, T., Miyake, K., Kimoto, M., & Takatsu, K. (1996). Cross-linking of CD38 induces tyrosine phosphorylation of Bruton's tyrosine kinase and expression of IL-5 receptor: Synergistic effect with IL-5. Proc. Natl. Acad. Sci. USA 92, 10879-10883.

Kinashi, T., Harada, N., Severinson, E., Tanabe, T., Sideras, P., Konishi, M., Azuma, C., Tominaga, A., Bergstedt-Lindqvist, S., Matsuda, F., Yaoita, Y., Takatsu, K., & Honjo, T. (1986). Cloning of complementary DNA encoding T-cell replacing factor and identity with B-cell growth factor II. Nature 324,70-73.

Kishimoto, T., Taga, T., & Akira, S. (1994). Cytokine signal transduction. Cell 76,253-262.

Kitamura, T., Sato, N., Arai, K., & Miyajima, A. (1991a). Expression cloning of the human IL-3 receptor cDNA reveals a shared b subunit for the human IL-3 and GM-CSF receptors. Cell 66,1165-1174.

Kitamura, T., Takaku, F., & Miyajima, A. (1991b). IL-1 up-regulates the expression of cytokine receptors on a factor-dependent human hemopoietic cell line, TF-1. Int. Immunol. 3, 571-577.

Koch, C. A., Anderson, D., Moran, M. F., Ellis, C., & Pawson, T. (1991). SH2 and SH3 domains: Elements that control interactions of cytoplasmic signaling proteins. Science 252,668-674.

Kodama, S., Tsuruoka, N., & Tsujimoto, M. (1991). Role of the C-terminus in the biological activity of human interleukin 5. Biochem. Biophys. Res. Commun. 178,514-519.

Kodama, S., Tsujimoto, M., Tsuruoka, N., Sugo, T., Endo, T., & Kobata, A. (1993). Role of sugar chains in the *in vitro* activity of recombinant human interleukin 5. Eur. J. Biochem. 211,903-908.

Koike, M., & Takatsu, K. (1994). IL-5 and its receptor: Which role do they play in the immune response? Int. Arch. Allergy. Immunol. 104,1-9.

Kopf, M., Brombacher, F., Hodgkin, P. D., Ramsey, A. J., Mibourne, E. A., Dai, W. J., Ovington, K. S., Behm, C. A., Kohler, G., Young, I. G., & Matthaei, K. I. (1996). IL-5 deficient mice have a developmental defect in CD5+ B-1 cells and lack eosinophilia but have normal antibody and cytotoxic T cell responses. Immunity 4, 15-24.

Kouro, T., Kikuchi, Y., Kanazawa, H., Hirokawa, K., Harada, N., Shiiba, M., Wakao, H., Takaki, S., & Takatsu, K. (1996). Critical proline residues of the cytoplasmic domain of the IL-5 receptor alpha chain and its function in IL-5 mediated activation of JAK kinase and STAT5. Int. Immunol. 8, 237-246.

Kung, T. T., Stelts, D. M., Zurcher, J. A., Adams, G. K., Egan, R. W., Kreutner, W., Watnick, A. S., Jones, H., & Chapman, R. W. (1995a). Involvement of IL-5 in a murine model of allergic pulmonary inflammation: Prophylactic and therapeutic effect of an anti-IL-5 antibody. Am. J. Respir. Cell. Mol. Biol. 13,360-365.

Kung, T. T., Stelts, D., Zurcher, Z. A., Jones, H., Umland, S. P., Kreutner, W., Egan, R. W., & Chapman, R. W. (1995b). Mast cells modulate allergic pulmonary eosinophilia in mice. Am. J. Respir. Cell. Mol. Biol. 12,404-409.

Larner, A. C., David, M., Feldman, G. M., Igarashi, H., Hackett, R. H., Webb, D. S., Sweitzer, S. M., Petricoin, E. F., & Finbloom, D. S. (1993). Tyrosine phosphorylation of DNA binding proteins by multiple cytokines. Science 261,1730-1733.

Le Beau, M. M., Lemons, R. S., Espinosa, R. 3d, Larson, R. A., Arai, N., & Rowley, J. D. (1989). Interleukin-4 and interleukin-5 map to human chromosome 5 in a region encoding growth factors and receptors and are deleted in myeloid leukemias with a del(5q). Blood 73,647-650.

Lee, H. J., Koyano-Nakagawa, N., Naito, Y., Nishida, J., Arai, N., Arai, K., & Yokota, T. (1993). cAMP activates the IL-5 promoter synergistically with phorbol ester through the signaling pathway involving protein kinase A in mouse thymoma line EL-4. J. Immunol. 151,6135-6142.

Le Gros, G., Ben-Sasson, S. Z., Seder, R. A., Finkelman, F. D., & Paul, W. E. (1990). Generation of interleukin 4 (IL-4)-producing cells *in vivo* and *in vitro*: IL-2 and IL-4 are required for *in vitro* generation of IL-4 producing cells. J. Exp. Med. 172,921-929.

Lemanske, R. F., & Kaliner, M. A. (1994). Late phase allergic reactions. In: *Allergy: Principles and Practices* (Middleton, E., Reed, C. E., Ellis, E. F., Adkinson, N. F., Yunginger, J. W., & Busse, W. W., Eds.). Mosby, St. Louis, 320-361.

Li, J., Cook, R., Dede, K., & Chaiken, I. (1996). Single chain human interleukin 5 and its asymmetric mutagenesis for mapping receptor binding sites. J. Biol. Chem. 271, 1817-1820.

Li, T., Tsukada, S., Satterthwaite, A., Havlik, M. H., Park, H., Takatsu, K., & Wittc, O. H. (1995). Activation of Bruton's tyrosine kinase (BTK) by a point mutation in its pleckstrin homology (PH) domain. Immunity 2,451-460.

Lopez, A. F., Begley, C. G., Williamson, D. J., Warren, D. J., Vadas, M. A., & Sanderson, C. J. (1986). Murine eosinophil differentiation factor: An eosinophil-specific colony-stimulating factor with activity for human cells. J. Exp. Med. 163,1085-1099.

Lopez, A. F., Sanderson, C. J., Gamble, J. R., Campbell, H. D., Young, I. G., & Vadas, M. A. (1988). Recombinant human interleukin 5 is a selective activator of human eosinophil function. J. Exp. Med. 167,219-224.

Lutticken, C., Wegenka, U. M., Yuan, J., Buschmann, J., Schindler, C., Ziemiecki, A., Harper, A. G., Wilks, A. F., Yasukawa, K., & Taga, T. (1994). Association of transcription factor APRF and protein kinase Jak1 with the interleukin-6 signal transducer gp130. Science 263,89-92.

MacDonald, D., Gordon, A. A., Kajitani, H., Enokihara, H., & Barrett, A. J. (1990). Interleukin-2 treatment-associated eosinophilia is mediated by interleukin-5 production. Br. J. Haematol. 76,168-173.

Mahanty, S., Abrams, J. S., King, C. L., Limaye, A. P., & Nutman, T. B. (1992). Parallel regulation of IL-4 and IL-5 in human helminth infections. J. Immunol. 148,3567-3571.

Mandler, R., Chu, C. C., Paul, W. E., Max, E. E., & Snapper, C. M. (1993). Interleukin 5 induces S mu-S gamma 1 DNA rearrangement in B cells activated with dextran-anti-IgD antibodies and interleukin 4: A three component model for Ig class switching. J. Exp. Med. 178,1577-1586.

Mano, H., Ishikawa, F., Nishida, J., Hirai, H., & Takaku, F. (1990). A novel protein-tyrosine kinase, tec, is preferentially expressed in liver. Oncogene 5,1781-1786.

Martinez, O. M., & Krams, S. M. (1993). Human IL-5 augments IL-2-dependent T cell proliferation. J. Immunol. 150,45A.

Martinez, O. M., Krams, S. M., Sterneck, M., Villanueva, J. C., Falco, D. A., Ferrell, L. D., Lake, J., Roberts, J. P., & Ascher, N. L. (1992). Intragraft cytokine profile during human liver allograft rejection. Transplant. 53,449-456.

Martinez, O. M., Ascher, N. L., Ferrell, L., Villanueva, J., Lake, J., Roberts, J. P., & Krams, S. M. (1993a). Evidence for a nonclassical pathway of graft rejection involving interleukin 5 and eosinophils. Transplant. 55,909-918.

Martinez, O. M., Krams, S. M., Villanueva, J. C., Ferrell, L., Lake, J., Roberts, J. P., & Ascher, N. L. (1993b). Intragraft eosinophilia and interleukin-5 mRNA accompany liver allograft rejection. Transplant. Proc. 25,126-127.

Martinez, O. M., Villanueva, J. C., Lake, J., Roberts, J. P., Ascher, N. L., & Krams, S. M. (1993c). IL-2 and IL-5 gene expression in response to alloantigen in liver allograft recipients and *in vitro*. Transplant. 55,1159-1166.

Matsuda, M. B., Mayer, B. G., & Hanafusa, H. (1991). Identification of domains of the v-crk oncogene product sufficient for association with phosphotyrosine-containing proteins. Mol. Cell. Biol. 11,1607-1613.

Matsumoto, R., Ando, M., Kohrogi, H., Araki, S., & Takatsu, K. (1992). Interleukin-5 levels of pleural fluid and serum samples in a patient with PIE syndrome. Chest 102,1296-1297.

Mauser, P. J., Pitman, A., Witt, A., Fernandez, X., Zurcher, J., Kung, T., Jones, H., Watnick, A. S., Egan, R. W., Kreutner, W., et al. (1993). Inhibitory effect of the TRFK-5 anti-IL-5 antibody in a guinea pig model of asthma. Am. Rev. Respir. Dis. 148,1623-1627.

Mauser, P. J., Pitman, A. M., Fernandez, X., Foran, S. K., Adams III, G. K., Kreutner, W., Egan, R. W., & Chapman, R. W. (1995). Effects of an antibody to IL-5 in a monkey model of asthma. Am. J. Respir. Crit. Care. Med. 152,467-472.

McKenzie, A. N., Barry, S. C., Strath, M., & Sanderson, C. J. (1991a). Structure-function analysis of interleukin-5 utilizing mouse/human chimeric molecules. EMBO J. 10,1193-1199.

McKenzie, A. N., Ely, B., & Sanderson, C. J. (1991b). Mutated interleukin-5 monomers are biologically inactive. Mol. Immunol. 28,155-158.

McNeil, H. P., & Austen, K. F. (1995). Biology of the Mast Cell. In: *Immunological Disease*, 5th ed. (Frank, M. M., Ausen, K. F., Claman, H. N., Unanue, E., Eds.). Little Brown, Boston, pp. 185-204.

Metwali, A., Elliott, D., Blum, A. M., & Weinstock, J. V. (1993). Granuloma eosinophils enhance IL-5 production by lymphocytes from mice infected with schistosoma mansoni. J. Immunol. 151,7048-7056.

Migita, M., Yamaguchi, N., Mita, S., Higuchi, S., Hitoshi, Y., Yoshida, Y., Tomonaga, M., Matsuda, I., Tominaga, A., & Takatsu, K. (1991). Characterization of the human IL-5 receptors on eosinophils. Cell. Immunol. 133,484-497.

Milburn, M. V., Hassell, A. M., Lambert, M. H., Jordan, S. R., Proudfoot, A. E., Graber, P., & Wells, T. N. (1993). A novel dimer configuration revealed by the crystal structure at 2.4 A resolution of human interleukin-5. Nature 363,172-176.

Mita, S., Harada, N., Naomi, S., Hitoshi, Y., Sakamoto, K., Akagi, M., Tominaga, A., & Takatsu, K. (1988). Receptors for T cell-replacing factor/interleukin 5. Specificity, quantitation, and its implication. J. Exp. Med. 168,863-878.

Mita, S., Tominaga, A., Hitoshi, Y., Sakamoto, K., Honjo, T., Akagi, M., Kikuchi, Y., Yamaguchi, N., & Takatsu, K. (1989). Characterization of high-affinity receptors for interleukin 5 on interleukin 5-dependent cell lines. Proc. Natl. Acad. Sci. USA 86,2311-2315.

Mita, S., Takaki, S., Tominage, A., & Takatsu, K. (1993). Comparative analysis of kinetics of binding and internalization of IL-5 murine IL-5 receptor of high and low affinity. J. Immunol. 151,6924-6932.

Miyajima, A. (1992). Molecular structure of the IL-3, GM-CSF and IL-5 receptors. Int. J. Cell. Cloning 10,126-134.

Miyajima, A., Mui, A. L., Ogorochi, T., & Sakamaki, K . (1993). Receptor for granulocyte-macrophage colony-stimulating factor, interleukin-3, and interleukin-5. Blood 82,1960-1974.

Miyatake, S., Otsuka, T., Yokota, T., & Lee, F. (1985). Structure of the chromosomal gene for granulocyte-macrophage colony stimulating factor: Comparison of the mouse and human gene. EMBO J. 4,2561-2568.

Miyatake, S., Shlomai, J., Arai, K., & Arai, N. (1991). Characterization of the mouse granulocyte-macrophage colony-stimulating factor (GM-CSF) gene promoter: Nuclear factors that interact with an element shared by three lymphokine genes—those for GM-CSF, interleukin-4 (IL-4), and IL-5. Mol. Cell. Biol. 11,5894-5901.

Morikawa, K., Oseko, F., Morikawa, S., Imai, K., & Sawada, M. (1993). Recombinant human IL-5 augments immunoglobulin generation by human B lymphocytes in the presence of IL-2. Cell. Immunol. 149,390-401.

Morla, A. O., Schreurs, J., Miyajima, A., & Wang, J. Y. (1988). Hematopoietic growth factors activate the tyrosine phosphorylation of distinct sets of proteins in interleukin-3-dependent murine cell lines. Mol. Cell. Biol. 8,2214-2218.

Morton, T., Li, J., Cook, R., & Chaiken, I. (1995). Mutagenesis in the C-terminal region of human interleukin-5 reveals a central patch for receptor alpha chain recognition. Proc. Natl. Acad. Sci. USA 92, 10879-10883.

Morton, T. A., Bennett, D. B., Appelbaum, E. R., Cusimano, D. M., Johanson, K. O., Matico, R. E., Young, P. R., Doyle, M., & Chaiken, I. M. (1994). Analysis of the interaction between human interleukin-5 and the soluble domain of its receptor using a surface plasmon resonance biosensor. J. Mol. Recognit. 7,47-55.

Mosmann, T. R. & Coffman, R. L. (1989). TH1 and TH2 cells: Different patterns of lymphokine secretion lead to different functional properties. Annu. Rev. Immunol. 7,145-173.

Mui, A. L., Kay, R. J., Humphries, R. K., & Krystal, G. (1992). Ligand-induced phosphorylation of the murine interleukin 3 receptor signals its cleavage. Proc. Natl. Acad. Sci. USA 89,10812-10816.

Murakami, M., Narazaki, M., Hibi, M., Yawata, H., Yasukawa, K., Hamaguchi, M., Taga, T., & Kishimoto, T. (1991). Critical cytoplasmic region of the interleukin 6 signal transducer gp130 is conserved in the cytokine receptor family. Proc. Natl. Acad. Sci. USA 88,11349-11353.

Murata, Y., Yamaguchi, N., Hitoshi, Y., Tominaga, A., & Takatsu, K. (1990). Interleukin 5 and interleukin 3 induce serine and tyrosine phosphorylations of several cellular proteins in an interleukin 5-dependent cell line. Biochem. Biophys. Res. Commun. 173,1102-1108.

Murata, Y., Takaki, S., Migita, M., Kikuchi, Y., Tominaga, A., & Takatsu, K. (1992). Molecular cloning and expression of the human interleukin 5 receptor. J. Exp. Med. 175,341-351.

Nagai, H., Yamaguchi, S., Inagaki, N., Tsuruoka, N., Hitoshi, Y., & Takatsu, K. (1993). Effect of anti-IL-5 monoclonal antibody on allergic bronchial eosinophilia and airway hyperresponsiveness in mice. Life. Sci. 53,PL243-PL247.

Nagasawa, M., Ohshiba, A., & Yata, J. (1991). Effect of recombinant interleukin 5 on the generation of cytotoxic T cells (CTL). Cell. Immunol. 133,317-326.

Nakajima, H., Iwamoto, I., Tomoe, S., Matsumura, R., Tomioka, H., Takatsu, K., & Yoshida, S. (1992). CD4+ T-lymphocytes and interleukin-5 mediate antigen-induced eosinophil infiltration into the mouse trachea. Am. Rev. Respir. Dis. 146,374-377.

Nakashima, Y., Mita, S., Takatsu, K., & Ogawa, M. (1993). Interleukin-5 induces tumor suppression by peritoneal exudate cells in mice. Cancer Immunol. Immunother. 37,227-232.

Naora, H., & Young, I. G. (1995). Comparison of the mechanism regulating IL-5, IL-4, and three other lymphokines genes in the Th2 clone D10.G4.1. Exp. Hematol. 23,597-602.

Nishinakamura, R., Nakayama, N., Hirabayashi, Y., Inoue, T., Aud, D., McNeil, T., Azuma, S., Yoshida, S., Toyoda, Y., Arai, K., Miyajima, A., & Murray, R. (1995). Mice deficient for the IL-3/GM-CSF/IL-5 βc exhibit lung pathology and impaired immune response, while β_{IL3} receptor-deficient mice are normal. Immunity 2, 211-222.

Noma, T., Mizuta, T., Rosen, A., Hirano, T., Kishimoto, T., & Honjo, T. (1987). Enhancement of the interleukin 2 receptor expression on T cells by multiple B-lymphotropic lymphokines. Immunol. Lett. 15,249-253.

Ogawa, M., Nishikawa, S., Ikuta, K., Yamamura, F., Naito, M., Takahashi, K., & Nishikawa, S. (1988). B cell ontogeny in murine embryo studied by a culture system with the monolayer of a stromal cell clone, ST2: Cell progenitor develops first in embryonal body rather than in yok sac. EMBO J. 7,1337-1343.

Okayama, H., Fushimi, T., Shimura, S., Sasaki, H., & Shirato, K. (1994). Allergens, IgE, mediators, inflammatory mechanisms. J. Allergy. Clin. Immunol. 93,1006-1012.

Okubo, Y., Hossain, M., M., Rai, R., Sato, E., Honda, T., Sekiguchi, M., Itoh, S., & Takatsu, K. (1995). Adhesion molecules on eosinophils in acute eosinophilic pneumonia. Am. J. Rev. Respr. Crit. Care Med. 151,1259-1262.

O'Neal, K. D., S. & hearer, W. T. (1995). Letter to the Editor. Clarification of the role of Pro-X-Pro sequence in IL-5 receptor signal transduction. Mol. Cell. Biol. 15,4657-4658.

O'Neal, K. D., & Yu-Lee, L. (1993). The proline-rich motif (PRM): A novel feature of the cytokine/hematopoietin receptor superfamily. Lymphokine. Cytokine Res. 12,309-312.

Owen, W. F., Rothenberg, M. E., Petersen, J., Weller, P. F., Silverstein, D., Sheffer, A. L., Stevens, R. L., Soberman, R. J., & Austen, K. F. (1989). Interleukin 5 and phenotypically altered eosinophils in the blood of patients with the idiopathic hypereosinophilic syndrome. J. Exp. Med. 170,343-348.

Owen, W. F., Jr., Petersen, J., Sheff, D. M., Folkerth, R. D., Anderson, R. J., Corson, J. M., Sheffer, A. L., & Austen, K. F. (1990). Hypodense eosinophils and interleukin 5 activity in the blood of patients with the eosinophilia-myalgia syndrome. Proc. Natl. Acad. Sci. USA 87,8647-8651.

Pandit, J., Bohm, A., Jancarik, J., Halenbeck, R., Koths, K., & Kim, S. (1992). Three-dimensional structure of dimeric human recombinant macrophage colony-stimulating factor. Science 258,1358-1362.

Pasternark, A., & Jansa, P. (1984). Local eosinophilia in stroma of tumors related to prognosis. Neoplasm 31,323-326.

Paul, C. C., Keller, J. R., Armpriester, J. M., & Baumann, M. A. (1990). Epstein-Barr virus transformed B lymphocytes produce interleukin-5. Blood 75,1400-1403.

Pazdrak, K., Stafford, S., & Alam, R. (1995). The activation of the Jak-STAT1 signaling pathway by IL-5 in eosinophils. J. Immunol. 155,397-402.

Pelicci, G., Lanfrancone, L., Grignani, F., McGlade, J., Cavallo, F., Forni, G., Nicoletti, I., Grignani, F., Pawson, T., & Pellicci, P. G. (1992). A novel transforming protein (SHC) with an SH2 domain is implicated in mitogenic signal transduction. Cell 70,93-104.

Plaetinck, G., Van der Heyden, J., Tavernier, J., Fache, I., Tuypens, T., Fischkoff, S., Fiers, W., & de Vos, R. (1990). Characterization of interleukin 5 receptors on eosinophilic sublines from human promyelocytic leukemia (HL-60) cells. J. Exp. Med. 172,683-691.

Plaut, M., Pierce, J. H., Watson, C. J., Hanley-Hyde, J., Nordan, R. P., & Paul, W. E. (1989). Mast cell lines produce lymphokines in response to cross-linkage of Fc epsilon RI or to calcium ionophores. Nature 339,64-67.

Pretlow, T. P., Keith, E. F., Cryar, A. K., Bartolucci, A. A., Pitts, A. M., Pretlow, T. G., Kimball, P. M., & Boohaker, E. A. (1983). Eosinophil infiltration of human colonic carcinomas as a prognostic indicators. Cancer Res. 43,2997-3000.

Rawlings, D., Saffran, D. C., Tsukada, S., Largaespada, D. A., Grimaldi, J. C., Cohen, L., & Mohr, R. N. (1993). Mutation of unique region of Bruton's tyrosine kinase in immunodeficient Xid mice. Science 261,358-360.

Renz, H., Smith, H. R., Henson, J. E., Ray, B. S., Irvin, C. G., & Gelfand, E. W. (1992). Aerosolized antigen exposure without adjuvant causes increased IgE production and increased airway responsiveness in the mouse. J. Allergy. Clin. Immunol. 89,1127-1138.

Rivoltini, L., Viggiano, V., Spinazze, S., Santoro, A., Colombo, M. P., Takatsu, K., & Parmiani, G. (1993). *In vitro* anti-tumor activity of eosinophils from cancer patients treated with subcutaneous administration of interleukin 2. Role of interleukin 5. Int. J. Cancer 54,8-15.

Robinson, D. S., Hamid, Q., Ying, S., Tsicopoulos, A., Barkans, J., Bentley, A. M., Corrigan, C., Durham, S. R., & Kay, A. B. (1992). Predominant TH2-like bronchoalveolar T-lymphocyte population in atopic asthma. N. Engl. J. Med. 326,298-304.

Rolink, A. G., Thalmann, P., Kikuchi, Y., & Erdei, A. (1990). Characterization of the interleukin 5-reactive splenic B cell population. Eur. J. Immunol. 20,1949-1956.

Rossi, G. A., Crimi, E., Lantero, S., Gianiorio, P., Oddera, S., Crimi, P., & Brusasco, V. (1991). Late-phase asthmatic reaction to inhaled allergen is associated with early recruitment of eosinophils in the airways. Am. Rev. Respir. Dis. 144,379-383.

Rozwarski, D. A., Gronenborn, A. M., Clore, G. M., Bazan, J. F., Bohm, A., Wlodawer, A., Hatada, M., & Karplus, P. A. (1994). Structural comparison among the short-chain helical cytokines. Structure 2,159-173.

Sakamaki, K., Miyajima, I., Kitamura, T., & Miyajima, A. (1992). Critical cytoplasmic domains of the common beta subunit of the human GM-CSF, IL-3 and IL-5 receptors for growth signal transduction and tyrosine phosphorylation. EMBO J. 11,3541-3549.

Samoszuk, M. & Nansen, L. (1990). Detection of interleukin-5 messenger RNA in Reed-Sternberg cells of Hodgkin's disease with eosinophilia. Blood 75,13-16.

Sasaki, O., Sugaya, H., Ishida, K., & Yoshimura, K. (1993). Ablation of eosinophils with anti-IL-5 antibody enhances the survival of intracranial worms of Angiostrongylus cantonensis in the mouse. Parasite. Immunol. 15,349-354.

Sato, N., Sakamaki, K., Terada, N., Arai, K., & Miyajima, A. (1993). Signal transduction by the high-affinity GM-CSF receptor: Two distinct cytoplasmic regions of the common b subunit responsible for different signaling. EMBO J. 12,4181-4189.

Sato, S., Katagiri, T., Takaki, S., Kikuchi, Y., Hitoshi, Y., Yonehara, S., Tsukada, S., Kitamura, D., Watanabe, T., Witte, O., & Takatsu, K. (1994). IL-5 receptor-mediated tyrosine phosphorylation of SH2/SH3-containing proteins and activation of Bruton's Tyrosine and Janus 2 Kinases. J. Exp. Med. 180,2101-2111.

Satoh, T., Nakafuku, M., Miyajima, A., & Kaziro, Y. (1991). Involvement of p21ras protein in signal-transduction pathways from interleukin 2, interleukin 3 and granulocyte/macrophage colony-stimulating factor, but not from interleukin 4. Proc. Natl. Acad. Sci. USA 88,3314-3318.

Schumacher, J. H., O'Garra, A., Shrader, B., van-Kimmenade, A., Bond, M. W., Mosmann, T. R., & Coffman, R. L. (1988). The characterization of four monoclonal antibodies specific for mouse IL-5 and development of mouse and human IL-5 enzyme-linked immunosorbent. J. Immunol. 141,1576-1581.

Sehmi, R., Wardlaw, A. J., Cromwell, O., Kurihara, K., Waltmann, P., & Kay, A. B. (1992). Interleukin-5 selectively enhances the chemotactic response of eosinophils obtained from normal but not eosinophilic subjects. Blood 79,2952-2959.

Senda, T., Shimazu, T., Matsuda, S., Kawano, G., Shimizu, H., Nakamura, K. T., & Mitsui, Y. (1992). Three-dimensional crystal structure of recombinant murine interferon-beta. EMBO J. 11,3193-3201.

Shanafelt, A. B., Miyajima, A., Kitamura, T., & Kastelein, R. A. (1991). The amino-terminal helix of GM-CSF and IL-5 governs high affinity binding to their receptors. EMBO J. 10,4105-4112.

Sher, A., Coffman, R. L., Hieny, S., & Cheever, A. W. (1990). Ablation of eosinophil and IgE responses with anti-IL-5 or anti-IL-4 antibodies fails to affect immunity against Schistosoma mansoni in the mouse. J. Immunol. 145,3911-3916.

Shirai, T., Hirose, S., Okada, T., & Nishimura, N. (1991). CD5+ B cells in autoimmune disease and lymphoid malignancy. Clin. Immunol. Immunopathol. 59, 173-186.

Shuai, K., Stark, G. R., Prezioso, V. R., Kerr, I. M., & Darnell, J. E. (1993). A single phosphotyrosine residue of Stat91 required for gene activation by interferon-γ. Science 261,1736-1739.

Siliciano, J. D., Morrow, T. A., & Desiderio, S. V. (1992). *itk*, a T-cell-specific tyrosine kinase gene inducible by interleukin-2. Proc. Natl. Acad. Sci. USA 89,11194-11198.

Silvennoinen, O., Witthuhn, B. A., Quelle, F. W., Cleveland, J. L., Yi, T., & Ihle, J. N. (1993). Structure of the murine Jak2 protein-tyrosine kinase and its role in interleukin 3 signal transduction. Proc. Natl. Acad. Sci. USA 90,8429-8433.

Smith, M. R., DeGudicubus, S. J., & Stacey, D. W. (1986). Requirement for c-ras proteins during viral oncogene transformation. Nature 320,540-543.

Sonoda, Y., Arai, N., & Ogawa, M. (1989). Humoral regulation of eosinophilopoiesis *in vitro*: Analysis of the targets of interleukin-3, granulocyte/macrophage colony-stimulating factor (GM-CSF), and interleukin-5. Leukemia 3,14-18.

Sprang, S. R. & Bazan, J. F. (1993). Cytokine structural taxonomy and mechanisms of receptor engagement. Cur. Opin. Struct. Biol. 3,815-827.

Stahl, N., Boulton, T. G., Farruggella, T., Ip, N. Y., Davis, S., Witthuhn, B. A., Quelle, F. W., Silvennoinen, O., Barbieri, G., & Pellegrini, S. (1994). Association and activation of Jak-Tyk kinases by CNTF-LIF-OSM-IL-6 b receptor components. Science 263,92-95.

Taga, T., Hibi, M., Hirata, Y., Yamasaki, K., Yasukawa, K., Matsuda, T., Hirano, T., & Kishimoto, T. (1989). Interleukin-6 triggers the association of its receptor with a possible signal transducer, gp130. Cell 58,573-581.

Takagi, M., Hara, T., Ichihara, M., Takatsu, K., & Miyajima, A. (1995). Multi-colony stimulating activity of interleukin-5 (IL-5) on hematopoietic progenitors from transgenic mice that express IL-5 receptor alpha subunit constitutively. J. Exp. Med. 181,889-899.

Takahashi, M., Yoshida, M. C., Satoh, H., Hilgers, J., Yaoita, Y., & Honjo, T. (1989). Chromosomal mapping of the mouse IL-4 and human IL-5 genes. Genomics 4,47-52.

Takaki, S., Tominaga, A., Hitoshi, Y., Mita, S., Sonoda, E., Yamaguchi, N., & Takatsu. (1990). Molecular cloning and expression of the murine interleukin-5 receptor. EMBO J. 9,4367-4374.

Takaki, S., Mita, S., Kitamura, T., Yonehara, S., Yamaguchi, N., Tominaga, A., Miyajima, A., & Takatsu, K. (1991). Identification of the second subunit of the murine interleukin-5 receptor: Interleukin-3 receptor-like protein, AIC2B is a component of the high affinity interleukin-5 receptor. EMBO J. 10,2833-2838.

Takaki, S., Murata, Y., Kitamura, T., Miyajima, A., Tominaga, A., & Takatsu, K. (1993). Reconstitution of the functional receptors for murine and human interleukin 5. J. Exp. Med. 177,1523-1529.

Takaki, S., Kanazawa, H., Shiiba, M., & Takatsu, K. (1994). A critical cytoplasmic domain of the interleukin-5 (IL-5) receptor alpha chain and its function in IL-5-mediated growth signal transduction. Mol. Cell. Biol. 14,7404-7413.

Takatsu, K. (1995) Interleukin-5 and its receptor system: From genes to disease. R.G. Landes Company, Austin, TX.

Takatsu, K., Tanaka, K., Tominaga, A., Kumahara, Y., Hamaoka, T. (1980) Antigen-induced T-cell replacing factor (TRF). III. Establishment of T cell hybridoma producing TRF and analysis of released TRF. J. Immunol. 125, 2646-2653.

Takatsu, K., Kikuchi, Y., Takahashi, T., Honjo, T., Matsumoto, M., Harada, N., Yamaguchi, N., & Tominaga, A. (1987). Interleukin 5, a T-cell-derived B-cell differentiation factor also induces cytotoxic T lymphocytes. Proc. Natl. Acad. Sci. USA 84,4234-4238.

Takatsu, K., Tominaga, A., Harada, N., Mita, S., Matsumoto, M., Takahashi, T., Kikuchi, Y., & Yamaguchi, N. (1988). T cell-replacing factor (TRF)/interleukin 5 (IL-5): Molecular and functional properties. Immunol. Rev. 102,107-135.

Takatsu, K., Takaki, S., Hitoshi, Y., Mita, S., Katoh, S., Yamaguchi, N., & Tominaga, A. (1992). Cytokine receptors on Ly-1 B cells. IL-5 and its receptor system. Ann. N. Y. Acad. Sci. 651,241-258.

Tanabe, T., Konishi, M., Mizuta, T., Noma, T., & Honjo, T. (1987). Molecular cloning and structure of the human interleukin-5 gene. J. Biol. Chem. 262,16580-16584.

Tavernier, J., de Vos, R., Cornelis, S., Tuypens, T., Van-der-Heyden, J., Fiers, W., & Plaetinck, (1991). A human high affinity interleukin-5 receptor (IL5R) is composed of an IL5-specific alpha chain and a beta chain shared with the receptor for GM-CSF. Cell 66,1175-1184.

Tavernier, J., Tuypens, T., Plaetinck, G., Verhee, A., Fiers, W., & de Vos, R. (1992). Molecular basis of the membrane-anchored and two soluble isoforms of the human interleukin 5 receptor alpha subunit. Proc. Natl. Acad. Sci. USA 89,7041-7045.

Tavernier, J., Tuypens, T., Verhee, A., Plaetinck, G., de Vos, R., Van der Heyden, J., Guisez, Y., & Oefner, C. (1995). Identication of receptor-binding domains on human interleukin 5 and design of an interleukin 5-derived receptor antagonist. Proc. Natl. Acad. Sci. USA 92,5194-5198.

Tepper, R. I., Coffman, R. L., & Leder, P. (1992). An eosinophil-dependent mechanism for the antitumor effect of interleukin-4. Science 257,548-551.

Tepper, T. I., Pattengale, P. K., & Leder, P. (1989). Murine interleukin-4 displays potent anti-tumor activity *in vivo*. Cell 57,503-512.

Thomas, J. D., Sideras, P., Smith, C. I. E., Vorechovsky, I., Chapman, V., & Paul, W. E. (1993). Colocalization of X-linked agamma-globulinemia and X-linked immunodeficiency genes. Science 261,355-361.

Thompson, C. B. (1995). Distinct roles for the costimulatory ligands B7-1 and B7-2 in T helper cells differentiation? Cell 81,979-982.

Till, S., Li, B., Durham, S., Humbert, M., Assoufi, B., Huston, D., Dickason, R., Jeannin, P., Lau, A. B., & Corrigan, C. (1995). Secretion of eosinophil-active cytokines by bronchoalveolar lavage CD4+ and CD8+ T cells lines in atopic asthmatics and atopic and non-atopic controls. Eur. J. Immunol. 25,2727-2731.

Tominaga, A., Matsumoto, M., Harada, N., Takahashi, T., Kikuchi, Y., & Takatsu, K. (1988). Molecular properties and regulation of mRNA expression for murine T cell-replacing factor/IL-5. J. Immunol. 140,1175-1181.

Tominaga, A., Mita, S., Kikuchi, Y., Hitoshi, Y., Takatsu, K., Nishikawa, S., & Ogawa, M. (1989). Establishment of IL-5-dependent early B cell lines by long-term bone marrow cultures. Growth. Factor 1,135-146.

Tominaga, A., Takahashi, T., Kikuchi, Y., Mita, S., Naomi, S., Harada, N., Yamaguchi, N., & Takatsu, K. (1990). Role of carbohydrate moiety of IL-5. Effect of tunicamycin on the glycosylation of IL-5 and the biologic activity of deglycosylated IL-5. J. Immunol. 144,1345-1352.

Tominaga, A., Takaki, S., Koyama, N., Katoh, S., Matsumoto, R., Migita, M., Hitoshi, Y., Hosoya, Y., Yamauchi, S., Kanai, Y., et al. (1991). Transgenic mice expressing a B cell growth and differentiation factor gene (interleukin 5) develop eosinophilia and autoantibody production. J. Exp. Med. 173,429-437.

Trinchieri, G. (1989). Biology of human natural killer cells. Adv. Immunol. 47,187-376.

Tsukada, S., Saffran, D. C., Rawlings, D. J., Parolini, O., Allen, R. C., Klisak, I., Sparkes, R. S., Kubagawa, H., Mohandas, T., & Quan, S. (1993). Deficient expression of a B cell cytoplasmic tyrosine kinase in human X-linked agammaglobulinemia. Cell 72,279-290.

Tsukada, S., Rawlings, D. J., & Witte, O. N. (1994). Role of Bruton's tyrosine kinase in human and murine B cell immunodeficiency. Curr. Opin. Immunol. 6,623-630.

Tsuruoka, N., Funakoshi, K., Kodama, S., & Tsujimoto, M. (1990). Interaction of interleukin-5 with its receptors on murine leukemic BCL1 cells and its implication in biological activity. Cell. Immunol. 125,354-362.

Tsuruoka, N., Yamashiro, K., & Tsujimoto, M. (1993). Purification of soluble murine interleukin 5 (IL-5) receptor alpha expressed in Chinese hamster ovary cells and its action as an IL-5 antagonist. Arch. Biochem. Biophys. 307,133-137.

Ullrich, A. & Schlessinger, J. (1990). Signal transduction by receptors with tyrosine inase activity. Cell 61,203-211.

Umland, S. P., Go, N. F., Cupp, J. E., & Howard, M . (1989). Responses of B cells from autoimmune mice to IL-5. J. Immunol. 142,1528-1535.

Urban, J. F., Madden, K. B., Svetic, A., Cheever, A., Trotta, P. P., Gause, W. C., Katona, I. M., & Finkelman, F. D. (1992). The importance of Th2 cytokines in protective immunity to nematodes. Immunol. Rev. 127,205-220.

Urban, J. F., Katona, I. M., Paul, W. E., & Finkelman, F. D. (1991). Interleukin 4 is important in protective immunity to a gastrointestinal nematode infection in mice. Proc. Natl. Acad. Sci. USA 88,5513-5517.

Van der Brugger, T., Caldenhover, E., Kanters, D., Coffer, P., Raaijimakers, J. A., Lammers, J. W., & Koenderman, L. (1995). IL-5 signaling in human eosinophils involves JAK2 tyrosine kinase and Stat1 α. Blood 85,1442-1448.

Van Oosterhout, A. J. M., Van Ark, I., Folkerts, G., Van der Linde, H. J., Savelkoul, H. F. J., Verheyen, K. C. P., & Nijkamp, F. P. (1995). Antibody to interleukin-5 inhibits virus-induced airway hyperresponsiveness to histamine in guinea pigs. Am. J. Respir. Crit. Care. Med. 151,177-183.

Van Straaten, J. F. M., Dokter, W. H. A., Stulp, B. K., & Vellenga, E. (1994). The regulation of interleukin 5 and interleukin 3 gene expression in human T cells. Cytokine 6,229-234.

Velazquez, L., Fellous, M., Stark, G. R., & Pellegrini, S. (1992). A protein tyrosine kinase in the interferon α/β signaling pathway. Cell 70,313-322.

Vetrie, D., Vorechovsky, I., Sideras, P., Holland, J., Davies, A., Flinter, F., Hammarstrom, L., Kinnon, C., Levinsky, R., & Bobrow, M. (1993). The gene involved in X-linked agammaglobulinaemia is a member of the src family of protein-tyrosine kinases. Nature 361,226-233.

Walsh, G. M., Hartnell, A., Wardlaw, A. J., Kurihara, K., Sanderson, C. J., & Kay, A. B. (1990). IL-5 enhances the in vitro adhesion of human eosinophils, but not neutrophils, in a leucocyte integrin (CD11/18)-dependent manner. Immunol. 71,258-265.

Walsh, G. M., Wardlaw, A. J., Hartnell, A., Sanderson, C. J., & Kay, A. B. (1991). Interleukin-5 enhances the in vitro adhesion of human eosinophils, but not neutrophils, in a leucocyte integrin (CD11/18)-dependent manner. Int. Arch. Allergy. Appl. Immunol. 94,174-178.

Walter, M. R., Cook, W. J., Zhao, B. G., Cameron, R. P., Jr., Ealick, S. E., Walter, R. L., Jr., Reichert, P., Nagabhushan, T. L., Trotta, P. P., & Bugg, C. E. (1992). Crystal structure of recombinant human interleukin-4. J. Biol. Chem. 267,20371-20376.

Wang, Y., Campbell, H. D., & Young, I. G. (1993). Sex hormones and dexamethasone modulate interleukin-5 gene expression in T lymphocytes. J. Steroid. Biochem. Mol. Biol. 44,203-210.

Warren, H. S., Kinnear, B. F., Phillips, J. H., & Lanier, L. L. (1995). Production of IL-5 by human NK cells and regulation of IL-5 secretion by IL-4, IL-10, and IL-12. J. Immunol. 154,5144-5152.

Watanabe, A., Mishima, H., Renzi, P. M., Xu, L.-J., Hamid, Q., & Martin, J. G. (1995). Transfer of allergic airway responses with antigen-primed CD4+ but not CD8+ T cells in brown norway rats. J. Clin. Invest. 96,1303-1310.

Watanabe, Y., Kitamura, T., Hayashida, K., & Miyajima, A. (1992). Monoclonal antibody against the common beta subunit (beta c) of the human interleukin-3 (IL-3), IL-5, and granulocyte-macrophage colony-stimulating factor receptors shows upregulation of beta c by IL-1 and tumor necrosis factor-alpha. Blood 80,2215-2220.

Webb, C. F., Das, C., Coffman, R. L., & Tucker, P. W. (1989). Induction of immunoglobulin mu mRNA in a B cell transfectant stimulated with interleukin-5 and a T-dependent antigen. J. Immunol. 143,3934-3939.

Wierenga, E. A., Snoek, M., de Groot, C., Chretien, L., Bos, J. D., Jansen, H. M., & Kapsenberg, M. I. (1990). Evidence for compartmentalization of functional subsets of CD2+ T lymphocytes in atopic patients. J. Immunol. 144,4651-4656.

Wilson, S. D., Billings, R. P., D'Eustachio, P., Fournier, R. E., Geissler, E., Lalley, P. A., Burd, P. R., Housman, D. E., Taylor, B. A., & Dorf, M. E. (1990). Clustering of cytokine gene on mouse chromosome 11. J. Exp. Med. 171,1301-1314.

Witthuhn, B. A., Quelle, F. W., Silvenninen, O., Yi, T., Tang, B., Miura, O., & Ihle, J. N. (1993). JAK2 associates with the erythropoietin receptor and is erythropoietin. Cell 74,227-236.

Witthuhn, B. A., Silvennoinen, O., Miura, O., Lai, K. S., Cwik, C., Liu, E. C., & Ihle, J. N. (1994). Involvement of the Jak-3 Janus kinase in signalling by interleukin 2 and 4 in lymphoid and myeloid cells. Nature 370,153-157.

Woodcock, J. M., Zacharakis, B., Plaetinck, G., Bagley, C. J., Qiyu, S., Hercus, T. R., Tavernier, J., & Lopez, A. F. (1994). Three residues in the common beta chain of the human GM-CSF, IL-3 and IL-5 receptors are essential for GM-CSF and IL-5 but not IL-3 high affinity binding and interact with Glu21 of GM-CSF. EMBO J. 13,5176-5185.

Wu, H. K., Hirai, H., Inamori, K., Kitamura, K., & Takaku, F. (1992). Anti-tumor effects of interleukin-4 and interleukin-5 against mouse B cell lymphoma and possible mechanisms of their action. Jpn. J. Cancer Res. 83,200-210.

Yamagata, T., Nishida, J., Sakai, R., Tanaka, T., Honda, H., Hirano, N., Mano, H., Yazaki, Y., & Hirai, H. (1995). Of the GATA-binding proteins, only GATA-4 selectively regulates the human interleukin-5 gene promoter in interleukin-5-producing cells which express multiple GATA-binding proteins. Mol. Cell. Biol. 15,3830-3839.

Yamaguchi, N., Takahashi, T., Harada, N., & Takatsu, K. (1989). Mechanisms of the interleukin 5-induced differentiation of B cells. J. Biochem. (Tokyo). 106,837-843.

Yamaguchi, S., Okubo, Y., Hossain, M., Fujimoto, K., Honda, T., Kubo, K., Sekiguchi, M., & Takatsu, K. (1995). IL-5 predominant in bronchoalveolar lavage fluid and peripheral blood in a patient with acute eosinophil pneumonia. Internal Med. 34, 65-68.

Yamaguchi, Y., Hayashi, Y., Sugama, Y., Miura, Y., Kasahara, T., Kitamura, S., Torisu, M., Mita, S., Tominaga, A., & Takatsu, K. (1988). Highly purified murine interleukin 5 (IL-5) stimulates eosinophil function and prolongs in vitro survival. IL-5 as an eosinophil chemotactic factor. J. Exp. Med. 167,1737-1742.

Yamaguchi, Y., Suda, T., Ohta, S., Tominaga, K., Miura, Y., & Kasahara, T. (1991). Analysis of the survival of mature human eosinophils: Interleukin-5 prevents apoptosis in mature human eosinophils. Blood 78,2542-2547.

Yasukawa, K., Saito, T., Fukanaga, T., Sekimori, Y., Koishihara, Y., Fukui, H., Osugi, Y., Matsuda, T., Yawata, H., Hirano, T., Taga, T., & Kishimoto, T. (1990). Purification and characterization of soluble human interleukin-6 receptor expressed in CHO cells. J. Biochem. (Tokyo). 108,673.

Ying, S., Durham, S. R., Barkans, J., Masuyama, K., Jacobson, M., Rak, S., Lowhagen, O., Moqbel, R., Kay, A. B., & Hamid, Q. A. (1993). T cells are the principal source of interleukin-5 mRNA in allergen-induced rhinitis. Am. J. Respir. Cell. Mol. Biol. 9,356-360.

Yoshida, T., Ikuta, I., Sugaya, H., Maki, K., Takagi, M., Kanazawa, H., Sunaga, S., Kinashi, T., Yoshimura, K., Miyazaki, J-I., Takaki, S., & Takatsu, K. (1996). Defective B-1 cell development and impaired immunity against *Angiostrongylus cantonensis* in IL-5Rα deficient mice. Immunity 4, 483-494.

Yuan, D., Dang, T., & Sanderson, C. (1990). Regulation of Ig H chain gene transcription by IL-5. J. Immunol. 145,3491-3496.

INTERLEUKIN-6

Chun-fai Lai and Heinz Baumann

Growth Factors and Cytokines in Health and Disease
Volume 2A, pages 201-256.

ISBN: 0-7623-0117-1

I. INTRODUCTION

Interleukin(IL)-6 is a multifunctional cytokine with a broad array of biological activities. The functional pleiotropy of IL-6 is best reflected in the history of its discovery. IL-6 was originally identified as a T cell-derived lymphokine which promoted the maturation of B cells into antibody-producing cells and hence named B cell differentiation factor (BCDF) or B cell stimulatory factor 2 (BSF-2; Hirano et al., 1985). Investigation on interferons led to the isolation of a soluble factor with antiviral activity. This factor was distinct from interferons α/α and γ, and termed interferon- (IFN-) β_2 or 26-kD protein (Weissenbach et al., 1980). In an independent study, hybridoma growth factor (HGF) was purified from the human endothelial culture supernatant based on its strong growth-promoting activity for hybridoma cells (Astaldi et al., 1980). In liver, systemic injury and trauma significantly alter the levels of a set of plasma proteins collectively termed acute-phase proteins (APPs). Monocytes and macrophages release a substance, the hepatocyte stimulating factor (HSF), which can induce the expression of the APP genes (Ritchie and Fuller, 1983). Based on other effects of IL-6 it is also called monocyte-granulocyte inducer type 2 (MGI-2; Shabo et al., 1988), T cell-activating factor (TAF; Garman and Raulet, 1987), cytotoxic T cell differentiation factor (CDF; Takai et al., 1988), hybridoma/plasmacytoma growth factor (HPGF; Van Damme et al., 1987a), plasmacytoma growth factor (PCT-GF; Nordan and Potter, 1986), or interleukin hybridoma plasmacytoma 1 (IL-HP1; Van Snick et al., 1986). Later, protein sequencing and cDNA cloning show that all of these factors actually refer to the same molecule, IL-6 (Van Damme et al., 1987b, 1988; Shabo et al., 1988; Simpson et al., 1988; Takai et al., 1988; Van Snick et al., 1988). Based on these biological responses, several bioassay systems have been established for determining the activity of IL-6.

The single-copy gene encoding human IL-6 contains five exons and has been mapped to chromosome 7p21 (Yasukawa et al., 1987) whereas the mouse homolog is located in the proximal region of chromosome 5 (Mock et al., 1989). The gene organization as well as the protein sequence are very similar to that of the granulocyte colony-stimulating factor (G-CSF) suggesting that they are derived from a common ancestral gene (Yasukawa et al., 1987). The cDNA for human IL-6 contains 212 amino acids including a 28-amino acid signal peptide (Hirano et al., 1986) whereas the mouse and rat homologs contain 211 amino acids with 24 amino acids in the signal peptide (Van Snick et al., 1988; Northemann et al., 1989). Human and murine IL-6 share 42% homology in amino acid sequence. The secreted form of IL-6 is a 20-30 kDa protein depending on the sources and purification methods. The molecular size is affected by post-translational modifications including phosphorylation and glycosylation (May et al., 1988a, 1988b; Gross et al., 1989).

The tertiary structure of IL-6 consists of four antiparallel α-helices arranged in the "up-up-down-down" sequence (Bazan, 1990a). The same structure is also found in the related growth hormone (GH), leukaemia inhibitory factor (LIF), erythropoietin (EPO), prolactin (PRL), and G-CSF (Abdel-Beguid et al., 1987; Bazan, 1990a; Boulay and Paul, 1993; Robinson et al., 1994). Human IL-6 acts on both the human and murine cells but murine IL-6 proved to be inactive on human cells. The N-terminal 1-28 amino acids in the mature human IL-6 protein are not required for biological activity (Brakenhoff et al., 1989). The carboxyl terminal and Gly^{77} -Glu^{95} are involved in the binding to the IL-6 ligand-binding subunit, IL-6Rα (Leebeek et al., 1992). More detailed analysis showed that the region containing residues 50-55 is the most critical for the interaction (Ehlers et al., 1995). Double mutations of Gln^{159} and Thr^{162} to Glu and Pro, respectively, greatly reduce the interaction of IL-6Rα and gp130 but only slightly affect the ligand binding (Brakenhoff et al., 1994). Similarly, exchange of the residues 43-55 on human IL-6 with murine corresponding region results in only a fivefold decrease in IL-6 binding but significant inhibition in the complex formation between IL-6Rα and gp130 (Ehlers et al., 1994). The chimeric human/murine IL-6 protein has no biological activity on human hepatoma HepG2 cells, but is able to induce the proliferation of human myeloma cell line, XG-1 at a 100-fold higher concentration. Introduction of both F170L and S176A mutations into the chimeric IL-6 protein completely abolished its activity on XG-1 cells (Ehlers et al., 1995).

IL-6 is produced by a great variety of cell types in the presence of appropriate stimulants (reviewed by Monique et al., 1990 and Akira et al., 1993). The IL-6 producers include fibroblasts, monocytes/macrophage, endothelial cells, keratinocytes, B cells, T cells, granulocytes, mast cells, mesangial cells, glial cells, osteoblasts, astrocytes, bone marrow stromal cells, Kupffer cells, smooth muscle cells, intestinal epithelial cells, anterior pituitary cells, adrenal gland cells, pancreatic β cells, and myocytes. Following inflammation, the major sources of circulating IL-6 are monocytic and stromal cells. In some tumor cell lines, IL-6 is produced constitutively (Biondi et al., 1989; Jourdan et al., 1990; Watson et al., 1990). The IL-6 inducers include cytokines (e.g., IL-1, IL-2, IL-4, IL-6, tumor necrosis factor [TNF], transforming growth factor [TGF-β], and interferons [IFN]), growth factors (e.g., platelet-derived growth factor [PDGF], granulocyte-macrophage colony-stimulating factor [GM-CSF] and CSF-1), hormones (prostaglandin E1 and E2, bradykinin, parathyroid hormone, angiotensin II and adrenocorticotrophic hormone [ACTH]), cAMP agonists (e.g., forskolin, dibutylyl-cAMP, isobutylmethylxanthine, and cholera toxin), protein kinase C (PKC) activators (e.g., 12-*O*-tetradecanoyl-phorbol-13-acetate [PMA] and diacylglycerol), calcium ionophore A-23187, viral infection, sepsis, and trauma. Other IL-6-inducing agents are LPS, cycloheximine, actinomycin D, serum, ConA, vasoactive intestinal peptide (VIP), and anti-C3 antibody. It is noteworthy that the effects of these stimuli on IL-6 production may be tissue-specific. For example, IL-4 suppresses IL-6 production in monocytes (Cheung et al., 1990; Lee et al., 1990; te Velde et al., 1990) but is stimulatory in endothelial cells and B cells (Smeland et al., 1989; Howells et al., 1991). Conceivably, the cellular context plays a key role in controlling the IL-6 gene promoter.

In the past few years, various excellent reviews have been published summarizing many different aspects of IL-6 biology ranging from biosynthesis and genetics to cell physiology and *in vivo* activities (Kishimoto, 1989; Monique et al., 1990; Van Snick, 1990; Akira and Kishimoto, 1992; Lee, 1992; Yoshizaki et al., 1992; Akira et al., 1993; Kushner, 1993; Baumann and Gauldie, 1994; Klein et al., 1995). This review will present some highlights of these aspects and focus on the current understanding in the gene regulation by IL-6 and the signaling mechanisms involved.

II. REGULATION OF IL-6 GENE EXPRESSION

Analysis of the IL-6 gene promoter has identified several potential regulatory elements including the multi-response element (MRE) and glucocorticoid response element (GRE), as well as the binding sites for NF-IL6, NF-κB, and AP-1 complex (Akira and Kishimoto, 1992; Akira et al., 1992). The MRE is located at -173 to -151 and is responsive to IL-1, TNF, and serum (Ray et al., 1989). It also confers responsiveness to forskolin and PMA. A cAMP response element (CRE) is also identified within the MRE sequence. Overlapping with the MRE at -158 to -145 is a 14 bp palindromic sequence representing a NF-IL6 (also known as C/EBPβ, LAP, IL-6DBP, or AGP/EBP) binding site for the response to IL-1, IL-6, TNF, and LPS (Akira and Kishimoto, 1992). NF-IL6 has been demonstrated to bind to the promoters of other cytokine genes such as IL-1, IL-8, TNFα, and G-CSF.

The region at -73 to -64 resembles the NF-κB consensus sequence and binds purified NF-κB *in vitro* (Akira & Kishimoto, 1992; Akira et al., 1992). Since NF-κB is prominently activated by IL-1 and TNF (Thanos and Maniatis, 1995), it is strongly suggested that NF-κB participates in the IL-6 induction by these cytokines (Libermann and Baltimore, 1990; Zhang, et al., 1990). The importance of NF-κB *in vivo* is confirmed by targeted disruption of the p50 subunit in mice (Sha et al., 1995). The p50-deficient mice show impaired IL-6 production from LPS-stimulated resting peritoneal macrophages. There is evidence showing that NF-κB and NF-IL6 are interacting to modulate the promoter activity of IL-6 gene. NF-κB and NF-IL6 act synergistically resulting in a strong activation of the transfected IL-6 promoter-reporter construct (Matsusaka et al., 1993). It is observed that NF-IL6 associates directly with NF-κB p65 subunit via its basic leucine-zipper domain and the Rel homology domain of p65.

The tumor suppressor proteins p53 and Rb inhibit the induction of IL-6 promoter by serum in HeLa cells whereas mutated p53 and Rb proteins increase the IL-6 production in neoplastic cells (Santhanam et al., 1991). In c-*fos* gene Rb repression is mediated via the retinoblastoma control element (RCE; Robbins et al., 1990). A homologous region is found in the IL-6 promoter at -128 to -101. Viral infection causes enhanced production of IL-6. Viral proteins such as SV40 large T antigen, adenovirus E1B protein, and human papillomavirus proteins are known to interact with p53 and Rb and, hence, may be involved in the dysregulation of the IL-6 promoter. It has also been reported that HTLV-1 p40 Tax protein and HIV-1 Tat protein can transactivate the gene expression of IL-6 (Gibellini et al., 1994; Mori et al., 1994; Rautonen et al., 1994; Yamashita et al., 1994). The effect of Tax protein is mediated via the NF-κB site (Mori et al., 1994).

Glucocorticoids mediate a prominent inhibition on IL-6 promoter (Helfgott et al., 1987). However, the mechanism of the steroid effects on IL-6 gene expression is still not well understood. GRE consensus sequences are identified upstream of the MRE. Similarly, IL-4 can suppress the IL-6 production in monocytes. It has been shown that NF-κB physically associates with the glucocorticoid receptor (Ray and Prefontaine, 1994). The inhibitory effect of IL-4 has been ascribed to a decrease in the nuclear NF-κB levels (Donnelly et al., 1993). Indeed, down-regulation of NF-κB correlates with the inhibition of IL-6 production mediated by glucocorticoids and IL-4. However, an alternative mode of IL-4 regulation of cytokine genes independent of NF-κB has also been proposed (Wang, P., et al., 1995).

III. BIOLOGICAL FUNCTIONS

A. Acute-phase Response

During inflammation or systemic injury, a drastic change in the expression of APPs occurs in the liver (Baumann, 1990; Kushner, 1993; Baumann and Gauldie, 1994). APPs may be up- or down-regulated by inflammatory signals and are referred to as positive or negative acute-phase reactants, respectively. The change in the expression levels may vary from as low as 50% to as high as several thousand-fold and may be transient followed by a rapid recovery or persistent as in chronic diseases. There are interspecies differences in the acute-phase response. For example, the plasma level of α_2-macroglobulin (α_2-MG) is elevated during inflammation in rats but no significant change is observed in human. Although the functions of many APPs are well characterized, some such as α_1-acid glycoprotein (AGP) and serum amyloid A (SAA), have not been precisely determined. In general, APPs perform protective functions for the organism to restore normal homeostasis and include blood coagulation, inhibition of proteases, removal of toxic oxidative metabolites, activation of host defense mechanism, and wound healing.

The expressions of most APPs reach maximal levels within 12-24 after the initiation of inflammatory response followed by a decline characteristic to specific proteins. However, the induction of different APP genes may vary kinetically. Genes such as fibrinogens (Fbs), hemopexin (Hpx), and haptoglobin (Hp) are activated as early as 30 mins after stimulation whereas the expression of the others, for example, AGP, is observed only after several hours (Baumann et al., 1992a). The transcriptional activation of the APP genes is well coordinated by cytokines and glucocorticoids. IL-1- and IL-6-type cytokines have been identified as the key regulators of the hepatic acute-phase response. APP genes are classified into two groups (Baumann, 1990). Type I APP genes require the combination of IL-1, IL-6, and glucocorticoids for maximal stimulation whereas Type II genes depend on IL-6 and glucocorticoids only. Induction of acute-phase response in human or animals by endotoxin results in an elevation of biologically active IL-6 in serum as high as 10 nM within 2 h (Coulie et al., 1987). This level is considered to be sufficient to explain the IL-6 effects on liver cells. In fact, IL-6 was named hepatocyte-stimulating factor due to its ability to induce most of the major APPs in hepatocytes (Andus et al., 1987). Rats injected with recombinant IL-6 show a typical acute-phase reaction similar to that induced by turpentine (Geiger et al., 1988).

Besides the regulation of APP genes, IL-6 also participates in the control of thermogenesis (Lemay et al., 1990) and release of ACTH and glucocorticoids (Naitoh et al., 1988). Intracerebroventricular injection of IL-6 can induce fever in animals (Helle et al., 1988; Lemay et al., 1990; Lesnikov et al., 1991; Sakata et al., 1991). The pyrogenic response of IL-6 was suggested to be mediated by prostaglandins (PGs). It has been shown that central or peripheral injection of IL-6 increases PGE_2 levels in cerebrospinal fluid (CSF) and this effect is diminished by the inhibitors of cyclogenase (Dinarello et al., 1991). However, it is still not clear whether the effect of IL-6 is direct or indirect. IL-6 can also increase the level of corticotrophin-releasing hormone (CRH) that serves as an alternate pathway for the regulation of thermogenesis (Navarra et al., 1991). IL-6 may affect the hypothalamic-pituitary-adrenal axis in the release of ACTH and glucocorticoids through the production of CRH. It is also conceivable that IL-6 may act directly on the pituitary and adrenal glands but experimental evidence to support this pathway is yet to be established. In addition, there is evidence showing that IL-6 acts synergistically with IL-1 to modulate the neuroendocrine response (Neta et al., 1992).

B. Hematopoiesis

Blood cells in circulation have a limited life span and thus require continuous renewal. Hematopoiesis is a complex process well controlled by a cytokine network that regulates the proliferation and differentiation of stem cells and committed progenitor cells into mature blood cells (Metcalf, 1989; Ogawa, 1993). Under physiological conditions, about 0.5% of the bone marrow cells produce IL-6 (Shirai et al., 1993). The major source of IL-6 in bone marrow is macrophages. IL-6 receptors are detectable on a large proportion of bone marrow blast cells expressing

CD34 antigen and other hematopoietic cells indicating an importance role of IL-6 in the development of blood cells. IL-6 alone has minimal stimulatory effects on the hematopoietic cells. However, in combination with other cytokines, it produces a prominent synergistic effect. IL-6 synergizes with IL-3 to support the proliferation of murine and human multipotent hematopoietic progenitor cells (Ikebuchi et al., 1987; Hoang et al., 1988; Koike et al., 1988). Similar synergism is observed between IL-1 and IL-3 but this effect cannot be reproduced in purified human bone marrow cells suggesting that the effect of IL-1 may be indirect (Ogawa, 1993). Since IL-1 is known to be a potent inducer of other cytokines, synergism of IL-1 and IL-3 is likely to be mediated via the production of other factors such as IL-6.

IL-6 has a weak proliferative effect on granulocyte/macrophage progenitors in mice. However, it synergizes with GM-CSF and macrophage colony-stimulating factor (M-CSF) effectively in promoting the formation of granulocyte precursors and macrophage colonies (Hoang et al., 1988; Bot et al., 1989). Stem-cell factor (SCF; also known as steel factor, SLF, mast cell growth factor, MGF, or *c- kit* ligand) is an essential growth factor for hematopoietic stem cells (Huang et al., 1990; Williams, et al., 1990; Zsebo et al., 1990). IL-6 interacts with SCF to support the formation of multipotential blast cells and multilineage colonies (Ogawa, 1993). The combination of IL-6 and SCF can also maintain B-lymphoid potentials of the lymphohematopoietic progenitors.

The cytokine critical for thrombopoiesis, a process of megakaryocyte formation and platelet production, is known as c-*mpl* ligand or thrombopoietin (TPO; Bartley et al., 1994; de Sauvage et al., 1994; Kaushansky et al., 1994; Lok et al., 1994; Wendling et al., 1994). IL-6 has been observed to affect thrombopoiesis before TPO is identified. In synergy with SCF, IL-6 expands the committed myeloid progenitor population (Williams et al., 1992). It increases the number of immature megakaryocytes and promotes their maturation to megakaryocytes in the presence of IL-3. The actions of IL-6 on megakaryocytopoiesis *in vivo* was examined. In mice injected with IL-6, the megakaryocytes undergo morphological and cytoplasmic changes resulting in enhanced maturation and platelet production (Williams et al., 1992). However, IL-6 does not stimulate the release of platelets. These results indicate that IL-6 mainly acts on the immature megakaryocytes but has little effects on the mature megakaryocytes.

In contrast to its proliferative effects, IL-6 was reported to exert anti-proliferative effects on human histiocytic U937 cells and mouse myeloid M1 cells (Miyaura et al., 1988; Chiu and Lee, 1989). The inhibition of growth was accompanied by differentiation-related morphological changes. In M1 cells, IL-6 induces the phagocytic activity and macrophage-like morphological changes.

Sequential administration of IL-6 for seven days followed by G-CSF for seven days in newborn rats result in the increase of peripheral neutrophil count and expansion of the granulocyte-macrophage colony-forming cells in bone marrow, liver, and spleen (Cairo et al., 1991). However, simultaneous IL-6 and G-CSF in mice for four days was reported to attenuate the stimulatory effects of G-CSF on the number of GM-CFC and Meg-CFC (Pojda et al., 1992).

IL-6 confers radioprotection on mice and Rhesus monkeys (Patchen et al., 1993; MacVittie et al., 1994). Administration of IL-6 alone can significantly reduce the severity of the radiation-induced anemia which resolved more quickly after exposure. It is at least partly due to the enhancing effects of IL-6 on the multilineage recovery from hemopoietic aplasia induced by sublethal radiation (Patchen et al., 1991).

C. Immunomodulation

Receptors for IL-6 are not detectable on resting B cells. However, upon activation, B cells express IL-6R to a level of about 500 per cell (Kishimoto, 1989). IL-6 can induce the antibody production of B cells (Hirano et al., 1985, 1986). Anti-IL-6 antibody inhibits the production of immunoglobulin (Ig) by peripheral blood monoclear cells induced by mitogens. IL-2 plays an essential role in regulating the immune system by stimulating the proliferation and cytokine production of activated T cells. IL-6 induces the expression of IL-2R in some T cell lines and thymocytes (Noma et al., 1987; Le et al., 1988), and acts as a secondary signal for IL-2 production (Garman et al., 1987; Houssiau et al., 1988). In the presence of IL-2, IL-6 stimulates the proliferation and differentiation of cytotoxic T lymphocytes (CTL; Garman et al., 1987; Lotz et al., 1988; Okada et al., 1988; Takai et al., 1988; Uyttenhove et al., 1988; Renauld et al., 1989).

The effects of IL-6 on the natural killer (NK) cells are observed in peripheral blood monoclear cells (PBMC)/ peripheral blood lymphocytes (PBL) which show increased cytotoxicity in the presence of IL-6 (Okada ct al., 1988; Renauld et al., 1989). The enhanced cytotoxicity is attributed to the endogenous production of IL-2 (Luger et al., 1989). Studies using purified NK cells indicate that IL-6 has a direct effect in enhancing the NK activity but not lymphokine-activated killer activity (LAK) of the CD3$^-$ large granular lymphocytes (LGL; Smyth and Ortaldo, 1991). In contrast to these results, no effect of IL-6 was observed in the CD16$^+$ NK subset (Gerosa et al., 1992). In human NK CD3$^-$CD56$^+$ cells, IL-6 induced low levels of cell proliferation and expression of activation antigens including CD25, CD54, CD69, and CD71 (Rabinowich et al., 1993). In addition, the stimulated cells exhibit increased integrin-mediated adhesion. These effects can be partially inhibited by neutralizing antibody to TNF-α or TNF-β.

IL-6 is a potent growth factor for myeloma and plasmacytoma (Astaldi et al., 1980; Nordan and Potter, 1986; Van Snick et al., 1986; Van Damme et al., 1987b). The tumorigenicity of an IL-6-dependent plasmacytoma cell line is greatly enhanced by transfecting cells with an expression vector for IL-6 (Vink et al., 1990). Proliferation of the plasmacytoma is inhibited by monoclonal antibody blocking the binding of IL-6 to its receptors. Injection of the same antibody effectively protects mice after tumor challenge. Similar observations are obtained in the studies of human myeloma *in vitro*. The growth of myeloma cells is enhanced in the presence of IL-6 but inhibited by anti-IL-6 antibody (Kawano et al., 1988).

Autoimmune diseases are associated with the elevation of Ig levels, increase in anti-viral antibody, and the presence of autoantibodies. The serum levels of IL-6 are increased in patients with autoimmune diseases such as rheumatoid arthritis, systemic lupus erythematosus, Castleman's disease, and AIDS implicating the involvement of IL-6 in these pathological conditions (Yoshizaki et al., 1992). MRL/lpr mice spontaneously develop hypergammaglobulinemia, anti-Ig antibody and plasmacytosis. Symptoms of arthritis and glomerulonephritis are also observed. The levels of APPs are elevated. An age-associated increase in serum levels of IL-6 is detected in these mice suggesting a link between IL-6 and the autoimmune diseases (Tang et al., 1991).

D. Nervous System

During viral infection of the central nervous system (CNS) and brain injury, the microglial cells and astrocytes are stimulated to produce IL-6 (Frei et al., 1989). Centrally injected LPS induces the expression of IL-6 in the cerebrospinal fluid 2 h after treatments and the increased IL-6 level is maintained for about 12 h (De Simoni et al., 1995). IL-6 induces the differentiation of rat pheochromocytoma cell line, PC12 cells into neural cells (Satoh et al., 1988). It also acts on the astrocytes to promote their proliferation and secretion of nerve growth factors (NGF; Frei et al., 1989), as well as on the cultured cholinergic neurons to maintain their survival and dopamine synthesis (Hama et al., 1989). IL-6 is produced constitutively in the Schwann cell line, iSC, when cocultured with PC12 cells (Bolin et al., 1995). In the sciatic nerve crush model of Schwann cell activation *in vivo* , IL-6 is up-regulated at mRNA level within 12 h after crush injury (Bolin et al., 1995). It suggests that the IL-6 production may be part of the mechanism by which Schwann cells facilitate the regeneration of peripheral nerve. Recent studies propose that IL-6 may be a mediator between peripheral tissues and CNS for the inflammatory response to systemic injury and viral infection. The levels of bioactive IL-1 and TNF-α are found to be insignificant in blood circulation, except after the most severe trauma (Hopkins and Rothwell, 1995). By contrast, IL-6 is increased substantially during the development of fever and is readily transported into the brain by a yet to be defined mechanism. Therefore, IL-6 is likely to be the messenger molecule responsible for the modulation of the neuroimmune response.

IL-6 stimulates the secretion of anterior pituitary hormones including PRL, GH, and luteinizing hormone (Spangelo et al., 1989). It has been shown that IL-6 has a dual role in regulating norepinephrine release from the myenteric plexus in rats (Ruhl et al., 1994). At low concentration, IL-6 augments the norepinephrine release whereas it exerts an inhibitory effect at high concentration. Furthermore, IL-6 acts synergistically with IL-1α to suppress the norepinephrine release.

E. Development

IL-6 is expressed in the yolk sac of developing mouse embryo. By using PCR technique, mRNA for IL-6 is detected in preimplantation blastocysts 3.5 days prior to the onset of hemopoiesis (Murray et al., 1990). The expression of IL-6 before hemopoiesis suggests that IL-6 may be involved in the growth and development of the trophoblasts or embryonic stem cells. IL-6 is produced and subjected to temporal regulation throughout the *in vitro* differentiation process of the ES cells to hematopoietic precursors (Schmitt et al., 1991).

In vivo , the expression of IL-6 undergoes a cyclic change during the estrous cycle (De et al., 1992b). While uterus from ovariectomized mice contains little or no cytokine activity, treatments with estrogen plus progesterone in the ovariectomized mice augment the production of

IL-6. During pregnancy in mice, IL-6 is detectable in days 9-18 and decreased to a low level near parturition (De et al., 1992a). The uterine stromal cells and polarized epithelial cells secrete IL-6 in cell culture (Jacobs et al., 1992). One possible function of IL-6 in uterus may be the indirect effect on angiogenesis. IL-6 may act as an autocrine factor on trophoblast for hCG release that leads to the production of progesterone in placenta (Nishino et al., 1990). The stimulation of angiogenesis by progesterone is important in maintaining the pregnancy state of the uterus.

Although the functions of IL-6 in development are still not clear (see also experimental results from IL-6 deficient mice discussed below), its expression in placenta as well as the known functions of IL-6 in hematopoiesis and modulation of the immune system suggest that IL-6 may be involved in embryonic development and fetal-maternal interaction.

F. Bone Resorption

The process of bone remodeling is regulated by cytokines including IL-1, IL-6, and TNF. IL-6 is produced by osteoblasts stimulated by parathyroid hormone, IL-1, TNF, and LPS (Feyen et al., 1989; Lowik et al., 1989; Ishimi et al., 1990). High level expression of IL-6 in mice caused hypercalcemia (Black et al., 1991). Activity of the osteoclasts is stimulated by IL-6. Due to the estrogen loss, ovariectomized mice overproduce IL-6 and show enhanced osteoclast development (Jilka et al., 1992). Neutralizing antibody to IL-6 can inhibit the enhanced osteoclast activity strongly suggesting a role of IL-6 in osteoporosis. IL-4 is found to inhibit bone resorption in an *ex vivo* model of rheumatoid arthritis and it correlates with the reduction of IL-6 expression in bone cells (Miossec et al., 1994).

IV. IL-6 TRANSGENIC AND KNOCK-OUT MICE

The functions of IL-6 have been explored by using *in vitro* systems and the data extrapolated to *in vivo* conditions. However, the interpretation of the *in vivo* effects is always complicated by the functional redundancy of cytokines such as LIF, oncostatin M (OSM), IL-11, and probably CNTF for IL-6. Transgenic mice proved to be a very useful experimental tool for identifying the biological functions of a specific gene product *in vivo* . By overexpressing IL-6 in particular tissues, the tissue-specific effects of IL-6 were studied. Transgenic mice with IL-6 overexpressed in the pancreatic islet β cells exhibit a reduction in body weight but are normal in the production of insulin and glucose tolerance (Campbell et al., 1994). Nevertheless, a remarkable alteration in the islet cells and surrounding tissues, as well as increased number of plasma cells in the peri-islet infiltrates are observed. These phenomena are similarly observed in patients with autoimmune diabetes. To explore the role of IL-6 in the nervous system, the gene encoding IL-6 is overexpressed in CNS of mice under the control of the glial fibrillary acidic protein gene promoter (Campbell et al., 1993). The transgenic animals develop severe neurologic diseases including neuro-degeneration, astrocytosis, and angiogenesis suggesting direct or indirect actions of IL-6 on the CNS. In addition, the serum levels of the APP are elevated. The adverse effects of IL-6 overexpression in the airway epithelial cells are relatively mild. Lymphocytic infiltration surrounding large and mid-sized airways occurs but it does not alter the basal respiratory resistance (DiCosmo et al., 1994). However, the airway sensitivity to methacholine is decreased. Interestingly, excessive expression of IL-6 in the basal cells of the stratified squamous epithelium of skin causes abnormalities including smaller cell size and retarded hair growth (Turksen et al., 1992). In addition, the stratum corneum is increased in thickness that may be related to the enhanced protection form injury or infection.

In contrast to the overexpression restricted to local tissues, studies were performed to investigate the systemic response in the presence of high serum level of IL-6. The IL-6 gene is inserted downstream either to the promoter containing the human Ig heavy chain enhancer (Turksen et al., 1992) or to the murine metallothionein-I gene promoter (Fattori et al., 1994b). Similar effects are observed. The initial activation of APP genes is followed by polyclonal hypergammaglobulinemia. A massive plasmacytosis is developed in spleen, lymph nodes and thymus. The proliferation of megakaryocytes is enhanced. Severe kidney damage results from mesangio-proliferative glomerulonephritis, probably due to the serum protein overload. Another approach to study the *in vivo* functions of IL-6 is by the use of human adenovirus as an transient expression vector. After intraperitoneal injection, the liver, spleen, and peritoneum are the major sites of expression. The serum level of IL-6 is increased after infection and splenomegaly with expansion of B and T cells is also observed. Intratracheal administration leads to the expression of IL-6 in lung and bronchus. The overexpression in lung causes prominent local lymphocytic hyperplasia with significant increase in the number of $CD3^+CD8^+$ cytotoxic T cells and $CD3^+CD4^+$ helper T cells. Notably, these changes are reversible as the expression level of IL-6 decline with time (Braciak et al., 1993; Xing et al., 1994).

The *in vivo* significance of IL-6 was further assessed in IL-6-deficient mice in which the IL-6-encoding gene was disrupted by homologous recombination (Kopf et al., 1994; Poli et al., 1994). Surprisingly, the knock-out mice are fertile and develop normally. They have normal life expectancy if left unchallenged in a pathogen-free environment. The seeming lack of IL-6 effects may be attributed to the substitute actions of other cytokines. Although interbreeding of these mice results in the normal number of siblings, the frequency of pregnancy is somewhat reduced. The number of thymocytes and peripheral T cells is decreased by 20-40%, even though the expression of the surface markers is normal. The IgG production by B cells is within normal ranges. The acute-phase response to tissue damage, Gram-positive bacteria (*Listeria monocytogenes*), or viral infection is severely impaired. Interestingly, the response to LPS and Gram-negative bacteria is only slightly affected indicating that the signaling pathway of LPS is different from that elicited by tissue injury (Fattori et al., 1994a; Kopf et al., 1994). TNF can induce lethal shock in mice. IL-6 was suspected to be the mediator since TNF strongly augments its expression. In the IL-6-deficient mice, the lethal response is identical to that exhibited by the wild-type littermates ruling out a critical role of IL-6 in the TNF-mediated lethality (Libert et al., 1994).

On the other hand, IL-6 deficiency caused altered control of proliferation and differentiation of hematopoietic cells resulting in the abnormal levels of committed progenitors of different lineages (Bernad et al., 1994). The number of the spleen colony-forming units (CFU-S)d12 and preCFU-S is decreased and the functionality of the long-term repopulating stem cells (LTRSC) is attenuated. The mucosal IgA-producing cells are reduced in number and the response to antigen challenge diminished. Ectopically supplied IL-6 restores the mucosal immune response in the lung to normal values demonstrating an exclusive role of IL-6 in this regulatory system. A striking observation is the protection of the IL-6-deficient female from bone loss (Poli et al., 1994). Ovariectomized mice do not undergo any change in either bone mass or bone remodeling rates which occur in normal mice. What still needs to be addressed is the contribution of other genetic loci to the process of bone remodeling since the knockout mice are not inbred and thus heterogenous in genetic background. Nevertheless, the results clearly indicate that IL-6 plays a role in bone turnover and may contribute to the bone resorption caused by estrogen deficiency.

V. IL-6 RECEPTOR SYSTEM

Recent advances in molecular cloning of the cytokine receptor subunits has uncovered a receptor superfamily termed hematopoietin receptor family (Bazan, 1990b; Cosman, 1993). Most of the cytokine receptors identified to date belong to this family which is characterized by a conserved extracellular domain structure referred to as the hematopoietin domain. This domain consists of approximately 200 amino acids which can be divided into two units. The N-terminal half contains four positionally conserved cysteine residues and the adjacent carboxyl-terminal half a WSXWS (W: Trp, S: Ser, X: any amino acid) motif. The hematopoietin domain may be duplicated in certain family members such as LIFR, IL-3Rβ, and c-mpl. Additional homology regions such as the Ig-like and fibronectin domains are also recognized in the extracellular portions of some but not all of the family members. By contrast, the intracellular regions of the cytokine receptors share only very limited similarity. Functional analyses of the hematopoietin receptors led to the discovery that different ligand-binding (or α-) subunits may share a common signal-transducing (or β/γ-) subunit. Based on the common signal-transducing subunit, the receptors have been divided into several subgroups (Miyajima et al., 1992; Taga and Kishimoto, 1992, 1995; Kishimoto et al., 1994; Sato and Miyajima, 1994). IL-6-type receptors share gp130 and include receptors for IL-6, LIF, OSM, CNTF, IL-11, and possibly IL-12. IL-2-type receptors utilize IL-2Rg chain and consist of receptors for IL-2, IL-4, IL-7, IL-9, IL-13, and IL-15. IL-3Rβ is shared among IL-3R, IL-5R, and GM-CSFR. The receptors for G-CSF, GH, PRL, EPO, and possibly TPO, function as homodimers. Upon ligand binding, the α chain interacts with the β subunit and triggers the downstream signaling events. Although the ligand binding is highly specific for each receptor, few exceptions are noted in which the same receptor complex is recognized by more than one cytokine. For instance, LIF, CNTF, and OSM bind to the complex formed by LIFRα (Bauman et al., 1993; Gearing et al., 1992) and gp130, IL-2, and IL-15 to IL-2Rβ·γ complex (Giri et al., 1994) and, IL-4 and IL-13 to IL-4Rα·IL-2Rγ complex (Obiri et al., 1995; Smerz-Bertling and Duschi, 1995; Vita et al., 1995).

IL-6R is composed of two polypeptide subunits, a 80 kD IL-6-binding protein (IL-6Rα) (Yamasaki et al., 1988) and gp130 (Hibi et al., 1990). In human hepatoma HepG2 cells, IL-6Rα and gp130 have a half-life of

2-3 h unaffected by IL-6 (Gerhartz et al., 1994). IL-6 binds to IL-6Rα with low affinity (K_d = 500 pm) and to the IL-6Rα-gp130 complex with high affinity (K_d = 30-50 pm; Zohlnhofer et al., 1992). The functional ligand-receptor complex formed under high-affinity conditions has recently been characterized to contain two molecules each of the IL-6Rα, gp130, and IL-6 (Ward et al., 1994). The cDNA for the human IL-6Rα encodes 468 amino acids including a signal peptide of 19 residues. It has a short cytoplasmic domain with 82 amino acids and six potential N-glycosylation sites. The signal-transducing subunit, gp130 is a glycoprotein of 130 kDa. The human gp130 cDNA encodes 918 amino acids in which 277 residues are located in the cytoplasmic region of the receptor. IL-6Rα is structurally similar to the IL-11Rα whereas gp130 shares a high degree of homology with LIFRα, OSMRα, G-CSFR, and an IL-12R subunit. It has been reported that gp130 can bind OSM at low affinity (Sporeno et al., 1994). LIFRα interacts with gp130 to constitute LIFR (Gearing et al., 1992). The LIFRα·gp130 complex, in the presence of CNTFRα, forms high-affinity binding site for CNTF (Davis et al., 1993). No interaction of IL-6Rα with either LIFRα or LIFRα·gp130 complex has been observed.

The region containing amino acids 249-264 in the extracellular domain of IL-6Rα is noted to be important for the interaction of the α chain and gp130 (Grube and Cochrane, 1994). Peptides representing this region inhibit the mitogenesis of the IL-6-dependent B9 cells and the expression of APP genes in HepG2 cells. The cytoplasmic domain of the IL-6Ra is not required for signaling (Taga et al., 1989). By contrast, gp130 is indispensable for the signal transduction of the IL-6-type cytokines. Monoclonal antibody against gp130 inhibits specifically the actions of IL-6, LIF, OSM, IL-11, and CNTF (Taga et al., 1992). The IL-6-type cytokines have similar but not identical biological activities in regulating APP genes, hematopoiesis, and neuronal differentiation. The CNTFR, although restricted to the neural cells, is able to stimulate the synthesis of APP when presented as soluble form or expressed in hepatoma cells by transfection (Baumann et al., 1993). The redundancy has been ascribed, at least in part, to the recruitment of gp130 as the common subunit. Knock-out mice deficient for gp130 die between 12.5 days postcoitum and term with myocardial and hematological disorders (Yoshida et al., 1996).

The soluble form of IL-6R (sIL-6R) is detected in the serum and urine of healthy individuals (Novick et al., 1989; Frieling et al., 1994). The expression level of sIL-6R is increased in patients with multiple myeloma and HIV infection (Honda et al., 1992; Gaillard et al., 1993).

In some myeloma cell lines sIL-6R is released spontaneously (Nakajima et al., 1992). The mechanism responsible for the production of sIL-6R is not fully understood. sIL-6R can be generated by limited proteolysis of the membrane-bound α chain (or shedding) at Gln357/Asp358 (Mullberg et al., 1994). This process is enhanced by PMA, an activator of PKC (Mullberg et al., 1992). Mutation analysis has shown that single amino acid change adjacent to the cleavage site decreased shedding of sIL-6R up to fivefold whereas deletion of five or 10 amino acid residues completely abolished the formation of sIL-6R (Mullberg et al., 1994). The mRNA encoding only the extracellular domain of IL-6Rα is detected suggesting that alternate splicing may be involved (Lust et al., 1992). The extracellular domain of IL-6Rα retains the ability to bind IL-6 and interact with gp130. In the presence of IL-6, sIL-6R-IL-6 complex can activate the synthesis of APP in hepatoma cells (Mackiewicz et al., 1992), promote the osteoclast formation (Tamura et al., 1993), and substitute for LIF to sustain self-renewal of pluripotential ES cells (Yoshida et al., 1994). *In vivo*, mice injected with B-78 melanoma cells transfected with both IL-6 and sIL-6R expression vectors are protected from tumor formation (Mackiewicz et al., 1995a). The effect was more prominent than with IL-6 alone indicating that sIL-6R is biologically active *in vivo*. The sIL-6R may provide a means by which IL-6 activates gp130-expressing cells normally not responsive to the ligand (Guschin et al., 1995). The soluble form of gp130 is detected in the ascitic fluids of tumor-bearing mice (Matsuda and Hirano, 1994a) and it acts as an antagonist to the IL-6 signal (Narazaki et al., 1993). It is noteworthy that the soluble forms of receptors are not uncommon among hematopoietin and other cytokine receptors (Fernandez-Botran, 1991; Heaney and Golde, 1993). Depending on the physiologic actions, soluble receptor form represents a general mechanism by which the cytokine activity is modulated (as antagonist) or the range of target cells is extended (as agonist).

VI. IL-6 RECEPTOR SIGNALING

Upon IL-6 stimulation, a number of intracellular proteins including gp130 are immediately phosphorylated on tyrosine, serine, and threonine residues (Murakami et al., 1993; Boulton et al., 1994). Since gp130 does not have intrinsic kinase activity, the phosphorylation of gp130 is proposed to be mediated by the receptor-associated kinases. The gp130 is maintained in the phosphorylated state for 60 min and then returns to the basal level (Wang and Fuller, 1994). There is a time lag of about 2 h before phosphorylated gp130 is detected again on cell surface and the re-activation process is blocked by protein synthesis inhibitor. These results suggest that phosphorylated gp130 is internalized and degraded in the cytoplasm. It may be responsible for the down-regulation of the IL-6 signal. A 10 amino acid domain, TQPLLDSEER, in the cytoplasmic region has been demonstrated to be important for the ligand-induced internalization of gp130 (Dittrich et al., 1994).

Among the signal-transducing receptor subunits for the IL-6-type cytokines, gp130, LIFR and OSMR, as well as the G-CSFR, three conserved sequence motifs are identified in the cytoplasmic domains. They are referred to as box 1, 2, and 3 (Asao et al., 1993). The single box 1 motif is a proline rich region (PXXPXP) whereas the box 2 represents an acidic domain (VSVVEIE) approximately 20 amino acids carboxyl terminal to box 1. The motifs of box 1 and 2 are also found in similar relative positions in other hematopoietin receptor subunits such as EPOR, GHR, IL-2Rβ, IL-3Rβ, and IL-4Rα (Ihle and Kerr, 1995). The box 3 motif (YXXQ) appears to be restricted to the IL-6-type cytokine receptor subunits and occurs in multiple copies in the carboxyl-terminal half of the cytoplasmic domain (Stahl et al., 1995).

A. Three Distinct IL-6 Signaling Pathways

The signaling function of gp130 has been defined in separate cell systems based on the activation of proliferation in hematopoietic cells and activation of gene transcription via gene elements in differentiated cells (Murakami et al., 1991; Asao et al., 1993; Lai et al., 1995b). Several types of gene elements responsive to IL-6 has been identified in the promoter regions of various differentiated genes. One type of elements which are restricted to the regulation by IL-6-type cytokines have been termed IL-6 response element (IL-6RE) and contain the consensus core sequence, CTGGGAA (Wegenka et al., 1993). These IL-6REs are found in the APP genes including a2-MG (Hattori et al., 1990; Hocke et al., 1992), Hpx (Immenschuh et al., 1994), C-reactive protein (CRP; Ganter et al., 1989), thiostatin (Chen and Liao, 1993), AGP (Won and Baumann, 1990), and Hp (Oliviero and Cortese, 1989; Marinkovic and Baumann, 1990) as well as α-, β- and γ-Fbs (Fowlkes et al., 1984; Anderson et al., 1993; Dalmon et al., 1993). These elements are sufficient to confer an IL-6 response onto a gene linked in *cis*.

The conserved motifs of box 1-3 are critical for IL-6 signaling. The first 133 amino acids in the cytoplasmic region are minimally required for all the normal functions of gp130. Mutant receptors form with the membrane-proximal region containing box 1 and 2 only, is still able to support IL-6-dependent growth when expressed in BAFB03 cells (Murakami et al., 1991), and control gene expression via the CytRE (Baumann et al., 1994), but fails to activate genes through IL-6RE (Lai et al., 1995b). Mutation of box 3 inactivates the minimal gp130 but not the full-length receptor suggesting that functionally redundant box 3 sequences are present (Lai et al., 1995b). Further study identifies four box 3-equivalent sequences in gp130 (Stahl et al., 1995). Similar duplication of box 3-related sequences was also observed in the LIFRα chain but not in G-CSFR (Lai et al., 1995a). As noted for both LIFRα and gp130, the single tyrosine residue in box 3 motif is critical for gene regulation via the IL-6RE. Mutation of this tyrosine residue to alanine completely abolished the transcriptional activation by IL-6 (Lai et al., 1995b).

As predicted from data defining the gp130 domain necessary for the regulation of cell growth, the cytoplasmic region containing box 1 and box 2 is sufficient to regulate the *sis*-inducible element (SIE) found in the early response genes, *c-fos* (Wagner et al., 1990) and *Jun*B (Lai and Baumann, unpublished observation). The importance of box 1 and box 2 motifs in proliferation and gene regulation are supported by similar studies on GHR (Colosi et al., 1993), EPOR (D'Andrea et al., 1991), G-CSFR (Fukunaga et al., 1993), IL-2Rβ (Hatakeyama et al., 1989), and GM-CSFRβ (Sato et al., 1993). The truncated gp130 to the membrane-proximal box 1 leads to the loss of proliferative control but partly retains the ability to stimulate gene expression through the hematopoietin receptor response element (HRRE) and the gamma-interferon activating site (GAS) of the FcRγ1 gene (Lai and Baumann, unpublished observation) in hepatoma cells and fibroblasts. Endogenous genes that are specifically regulated by the box 1 motif have not been identified yet. The results indicate that IL-6R triggers at least three types of gene regulatory signals which depend on three functional receptor motifs (Figure 1). Box 1 seems to be the most critical since all signaling activities cease with mutation of the proline residues in this motif (Murakami et al., 1991; Lai et al., 1995b).

B. JAKs and STATs

The biochemical link from gp130 to thc phosphorylated downstream substrates has been attributed to the receptor-associated protein kinases including the *Janus* kinases, JAKs, and the *src*-related kinases. JAKs

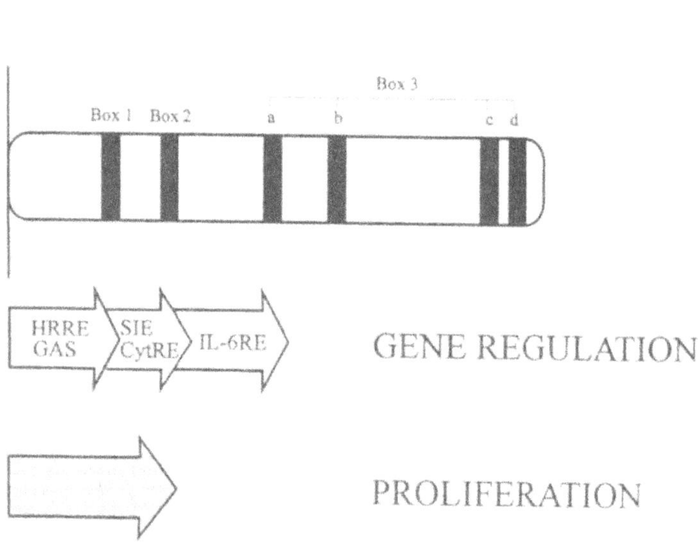

Figure 1. Schematic diagram indicating the minimal cytoplasmic domains of gp130 required for specific IL-6 responses (drawing not to scale).

were originally cloned by a PCR approach designed to identify novel kinases (Wilks et al., 1991). Four members of this family, JAK1, JAK2, JAK3, and TYK2 have been cloned and mapped in the human genome (Firmbach-Kraft et al., 1990; Wilks et al., 1991; Kawamura et al., 1994; Rane and Reddy, 1994; Witthuhn et al., 1994). While the expression of JAK1, JAK2, and TYK2 is ubiquitous, JAK3 is restricted to the myeloid and lymphoid cells. All JAK members are related as evident from the five homology domains (JH domains) in the amino terminus. There are no identifiable SH2 or SH3 domains in JAKs suggesting that these kinases may interact with the receptor and other signaling proteins by a novel mechanism. Unlike other kinases, JAKs have two kinase domains at the carboxyl portion. The carboxyl-terminal domain has catalytic functions whereas the more N-terminal one is predicted to be a non-functional kinase domain due to the lack of certain critical amino acids. However, both domains are indispensable for kinase activity (Velazuez et al., 1995).

The importance of JAKs in signaling by many cytokines is recognized based on the observations that cytokine treatments specifically activate the enzymatic activities of JAK members. JAK1 and JAK2 are involved in the signaling of most of the cytokines whereas JAK3 is confined to the signaling by receptors sharing IL-2γ chain such as IL-2R, IL-4R,

IL-7R, and IL-9R (Taniguchi, 1995). JAK1, JAK2, and TYK2 have been shown to associate with gp130 and activated upon IL-6 induction (Lütticken et al., 1994; Stahl et al., 1994) and JAK1 has been demonstrated to be the most critical in IL-6 signaling (Guschin et al., 1995). The functional roles of JAKs have been demonstrated in mutant cell lines which are unresponsive to IFNα/β or -γ. Complementation by transfecting cDNA for JAKs successfully restore their responsiveness (Muller et al., 1993; Watling et al., 1993). Furthermore, the expression of kinase-deficient JAK2 mutant which function in dominant-negative fashion effectively inhibits the EPO-dependent proliferation of a growth factor-dependent cell line transfected with EPOR (Zhuang et al., 1994).

The potential downstream substrates of JAKs are a group of protein factors first observed in the study of interferon signaling (Darnell et al., 1994). They are latent in the cytoplasm, activated by phosphorylation,and then translocated to the nucleus. Based on the feature that the same protein is responsible for both signal transduction and transcriptional regulation, they are termed Signal Transducer and Activator of Transcription (STAT; Shuai et al., 1993). Molecular cloning of the STAT proteins indicate that they are related proteins and belong to one gene family. To date, six members of the STAT family have been cloned, namely STAT 1-6 (Ihle and Kerr, 1995). Additional members have been proposed to exist (Ihle and Kerr, 1995). Each cloned member contains a putative SH2 and SH3 domain. The SH2 domain is important for the association of the STAT proteins to the tyrosine-phosphorylated receptor (Heim et al., 1995; Stahl et al., 1995) as well as the formation of homo- or heterodimers between STAT members (Shuai et al., 1994). IL-6 stimulates the phosphorylation of STAT3 at the tyrosine (Zhong et al., 1994) and serine residues (Lütticken et al., 1995; Zhang, X. et al., 1995). While the tyrosine phosphorylation is required for dimerization and DNA-binding, the phosphorylation of the serine residues is suggested to be necessary for the transactivation function of STAT proteins (Lütticken et al., 1995; Wen et al., 1995). Mutation of a single tyrosine residue carboxyl-terminal to the SH2 domain completely abolished the ligand-induced dimer formation and thus DNA-binding activity (Shuai et al., 1993; Gouilleux et al., 1994).

The activation of STATs has been shown to be the immediate early event that occurs in the signaling of many different cytokines. STAT3 (or APRF; Akira et al., 1994; Zhong et al., 1994) which binds to the APRE of the APP genes and is assumed to be the principal regulator of the APP gene expression (Wegenka et al., 1993), is similarly activated following treatments with IL-6-type cytokines (Raz et al., 1994; Lai et

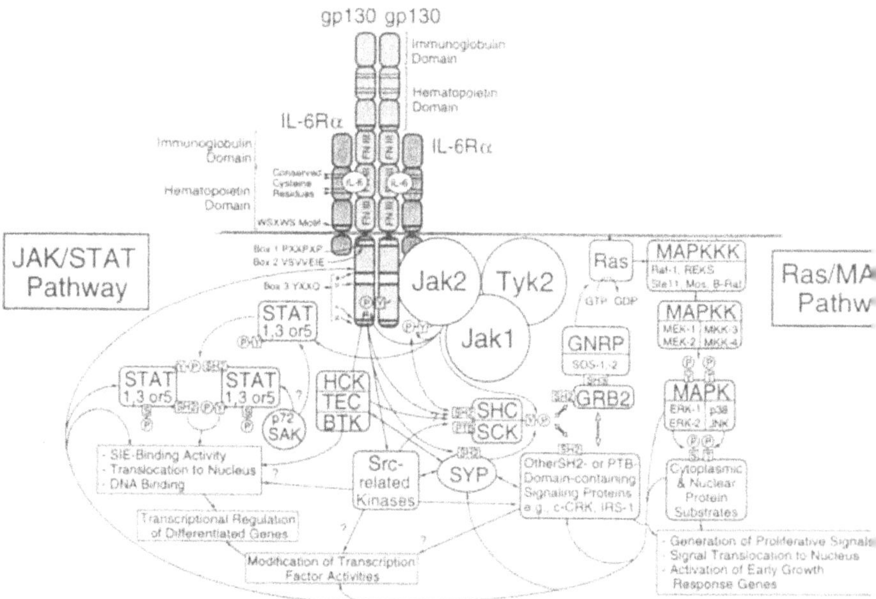

2. Working model integrating the known signal transduction pathways and the proposed signaling molec
nse to IL-6. Two signaling pathways are considered to participate: the JAK/STAT pathway for the differenti
ıs/MAPK pathway *for the control of cell proliferation. However, these two pathways have considerable over.*
indicated signaling proteins are also part of the signaling pathways activated by other effectors such as
tors, growth factor tyrosine kinase receptors or cytokine receptors of the IL-1/TNF receptor families. Different
IL-6 response are conceivable. For details, see text description. The model does not take into consideration
tion of IL-6 with IL-6Rα and gp130 is artistically inaccurate; IL-6 should contact both receptor subunits at the

Figure 3. Stimulation of SIE-binding activities. Rat hepatoma H-35 cells were treated with 100 ng/ml IL-6 for the indicated length of time. Whole cell extracts were prepared and relative amounts of activated STAT proteins were determined by EMSA using SIE as a probe. The positions of the STAT proteins - SIE complexes, SIF-A, -B, and -C are indicated. SIF-A and SIF-C are formed by STAT3 and STAT1 homodimers respectively whereas SIF-B by the heterodimer of STAT1 and STAT3

al., 1995a). In addition, STAT3 can be activated by cytokines including G-CSF, TPO, and GH, all of which are not associated with APP gene regulation (Gronowski et al., 1995; Lai et al., 1995a; Morella et al., 1995). IL-6 increases the mRNA levels of STAT1, STAT3, and STAT5 (or mammary gland factor—MGF; Ripperger et al., 1995). STAT3 and STAT5 have been sown to be transcription factors participating in separate IL-6 signaling pathways (Lai et al., 1995c). Since STATs are activated not only by cytokines but also by growth factors like EGF and PDGF, they are proposed to account for the overlapping activities of cytokines and growth factors (Sadowski et al., 1993).

C. IL-6 Signaling by JAK/STAT Pathway

The prominent activation of JAKs and STATs by IL-6 has suggested that the transcriptional regulation of the IL-6 responsive genes is causally linked to these protein factors. Upon stimulation, the DNA-binding activity of STAT1 and STAT3 are induced and can be detected by the electrophoretic mobility-shift assay (EMSA) using SIE or related sequences as the binding substrate (Figure 3). The SIE-STAT complexes are referred to as the sis-inducible factors (SIFs) and have distinct electrophoretic mobilities separating into SIF-A, -B, and -C (Sadowski et al., 1993). SIF-A and SIF-C consists of homodimers of STAT3 and STAT1, respectively, whereas SIF-B represent the heterodimer of STAT1 and STAT3. In hepatoma cells, IL-6 activates STAT proteins in a biphasic manner. The activation of STAT proteins is at maximal levels within 15 min and yields all three SIF forms indicating the activation of both STAT1 and STAT3. The activation is followed by a rapid decline of the DNA-binding activities. Only STAT3 but not STAT1 DNA-binding re-appears after an hour and persists for 24 h. The second phase requires *de novo* protein synthesis (Lai and Baumann, unpublished observation). Notably, the activation of STAT1 only occurs at high concentration of IL-6 (>10 ng/ml; Sadowski et al., 1993) and is very transient. The kinetics of STAT activation correlates only in part with the kinetics of APP genes regulation by IL-6 in the same cells. Most notable is the temporal difference with the delayed transcriptional activation of the "late" APP genes. Nevertheless, the results demonstrate that gp130 is capable of recruiting multiple STAT members.

The region of gp130 necessary for STAT3 activation has been determined and the box 3 motif is found to be indispensable. Mutation of the tyrosine residue in box 3 of the minimal gp130 construct (containing 133 amino acids in cytoplasmic region) dramatically abolished the DNA-binding activity of STAT proteins induced by IL-6 (Lai et al., 1995b). Therefore, the activation of STAT proteins seems to correlate with the transcriptional regulation via the IL-6RE. However, the functional role of STAT3 as transcription factor in IL-6 signaling pathway has been demonstrated to be essential for terminal differentiation and growth arrest in myeloid leukemia MI cells (Minami et al., 1996). It has been shown that STATs associate with a 72 kDa protein with tyrosine kinase activity (p72 SAK) suggesting the presence of additional signaling molecules in the STAT pathway (Matsuda and Hirano, 1994b).

The current model has proposed that the recruitment of STATs to the receptor occurs via the interaction of the SH2 domain of the STAT proteins and phosphotyrosine residue on the receptor. The box 3 motif is a potential STAT3 binding site explaining why it is necessary for the regulation of IL-6RE. However, it was found that box 3 is dispensable for STAT activation in the presence of high expression levels of JAK2 or TYK2 (Lai et al., 1995b). A possible alternative binding and activation of STAT3 may occur through other receptor-bound proteins such as JAKs or the adaptor proteins, IRS-1 and Shc. The box 3 motif of gp130 apparently provides a high-affinity STAT binding site which facilitates the interaction of the STAT proteins with the kinases. Furthermore, mutant GHR in which all tyrosine residues of the cytoplasmic domains have been removed, still supports the activation of STATs (Wang et al., 1995). These results argue that the requirement of a receptor binding site for STATs may not be obligatory for their activation, but alternative pathways using non-receptor molecules as interaction sites through SH2-phosphotyrosine association exist.

The phosphorylation of STATs is believed to be mediated by JAKs. The box 1 motif in gp130 as in other hematopoietin receptors has been demonstrated to be an interaction site for JAK2 (Tanner et al., 1995). The ability of JAKs to activate STATs via gp130 is verified by the overexpression of JAK2 or TYK2. As indicated recently (Ihle and Kerr, 1995), the activation process does not show any specificity toward STAT1 or STAT3. Moreover, similar non-discriminating activation is observed with other hematopoietin receptors. The results suggest that the recruitment of STAT proteins to the receptor is not a highly specific process as predicted (Heim et al., 1995).

D. IL-6 Signaling by STAT-independent Pathway

The gp130 mutant without box 3, although unable to mediate the activation of IL-6RE, is capable of regulating gene expression via other elements such as HRRE and SIE as well as stimulating cell proliferation (Figure 1). These processes seem to occur without detectable STAT activation suggesting the existence of IL-6 regulated pathways independent of STAT proteins. IL-6 treatment leads to the activation of mitogen-activated protein kinase (MAPK) pathway in the IL-6-dependent B cell line, AF10, and fibroblasts (Kumar et al., 1994). IL-6 stimulates the tyrosine phosphorylation of gp130 and the receptor-associated proteins including JAK1, JAK2 and the adaptor protein, $p52^{Shc}$. The phosphorylation of Shc causes the association with Grb2 mediated by the SH2 domain. Grb2 is complexed with the guanylate nucleotide exchange factor hSOS via its SH3 domains and activates the membrane-bound Ras protein. The GTP-bound Ras in turn stimulates the downstream signaling molecules including Raf-1 (or MAPK kinase kinase), MAPK kinase, and MAPK. Substrates of the activated MAPK include the ribosomal protein S6 serine/threonine kinase (Sturgill et al., 1988; Chen et al., 1992), MAPKAP kinase 2 (Stokoe et al., 1992), heat shock protein (Hsp) 27 (Belka et al., 1995), and transcription factors such as NF-IL6 (Nakajima et al., 1993), c-*myc* (Seth et al., 1991), c-*jun* (Pulverer et al., 1991), and $p62^{TCF}$ (Gille et al., 1992). It is still not clear whether JAKs are responsible for the initial phosphorylation of Shc and whether Shc directly binds to gp130 through the PTB domain (Kavanaugh and Williams, 1994) or to other receptor-associated proteins via the SH2 domain.

Besides JAKs, other kinases are also implicated in IL-6 signaling. The *src*-related tyrosine kinase, Hck is physically associated with gp130 in ES cells (Ernst et al., 1994). The roles of the *src*-related kinases in cytokine signaling are also demonstrated in IL-2R to which Lck, Fyn, and Syk are associated (Horak et al., 1991; Kobayashi et al., 1993). In the IL-3-dependent murine B cell line, BAFB03, Btk and Tec tyrosine kinases were found to bind constitutively to the cytoplasmic region of gp130 and be activated by IL-6 plus soluble IL-6Rα (Matsuda et al., 1995). Btk, and Tec are members of a subfamily of cytoplasmic tyrosine kinases containing a novel pleckstrin homology (PH) domain in the amino-terminal region. Btk is involved in the lymphopoiesis of B cells (Rawlings et al., 1993; Thomas et al., 1993; Tsukada et al., 1993; Vetrie et al., 1993) whereas Tec is expressed at a high level in cells of hematopoietic lineages and is involved in the IL-3 and SCF signaling in the control of proliferation (Tang et al., 1994; Mano et al., 1995). The interactions of Btk and Tec with the *src*-related kinases are observed. Btk interacts with Fyn, Lyn and Hck via their SH3 domains (Cheng et al., 1994) whereas, Tec binds to Lyn at the N-terminal unique domain (Mano et al., 1994). These results reveal a complex network of kinases in the cytoplasm contributing to an IL-6 signaling pathway independent of JAK/STAT proteins.

The transcription factor, NF-IL6 (C/EBP) is strongly upregulated at mRNA level by LPS and the inflammatory cytokines such as IL-1, IL-6, and TNF which also activate the DNA-binding activity of NF-IL6 (Akira and Kishimoto, 1992; Natsuda et al., 1992). In the mouse liver, the mRNA level of NF-IL6 is drastically increased within 15 min after IL-6 induction. The expression reached a maximum between 1 and 2 h and then returned to the basal level by 6 h. The enhanced expression of NF-IL6 coincides with the regulation of the APP genes and the macrophage differentiation of mouse myeloid leukemia M1 cells by IL-6 (Natsuda et al., 1992). Concomitantly, the expression of Hp is stimulated within 30 min with a maximum at 3 h. The increased level persists up to 6 h. The binding sites for NF-IL6 (also referred to as the Type I IL-6RE) have been identified in the promoters of the APP genes including AGP, Hp, Hpx, complement C-3, CRP, and T kininogen, as well as the α-, β- and γFbs (Akira and Kishimoto, 1992). Overexpression of NF-IL6 augments the IL-6 induction of APP gene promoters (Baumann et al., 1991, 1992b). NF-IL6 is able to interact with other transcription factors. NF-IL6 forms heterodimers with NF-IL6β (or C/EBPδ) resulting in a synergistic effect on transactivation (Kinoshita et al., 1992). It also interacts with NF-κB in regulating the SAA and complement C3 genes (Shimizu and Yamamoto, 1994; Ray et al., 1995). It has been shown that NF-IL6 associates with Fos and Jun of the AP-1 family through the basic leucine zipper (bZIP) region (Hsu et al., 1994; Klamfer et al., 1994). Additionally, NF-IL6 and the glucocorticoid receptor act synergically to activate the transcription of AGP gene via direct protein-protein interaction (Nishio et al., 1993).

Numerous studies suggest that IL-6-regulated gene expression is the result of the interaction of multiple transcription factors, ubiquitous or cell specific (Figure 2). For instance, the regulation of human βFb gene promoter by IL-6 is critically controlled by the IL-6RE as well as C/EBP and HNF-1 binding sites. Mutation in any of these elements attenuates the IL-6 response (Dalmon et al., 1993). Given the possibility of cross-talk between signaling pathways and interaction of gene regulatory factors, it is conceivable that various pathways may converge at different levels of the regulatory loops (David et al., 1995; Cao et al., 1996). Therefore, the post-receptor components of individual pathway need to be determined before we can fully understand the actions of IL-6.

VII. IL-6 IN DISEASES AND CLINICAL APPLICATIONS

Elevated IL-6 levels in serum are observed in various disease states and the pathological conditions resemble the known biological activities of IL-6. Considering the facts that IL-6 can be produced by many cell types, whether normal or malignant, that its expression is induced by a broad array of inflammatory mediators and that IL-6 acts as part of an intricate cytokine network, the precise sources of systemic IL-6 are difficult to assess and the specific roles of IL-6 in the cause and progression of diseases may not be apparent. The levels of IL-6 are increased in acute inflammatory disorders (e.g., meningitis, acute hepatitis, transplant rejection, and trauma), chronic inflammatory diseases (e.g., cachexia and psoriasis), autoimmune diseases (e.g., rheumatoid diseases, Castleman's disease, systemic lupus erythematosus, and insulin-dependent diabetes mellitus), tumor (e.g., multiple myeloma/plasmacytoma, Kaposi's sarcoma—KS, cardiac myxoma and renal cell carcinoma), and acquired immunodeficiency syndrome (AIDS).

In patients with multiple myeloma, the tumor cells proliferate and accumulate in the bone marrow. Bone marrow cells freshly isolated from patients with multiple myeloma express IL-6R and produce IL-6 spontaneously (Kawano et al., 1988). In addition, high serum levels of IL-6 are detected in these patients (Bataille et al., 1989). IL-6 is a growth factor for the human myeloma and murine plasmacytoma cells *in vitro* (Van Damme et al., 1987b; Kawano et al., 1988). The *in vitro* growth is inhibited by anti-IL-6 antibody strongly suggesting that IL-6 is an essential growth factor for myeloma cells. It is still controversial whether IL-6 serves as an autocrine or paracrine factor for myeloma cells. IL-6 acts in an autocrine manner in the myeloma cell line U266 (Schwab et al., 1991). However, in another study, IL-6 mRNA expression can only be detected in bone marrow stromal cells but not purified myeloma cells (Portier et al., 1991). Furthermore, myeloma cell lines which are completely dependent on exogenous IL-6 can be established from patients with extramedullary proliferation (Shimizu et al., 1989; Zhang et al., 1994). These results suggest that IL-6 may function as a paracrine factor. Transfection of IL-6 cDNA and subsequent overexpression of IL-6 in IL-6-dependent B cell hybridoma and murine plasmacytoma cells enhance their tumorigenicity (Vink et al., 1990). Antibodies against IL-6 or its receptor can effectively suppress this stimulatory effect. Anti-IL-6 antibody has been tested in a clinical trial in which the antibody is injected intravenously to a myeloma patient daily for two months (Klein et al., 1991). The clinical status of the patient has been improved. The proliferation of the myeloma is suppressed and a reduction of calcium, monoclonal IgG and CRP levels in serum is observed.

The symptoms found in patients with autoimmune diseases suggest a relation of IL-6 with these diseases. In autoimmune diseases, increased serum levels of Ig and autoantibodies are generally observed in the patients. In addition, inflammatory response including elevation of APP and fever is frequently associated with the autoimmune diseases. Increased IL-6 levels can be detected in the synovial fluid and serum from patients with rheumatoid arthritis (RA; Hirano et al., 1988; Houssiau et al., 1988; Bhardwaj et al., 1989; Guerne et al., 1989). At the affected tissue, the production of IL-6 from mononuclear cells, T cells, B cells, synoviocytes, and chondrocytes is detected. Moreover, the IL-6-inducing cytokines such as IL-1 and TNF are found in the synovial fluid (Firestein et al., 1990). These cytokines, together with IL-6 may be responsible for the hypertrophy, formation of pannus and bone destruction occurred in the joints of the RA patients. Castleman's disease is a benign lymphoproliferative disease characterized by the enlargement of lymph nodes with infiltration of plasma cells. Patients have symptoms including polyclonal hypergammaglobulinemia, fever, anemia as well as increased APP expression and platelets. Enhanced production of IL-6 is detected in the B blastoid cells in the germinal center of the enlarged lymph nodes (Yoshizaki et al., 1989). The level of IL-6 is consistent with the severity of the symptoms suggesting that IL-6 plays a role in the pathogenesis of this disease.

Cachexia is often resulted from chronic infection and symptoms including weight loss, muscle wasting and alteration in protein and lipid metabolism are observed. The contributions of IL-6 to cachexia is confirmed in C-26.IVX-bearing mice (Strassmann et al., 1992). The serum level of IL-6 which is believed to be derived from the tumor as a result of stimulation by infiltrated inflammatory cells, correlates with the development of cachexia in these mice. When the primary tumors were resected, mice gained weight with a concomitant decline of IL-6 level in serum. More importantly, antibody to IL-6 was able to prevent the development of the cachectic syndrome supporting the notion that IL-6 acts in part as an indirect mediator of the proinflammatory cytokines, IL-1β and TNFα.

The increase of IL-6 level is detected in plasma (Breen et al., 1990; Honda et al., 1990; de Wit et al., 1991; Rautonen et al., 1991; Zauli et al., 1993), CSF (Gallo et al., 1989; Laurenzi et al., 1990) and lymph nodes (Emilie et al., 1990; Boyle et al., 1993) of the HIV-infected patients. IL-6 is produced constitutively from the PBMC, lung alveolar macrophage, B lymphocytes, and intestinal mononuclear cells of the patients (Rieckmann et al., 1991a, 1991b; Boue et al., 1992; Twigg et al., 1992; Steffen et al., 1993). The augmented IL-6 expression is correlated with the increase of Ig production in plasma and CSF, as well as the enhanced expression of ICAM-1 adhesion molecules (Gallo et al.,

1989; Birx et al., 1990; Rautonen et al., 1991; Most et al., 1993). It is found that IL-6 promotes the shedding of IL-2R in T cells *in vitro* and it may, at least in part, be responsible for the attenuated proliferative response of T cells during HIV infection (Honda et al., 1990; Tsunetsugu-Yokota and Honda, 1990). The envelop proteins, gp120 and gp160 of HIV-1 can increase the IL-6 production from PBMC and $CD4^{+}$ T cell clones (Oyaizu et al., 1991). In addition, the viral *trans*-acting factors, Tat and Tax can enhance the expression of IL-6 (Gibellini et al., 1994; Mori et al., 1994; Rautonen et al., 1994; Yamashita et al., 1994). IL-6 induces the expression of HIV in infected primary astrocytes, monocytic cell line THP-1, and a neuroblastoma-derived cell line (Tsunetsugu-Yokota and Honda, 1990; Swingler et al., 1992; Vesanen et al., 1992). In primary macrophages, IL-6 acts in synergy with TNFα to potentiate the replication efficiency of HIV (Poli et al., 1990). This stimulatory effect is mediated by both transcriptional and post-transcriptional mechanisms. IL-6 also synergizes with IL-1 to upregulate the virus expression in the promonocytic cell line, U1 (Poli et al., 1994). Kaposi's sarcoma (KS) is the most frequent tumor associated with AIDS. Cell lines derived from AIDS-KS express IL-6R and secrete IL-6 constitutively (Ensoli et al., 1989; Miles et al., 1990; Corbeil et al., 1991). IL-6 antisense oligonucletides can inhibit the IL-6 production by these cells as well as their proliferation. Interestingly, these oligonucleotides also suppress the mitogenic effects of OSM which is a strong growth-promoting factor for the AIDS-KS cells (Miles et al., 1992; Nair et al., 1992). Notably, transgenic mice expressing *tat* gene develop KS-like disease (Vogel et al., 1988) and Tat protein can stimulate the IL-6 production as well as the proliferation of the KS cells (Ensoli et al., 1990).

Based on the observations that many pathological conditions are associated with the overproduction of IL-6, a number of different strategies have been developed to inhibit the IL-6 activity. These include IL-6 synthesis inhibitors, anti-IL-6 antibodies, antibodies against IL-6 receptor subunits, and, more recently, IL-6 mutant proteins. Corticosteroids, estrogens, retinoid acids, IFNα, and IL-4 are known to act as inhibitors of IL-6 expression (Helfgott et al., 1987; Tabibzadeh et al., 1989; Sidell et al., 1991). For example, IL-4 inhibits the cellular production of proinflammatory cytokines in RA synovium (Miossec et al., 1992) and acts as a potent inhibitor of bone resorption *in vivo* (Watanabe et al., 1990). Neutralizing antibody to IL-6 proved to be effective in the suppression of myeloma growth in patients with multiple myeloma (Suzuki et al., 1992). It also improves the clinical conditions of RA patients (Wendling et al., 1993) and attenuates the endotoxin-induced activation of coagulation in experimental endotoxemia in chimpanzees (van der Poll et al., 1994). The effects of the anti-IL-6R antibody have been tested *in vivo* . Mice administrated with specific antibody to the IL-6R show reduced growth of inoculated plasmacytoma/myeloma cells (Vink et al., 1990). Moreover, the antibody inhibits the increase of platelet induced by IL-6.

A novel approach aims at designing mutant IL-6 protein with antagonist function. The regions on the IL-6 molecule important for the interaction with receptor subunits have been identified. These variant proteins retain the binding affinity to the IL-6Rα but reduce the ability of the IL-6Rα to associate with gp130 and hence predicted to interrupt the signaling process (Ehlers et al., 1994, 1995). Although they have not been tested in vivo, they are believed to be of potential therapeutic benefits. An alternative approach to suppress endogenous IL-6 action is the design of pharmacological reagents inhibiting the signaling mechanism by the IL-6Rα-gp130 complex via protein tyrosine kinases (Levitzki and Gazit, 1995).

Although emphasis has been made on the inhibition of IL-6 actions as part of the treatments. IL-6 may also be a beneficial factor. For instance, the stimulatory effects in the course of acute inflammatory process are necessary for reestablishing and maintaining homeostasis. Moreover, as discussed above, IL-6 has been noted to have a functional role in hematopoiesis and immune response. Hence, it has been tried to boost the anti-tumor response. In a phase I-II study of 20 patients with breast carcinoma or nonsmall cell lung cancer, recombinant human IL-6 was administered by continuous intravenous infusion followed by subcutaneous administration for six days (van Gameren et al., 1994). Patients exhibited an increase in platelet counts as well as the numbers of neutrophils, monocytes, T cells, and NK cells. Reduced total cholesterol levels and dose-dependent increase of CRP and SAA in plasma were also observed. These results indicate the potential clinical use of IL-6 as a stimulatory agent in leukopoiesis and thrombopoiesis. The applications of IL-6 as a therapeutic agent and the results of clinical trials are also discussed by Mackiewicz et al. (1995b).

VIII. CONCLUSIONS

IL-6 is a typical pleiotropic cytokine that elicits multiple physiologic responses. Its biological activities have been extensively explored by using different *in vitro* systems. Studies on IL-6 transgenic mice which overexpress the cytokine have demonstrated the *in vivo* significance of IL-6 in the regulation of hematopoiesis and acute-phase response, as well as the modulation of the immune and nervous systems. Although the effects of IL-6 on the other systems are relatively clear, its impacts on CNS and embryonic development are still poorly understood. The minor defects observed in the IL-6-deficient mice indicate that most functions of IL-6 can be substituted by other cytokines. Nevertheless, the study has suggested an exclusive role of IL-6 in osteoporosis.

The structure/function analysis of gp130 has established a relationship between receptor domains and the IL-6-regulated events. Recent progress in cytokine signaling has advanced our understanding in the functional redundancy among members of the cytokine family. The fact that receptors for the IL-6-type cytokines (IL-6, LIF, OSM, CNTF, IL-11) share gp130 as the common signal-transducing subunit provides a basis for the redundancy among these cytokines. IL-6R, like many other hematopoietin receptors, mediates the activation of the JAK/STAT pathway which may allow cross-talk within the cytokine network, as well as with the actions of growth factors and endocrine system. To date, the contributions of the JAK/STAT proteins and the STAT-independent pathways mediated by kinases including Hck, Tec, Btk, and MAPK, to the IL-6 regulation of cell proliferation, differentiation, and gene expression are still not precisely defined. Furthermore, the roles of phosphatase such as Syp (or PTP1D) which is activated upon IL-6 stimulation (Boulton et al., 1994; Stahl et al., 1995) are largely unaddressed at the present. It will remain to be the focus of attention for investigators in these fields.

An elevated serum level of IL-6 is observed in cancer and a number of inflammatory and immune diseases. The increase in IL-6 level correlates with the pathogenesis of these diseases. Based on the knowledge of the cytokine and receptor structure, ligand-receptor interaction, and cell activation process, therapeutic methods to intervene or potentiate the IL-6 effects are under development. It is anticipated that by elucidating the downstream signaling events of IL-6, more specific and effective therapeutic drugs for the treatments of IL-6-related diseases can be designed and applied.

ACKNOWLEDGMENTS

Research in authors' laboratory was supported by National Institute of Health Grants CA26122 and DK 33886. We thank Karen K. Morella and Yanping Wang for their excellent technical assistance.

REFERENCES

Abdel-Beguid, S. S., Shieh, H. S., Smith, W. W., Dayringer, H. E., Violand, B. N., & Bentle, L. A. (1987). Three-dimensional structure of a genetically engineered variant of porcine growth hormone. Proc. Natl. Acad. Sci. USA 84, 6434-6437.

Akira, S., & Kishimoto, T. (1992). IL-6 and NF-IL6 in acute-phase response and viral infection. Immunol. Rev. 127, 25-50.

Akira, S., Isshiki, H., Nakajima, T., Kinoshita, S., Nishio, Y., Natsuka, S., & Kishimoto, T. (1992). Regulation of expression of the interleukin 6 gene: Structure and function of the transcription factor NF-IL6. Ciba Foundation Symposium 167, 47-67.

Akira, S., Taga, T., & Kishimoto, T. (1993). Interleulin-6 in biology and medicine. Adv. Immunol. 54, 1-78.

Akira, S., Nishio, Y., Inoue, M., Wang, X.-J., Wei, S., Matsusaka, T., Yoshida, K., Sudo, T., Naruto, M., & Kishimoto, T. (1994). Molecular cloning of APRF, a novel IFN-stimulated gene factor 3 p91-related transcription factor involved in the gp130-mediated signaling pathway. Cell 77, 63-71.

Anderson, G. M., Shaw, A. R., & Shafer, J. A. (1993). Functional characterization of promoter elements involved in regulation of human β-fibrinogen expression. Evidence for binding of novel activator and repressor proteins. J. Biol. Chem. 268, 22650-22655.

Andus, T., Geiger, T., Hirano, T., Northoff, H., Ganter, U., Bauer, J., Kishimoto, T., & Heinrich, P. C. (1987). Recombinant human B cell stimulatory factor 2 (BSF-2/IFN-β2) regulates β-fibrinogen and albumin mRNA levels in Fao-9 cells. FEBS Lett. 221, 18-22.

Asao, H., Takeshita, T., Ishii, N., Kumaki, S., Nakamura, M., & Sugamura, K. (1993). Reconstitution of functional interleukin 2 receptor complexes on fibroblastoid cells: Involvement of the cytoplasmic domain of the γ chain in two distinct signaling pathways. Proc. Natl. Acad. Sci. USA 90, 4127-4131.

Astaldi, G. C., Janssen, M. C., Lansdorp, P., Willems, C., Zeijlemaker, W. P., & Oosterhof, F. (1980). Human endothelial culture supernatant (HECS): A growth factor for hybridomas. J. Immunol. 125, 1411-1414.

Bartley, T. D., Bogenberger, J., Hunt, P., Li, Y.-S., Lu, H.-S., Martin, F., Chang, M.-S., Samal, B., Nichol, J. L., Swift, S., Johnson, M. J., Hsu, R.-Y., Parker, V. P., Suggs, S., Skrine, J. D., Merewether, L. A., Clogston, C., Hsu, E., Hokom, M. M., Hornkohl, A., Choi, E., Pangelinan, M., Sun, Y., Mar, V., McNinch, J., Simonet, L., Jacobsen, F., Xie, C., Shutter, J., Chute, H., Basu, R., Selander, L., Trollinger, D., Sieu, L., Padilla, D., Trail, G., Elliott, G., Izumi, R., Covey, T., Crouse, J., Garcia, A., Xu, W., Del Castillo, J., Biro, J., Cole, S., Hu, M. C.-T., Pacifici, R., Ponting, I., Saris, C., Wen, D., Yung, Y. P., Lin, H., & Bosselman, R. A. (1994). Identification and cloning of a megakaryocyte growth and development factor that is a ligand for the cytokine receptor Mpl. Cell 77, 1117-1124.

Bataille, R., Jourdan, M., Zhang, X. G., & Klein, B. (1989). Serum levels of interleukin 6, a potent myeloma cell growth factor, as a reflect of disease severity in plasma cell dyscrasias. J. Clin. Invest. 84, 2008-2011.

Baumann, H. (1990). Regulation of hepatic acute phase plasma protein genes by hepatocyte stimulating factors and other mediators of inflammation. Mol. Biol. Med. 7, 147-159.

Baumann, H., & Gauldie, J. (1994). The acute phase response. Immunology Today 15, 74-80.

Baumann, H., Jahreis, G. P., Morella, K. K., Won, K.-A., Pruitt, S. C., Jones, V. E., & Prowse, K. R. (1991). Transcriptional regulation through cytokine and glucocorticoid response element of rat acute phase plasma protein genes by C/EBP and JunB. J. Biol. Chem. 266, 20390-20399.

Baumann, H., Marinkovic-Pajovic, S., Won, K.-A., Jones, V. E., Campos, S. P., Jahreis, G. P., & Morella, K. K. (1992a). The action of interleukin-6 and leukaemia inhibitory factor on liver cells. Ciba Foundation Symposium 167, 100-124.

Baumann, H., Morella, K. K., Campos, S. P., Cao, Z., & Jahreis, G. P. (1992b). Role of CAAT-enhancer binding protein isoforms in the cytokine regulation of acute-phase plasma protein genes. J. Biol. Chem. 267, 19744-19751.

Baumann, H., Ziegler, S. F., Mosley, B., Morella, K. K., Pajovic, S., & Gearing, D. P. (1993). Reconstitution of the response to leukemia inhibitory factor, oncostatin M, and ciliary neurotrophic factor in hepatoma cells. J. Biol. Chem. 268, 8414-8417.

Baumann, H., Gearing, D., & Ziegler, S. F. (1994). Signaling by the cytoplasmic domain of hematopoietin receptors involves two distinguishable mechanisms in hepatic cells. J. Biol. Chem. 269, 16297-16304.

Bazan, J. F. (1990a). Haemopoietic receptors and helical cytokines. Immunology Today 11, 350-354.

Bazan, J. F. (1990b). Structural design and molecular evolution of a cytokine receptor superfamily. Proc. Natl. Acad. Sci. USA 87, 6934-6938.

Belka, C., Ahlers, A., Sott, C., Gaestel, M., Herrmann, F., & Brach, M. A. (1995). Interleukin (IL)-6 signaling leads to phosphorylation of the small heat shock protein (Hsp)27 through activation of the MAP kinase and MAPKAP kinase 2 pathway in monocytes and monocytic leukemia cells. Leukemia 9, 288-294.

Bernad, A., Kopf, M., Kulbacki, R., Weich, N., Koehler, G., & Gutierrez-Ramos, J. C. (1994). Interleukin-6 is required *in vivo* for the regulation of stem cells and committed progenitors of the hematopoietic system. Immunity 1, 725-731.

Bhardwaj, N., Santhanam, U., Lau, L. L., Tatter, S. B., Ghrayeb, J., Rivelis, M., Steinman, R. M., Sehgal, P. B., & May, L. T. (1989). IL-6/IFN-β2 in synovial effusions of patients with rheumatoid arthritis and other arthritides. Identification of several isoforms and studies of cellular sources. J. Immunol. 143, 2153-2159.

Biondi, A., Rossi, V., Bassan, R., Barbui, T., Bettoni, S., Sironi, M., Mantovani, A., & Rambaldi, A. (1989). Constitutive expression of the interleukin-6 gene in chronic lymphocytic leukemia. Blood 73, 1279-1284.

Birx, D. L., Redfield, R. R., Tencer, K., Fowler, A., Burke, D. S., & Tosato, G. (1990). Induction of interleukin-6 during human immunodeficiency virus infection. Blood 76, 2303-2310.

Black, K., Garrett, I. R., & Mundy, G. R. (1991). Chinese hamster ovarian cells transfected with the murine interleukin-6 gene cause hypercalcemia as well as cachexia, leukocytosis and thrombocytosis in tumor-bearing nude mice. Endocrinology 128, 2657-2659.

Bolin, L. M., Verity, A. N., Silver, J. E., Shooter, E. M., & Abrams, J. S. (1995). Interleukin-6 production by Schwann cells and induction in sciatic nerve injury. J. Neurochem. 64, 850-858.

Bot, F. J., van Eijk, L., Broeders, L., Aarden, L. A., & Lowenberg, B. (1989). Interleukin-6 synergizes with M-CSF in the formation of macrophage colonies from purified human marrow progenitor cells. Blood 73, 435-437.

Boue, F., Wallon, C., Goujard, C., Barre-Sinoussi, F., Galanaud, P., & Delfraissy, J. F. (1992). HIV induces IL-6 production by human B lymphocytes. Role of IL-4. J. Immunol. 148, 3761-3767.

Boulay, J.-L., & Paul, W. E. (1993). Hematopoietin sub-family classification based on size, gene organization and sequence homology. Curr. Biol. 3, 573-581.

Boulton, T. G., Stahl, N., & Yancopoulos, G. D. (1994). Ciliary neurotrophic factor/leukemia inhibitory factor/interleukin 6/oncostatin M family of cytokines induces tyrosine phosphorylation of a common set of proteins overlapping those induced by other cytokines and growth factors. J. Biol. Chem. 269, 9-17.

Boyle, M. J., Berger, M. F., Tschuchnigg, M., Valentine, J. E., Kennedy, B. G., Divjak, M., Cooper, D. A., Turner, J. J., Penny, R., & Sewell, W. A. (1993). Increased experssion of interferon-gamma in hyperplastic lymph nodes from HIV-infected patients. Clin. Exp. Immunol. 92, 100-105.

Braciak, T. A., Mittal, S. K., Graham, F. L., Richards, C. D., & Gauldie, J. (1993). Construction of recombinant human type 5 adenoviruses expressing rodent IL-6 genes. An approach to investigate *in vivo* cytokine function. J. Immunol. 151, 5145-5153.

Brakenhoff, J. P., Hart, M., & Aarden, L. A. (1989). Analysis of human IL-6 mutants expressed in *Escherichia coli*. Biologic activities are not affected by deletion of amino aicds 1-28. J. Immunol. 143, 1175-1182.

Brakenhoff, J. P., de Hon, F. D., Fontaine, V., ten Boekel, E., Schooltink, H., Rose-John, S., Heinrich, P. C., Content, J., & Aarden, L. A. (1994). Development of a human interleukin-6 receptor antagonist. J. Biol. Chem. 269, 86-93.

Breen, E. C., Rezai, A. R., Nakajima, K., Beall, G. N., Mitsuyasu, R. T., Hirano, T., Kishimoto, T., & Martinez-Mada, O. (1990). Infection with HIV is associated with elevated IL-6 levels and production. J. Immunol. 144, 480-484.

Cairo, M. S., Plunkett, J. M., Nguyen, A., Clark, S., & van de Ven, C. (1991). Sequential administration of interleukin-6 and granulocyte-colony stimulating factor in newborn rats: Modulation of newborn granulopoiesis and thrombopoiesis. Pediatr. Res. 30, 554-559.

Campbell, I. L., Abraham, C. R., Masliah, E., Kemper, P., Inglis, J. D., Oldstone, M. B., & Mucke, L. (1993). Neurologic disease induced in transgenic mice by cerebral overexpression of interleukin 6. Proc. Natl. Acad. Sci. USA 90, 10061-10065.

Campbell, I. L., Hobbs, M. V., Dockter, J., Oldstone, M. B., & Allison, J. (1994). Islet inflammation and hyperplasia induced by the pancreatic islet-specific overexpression of interleukin-6 in transgenic mice. Am. J. Pathol. 145, 157-166.

Cao, X., Tay, A., Guy, F. R., & Tan, Y. H. (1996). Activation and association of Stat3 with Src in v-Src-transformed cell lines. Mol. Cell. Biol. 16, 1595-1603.

Chen, H.-M., & Liao, W. S. L. (1993). Differential acute-phase response of rat kininogen genes involves type I and type II interleukin-6 response elements. J. Biol. Chem. 268, 25311-25319.

Chen, R. H., Sarnecki, C., & Blenis, J. (1992). Nuclear localization and regulation of *erk*- and *rsk*-encoded protein. Mol. Cell. Biol. 12, 915-927.
Cheng, G., Ye, Z. S., & Baltimore, D. (1994). Binding of Bruton's tyrosine kinase to Fyn, Lyn, or Hck through a Src homology 3 domain-mediated interaction. Proc. Natl. Acad. Sci. USA 91, 8152-8155.
Cheung, D. L., Hart, P. H., Vitti, G. F., Whitty, G. A., & Hamilton, J. A. (1990). Contrasting effects of interferon-γ and interleukin-4 on the interleukin-6 activity of stimulated human monocytes. Immunology 71, 70-75.
Chiu, C. P., & Lee, F. (1989). IL-6 is a differentiation factor for M1 and WEHI-3B myeloid leukemic cells. J. Immunol. 142, 1909-1915.
Colosi, P., Wong, K., Leong, S. R., & Wood, W. I. (1993). Mutational analysis of the intracellular domain of the human growth hormone receptor. J. Biol. Chem. 268, 12617-12623.
Corbeil, J., Evans, L. A., Vasak, E., Cooper, D. A., & Penny, R. (1991). Culture and properties of cells derived from Kaposi sarcoma. J. Immunol. 146, 2972-2976.
Cosman, D. (1993). The hematopoietin receptor superfamily. Cytokine 5, 95-106.
Coulie, P. G., Cayphas, S., Vink, A., Uyttenhove, C., & Van Snick, J. (1987). Interleukin-HP1-related hybridoma and plasmacytoma growth factors induced by lipopolysaccharide *in vivo*. Eur. J. Immunol. 17, 1217-1220.
Dalmon, J., Laurent, M., & Courtois, G. (1993). The human β fibrinogen promoter contains a hepatocyte nuclear factor 1-dependent interleukin-6-responsive element. Mol. Cell. Biol. 13, 1183-1193.
D'Andrea, A. D., Yoshimura, A., Youssoufian, H., Zon, L. I., Koo, J.-W., & Logish, H. F. (1991). The cytoplasmic region of the erythropoietin receptor contains nonoverlapping positive and negative growth regulatory domains. Mol. Cell. Biol. 11, 1980-1987.
Darnell, J. E., Jr., Kerr, I. M., & Stark, G. R. (1994). Jak-STAT pathways and transcriptional activation in response to IFNs and other extracellular signaling proteins. Science 264, 1415-1421.
David, M., Petricoin, E., III, Benjamin, C., Pine, R., Weber, M. J., & Larner, A. C. (1995). Requirement for MAP kinase (ERK2) activity in interferon α- and interferon β-stimulated gene expression through STAT proteins. Science 269, 1721-1723.
Davis, S., Aldrich, T. H., Stahl, N., Pan, L., Taga, T., Kishimoto, T., Ip, N. Y., & Yancopoulos, G. D. (1993). LIFR and gp130 as heterodimerizing signal transducers of the tripartite CNTF receptor. Science 260, 1805-1808.
De, M., Sanford, T. H., & Wood, G. W. (1992a). Detection of interleukin-1, interleukin-6, and tumor necrosis factor-alpha in the uterus during the second half of pregnancy in the mouse. Endocrinology 131, 14-20.
De, M., Sanford, T. R., & Wood, G. W. (1992b). Interleukin-1, interleukin-6, and tumor necrosis factor alpha are produced in the mouse uterus during the estrous cycle and are induced by estrogen and progesterone. Dev. Biol. 151, 297-305.
de Sauvage, F. J., Hass, P. E., Spencer, S. D., Malloy, B. E., Gurney, A. L., Spencer, S. A., Darbonne, W. C., Henzel, W. J., Wong, S. C., Kuang, W.-J., Oles, K. J., Hultgren, B., Solberg, L., Jr., Goeddel, D. V., & Eaton, D. L. (1994). Stimulation of megakaryocytopoiesis and thrombopoiesis by the c-Mpl ligand. Nature 369, 533-538.
De Simoni, M. G., Del Bo, R., De Luigi, A., Simard, S., & Forloni, G. (1995). Central endotoxin induces different patterns of interleukin(IL)-1β and IL-6 messenger

ribonucleic acid expression and IL-6 secretion in the brain and periphery. Endocrinology 136, 897-902.

de Wit, R., Raasveld, M. H., ten Berge, R. J., van der Wouw, P. A., & Bakker, P. J. (1991). Interleukin-6 concentrations in the serum of patients with AIDS-associated Kaposi's sarcoma during treatment with interferon-alpha. J. Intern. Med. 229, 539-542.

DiCosmo, B. F., Geba, G. P., Picarella, D., Elias, J. A., Rankin, J. A., Stripp, B. R., Whitsett, J. A., & Flavell, R. A. (1994). Airway epithelial cell expression of interleukin-6 in transgenic mice. Uncoupling of airway inflammation and bronchial hyperreactivity. J. Clin. Invest. 94, 2028-2035.

Dinarello, C. A., Cannon, J. G., Mancilla, J., Bishai, I., Lees, J., & Coceani, F. (1991). Interleukin-6 as an endogenous pyrogen: Induction of prostaglandin E2 in brain but not in peripheral blood mononuclear cells. Brain Res. 562, 199-206.

Dittrich, E., Rose-John, S., Gerhartz, C., Mullberg, J., Stoyan, T., Yasukawa, K., Heinrich, P. C., & Graeve, L. (1994). Identification of a region within the cytoplasmic domain of the interleukin-6 (IL-6) signal transducer gp130 important for ligand-induced endocytosis of the IL-6 receptor. J. Biol. Chem. 269, 19014-19020.

Donnelly, R. P., Crofford, L. J., Freeman, S. L., Buras, J., Remmers, E., Wilder, R. L., & Fenton, M. J. (1993). Tissue-specific regulation of IL-6 production by IL-4. Differential effects of IL-4 on nuclear factor-κB activity in monocytes and fibroblasts. J. Immunol. 151, 5603-5612.

Ehlers, M., Grotzinger, J., deHon, F. D., Mullberg, J., Brakenhoff, J. P., Liu, J., Wollmer, A., & Rose-John, S. (1994). Identification of two novel regions of human IL-6 responsible for receptor binding and signal transduction. J. Immunol. 153, 1744-1753.

Ehlers, M., de Hon, F. D., Bos, H. K., Horsten, U., Kurapkat, G., van De Leur, H. S., Grötzinger, J., Wollmer, A., Brakenhoff, J. P. J., & Rose-John, S. (1995). Combining two mutations of human interleukin-6 that affect gp130 activation results in a potent interleukin-6 receptor antagonist on human myeloma cells. J. Biol. Chem. 270, 8158-8163.

Emilie, D., Peuchmaur, M., Maillot, M. C., Crevon, M. C., Brousse, N., Delfraissy, J. F., Dormont, J., & Galanaud, P. (1990). Production of interleukins in human immunodeficiency virus-1-replicating lymph nodes. J. Clin. Invest. 86, 148-159.

Ensoli, B., Nakamura, S., Salahuddin, S. Z., Biberfeld, P., Larsson, L., Beaver, B., Wong-Staal, F., & Gallo, R. C. (1989). AIDS-Kaposi's sarcoma-derived cells express cytokines with autocrine and paracrine growth effects. Science 243, 223-226.

Ensoli, B., Barillari, G., Salahuddin, S. Z., Gallo, R. C., & Wong-Staal, F. (1990). Tat protein of HIV-1 stimulates growth of cells derived from Kaposi's sarcoma lesions of AIDS patients. Nature 345, 84-86.

Ernst, M., Gearing, D. P., & Dunn, A. R. (1994). Functional and biochemical association of Hck with the LIF/IL-6 receptor signal transducing subunit gp130 in embryonic stem cells. EMBO J. 13, 1574-1584.

Fattori, E., Cappelletti, M., Costa, P., Sellitto, C., Cantoni, L., Carelli, M., Faggioni, R., Fantuzzi, G., Ghezzi, P., & Poli, V. (1994a). Defective inflammatory response in interleukin 6-deficient mice. J. Exp. Med. 180, 1243-1250.

Fattori, E., Della Rocca, C., Costa, P., Giorgio, M., Dente, B., Pozzi, L., & Ciliberto, G. (1994b). Development of progressive kidney damage and myeloma kidney in interleukin-6 transgenic mice. Blood 83, 2570-2579.

Fernandez-Botran, R. (1991). Soluble cytokine receptors: Their role in immunoregulation. FASEB J. 5, 2567-2574.

Feyen, J. H., Elford, P., Di Padova, F. E., & Trechsel, U. (1989). Interleukin-6 is produced by bone and modulated by parathyroid hormone. J. Bone Miner. Res. 4, 633-638.

Firestein, G. S., Alvaro-Gracia, J. M., Maki, R., & Alvaro-Gracia, J. M. (1990). Quantitative analysis of cytokine gene expression in rheumatoid arthritis. J. Immunol. 144, 3347-3353.

Firmbach-Kraft, I., Byers, M., Shows, T., Dalla-Favera, R., & Krolewski, J. J. (1990). *tyk2*, prototype of a novel class of non-receptor tyrosine kinase genes. Oncogene 5, 1329-1336.

Fowlkes, D. M., Mullis, N. T., Comeau, C. M., & Crabtree, G. R. (1984). Potential basis for regulation of the coordinately expressed fibrinogen genes: Homology in the 5' flanking regions. Proc. Natl. Acad. Sci. USA 81, 2313-2316.

Frei, K., Malipiero, U. V., Leist, T. P., Zinkernagel, R. M., Schwab, M. E., & Fontana, A. (1989). On the cellular source and function of interleukin 6 produced in the central nervous system in viral diseases. Eur. J. Immunol. 19, 689-694.

Frieling, J. T., Sauerwein, R. W., Wijdenes, J., Hendriks, T., & van der Linden, C. J. (1994). Soluble interleukin 6 receptor in biological fluids from human orgin. Cytokine 6, 376-381.

Fukunaga, R., Ishizaka-Ikeda, E., & Nagata, S. (1993). Growth and differentiation signals mediated by different regions in the cytoplasmic domain of granulocyte colony-stimulating factor receptor. Cell 74, 1079-1087.

Gaillard, J. P., Bataille, R., Brailly, H., Zuber, C., Yasukawa, K., Attal, M., Maruo, N., Taga, T., Kishimoto, T., & Klein, B. (1993). Increased and highly stable levels of functional soluble interleukin-6 receptor in sera of patients with monoclonal gammopathy. Eur. J. Immunol. 23, 820-824

Gallo, P., Frei, K., Rordorf, C., Lazdins, J., Tavolato, B., & Fontana, A. (1989). Human immunodeficiency virus type 1 (HIV-1) infection of the central nervous system: An evaluation of cytokines in cerebrospinal fluid. J. Neuroimmunol. 23, 109-116.

Ganter, U., Arcone, R., Toniatti, C., Morrone, G., & Ciliberto, G. (1989). Dual control of C-reactive protein gene expression by interleukin-1 and interleukin-6. EMBO J. 8, 3773-3779.

Garman, R. D., & Raulet, D. H. (1987). Characterization of a novel murine T cell-activating factor. J. Immunol. 138, 1121-1129.

Garman, R. D., Jacobs, K. A., Clark, S. C., & Raulet, D. H. (1987). B-cell-stimulatory factor 2 (β2 interferon) functions as a second signal for interleukin 2 production by mature murine T cells. Proc. Natl. Acad. Sci. USA 84, 7629-7633.

Gearing, D. P., Comeau, M. R., Friend, D. J., Gimpel, S. D., Thut, C. J., McGourty, J., Brasher, K. K., King, J. A., Gillis, S., Mosley, B., Ziegler, S. F., & Cosman, D. (1992). The IL-6 signal transducer, gp130: An oncostatin M receptor and affinity converter for the LIF receptor. Science 255, 1434-1437.

Geiger, T., Andus, T., Klapproth, J., Hirano, T., Kishimoto, T., & Heinrich, P. C. (1988). Induction of rat acute-phase proteins by interleukin 6 *in vivo*. Eur. J. Immunol. 18, 717-721.

Gerhartz, C., Dittrich, E., Stoyan, T., Rose-John, S., Yasukawa, K., Heinrich, P. C., & Graeve, L. (1994). Biosynthesis and half-life of the interleukin-6 receptor and its signal transducer gp130. Eur. J. Biochem. 223, 265-274.
Gerosa, F., Tommasi, M., Carra, G., Gandini, G., Tridente, G., & Benati, C. (1992). Different sensitivity to interleukin 4 of interleukin 2- and interferon alpha-induced CD69 antigen experssion in human resting NK cells and CD3+, CD4-, CD8- lymphocytes. Cellular Immunology 141, 342-351.
Gibellini, D., Zauli, G., Re, M. C., Milani, D., Furlini, G., Caramelli, E., Capitani, S., & La Placa, M. (1994). Recombinant human immunodeficiency virus type-1 (HIV-1) Tat protein sequentially up-regulates IL-6 and TGF-beta 1 mRNA expression and protein synthesis in peripheral blood monocytes. Br. J. Haematol. 88, 261-267.
Gille, H., Sharrocks, A. D., & Shaw, P. E. (1992). Phosphorylation of transcription factor $p62^{TCF}$ by MAP kinase stimulates ternary complex formation at c-*fos* promoter. Nature 358, 414-417.
Giri, J. G., Ahdieh, M., Eisenman, J., Shanebeck, K., Grabstein, K., Kumaki, S., Namen, A., Park, L. S., Cosman, D., & Anderson, D. (1994). Utilization of the β and γ chains of the IL-2 receptor by the novel cytokine IL-15. EMBO J. 13, 2822-2830.
Gouilleux, F., Wakao, H., Mundt, M., & Groner, B. (1994). Prolactin induces phosphorylation of Tyr694 of Stat5 (MGF), a prerequisite for DNA binding and induction of transcription. EMBO J. 13, 4361-4369.
Gronowski, A. M., Zhong, Z., Wen, Z., Thomas, M. J., Darnell, J. E., Jr., & Rotwein, P. (1995). *In vivo* growth hormone treatment rapidly stimulates the tyrosine phosphorylation and activation of Stat3. Mol. Endo. 9, 171-177.
Gross, T., Andus, T., Castell, J., Wom Berg, D., Heinrich, P. C., & Gerok, W. (1989). O- and N-glycosylation lead to different molecular mass forms of human monocyte interleukin-6. FEBS Lett. 247, 323-326.
Grube, B. J., & Cochrane, C. G. (1994). Identification of a regulatory domain of the interleukin-6 receptor. J. Biol. Chem. 269, 20791-20797.
Guerne, P. A., Zuraw, B. L., Vaughan, J. H., Carson, D. A., & Lotz, M. (1989). Synovium as a source of interleukin 6 *in vitro*. Contribution to local and systemic manifestations of arthritis. J. Clin. Invest. 83, 585-592.
Guschin, D., Rogers, N., Briscoe, J., Witthuhn, B., Watling, D., Horn, F., Pellegrini, S., Yasukawa, K., Heinrich, P., Stark, G. R., Ihle, J. N., & Kerr, I. M. (1995). A major role for the protein tyrosine kinase JAK1 in the JAK/STAT signal transduction pathway in response to interleukin-6. EMBO J. 14, 1421-1429.
Hama, T., Miyamato, M., Tsukui, H., Nishio, C., & Hatanaka, H. (1989). Interleukin-6 as a neurotrophic factor for promoting the survival of cultured basal forebrain cholinergic neurons from postnatal rats. Neurosci. Lett. 104, 340-344.
Hatakeyama, M., Mori, H., Doi, T., & Taniguchi, T. (1989). A restricted cytoplasmic region of IL-2 receptor beta chain is essential for growth signal transduction but not for ligand binding and internalization. Cell 59, 837-845.
Hattori, M., Abraham, L. J., Northemann, W., & Fey, G. H. (1990). Acute-phase reaction induces a specific complex between hepatic nuclear proteins and the interleukin 6 response element of the rat α_2-macroglobulin gene. Proc. Natl. Acad. Sci. USA 87, 2364-2368.
Heaney, M. L., & Golde, D. W. (1993). Soluble hormone receptors. Blood 82, 1945-1948.

Heim, M. H., Kerr, I. M., Stark, G. R., & Darnell, J. E., Jr. (1995). Contribution of STAT SH2 groups to specific interferon signaling by the Jak-STAT pathway. Science 267, 1347-1349.

Helfgott, D. C., May, L. T., Sthoeger, Z., Tamm, I., & Sehgal, P. B. (1987). Bacterial lipopolysaccharide (endotoxin) enhances expression and secretion of β2 interferon by human fibroblasts. J. Exp. Med. 166, 1300-1309.

Helle, M., Brakenhoff, J. P., De Groot, E. R., & Aarden, L. A. (1988). Interleukin 6 is involved in interleukin 1-induced activities. Eur. J. Immunol. 18, 957-959.

Hibi, M., Murakami, M., Saito, M., Hirano, T., Taga, T., & Kishimoto, T. (1990). Molecular cloning and expression of an IL-6 signal transducer, gp130. Cell 63, 1149-1157.

Hirano, T., Taga, T., Nakano, N., Yasukawa, K., Kashiwamura, S., Shimizu, K., Nakajima, K., Pyun, K. H., & Kishimoto, T. (1985). Purification to homogeneity and characterization of human B-cell differentiation factor (BCDF or BSFp-2). Proc. Natl. Acad. Sci. USA 82, 5490-5494.

Hirano, T., Yasukawa, K., Harada, H., Taga, T., Watanabe, Y., Matsuda, T., Kashiwamura, S., Nakajima, K., Koyama, K., Iwamatsu, A., Tsunasawa, S., Saikiyama, F., Matsui, H., Takahara, Y., Taniguchi, T., & Kishimoto, T. (1986). Complementary DNA for a novel human interleukin (BSF-2) that induces B lymphocytes to produce immunogloblin. Nature 324, 73-76.

Hirano, T., Matsuda, T., Turner, M., Miyasaka, N., Buchan, G., Tang, B., Sato, K., Shimizu, M., Maini, R., Feldmann, M., & Kishimoto, T. (1988). Excessive production of interleukin 6/B cell stimulatory factor-2 in rheumatoid arthritis. Eur. J. Immunol. 18, 1797-1801.

Hoang, T., Haman, A., Goncalves, O., Wong, G. G., & Clark, S. C. (1988). Interleukin-6 enhances growth factor-dependent proliferation of the blast cells of acute myeloblastic leukemia. Blood 72, 823-826.

Hocke, G. M., Barry, D., & Fey, G. H. (1992). Synergistic action of interleukin-6 and glucocorticoids is mediated by the interleukin-6 response element of the rat α_2 macroglobulin gene. Mol. Cell. Biol. 12, 2282-2294.

Honda, M., Kitamura, K., Mizutani, Y., Oishi, M., Arai, M., Okura, T., Igarahi, K., Yasukawa, K., Hirano, T., Kishimoto, T., Mitsuyasu, R., Chermann, J.-C., & Tokunaga, T. (1990). Quantitative analysis of serum IL-6 and its correlation with increased levels of serum IL-2R in HIV-induced diseases. J. Immunol. 145, 4059-4064.

Honda, M., Yamamoto, S., Cheng, M., Yasukawa, K., Suzuki, H., Saito, T., Osugi, Y., Tokunaga, T., & Kishimoto, T. (1992). Human soluble IL-6 receptor: Its detection and enhanced release by HIV. J. Immunol. 148, 2175-2180.

Hopkins, S. J., & Rothwell, N. J. (1995). Cytokines and the nervous system I: Expression and recognition. Trends Neurosci. 18, 83-88.

Horak, I. D., Gress, R. E., Lucas, P. J., Horak, E. M., Waldmann, T. A., & Bolen, J. B. (1991). T-lymphocyte interleukin 2-dependent tyrosine protein kinase signal transduction involves the activation of $p56^{lck}$. Proc. Natl. Acad. Sci. USA 88, 1996-2000.

Houssiau, F. A., Coulie, P. G., Olive, D., & Van Snick, J. (1988). Synergistic activation of human T cells by interleukin 1 and interleukin 6. Eur. J. Immunol. 18, 653-656.

Houssiau, F. A., Devogelaer, J. P., Van Damme, J., de Deuxchaisnes, C. N., & Van Snick, J. (1988). Interleukin-6 in synovial fluid and serum of patients with rheumatoid arthritis and other inflammatory arthritides. Arthritis Rheum. 31, 784-788.

Howells, G., Pham, P., Taylor, D., Foxwell, B., & Feldmann, M. (1991). Interleukin 4 induces interleukin 6 production by endothelial cells: Synergy with interferon-γ. Eur. J. Immunol. 21, 97-101.

Hsu, W., Kerppola, T. K., Chen, P. L., Curran, T., & Chen-Kiang, S. (1994). *Fos* and *Jun* repress transcription activaiton by NF-IL6 through association at the basic zipper region. Mol. Cell. Biol. 14, 268-276.

Huang, E., Nocka, K., Beier, D. R., Chu, T. Y., Buck, J., Lahm, H. W., Wellner, D., Leder, P., & Besmer, P. (1990). The hematopoietic growth factor KL is encoded by the S1 locus and is the ligand of the *c-kit* receptor, the gene product of the W locus. Cell 63, 225-233.

Ihle, J. N., & Kerr, I. M. (1995). Jaks and Stats in signaling by the cytokine receptor superfamily. Trends Genet. 11, 69-74.

Ikebuchi, K., Wong, G. G., Clark, S. C., Ihle, J. N., Hirai, Y., & Ogawa, M. (1987). Interleukin 6 enhancement of interleukin 3-dependent proliferation of multipotential hemopoietic progenitors. Proc. Natl. Acad. Sci. USA 84, 9035-9039.

Immenschuh, S., Nagae, Y., Satoh, H., Baumann, H., & Muller- Eberhard, U. (1994). The rat and human hemopexin genes contain an identical interleukin-6 response element that is not a target of CAAT enhancer-binding protein isoforms. J. Biol. Chem. 269, 12654-12661.

Ishimi, Y., Miyaura, C., Jin, C. H., Akatsu, T., Abe, E., Nakamura, Y., Yamaguchi, A., Yoshiki, S., Matsuda, T., Hirano, T., Kishimoto, T., & Suda, T. (1990). IL-6 is produced by osteoblasts and induces bone resorption. J. Immunol. 145, 3297-3303.

Jacobs, A. L., Sehgal, P. B., Julian, J., & Carson, D. D. (1992). Secretion and hormonal regulation of interleukin-6 production by mouse uterine stromal and polarized epithelial cells cultured *in vitro*. Endocrinology 131, 1037-1046.

Jilka, R. L., Hangoc, G., Girasole, G., Passeri, G., Williams, D. C., Abrams, J. S., Boyce, B., Broxmeyer, H., & Manolagas, S. C. (1992). Increased osteoclast development after estrogen loss: Mediation by interleukin-6. Science 257, 88-91.

Jourdan, M., Bataille, R., Segulin, J., Zhang, X. G., Chaptal, P. A., & Klein, B. (1990). Constitutive production of interleukin-6 and immunologic features in cardiac myxomas. Arthritis Rheum. 33, 398-402.

Kaushansky, K., Lok, S., Holly, R. D., Broudy, V. C., Lin, N., Bailey, M. C., Forstrom, J. W., Buddle, M. M., Oort, P. J., Hagen, F. S., Roth, G. J., Papayannopoulou, T., & Foster, D. C. (1994). Promotion of megakaryocyte progenitor expansion and differentiation by the *c-Mpl* ligand thrombopoietin. Nature 369, 568-571.

Kavanaugh, W. M., & Williams, L. T. (1994). An alternative to SH2 domains for binding tyrosine-phosphorylated proteins. Science 266, 1862-1865.

Kawamura, M., McVicar, D. W., Johnston, J. A., Blake, T. B., Chen, Y.-Q., Lal, B. K., Lloyd, A. R., Kelvin, D. J., Staples, J. E., Ortaldo, J. R., & O'Shea, J. J. (1994). Molecular cloning of L-JAK, a Janus family protein-tyrosine kinase expressed in natural killer cells and activated leukocytes. Proc. Natl. Acad. Sci. USA 91, 6374-6378.

Kawano, M., Hirano, T., Matsuda, T., Taga, T., Horii, Y., Iwato, K., Asaoku, H., Tang, B., Tanabe, O., Tanaka, H., Kuramoto, A., & Kishimoto, T. (1988). Autocrine generation and requirement of BSF-2/IL-6 for human multiple myelomas. Nature 332, 83-85.

Kinoshita, S., Akira, S., & Kishimoto, T. (1992). A member of the C/EBP family, NF-IL6β, forms a heterodimer and transcriptionally synergizes with NF-IL6. Proc. Natl. Acad. Sci. USA 89, 1473-1476.

Kishimoto, T. (1989). The biology of interleukin-6. Blood 74, 1-10.

Kishimoto, T., Taga, T., & Akira, S. (1994). Cytokine signal transduction. Cell 76, 253-262.

Klamfer, L., Lee, T. H., Hsu, W., Vilcek, J., & Chen-Kiang, S. (1994). NF-IL6 and AP-1 cooperatively modulate the activation of the TSG-6 gene by tumor necrosis factor alpha and interleukin-1. Mol. Cell. Biol. 14, 6561-6569.

Klein, B., Wijdenes, J., Zhang, X. G., Jourdan, M., Boiron, J. M., Brochier, J., Liautard, J., Merlin, M., Clement, C., Morel-Fournier, B., Lu, Z.-Y., Mannoni, P., Sany, J., & Bataille, R. (1991). Murine anti-interleukin 6 monoclonal antibody therapy for a patient with plasma cell leukemia. Blood 78, 1198-1204.

Klein, B., Zhang, X.-G., Lu, Z.-Y., & Bataille, R. (1995). Interleukin-6 in human multiple myeloma. Blood 85, 863-872.

Kobayashi, N., Kono, T., Hatakeyama, M., Minami, Y., Miyazaki, T., Perlmutter, R. M., & Taniguchi, T. (1993). Functional coupling of *src*-family protein tyrosine kinases $p59^{fyn}$ and $p53/56^{lyn}$ with the interleukin 2 receptor: Implications for redundancy and pleiotropism in cytokine signal transduction. Proc. Natl. Acad. Sci. USA 90, 4201-4205.

Koike, K., Nakahata, T., Takagi, M., Kobayashi, T., Ishiguro, A., Tsuji, K., Naganuma, K., Okano, A., Akiyama, Y., & Akabane, T. (1988). Synergism of BSF-2/interleukin 6 and interleukin 3 on development of multipotential hemopoietic progenitors in serum-free culture. J. Exp. Med. 168, 879-890.

Kopf, M., Baumann, H., Freer, G., Freudenberg, M., Lamers, M., Kishimoto, T., Zinkernagel, R., Bluethmann, H., & Köhler, G. (1994). Impaired immune and acute-phase responses in interleukin-6-deficient mice. Nature 368, 339-342.

Kumar, G., Gupta, S., Wang, S., & Nel, A. E. (1994). Involvment of Janus kinases, $p52^{shc}$, Raf-1, and MEK-1 in the IL-6-induced mitogen-activated protein kinase cascade of a growth-responsive B cell line. J. Immunol. 153, 4436-4447.

Kushner, I. (1993). Regulation of the acute phase response by cytokines. Perspectives in Biology and Medicine 36, 611-622.

Lai, C.-F., Morella, K. K., Wang, Y., Kumaki, S., Gearing, D., Ziegler, S. F., Tweardy, D. J., Campos, S. P., & Baumann, H. (1995a). Function of hematopoietin receptor subunits in hepatic cells and fibroblasts. Ann. N. Y. Acad. Sci. 762,189-206.

Lai, C.-F., Ripperger, J., Morella, K. K., Wang, Y., Gearing, D. P., Fey, G. H., & Baumann, H. (1995b). Separate signaling mechanisms are involved in the control of STAT protein activation and gene regulation via the interleukin 6 response element by the box 3 motif of gp130. J. Biol. Chem. 270, 14847-14850.

Lai, C.-F., Ripperger, J., Morella, K. K., Wang, Y., Gearing, D. P., Horseman, N. D., Campos, S. P., Fey, G. H., & Baumann, H. (1995c). STAT3 and STAT5B are targets of two different signal pathways activated by hematopoietin receptors and control transcription via separate cytokine response elements J. Biol. Chem. 270, 23254-23257.

Laurenzi, M. A., Siden, A., Persson, M. A., Norkrans, G., Hagberg, L., & Chiodi, F. (1990). Cerebrospinal fluid interleukin-6 activity in HIV infection and inflammatory and noninflammatory diseases of the nervous system. Clin. Immunol. Immunopathol. 57, 233-241.

Le, J. M., Fredrickson, G., Reis, L. F., Diamantstein, T., Hirano, T., Kishimoto, T., & Vilcek, J. (1988). Interleukin 2-dependent and interleukin 2-independent pathways of regulation of thymocyte function by interleukin 6. Proc. Natl. Acad. Sci. USA 85, 8643-8647.

Lee, F. D. (1992). The role of interleukin-6 in development. Dev. Biol. 151, 331-338.

Lee, J. D., Swisher, S. G., Minehart, E. H., McBride, W. H., & Economou, J. S. (1990). Interleukin-4 downregulates interleukin-6 production in human peripheral blood monoclear cells. J. Leukoc. Biol. 47, 475-479.

Leebeek, F. W., Kariya, K., Schwabe, M., & Fowlkes, D. M. (1992). Identification of a receptor binding site in the carboxyl terminus of human interleukin-6. J. Biol. Chem. 267, 14832-14838.

LeMay, L. G., Vander, A. J., & Kluger, M. J. (1990). Role of interleukin 6 in fever in rats. Am. J. Physiol. 258, R798-R803.

Lesnikov, V. A., Efremov, O. M., Korneva, E. A., Van Demme, J., & Billiau, A. (1991). Fever produced by intrahypothalamic injection of interleukin-1 and interleuki-6. Cytokine 3, 195-198.

Levitzki, A., & Gazit, A. (1995). Tyrosine kinase inhibition: An approach to drug development. Science 267, 1782-1788.

Libermann, T. A., & Baltimore, D. (1990). Activation of interleukin-6 gene expression through the NF-κB. Mol. Cell. Biol. 10, 2327-2334.

Libert, C., Takahashi, N., Cauwels, A., Brouckaert, P., Bluethmann, H., & Fiers, W. (1994). Response of interleukin-6-deficient to tumor necrosis factor-induced metabolic changes and lethality. Eur. J. Immunol. 24, 2237-2242.

Lok, S., Kaushansky, K., Holly, R. D., Kuijper, J. L., Lofton-Day, C. E., Oort, P. J., Grant, F. J., Heipel, M. D., Burkhead, S. K., Kramer, J. M., Bell, L. A., Sprecher, C. A., Blumberg, H., Johnson, R., Prunkard, D., Ching, A. F. T., Mathews, S. L., Bailey, M. C., Forstorm, J. W., Buddle, M. M., Osborn, S. G., Evans, S. J., Sheppard, P. O., Presnell, S. R., O'hara, P. J., Hagen, F. S., Roth, G. J., & Foster, D. C. (1994). Cloning and expression of murine thrombopoietin cDNA and stimulation of platelet production *in vivo*. Nature 369, 565-568.

Lotz, M., Jirik, F., Kabouridis, P., Tsoukas, C., Hirano, T., & Kishimoto, T. (1988). B cell stimulating factor 2/interleukin 6 is a costimulant for human thymocytes and T lymphocytes. J. Exp. Med. 167, 1253-1258.

Lowik, C. W., van der Pluijm, G., Bloys, H., Hoekman, K., Bijvoet, O. L., Aarden, L. A., & Papapoulos, S. E. (1989). Parathyroid hormone (PTH) and PTH-like protein (PLP) stimulate interleukin-6 production by osteogenic cells: A possible role of interleukin-6 in osteoclastogenesis. Biochim. Biophys. Res. Commun. 162, 1546-1552.

Lütticken, C., Wegenka, U. M., Yuan, J., Buschmann, J., Schindler, C., Ziemiecki, A., Harpur, A. G., Wilks, A. F., Yasukawa, K., Taga, T., Kishimoto, T., Barbieri, G., Pellagrini, A., Sendtner, M., Heinrich, P. C., & Horn, F. (1994). Association of transcription factor APRF and protein kinase Jak1 with the interleukin-6 signal transducer gp130. Science 263, 89-92.

Lütticken, C., Coffer, P., Yuan, J., Schwartz, C., Caldenhoven, E., Schindler, C., Kruijer, W., Heinrich, P. C., & Horn, F. (1995). Interleukin-6-induced serine phosphorylation of transcription factor APRF: Evidence for a role in interleukin-6 target gene induction. FEBS Lett. 360, 137-143.

Luger, T. A., Krutmann, J., Kirnbauer, R., Urbanski, A., Schwarz, T., Klappacher, G., Kock, A., Micksche, M., Malejczyk, J., Schauer, E., May, L. T., & Sehgal, P. B. (1989). IFN-β2/IL-6 augments the activity of human natural killer cells. J. Immunol. 143, 1206-1209.

Lust, J. A., Donovan, K. A., Kline, M. P., Greipp, P. R., Kyle, R. A., & Maihle, N. J. (1992). Isolation of an mRNA encoding a soluble form of the human interleukin-6 receptor. Cytokine 4, 96-100.

Mackiewicz, A., Schooltink, H., Heinrich, P. C., & Rose-John, S. (1992). Complex of soluble human IL-6-receptor/IL-6 up-regulates expression of acute-phase proteins. J. Immunol. 149, 2021-2027.

Mackiewicz, A., Wiznerowicz, M., Roeb, E., Karczewska, A., Nowak, J., Heinrich, P. C., & Rose-John, S. (1995a). Soluble interleukin 6 receptor is biologically active *in vivo*. Cytokine 7, 142-149.

Mackiewicz, A., Wiznerowicz, M., Roeb, E., Nowak, J., Pawlowski, T., Baumann, H., Heinrich, P. C., & Rose-John, S. (1995b). Interleukin-6-type cytokines and their receptors for gene therapy of melanoma. Ann. N. Y. Acad. Sci. 762, 362-374.

MacVittie, T. J., Farese, A. M., Patchen, M. L., & Myers, L. A. (1994). Therapeutic efficacy of recombinant interleukin-6 (IL-6) alone and combined with recombinant human IL-3 in a nonhuman primate model of high-dose, sublethal radiation-induced marrow aplasia. Blood 84, 2515-2522.

Mano, H., Yazaki, Y., & Hirai, H. (1994). Tec protein-tyrosine kinase directly associates with Lyn protein-tyrosine kinase through its N-terminal unique domain. Oncogene 9, 3205-3211.

Mano, H., Yamashita, Y., Sato, K., Yazaki, Y., & Hirai, H. (1995). Tec protein-tyrosine kinase is involved in interleukin-3 signaling. Blood 85, 343-350.

Marinkovic, S., & Baumann, H. (1990). Structure, hormonal regulation, and identification of the interleukin-6- and dexamethasone-responsive element of the rat haptoglobin gene. Mol. Cell. Biol. 10, 1573-1583.

Matsuda, T., & Hirano, T. (1994a). Establishment of the ELISA for murine soluble gp130, a signal transducer for the IL-6 family cytokine, and its detection in the ascitic fluids of tumor-bearing mice. Biochim. Biophys. Res. Commun. 202, 637-642.

Matsuda, T., & Hirano, T. (1994b). Association of p72 tyrosine kinase with Stat factors and its activation by interleukin-3, interleukin-6, and granulocyte colony-stimulating factor. Blood 83, 3457-3461.

Matsuda, T., Takahashi-Tezuka, M., Fukada, T., Okuyana, Y., Fujitani, Y., Tsukada, S., Mano, H., Hirai, H., Witte, O. N., & Hirano, T. (1995). Association and activation of Btk and Tec tyrosine kinases by gp130, a signal transducer of the interleukin-6 family of cytokines. Blood 85, 627-633.

Matsusaka, T., Fujikawa, K., Nishio, Y., Mukaida, N., Matsushima, K., Kishimoto, T., & Akira, S. (1993). Transcription factors NF-IL6 and NF-κB synergistically activate transcription of the inflammatory cytokines. Proc. Natl. Acad. Sci. USA 90, 10193-10197.

May, L. T., Ghrayeb, J., Santhanam, U., Tatter, S. B., Sthoeger, Z., Helfgott, D. C., Chiorazzi, N., Grieninger, G., & Sehgal, P. B. (1988a). Synthesis and secretion of multiple forms of β2-interferon/B-cell differentiation factor 2/hepatocyte-stimulating factor by human fibroblasts and monocytes. J. Biol. Chem. 263, 7760-7766.

May, L. T., Santhanam, U., Tatter, S. B., Bhardwaj, N., Ghrayeb, J., Ghrayeb, J., & Sehgal, P. B. (1988b). Phosphorylation of secreted forms of human β2-interferon/hepatocyte stimulating factor/interleukin-6. Biochim. Biophys. Res. Commun. 152, 1144-1150.

Metcalf, D. (1989). The molecular control of cell division, differentiation commitment and maturation in haemopoietic cells. Nature 339, 27-30.

Miles, S. A., Martinez-Maza, O., Rezai, A., Magpantay, L., Kishimoto, T., Nakamura, S., Radka, S. F., & Linsley, P. S. (1992). Oncostatin M as a potent mitogen for AIDS-Kaposi's sarcoma-derived cells. Science 255, 1432-1434.

Miles, S. A., Rezai, A. R., Salazar-Gonzalez, J. F., Vander Meyden, M., Stevens, R. H., Logan, D. M., Mitsuyasu, R. T., Taga, T., Hirano, T., Kishimoto, T., & Martinez-Maza, O. (1990). AIDS Kaposi sarcoma-derived cells produce and respond to interleukin 6. Proc. Natl. Acad. Sci. USA 87, 4068-4072.

Minami, M., Inoue, M., Wei, S., Takeda, K., Matsumoto, M., Kishimoto, T., & Akira, S. (1996). STAT3 activation is a critical step in gp130-mediated terminal differentiation and growth arrest of a myeloid cell line. Proc. Natl. Acad. Sci. USA 93, 393-3966.

Miossec, P., Briolay, J., Dechanet, J., Wijdenes, J., Martinez-Valdez, H., & Banchereau, J. (1992). Inhibition of the production of proinflammatory cytokines and immunoglobulins by interleukin-4 in an *ex vivo* model of rheumatoid synovitis. Arthritis Rheum. 35, 874-883.

Miossec, P., Chomarat, P., Dechanet, J., Moreau, J. F., Roux, J. P., Delmas, P., & Banchereau, J. (1994). Interleukin-4 inhibits bone resorption through an effect on osteoclasts and proinflammatory cytokines in an *ex vivo* model of bone resorption in rheumatoid arthritis. Arthritis Rheum. 37, 1715-1722.

Miyajima, A., Hara, T., & Kitamura, T. (1992). Common subunits of cytokine receptors and the functional redundancy of cytokines. Trends Biochem. Sci. 17, 378-382.

Miyaura, C., Onozaki, K., Akiyama, Y., Taniyama, T., Hirano, T., Kishimoto, T., & Suda, T. (1988). Recombinant human interleukin 6 (B-cell stimulatory factor 2) is a potent inducer of differentiation of mouse myeloid leukemia cells (M1). FEBS Lett. 234, 17-21.

Mock, B. A., Nordan, R. P., Justice, M. J., Kozak, C., Jenkins, N. A., Copeland, N. G., Clark, S. C., Wong, G. G., & Rudikoff, S. (1989). The murine IL-6 gene maps to the proximal region of chromosome 5. J. Immunol. 142, 1372-1376.

Monique, C., Wolvekamp, J., & Marquet, R. L. (1990). Interleukin-6: Historical background, genetics and biological significance. Immunol. Lett. 24, 1-10.

Morella, K. K., Bruno, E., Kumaki, S., Lai, C.-F., Fu, J., Wang, H.-M., Murray, L., Hoffman, R., Timour, M., Bénit, L., Gisselbrecht, S., Zhuang, H., Wojchowski, D. M., Baumann, H., & Gearing, D. P. (1995). Signal transduction by the receptors for thrombopoietin (c-*mpl*) and interleukin-3 in hematopoietic and non-hematopoietic cells. Blood 86, 557-571.

Mori, N., Shirakawa, F., Shimizu, H., Murakami, S., Oda, S., Yamamoto, K., & Eto, S. (1994). Transcriptional regulation of the human interleukin-6 gene promoter in human T cell leukemia virus type I-infected T cell lines: Evidence for the involvement of NF-κB. Blood 84, 2904-2911.

Most, J., Zangerle, R., Herold, M., Fuchs, D., Wachter, H., Fritsch, P., & Dierich, M. P. (1993). Elevated concentrations of circulating intercellular adhesion molecule 1 (ICAM-1) in HIV-1 infection. J. AIDS 6, 221-226.

Mullberg, J., Schooltink, H., Stoyan, T., Heinrich, P. C., & Rose-John, S. (1992). Protein kinase C activity is rate limiting for shedding of the interleukin-6 receptor. Biochim. Biophys. Res. Commun. 189, 794-800.

Mullberg, J., Oberthur, W., Lottspeich, F., Mehl, E., Dittrich, E., Graeve, L., Heinrich, P. C., & Rose-John, S. (1994). The soluble human IL-6 receptor. Mutational characterization of the proteolytic cleavage site. J. Immunol. 152, 4958-4968.

Muller, M., Briscoe, J., Laxton, C., Guschin, D., Ziemiecki, A., Silvennoninen, O., Harpur, A. G., Barbieri, G., Witthuhn, B. A., Schindler, C., Pellegrini, S., Wilks, A. F., Ihle, H. N., Stark, G. R., & Kerr, I. M. (1993). The protein tyrosine kinase JAK1 complements defects in interferon-α/β and -γ signal transduction. Nature 366, 129-135.

Murakami, M., Narazaki, M., Hibi, M., Yawata, H., Yasukawa, K., Hamaguchi, M., Taga, T., & Kishimoto, T. (1991). Critical cytoplasmic region of the interleukin 6 signal transducer gp130 is conserved in the cytokine receptor family. Proc. Natl. Acad. Sci. USA 88, 11349-11353.

Murakami, M., Hibi, M., Nakagawa, N., Nakagawa, T., Yasukawa, K., Yamanishi, K., Taga, T., & Kishimoto, T. (1993). IL-6-induced homodimerization of gp130 and associated activation of a tyrosine kinase. Science 260, 1808-1810.

Murray, R., Lee, F., & Chiu, C. P. (1990). The genes for leukemia inhibitory factor and interleukin-6 are expressed in mouse blastocysts prior to the onset of hemopoiesis. Mol. Cell. Biol. 10, 4953-4956.

Nair, B. C., DeVico, A. L., Nakamura, S., Copeland, T. D., Chen, Y., Patel, A., O'Neil, T., Oroszlans, S., Gallo, R. C., & Sarngadharan, M. G. (1992). Identification of a major growth factor for AIDS-Kaposi's sarcoma cell as oncostatin M. Science 255, 1430-1432.

Naitoh, Y., Fukata, J., Tominaga, T., Nakai, Y., Tamai, S., Mori, K., & Imura, H. (1988). Interleukin-6 stimulates the secretion of adrenocorticotropic hormone in conscious, freely-moving rats. Biochim. Biophys. Res. Commun. 155, 1459-1463.

Nakajima, T., Yamamoto, S., Cheng, M., Yasukawa, K., Hirano, T., Kishimoto, T., Tokunaga, T., & Honda, M. (1992). Soluble interleukin-6 receptor is released from receptor-bearing cell lines *in vitro*. Jpn. J. Cancer Res. 83, 373-378.

Nakajima, T., Kinoshita, S., Sasagawa, T., Sasaki, K., Naruto, M., Kishimoto, T., & Akira, S. (1993). Phosphorylation at threonine-235 by a ras-dependent mitogen-activated protein kinase cascade is essential for transcription factor NF-IL6. Proc. Natl. Acad. Sci. USA 90, 2207-2211.

Narazaki, M., Yasukawa, K., Saito, T., Ohsugi, Y., Fukui, H., Koishihara, Y., Yancopoulos, G. D., Taga, T., & Kishimoto, T. (1993). Soluble forms of the interleukin-6 signal-transducing receptor component gp130 in human serum possessing a potential to inhibit signals through membrane-anchored gp130. Blood 82, 1120-1126.

Natsuda, S., Akira, S., Nishio, Y., Hashimoto, S., Sugita, T., Isshiki, H., & Kishimoto, T. (1992). Macrophage differentiation-specific expression of NF-IL6, a transcription factor for interleukin-6. Blood 79, 460-466.

Navarra, P., Tsagarakis, S., Faria, M. S., Rees, L. H., Besser, G. M., & Grossman, A. B. (1991). Interleukin-1 and -6 stimulate the release of corticotropin-releasing hormone-41 from rat hypothalamus *in vitro* via the eicosanoid cyclooxygenase pathway. Endocrinology 128, 37-44.

Neta, R., Perlstein, R., Vogel, S. N., Ledney, G. D., & Abrams, J. (1992). Role of interleukin 6 (IL-6) in protection from lethal irradiation and in endocrine response to IL-1 and tumor necrosis factor. J. Exp. Med. 175, 689-694.

Nishino, E., Matsuzaki, N., Masuhiro, K., Kameda, T., Taniguchi, T., Takagi, T., Saji, F., & Tanizawa, O. (1990). Trophoblast-derived interleukin-6 (IL-6) regulates human chorionic gonadotropin release through IL-6 receptor on human trophoblasts. J. Clin. Endocrinol. Metabol. 71, 436-441.

Nishio, Y., Isshiki, H., Kishimoto, T., & Akira, S. (1993). A nuclear factor for interleukin-6 expression (NF-IL6) and the glucocorticoid receptor synergistically activate transcription of the rat alpha 1-acid glycoprotein gene via direct protein-protein interaction. Mol. Cell. Biol. 13, 1854-1862.

Noma, T., Mizuta, T., Rosen, A., Hirano, T., Kishimoto, T., & Honjo, T. (1987). Enhancement of the interleukin 2 receptor expression on T cells by multiple B-lymphotropic lymphokines. Immunol. Lett. 15, 249-253.

Nordan, R. P., & Potter, M. (1986). A macrophage-derived factor required by plasmacytomas for survival and proliferation *in vitro*. Science 233, 566-569.

Northemann, W., Braciak, T. A., Hattori, M., Lee, F., & Fey, G. H. (1989). Structure of the rat interleukin 6 gene and its expression in macrophage-derived cells. J. Biol. Chem. 264, 16072-16082.

Novick, D., Engelmann, H., Wallach, D., & Rubinstein, M. (1989). Soluble cytokine receptors are present in normal human urine. J. Exp. Med. 170, 1409-1414.

Obiri, N. I., Debinski, W., Leonard, W. J., & Puri, R. K. (1995). Receptor for interleukin 13: interaction with interleukin 4 by a mechanism that does not involve the common chain shared by the receptors for interleukins 2, 4, 7, 9, and 15. J. Biol. Chem. 270, 8797-8804.

Ogawa, M. (1993). Differentiation and proliferation of hematopoietic cells. Blood 81, 2844-2853.

Okada, M., Kitahara, M., Kishimoto, S., Matsuda, T., Hirano, T., & Kishimoto, T. (1988). IL-6/BSF-2 functions as a killer helper factor in the *in vitro* induction of cytotoxic T cells. J. Immunol. 141, 1543-1549.

Oliviero, S., & Cortese, R. (1989). The human haptoglobin gene promoter: Interleukin-6-response elements interact with a DNA-binding protein induced by interleukin-6. EMBO J. 8, 1145-1151.

Oyaizu, N., Chirmule, N., Ohnishi, Y., Kalyanaraman, V. S., & Pahwa, S. (1991). Human immunodeficiency virus type 1 envelop glycoproteins gp120 adn gp160 induce interleukin-6 production in CD4+ T-cell clones. J. Virol. 65, 6277-6282.

Patchen, M. L., MacVittie, T. J., Williams, J. L., Schwartz, G. N., & Souza, L. M. (1991). Administration of interleukin-6 stimulates multilineage hematopoiesis and accelerates recovery from radiation-induced hematopoietic depression. Blood 77, 472-480.

Patchen, M. L., Fischer, R., & MacVittie, T. J. (1993). Effects of combined administration of interleukin-6 and granulocyte colony-stimulating factor on recovery from radiation-induced hemopoietic aplasia. Exp. Hematol. 21, 338-344.

Pojda, Z., Tanaka, K., Aoki, Y., & Tsuboi, A. (1992). *In vivo* and *in vitro* interaction between interleukin 6 and granulocyte colony-stimulating factor in the regulaiton of murine hematopoiesis. Exp. Hematol. 20, 874-878.

Poli, G., Bressler, P., Kinter, A., Duh, E., Timmer, W. C., Rabson, A., Justement, J. S., Stanley, S., & Fauci, A. S. (1990). Interleukin 6 induces human immunodeficiency virus expression in infected monocytic cells alone and in synergy with tumor necrosis factor alpha by transcriptional and post-transcriptional mechanisms. J. Exp. Med. 172, 151-158.

Poli, G., Kinter, A. L., & Fauci, A. S. (1994a). Interleukin 1 induces expression of the human immunodeficiency virus alone and in synergy with interleukin 6 in chronically infected U1 cells: Inhibition of inductive effects by the interleukin 1 receptor antagonist. Proc. Natl. Acad. Sci. USA 91, 108-112.

Poli, V., Balena, R., Fattori, E., Markatos, A., Yamamoto, M., Tanaka, H., Ciliberto, G., Rodan, G. A., & Costantini, F. (1994b). Interleukin-6 deficient mice are protected from bone loss caused by estrogen depletion. EMBO J. 13, 1189-1196.

Portier, M., Rajzbaum, G., Zhang, X. G., Attal, M., Rusalen, C., Wijdenes, J., Mannoni, P., Maraninchi, D., Piechaczyk, M., Bataille, R., & Klein, B. (1991). *In vivo* interleukin 6 gene expression in the tumoral environment in multiple myeloma. Eur. J. Immunol. 21, 1759-1762.

Pulverer, B. J., Kyriakis, J. M., Avruch, J., Nikolakaki, E., & Woodgett, J. R. (1991). Phosphorylation of *c-jun* mediated by MAP kinases. Nature 353, 670-674.

Rabinowich, H., Sedlmayr, P., Herberman, R. B., & Whiteside, T. L. (1993). Response of human NK cells to IL-6 alterations of the cell surface phenotype, adhesion to fibronectin and laminin, and tumor necrosis factor α/β secretion. J. Immunol. 150, 4844-4855.

Rane, S. G., & Reddy, E. P. (1994). JAK3: A novel JAK kinase associated with terminal differentiation of hematopoietic cells. Oncogene 9, 2415-2423.

Rautonen, J., Rautonen, N., Martin, N. L., Philip, R., & Wara, D. W. (1991). Serum interleukin-6 concentrations are elevated and associated with elevated tumor necrosis factor-α and immunoglobulin G and A concentrations in children with HIV infection. AIDS 5, 1319-1325.

Rautonen, J., Rautonen, N., Martin, N. L., & Wara, D. W. (1994). HIV type 1 Tat protein induces immunoglobulin and interleukin 6 synthesis by uninfected peripheral blood mononuclear cells. AIDS Res. Hum. Retroviruses 10, 781-785.

Rawlings, D. J., Saffran, D. C., Tsukada, S., Largaespada, D. A., Grimaldi, J. C., Cohen, L., Mohr, R. N., Bazan, J. F., Howard, M., Copeland, N. G., Jenkins, N. A., & Witte, O. N. (1993). Mutation of unique region of Bruton's tyrosine kinase in immunodeficient XID mice. Science 261, 358-361.

Ray, A., & Prefontaine, K. E. (1994). Physical association and functional antagonism between the p65 subunit of transcription factor NF-κB and the glucocorticoid receptor. Proc. Natl. Acad. Sci. USA 91, 752-756.

Ray, A., Sassone-Corsi, P., & Sehgal, P. B. (1989). A multiple cytokine- and second messenger-responsive element in the enhancer of the human interleukin-6 gene: Similarities with *c-fos* gene regulation. Mol. Cell. Biol. 9, 5537-5547.

Ray, A., Hannink, M., & Ray, B. K. (1995). Concerted participation of NF-κB and C/EBP heteromer in lipopolysaccharide induction of serum amyloid A gene expression in liver. J. Biol. Chem. 270, 7365-7374.

Raz, R., Durbin, J. E., & Levy, D. E. (1994). Acute phase response factor and additional members of the interferon-stimulated gene factor 3 family integrate diverse signals from cytokines, interferons, and growth factors. J. Biol. Chem. 269, 24391-24395.

Renauld, J. C., Vink, A., & Van Snick, J. (1989). Accessory signals in murine cytolytic T cell responses. Dual requirement for IL-1 and IL-6. J. Immunol. 143, 1894-1898.

Rieckmann, P., Poli, G., Fox, C. H., Kehrl, J. H., & Fauci, A. S. (1991a). Recombinant gp120 specifically enhances tumor necrosis factor-alpha production and Ig secretion in B lymphocytes from HIV-infected individuals but not from seronegative donors. J. Immunol. 147, 2922-2927.

Rieckmann, P., Poli, G., Kehrl, J. H., & Fauci, A. S. (1991b). Activated B lymphocytes from human immunodeficiency virus-infected individuals induce virus expression in infected T cells and a promonocytic cell line, U1. J. Exp. Med. 173, 1-5.

Ripperger, J. A., Fritz, S., Richter, K., Hocke, G. M., Lottspeich, F., & Fey, G. H. (1995). Transcription factors Stat3 and Stat5b are present in rat liver nuclei late in an acute phase response and bind interleukin-6 response elements. J. Biol. Chem. 270, 29998-30006.

Ritchie, D. G., & Fuller, G. M. (1983). Hepatocyte-stimulating factor: A monocyte-derived acute-phase regulatory protein. Ann. N. Y. Acad. Sci. 408, 490-502.

Robbins, P. D., Horowitz, J. M., & Mulligan, R. C. (1990). Negative regulation of human *c-fos* expression by the retinoblastoma gene product. Nature 346, 668-671.

Robinson, R. C., Grey, L. M., Staunton, D., Vankelecom, H., Vernallis, A. B., Moreau, J.-F., Stuart, D. I., Heath, J. K., & Jones, E. Y. (1994). The crystal structure and biological function of leukemia inhibitory factor: Implications for receptor binding. Cell 77, 1101-1116.

Ruhl, A., Hurst, S., & Collins, S. M. (1994). Synergism between interleukin 1 beta and 6 on noradrenergic nerves in rat myenteric plexus. Gastroenterology 107, 993-1001.

Sadowski, H. B., Shuai, K., Darnell, J. E., Jr., & Gilman, M. Z. (1993). A common nuclear signal transduction pathway activated by growth factor and cytokine receptors. Science 261, 1739-1744.

Sakata, Y., Morimoto, A., Long, N. C., & Murakami, N. (1991). Fever and acute-phase response induced in rabbits by intravenous and intracerebroventricular injection of interleukin 6. Cytokine 3, 199-203.

Santhanam, U., Ray, A., & Sehgal, P. B. (1991). Repression of the interleukin 6 gene promoter by p53 and the retinoblastoma susceptibility gene product. Proc. Natl. Acad. Sci. USA 88, 7605-7609.

Sato, N., & Miyajima, A. (1994). Multimeric cytokine receptors: common versus specific functions. Curr. Opin. Cell Biol. 6, 174-179.

Sato, N., Sakamaki, K., Terada, N., Arai, K., & Miyajima, A. (1993). Signal transduction by the high-affinity GM-CSF receptor: Two distinct cytoplasmic regions of the common subunit responsible for different signaling. EMBO J. 12, 4181-4189.

Satoh, T., Nakamura, S., Taga, T., Matsuda, R., Hirano, T., Kishimoto, T., & Kaziro, Y. (1988). Induction of neuronal differentiation in PC12 cells by B-cell stimulatory factor 2/interleukin 6. Mol. Cell. Biol. 8, 3546-3549.

Schmitt, R. M., Bruyns, E., & Snodgrass, H. R. (1991). Hematopoietic development of embryonic stem cells *in vitro*: Cytokine and receptor gene expression. Genes Dev. 5, 728-740.

Schwab, G., Siegall, C. B., Aarden, L. A., Neckers, L. M., & Nordan, R. P. (1991). Characterization of an interleukin-6-mediated autocrine growth loop in the human multiple myeloma cell line, U266. Blood 77, 587-593.

Seth, A., Alvarez, E., Gupta, S., & Davis, R. J. (1991). A phosphorylation site located in the NH_2-terminal domain of c-Myc increases transactivation of gene expression. J. Biol. Chem. 266, 23521-23524.

Sha, W. C., Liou, H.-C., Tuomanen, E. I., & Baltimore, D. (1995). Targeted disruption of the p50 subunit of NF-κB leads to multifocal defects in immune responses. Cell 80, 321-330.

Shabo, Y., Lotem, J., Rubinstein, M., Revel, M., Clark, S. C., Wolf, S. F., Kamen, R., & Sachs, L. (1988). The myeloid blood cell differentiation-inducing protein MGI-2A is interleukin-6. Blood 72, 2070-2073.

Shimizu, H., & Yamamoto, K. (1994). NF-κB and C/EBP transcription factor families synergistically function in mouse serum amyloid A gene expression induced by inflammatory cytokines. Gene 149, 305-310.

Shimizu, S., Yoshioka, R., Hirose, Y., Sugai, S., Tachibana, J., & Konda, S. (1989). Establishment of two interleukin 6 (B cell stimulatory factor 2/interferon beta 2)-dependent human bone marrow-derived myeloma cell lines. J. Exp. Med. 169, 339-344.

Shirai, A., Holmes, K., & Klinman, D. (1993). Detection and quantitation of cells secreting IL-6 under physiologic conditions in BALB/c mice. J. Immunol. 150, 793-799.

Shuai, K., Ziemiecki, A., Wilks, A. F., Harpur, A. G., Sadowski, H. B., Gilmam, M. Z., & Darnell, J. E. (1993). Polypeptide signalling to the nucleus through tyrosine phosphorylation of Jak and Stat proteins. Nature 366, 580-583.

Shuai, K., Horvath, C. M., Huang, L. H. T., Qureshi, S. A., Cowburn, D., & Darnell, J. E., Jr. (1994). Interferon activation of the transcription factor stat91 involves dimerization through SH2-phosphotyrosyl peptide interactions. Cell 76, 821-828.

Sidell, N., Taga, T., Hirano, T., Kishimoto, T., & Saxon, A. (1991). Retinoic acid-induced growth inhibition of a human myeloma cell line via down-regulation of IL-6 receptors. J. Immunol. 146, 3809-3814.

Simpson, R. J., Moritz, R. L., Rubira, M. R., & Van Snick, J. (1988). Murine hybridoma/plasmacytoma growth factor. Complete amino-acid sequence and relation to human interleukin-6. Eur. J. Biochem. 176, 187-197.

Smeland, E. B., Blomhoff, H. K., Funderud, S., Shalaby, M. R., & Espevik, T. (1989). Interleukin 4 induces selective production of interleukin 6 from normal human B lymphocytes. J. Exp. Med. 170, 1463-1468.

Smerz-Bertling, C., & Duschi, A. (1995). Both interleukin 4 and interleukin 13 induce tyrosine phosphorylation of the 140-kDa subunit of the interleukin 4 receptor. J. Biol. Chem. 270, 966-970.

Smyth, M. J., & Ortaldo, J. R. (1991). Comparison of the effect of IL-2 and IL-6 on the lytic activity of purified human peripheral blood large granular lymphocytes. J. Immunol. 146, 1380-1384.

Spangelo, B. L., Judd, A. M., Isakson, P., & MacLeod, R. M. (1989). Interleukin-6 stimulates anterior pituitary hormone release in vitro. Endocrinology 125, 575-577.

Sporeno, E., Paonessa, G., Salvati, A. L., Graziani, R., Delmastro, P., Ciliberto, G., & Toniatti, C. (1994). Oncostatin M binds directly to gp130 and behaves as interleukin-6 antagonist on a cell line expressing gp130 but lacking functional oncostatin M receptors. J. Biol. Chem. 269, 10991-10995.

Stahl, N., Boulton, T. G., Farruggella, T., Ip, N. Y., Davis, S., Witthuhn, B. A., Quelle, F. W., Silvennoinen, O., Barbieri, G., Pellegrini, S., Ihle, J. N., & Yancopoulos, G. D. (1994). Association and activation of Jak-Tyk kinases by CNTF-LIF-OSM-IL-6β receptor components. Science 263, 92-95.

Stahl, N., Farruggella, T. J., Boulton, T. G., Zhong, Z., Darnell, J. E., Jr., & Yancopoulos, G. D. (1995). Choice of STATs and other substrates specified by modular tyrosine-based motifs in cytokine receptors. Science 267, 1349-1353.

Steffen, M., Reinecker, H. C., Petersen, J., Doehn, C., Pfluger, I., Voss, A., & Raedler, A. (1993). Differences in cytokine secretion by intestinal mononuclear cells, peripheral blood monocytes and alveolar macrophages from HIV-infected patients. Clin. Exp. Immunol. 91, 30-36.

Stokoe, D., Campbell, D. G., Nakielny, S., Hidaka, H., Leerers, S. J., Marshall, C., & Cohen, P. (1992). MAPKAP kinase 2: A novel protein kinase activated by mitogen-activated protein kinase. EMBO J. 10, 3985-3994.

Strassmann, G., Fong, M., Kenney, J. S., & Jacob, C. O. (1992). Evidence for the involvement of interleukin 6 in experimental cancer cachexia. J. Clin. Invest. 89, 1681-1684.

Sturgill, T. W., Ray, L. B., Erikson, E., & Maller, J. L. (1988). Insulin-stimulated MAP-2 kinase phosphorylates and activates ribosomal protein S6 kinase II. Nature 334, 715-718.

Suzuki, H., Yasukawa, K., Saito, T., Goitsuka, R., Hasegawa, A., Ohsugi, Y., Taga, T., & Kishimoto, T. (1992). Anti-human interleukin-6 receptor antibody inhibits human myeloma growth *in vivo*. Eur. J. Immunol. 22, 1989-1993.

Swingler, S., Easton, A., & Morris, A. (1992). Cytokine augmentation of HIV-1 LTR-driven gene expression in neural cells. AIDS Res. Hum. Retroviruses 8, 487-493.

Tabibzadeh, S. S., Santhanam, U., Sehgal, P. B., & May, L. T. (1989). Cytokine-induced production of IFN-β2/IL-6 by freshly explanted human endometrial stromal cells. Modulation by estradiol-17bβ. J. Immunol. 142, 3134-3139.

Taga, T., & Kishimoto, T. (1992). Cytokine receptors and signal transduction. FASEB J. 6, 3387-3396.

Taga, T., & Kishimoto, T. (1995). Signaling mechanisms through cytokine receptors that share signal transducing receptor components. Curr. Opin. Immunol. 7, 17-23.

Taga, T., Hibi, M., Hirata, Y., Yamasaki, K., Yasukawa, K., Matsuda, T., Hirano, T., & Kishimoto, T. (1989). Interleukin-6 triggers the association of its receptor with a possible signal transducer,gp130. Cell 58, 573-581.

Taga, T., Narazaki, M., Yasukawa, K., Saito, T., Miki, D., Hamaguchi, M., Davis, S., Shoyab, M., Yancopoulos, G. D., & Kishimoto, T. (1992). Functional inhibition of hematopoietic and neurotrophic cytokines by blocking the interleukin 6 signal transducer gp130. Proc. Natl. Acad. Sci. USA 89, 10988-11001.

Takai, Y., Wong, G. G., Clark, S. C., Burakoff, S. J., & Herrmann, S. H. (1988). B cell stimulatory factor-2 is involved in the differentiation of cytotoxic T lymphocytes. J. Immunol. 140, 508-512.

Tamura, T., Udagawa, N., Takahashi, N., Miyaura, C., Tanaka, S., Yamada, Y., Koishihara, Y., Ohsugi, Y., Kumaki, K., & Taga, T. (1993). Soluble interleukin-6 receptor triggers osteoclast formation by interleukin 6. Proc. Natl. Acad. Sci. USA 90, 11924-11928.

Tang, B., Matsuda, T., Akira, S., Nagata, N., Ikehara, S., Hirano, T., & Kishimoto, T. (1991). Age-associated increase in interleukin 6 in MRL/lpr mice. Int. Immunol. 3, 273-278.

Tang, B., Mano, H., Yi, T., & Ihle, J. N. (1994). Tec kinase associates with c-kit and is tyrosine phosphorylated and activated following stem cell factor binding. Mol. Cell. Biol. 14, 8432-8437.

Taniguchi, T. (1995). Cytokine signaling through nonreceptor protein tyrosine kinases. Science 268, 251-255.

Tanner, J. W., Chen, W., Young, R. L., Longmore, G. D., & Shaw, A. S. (1995). The conserved box 1 motif of cytokine receptors is required for association with JAK kinases. J. Biol. Chem. 270, 6523-6530.

te Velde, A. A., Hoijbens, R. J., Heije, K., de Vries, J. E., & Figdor, C. G. (1990). Interleukin-4 (IL-4) inhibits secretion of IL-1β, tumor necrosis factor alpha, and IL-6 by human monocytes. Blood 76, 1392-1397.

Thanos, D., & Maniatis, T. (1995). NF-κB: A lesson in family values. Cell 80, 529-532.

Thomas, J. D., Sideras, P., Smith, C. I., Vorechovsky, I., Chapman, V., & Paul, W. E. (1993). Colocalization of X-linked agammaglobulinemia and X-linked immunodeficiency genes. Science 261, 355-358.

Tsukada, S., Saffran, D. C., Rawlings, D. J., Parolini, O., Allen, R. C., Klisak, I., Sparkes, R. S., Kubagawa, H., Mohandas, T., Quan, S., Belmont, J. W., Cooper, M. D., Conley, M. E., & Witte, O. N. (1993). Deficient expression of a B cell cytoplasmic tyrosine kinase in human X-linked agammaglobulinemia. Cell 72, 279-290.

Tsunetsugu-Yokota, Y., & Honda, M. (1990). Effect of cytokines on HIV release and IL-2 receptor alpha expression in monocytic cell lines. J. AIDS 3, 511-516.

Turksen, K., Kupper, T., Degenstein, L., Williams, I., & Fuchs, E. (1992). Interleukin 6: Insights to its function in skin by overexpression in transgenic mice. Proc. Natl. Acad. Sci. USA 89, 5068-5072.

Twigg, H. L., Iwamoto, G. K., & Soliman, D. M. (1992). Role of cytokines in alveolar macrophage accessory cell function in HIV-infected individuals. J. Immunol. 149, 1462-1469.

Uyttenhove, C., Coulie, P. G., & Van Snick, J. (1988). T cell growth and differentiation induced by interleukin-HP1/IL-6, the murine hybridoma/plasmacytoma growth factor. J. Exp. Med. 167, 1417-1427.

Van Damme, J., Cayphas, S., Opdenakker, G., Billiau, A., & Van Snick, J. (1987a). Interleukin-1 and poly(rl).poly(rC) induce production of a hybridoma growth factor by human fibroblasts. Eur. J. Immunol. 17, 1-7.

Van Damme, J., Opdenakker, G., Simpson, R. J., Rubira, M. R., Cayphas, S., Vink, A., Billiau, A., & Van Snick, J. (1987b). Identification of the human 26-kD protein, interferon beta 2 (IFN-β2), as a B cell hybridoma/plasmacytoma growth factor induced by interleukin 1 and tumor necrosis factor. J. Exp. Med. 165, 914-919.

Van Damme, J., Van Beeumen, J., Decock, B., Van Snick, J., De Ley, M., & Billiau, A. (1988). Separation and comparison of two monokines with lymphocyte-activating factor activity: IL-1β and hybridoma growth factor (HGF).Identification of leukocyte-derived HGF as IL-6. J. Immunol. 140, 1534-1541.

van der Poll, T., Levi, M., Hack, C. E., ten Cate, H., van Deventer, S. J., Eerenberg, A. J., de Groot, E. R., Jansen, J., Gallati, H., B ll er, H. R., ten Cate, J. W., & Aarden, L. A. (1994). Elimination of interleukin 6 attenuates coagulation activation in experimental endotoxemia in chimpanzees. J. Exp. Med. 179, 1253-1259.

Van Gameren, M. M., Willemse, P. H., Mulder, N. H., Limburg, P. C., Groen, H. J., Vellenga, E., & de Vries, E. G. (1994). Effects of recombinant human interleukin-6 in cancer patients: A phase I-II study. Blood 84, 1434-1441.

Van Snick, J. (1990). Interleukin-6: An overview. Annu. Rev. Immunol. 8, 253-278.

Van Snick, J., Cayphas, S., Vink, A., Uyttenhove, C., Coulie, P. G., Rubira, M. R., & Simpson, R. J. (1986). Purification and NH_2-terminal amino acid sequence of a T-cell-derived lymphokine with growth factor activity for B-cell hybridomas. Proc. Natl. Acad. Sci. USA 83, 9679-9683.

Van Snick, J., Cayphas, S., Szikora, J. P., Renauld, J. C., Van Roost, E., Boon, T., & Simpson, R. J. (1988). cDNA cloning of murine interleukin-HP1: Homology with human interleukin 6. Eur. J. Immunol. 18, 193-197.

Velazuez, L., Mogensen, K. E., Barbieri, G., Fellous, M., Uze, G., & Pellegrini, S. (1995). Dinstint domains of the protein tyrosine kinase tyk2 required for binding of interferon-α/β and for sigal transduction. J. Biol. Chem. 270, 3327-3334.

Vesanen, M., Wessman, M., Salminen, M., & Vaheri, A. (1992). Activation of intergrated human immunodeficiency virus type 1 in human neuroblastoma cells by the cytokine tumor necrosis factor alpha and interleukin-6. J. Gen. Virol. 73, 1753-1760.

Vetrie, D., Vorechovsky, I., Sideras, P., Holland, J., Davies, A., Flinter, F., Hammarstrom, L., Kinnon, C., Levinsky, R., Bobrow, M., Smith, C. I. M., & Bentley, D. R. (1993). The gene involved in X-linked agammaglobulinaemia is a member of the src family of protein-tyrosine kinases. Nature 361, 226-233.

Vink, A., Coulie, P., Warnier, G., Renauld, J. C., Stevens, M., Donckers, D., & Van Snick, J. (1990). Mouse plasmacytoma growth *in vivo*: Enhancement by interleukin 6 (IL-6) and inhibition by antibodies directed against IL-6 or its receptor. J. Exp. Med. 172, 997-1000.

Vita, N., Lefort, S., Laurent, P., Caput, D., & Ferrara, P. (1995). Characterization and comparison of the interleukin 13 receptor with the interleukin 4 receptor on several cell types. J. Biol. Chem. 270, 3512-3517.

Vogel, J., Hinrichs, S. H., Reynolds, R. K., Luciw, P. A., & Jay, G. (1988). The HIV tat gene induces dermal lesions resembling Kaposi's sarcoma in transgenic mice. Nature 335, 606-611.

Wagner, B. J., Hayes, T. E., Hoban, C. J., & Cochran, B. H. (1990). The SIF binding element confers *sis*/PDGF inducibility onto the c-*fos* promoter. EMBO J. 9, 4477-4484.

Wang, P., Wu, P. W., Siegel, M. I., Egan, R. W., & Billah, M. M. (1995). Interleukin (IL)-10 inhibits nuclear factor $\kappa\beta$ (NF-κB) activation in human monocytes. J. Biol. Chem. 270, 9558-9563.

Wang, Y., & Fuller, G. M. (1994). Phosphorylation and internalization of gp130 occur after IL-6 activation of Jak2 kinase in hepatocytes. Mol. Biol. Cell 5, 819-828.

Wang, Y.-D., Wong, K., & Wood, W. I. (1995). Intracellular tyrosine residues of the human growth hormone receptor are not required for the signaling of proliferation or Jak-STAT Activation. J. Biol. Chem. 270, 7021-7024.

Ward, L. D., Howlett, G. J., Discolo, G., Yasukawa, K., Hammacher, A., Moritz, R. L., & Simpson, R. J. (1994). High affinity interleukin-6 receptor is a hexameric complex consisting of two molecules each of interleukin-6, interleukin-6 receptor, and gp130. J. Biol. Chem. 269, 23286-23289.

Watanabe, K., Tanaka, Y., Morimoto, I., Yahata, K., Zeki, K., Fujihira, T., Yamashita, U., & Eto, S. (1990). Interleukin-4 as a potent inhibitor of bone resorption. Biochim. Biophys. Res. Commun. 172, 1035-1041.

Watling, D., Guschin, D., Muller, M., Silvennoinen, O., Witthuhn, B. A., Quelle, F. W., Rogers, N. C., Schindler, C., Stark, G. R., Ihle, J. N., & Kerr, I. M. (1993). Complementation by the protein tyrosine kinase JAK2 of a mutant cell line defective in the interferon-gamma signal transduction pathway. Nature 366, 166-170.

Watson, J. M., Sensintaffar, J. L., Berek, J. S., & Martinez-Maza, O. (1990). Constitutive production of interleukin 6 by ovarian cancer cell lines and by primary ovarian tumor cultures. Cancer Research 50, 6959-6965.

Wegenka, U. M., Buschmann, J., Lütticken, C., Heinrich, P. C., & Horn, F. (1993). Acute-phase response factor, a nuclear factor binding to acute-phase response elements, is rapidly activated by interleukin-6 at the posttranslational level. Mol. Cell. Biol. 13, 276-288.

Weissenbach, J., Chernajovsky, Y., Zeevi, M., Shulman, L., Soreq, H., Nir, U., Wallach, D., Perricaudet, M., Tiollais, P., & Revel, M. (1980). Two interferon mRNAs in human fibroblasts: *In vitro* translation and Escherichia coli cloning studies. Proc. Natl. Acad. Sci. USA 77, 7152-7156.

Wen, Z., Zhong, Z., & Darnell, J. E., Jr. (1995). Maximal activation of transcription by Stat1 and Stat3 requires both tyrosine and serine phosphorylation. Cell 82, 241-250.

Wendling, F., Maraskovsky, E., Debili, N., Florindo, C., Teepe, M., Titeux, M., Methia, N., Breton-Gorius, J., Cosman, D., & Vainchenker, W. (1994). cMpl ligand is a humoral regulator of megakaryocytopoiesis. Nature 369, 571-574.

Wendling, D., Racadot, E., & Wijdenes, J. (1993). Treatment of severe rheumatoid arthritis by anti-interleukin 6 monoclonal antibody. J. Rheumatol. 20, 259-262.

Wilks, A. F., Harpur, A. G., Kurban, R. R., Ralph, S. J., & Zurcher, G. (1991). Two novel protein-tyrosine kinases, each with a second phosphotransferase-related catalytic domain, define a new class of protein kinase. Mol. Cell. Biol. 11, 2057-2065.

Williams, D. E., Eisenman, J., Baird, A., Rauch, C., Van Ness, K., March, C. J., Park, L. S., Martin, U., Mochizuki, D. Y., Boswell, H. S., Burgess, G. S., Cosman, D., & Lyman, S. D. (1990). Identification of a ligand for the c-*kit* proto-oncogene. Cell 63, 167-174.

Williams, N., Bertoncello, I., Jackson, H., Arnold, J., & Kavnoudias, H. (1992). The role of interleukin 6 in megakaryocyte formation, megakaryocyte development and platelet production. Ciba Foundation Symposium 167, 160-173.

Witthuhn, B. A., Silvennoinen, O., Miura, O., Lai, K. S., Cwik, C., Liu, E. T., & Ihle, J. N. (1994). Involvement of the Jak-3 Janus kinase in signalling by interleukin 2 and 4 in lymphoid and myeloid cells. Science 370, 153-157.

Won, K.-A., & Baumann, H. (1990). The cytokine response element of the rat α1-acid glycoprotein gene is a complex of several interacting regulatory sequences. Mol. Cell. Biol. 10, 3965-3978.
Xing, Z., Braciak, T., Jordana, M., Croitoru, K., Graham, F. L., & Gauldie, J. (1994). Adenovirus-mediated cytokine gene transfer at tissue sites. Overexpression of IL-6 induces lymphocytic hyperplasia in the lung. J. Immunol. 153, 4059-4069.
Yamasaki, T., Taga, T., Hirata, Y., Yawata, H., Kawanishi, Y., Seed, B., Taniguchi, T., Hirano, T., & Kishimoto, T. (1988). Cloning and expression of the human interleukin-6 (BSF-2/IFN beta 2) receptor. Science 241, 825-828.
Yamashita, I., Katamine, S., Moriuchi, R., Nakamura, Y., Miyamoto, T., Eguchi, K., & Nagataki, S. (1994). Transactivation of the human interleukin-6 gene by human T-lymphotropic virus type 1 Tax protein. Blood 84, 1573-1578.
Yasukawa, K., Hirano, T., Watanabe, Y., Muratani, K., Matsuda, T., Nakai, S., & Kishimoto, T. (1987). Structure and expression of human B cell stimulatory factor-2 (BSF-2/IL-6). EMBO J. 6, 2939-2945.
Yoshida, K., Taga, T., Saito, M., Suematsu, S., Kumanagoh, A., Tanaka, T., Fujiwara, H., Hirata, M., Yamagami, T., Nakahata, T., Hirabayashi, T., Yoneda, Y., Tanaka, K., Wang, W.-Z., Mori, C., Shiota, K., Yoshida, N., & Kishimoto, T. (1996). Targeted disruption of gp130, a common signal transducer for the interleukin 6 family of cytokines, leads to myocardial and hematological disorders. Proc. Natl. Acad. Sci. USA 93, 407-411.
Yoshida, K., Chambers, I., Nichols, J., Smith, A., Saito, M., Yasukawa, K., Shoyab, M., Taga, T., & Kishimoto, T. (1994). Maintenance of the pluripotential phenotype of embryonic stem cells through direct activation of gp130 sigalling pathways. Mech. Dev. 45, 163-171.
Yoshizaki, K., Matsuda, T., Nishimoto, N., Kuritani, T., Taeho, L., Aozasa, K., Nakahata, T., Kawai, H., Tagoh, H., Komori, T., Kishimoto, S., Hirano, T., & Kishimoto, T. (1989). Pathogenic significance of interleukin-6 (IL-6/BSF-2) in Castleman's disease. Blood 74, 1360-1367.
Yoshizaki, K., Kuritani, T., & Kishimoto, T. (1992). Interleukin-6 in autoimmune disorders. Semin. Immunol. 4, 155-166.
Zauli, G., Re, M. C., Gugliotta, L., Furlini, G., & La Placa, M. (1993). The elevatiion of circulating platelets after IFN-α therapy in HIV-1 seropositive thrombocytopenic patients correlates with increased plasma levels of IL-6. Microbiologica 16, 27-34.
Zhang, X., Blenis, J., Li, H.-C., Schindler, C., & Chen-Kiang, S. (1995). Requirement of serine phosphorylation for formation of STAT-promoter complexes. Science 267, 1990-1994.
Zhang, X. G., Gaillard, J. P., Robillard, N., Lu, Z. Y., Gu, Z. J., Jourdan, M., Boiron, J. M., Bataille, R., & Klein, B. (1994). Reproducible obtaining of human myeloma cell lines as a model for tumor stem cell study in human multiple myeloma. Blood 83, 3654-3663.
Zhang, Y. H., Lin, J. X., & Vilcek, J. (1990). Interleukin-6 induction by tumor necrosis factor and interleukin-1 in human fibroblasts involves activation of a nuclear factor binding to a κB-like sequence. Mol. Cell. Biol. 10, 3818-3823.
Zhong, Z., Wen, Z., & Darnell, J. E., Jr. (1994). Stat3: A STAT family member activated by tyrosine phosphorylation in response to epidermal growth factor and interleukin-6. Science 264, 95-98.

Zhuang, H., Patel, S. V., He, T.-C., Sonsteby, S. K., Niu, Z., & Wojchowski, D. M. (1994). Inhibition of erythropoietin-induced mitogenesis by a kinase-deficient form of JAK2. J. Biol. Chem. 269, 21411-21414.

Zohlnhofer, D., Graeve, L., Rose-John, S., Schooltink, H., Dittrich, E., & Heinrich, P. C. (1992). The hepatic interleukin-6 receptor. Down-regulation of the interleukin-6 binding subunit (gp80) by its ligand. FEBS Lett. 306, 219-222.

Zsebo, K. M., Wypych, J., McNiece, I. K., Lu, H. S., Smith, K. A., Karkare, S. B., Sachdev, R. K., Yuschenkoff, V. N., Birkett, N. C., Williams, L. R., Satyagal, V. N., Tung, W., Bosselman, R. A., Mendiaz, E. A., & Langley, K. E. (1990). Identification, purification, and biological characterization of hematopoietic stem cell factor from buffalo rat liver-conditioned medium. Cell 63, 195-201.

THE ROLE OF INTERLEUKIN-7 (IL-7) IN THE GENERATION AND REGULATION OF IMMUNE RESPONSES

David H. Lynch

Growth Factors and Cytokines in Health and Disease
Volume 2A, pages 257-275.

ISBN: 0-7623-0117-1

I. INTRODUCTION

Interleukin-7 (IL-7) was initially described as a cytokine produced by murine bone marrow stromal cells that promoted the growth of B cell progenitors *in vitro*. Further studies have demonstrated that, in addition to its effects on B lymphopoiesis, IL-7 affects thymopoiesis, myelopoiesis, and, more recently, neuronal cell survival. IL-7 also has profound immunomodulatory effects ranging from augmentation of immunotherapeutically effective T cell responses to tumor antigens to enhancing immune responses of mice to virally mediated pathologic conditions. IL-7 has also been found to affect immune responses to parasitic infections. Finally, the receptor for IL-7 is a heterodimeric complex comprised of a unique IL-7Rα-chain and a common γ-chain shared by both the IL-2R and IL-4R. Collectively, the data strongly suggest that IL-7 plays a pivotal role in both the ontogeny of the immune system and the regulation of immune reponses.

II. THE CYTOKINE

Interleukin 7 (IL-7) was initially described in 1988 as a 25 K MW protein (lymphopoietin-1) produced by murine bone marrow stromal cells that promoted the growth of B cell progenitors *in vitro* (Namen et al., 1988b). Expression cloning of mouse IL-7 resulted in the isolation of a 1,607 bp sequence that contained one predominant open reading frame of 462 bp (Namen et al., 1988a). The open reading frame encoded a 25 amino acid leader sequence followed by a 129 amino acid protein with a predicted relative molecular mass (M_r) of 14.9 K. Two N-linked glycosylation sites at amino acids 61 and 90 were also identified, which likely account for the difference between the predicted and observed molecular weights. The human IL-7 cDNA was subsequently isolated by cross-species hybridization from an SK-HEP-1 library and found to be 60% identical to mouse IL-7 at the amino acid level (Goodwin et al., 1989). The cDNA for human IL-7 contained a coding region of 534 bp which encoded a 25 amino acid leader sequence and a 152 amino acid mature protein with three potential N-linked glycosylation sites and a predicted Mr of 17.4 K. The difference in predicted M_r of mouse and human IL-7 was due to the inclusion of a 19 amino acid stretch in the human sequence (amino acid residues 96-114) not found in the mouse sequence. Interestingly, these amino acids are not required for biological function since their removal had no effect on the proliferative activity of human IL-7 on bone marrow stromal cells *in vitro* (Goodwin et al., 1989).

III. THE IL-7 RECEPTOR AND ASSOCIATED PROTEINS

IL-7 mediates its biologic activity by binding to specific cell-surface receptors. Initial studies have shown that binding of ^{125}I-IL-7 to the IL-7 dependent cell line IxN/2b was both rapid and saturable (Park et al., 1990). These studies have also demonstrated the presence of both high (Ka ~$1 \times 10^{10} M^{-1}$) and low (Ka ~$4 \times 10^{8} M^{-1}$) affinity binding sites. A combined total of 2,000-2,500 sites per cell was detected on these cells, of which only 15-20% were high affinity. Distribution studies have shown IL-7R on bone marrow cells (360/cell), bone marrow macrophages (3,900/cell), lymph node cells (860/cell), spleen cells (640/cell), and thymocytes (330/cell) but not on mature B cells (Park et al., 1990). Characterization of the IL-7R by affinity cross-linking has indicated the presence of two membrane proteins with M_r of approximately 75-78 K and 160 K, which most likely correspond to the monomeric and dimeric forms of the IL-7R, respectively (Park et al., 1990).

Initial cloning of both human and murine IL-7R revealed only a single cDNA whose extracellular domain is homologous to the hematopoietic receptor superfamily (Goodwin et al., 1990). Goodwin et al. (1990) also found that alternative splicing of the human gene results in three different mRNA species including a soluble form and two membrane-bound forms (one a full-length receptor and the other containing a truncated cytoplasmic domain). Interestingly, although both of the membrane-bound forms of this receptor bind IL-7 with high and low affinity, neither contains an obvious kinase domain. Studies by a number of different groups have since demonstrated that the functional form of the IL-7R is a heterodimeric protein composed of the cytokine-specific IL-7R protein and the IL-2Rγ chain (Noguchi et al., 1993; Page et al., 1993; Kawahara et al., 1994). The IL-2Rγ chain has also been found to be a component of several other receptor complexes, including receptors for the IL-2, IL-4, IL-9, and IL-15 (Takeshita et al., 1992; Russell et al., 1993; 1994; Giri et al., 1994; Schlossman et al., 1994), and thus is referred to as the common γ chain (γc). Further studies have shown the existence of three affinities of IL-7 binding: a low affinity class (~125 nM) detected on either untransfected COS cell or cells expressing only the γc subunit; a high affinity class (~250 pM) detected on cells transfected with IL-7R cDNA; and a slightly higher affinity class (~30-80 pM) detected on cells expressing both IL-7R and γc (Kondo et al., 1993; Noguchi et al., 1993).

Ligation of the IL-7R in lymphocyte progenitor cells results in rapid phosphorylation of tyrosine residues on a number of different proteins and inositol phospholipid turnover (Dibirdik et al., 1991; Uckun et al., 1991a, 1991b). One of the phosphorylated proteins is the 85kDa subunit (including both α and β isomers) of phosphatidyl inositol-3 kinase (PI-3 kinase; Dadi et al., 1994). Interestingly, the IL-7R is also a substrate for tyrosine phosphorylation, and IL-7 binding causes association of the IL-7R α chain with activated PI-3 kinase and phosphorylation of a single Tyr residue (Tyr 449) on the IL-7Rα chain (Venkitaraman and Cowling, 1994). The phosphotyrosine-dependent recognition motif surrounding Tyr 449 corresponds to a canonical p85 binding sequence, Tyr (P)-X-X-Met. Interestingly, although both the IL-2 and IL-4 receptor complexes contain the IL-2Rγ chain and activate PI-3 kinase, neither contains sequences resembling the recognition motifs for the SH2 domains of p85. Researchers speculate that the mechanism by which IL-7R activates PI-3 kinase is distinct from the mechanism used by the IL-2R and IL-4R.

The IL-7R has also been shown to interact with protein tyrosine kinases (PTK) of the *src* family (Hatakeyama et al., 1991; Venkitaraman and Cowling, 1992; Seckinger and Fogereau, 1994) and induces both *N-myc* and *C-myc* expression (Morrow et al., 1992; Seckinger et al., 1994). Recent studies have also indicated that activation of a T cell clone by IL-7 results in tyrosine phosphorylation of the JAK3 tyrosine kinase, but not JAK1, JAK2, or Tyk2 (Zeng et al., 1994). Two additional proteins reported to be phosphorylated after IL-7 stimulation are STAT 1α and the p44 MAP kinase, suggesting that the STAT (signal transducers and activators of transcription) and MAPK (mitogen-activated protein kinase) signal pathways may also be involved in IL-7-mediated signal transduction (Zeng et al., 1994).

IV. IL-7 IN B CELL LYMPHOPOIESIS

Although it is now known that IL-7 can mediate effects on a variety of cells, it was first identified and isolated based on its ability to promote the proliferation of B-cell precursors (Namen et al., 1988b). Early studies indicated that IL-7 supported the *in vitro* growth of $B220^+$, surface IgM^- pre-B cells from either normal bone marrow or long-term Whitlock-

Witte cultures, but did not drive differentiation of these cells into surface IgM^+ cells (Whitlock and Witte, 1982; Namen et al., 1988a, 1988b; Lee et al., 1989). However, subsequent analyses of subpopulations of $B220^+$, IgM^- B lineage cells in mouse bone marrow that hase been separated based on differential expression of CD43, BP-1, and HSA have indicated that both early ($CD43^+$, HSA^{dull}) and late ($CD43^+$, HSA^{bright}) pro-B cells are directly responsive to IL-7, while pre-B cells ($CD43^-$, HSA^{bright}) are not (Hardy et al., 1991). Interestingly, although purified pre-pro-B cells are not directly responsive to IL-7, the addition of IL-7-specific mAb to pre-pro-B cells cultured on stromal cell monolayers has resulted in a decrease in cell recovery. Thus, the data suggest that pre-pro-B cells requires both IL-7 and a contact-dependent signal for progression to a stage of differentiation in which they are capable of responding to IL-7 alone (early and late pro-B cell stage). Coupled with analyses of immunoglobulin gene rearrangements, the data indicate that IL-7-mediated signals are critical in promoting the proliferation of those B cell progenitors undergoing heavy chain rearrangements (Hardy et al., 1991).

The growth-promoting effects of IL-7 are also true for *in vivo* B lymphopoiesis. Administration of purified recombinant IL-7 to normal mice has stimulated significant increases in numbers of $B220^+$, surface IgM^- pre-B cells in spleen, lymph nodes, and bone marrow (Morrissey et al., 1991b). Mature surface IgM^+/IgD^+ B cells and mature T cells have also been found to increase in the spleen and lymph nodes of IL-7 treated mice. The effects were transient, as both B cell and T cell numbers returned to normal levels after cessation of IL-7 treatments. The importance of IL-7 in normal B lymphopoiesis *in vivo* has also been demonstrated. Treatment of mice with a neutralizing IL-7 specific mAb completely inhibited the development of B-cell progenitors from the pro-B cell stage forward (Grabstein et al., 1993). However, treatment of mice with the IL-7-specific mAb for up to three weeks did not result in the depletion of mature peripheral B cells, indicating that while IL-7 is an essential factor for the development of B lineage cells, it does not appear to be required for the maintenance of mature B cells *in vivo*.

Mice that are genetically deficient in the expression of the IL-7R have been produced by targeted disruption of the IL-7Rα chain gene (Peschon et al., 1994). Phenotypic analyses of bone marrow from these mice demonstrated the presence of pre-pro-B cells and early pro-B cells, but not late pro-B cells and pre-B cells, a finding consistent with the conclusion that B cell development up to, but not including, the pre-B cell state, requires the high affinity of IL-7R. Surprisingly, a low number of mature B cells (7-10% of normal) were detected in the spleens of these mice, suggesting that to a limited extent, B lymphocyte development can proceed in the absence of the high affinity IL-7R. It is tempting to speculate that this may be due to low affinity binding of IL-7 to the γc chain, which, as noted earlier, is a functional component of the high affinity IL-7R complex. Surprisingly, the block in B cell development in mice that are genetically deficient in the expression of IL-7 appears to be at the transition of pro-B cells to pre-B cells (von Freeden-Jeffry et al., 1995). Whether this is due to the influence of other cytokines that can also bind to and signal through the IL-7Ra chain is unknown.

V. IL-7 IN THYMOPOIESIS

IL-7 has also been shown to be expressed in both murine and human thymus tissue (Namen et al., 1988a; Goodwin et al., 1989; Wiles et al., 1992; Fernández et al., 1994), implying a role for this cytokine in the thymic development of T cells. IL-7 has been shown to induce thymocyte proliferation (Conlon et al., 1989; Watson et al., 1989), and administration of IL-7 to mice with cyclophosphamide-induced lymphopenia accelerated lymphocyte repopulation (Morrissey et al., 1991a). Evaluation of the effects of IL-7 on subpopulations of highly purified $CD3^-$, $CD4^-$, and $CD8^-$ mouse thymocytes *in vitro* demonstrated the differentiation of significant numbers of $CD4^+$ and/or $CD8^+$ cells within one day, and $CD3^+$ cells within 2-3 days (Vissinga et al., 1992). Although the principal subpopulations of $CD3^-$, $CD4^-$, and $CD8^-$ thymocytes induced to differentiate by IL-7 were contained in the $IL\text{-}2R^-$ subset, the data also indicate that IL-7 stimulates proliferation of $CD3^+$, $CD4^-$, $CD8^-$ thymocytes, but without inducing their differentiation. Culture of nylon-wool purified human thymocytes for 10-14 days in serum-free medium containing IL-7 has also been performed, and demonstrated to promote the generation of a population of large thymocytes expressing high density CD3, CD4, and/or CD8 and CD45RA (Vollger and Uittenbogaart, 1993). IL-7 has also been shown to promote differentiation of both T cell receptor (TcR)-αβ and TcR-γδ T cells by inducing TcR gene rearrangements (Appasamy et al., 1993; Muegge et al., 1993; Vicari et al., 1994). Whether this result is due to a direct effect of IL-7 or whether it acts by maintaining thymocyte survival long enough for an unknown second signal to initiate TcR gene rearrangement is not known at this time.

Experiments on the effects of ablation of IL-7 mediated signals on lymphocyte development have also been conducted. For example, treatment of mice with IL-7 specific mAb causes a three to fourfold reduction in thymic cellularity (Grabstein et al., 1993). Thymic cellularity in IL-7R$^{-/-}$ mice was found to vary between 0.1 and 10% of control values and to be composed primarily of CD4^{-}/CD8^{-} cells (Peschon et al., 1994). Further phenotypic evaluation of those thymocytes suggested that thymocyte development in IL-7R$^{-/-}$ mice is affected prior to expression of cell-surface CD3, CD4, and CD8 with expression of these cells interrupted at the HSA^{+}, CD44^{+}, and CD25^{-} stage, which correlates to a stage prior to the initiation of TcR β and γ chain rearrangement. These findings are supported by studies in which mice treated for extended periods of time with IL-7-specific mAb demonstrated a 99% reduction in thymic cellularity. In one study the principal depletion in thymocytes was in the CD3^{-}, CD4^{-}, and CD8^{-} precursors, and the data indicate interruption at the transition to the CD44^{+}, CD25^{+} stage of development. Similar reductions in thymic cellularity have also been reported in mice rendered genetically deficient in IL-7, although some variations have been reported in the distribution of CD4- and CD8-expressing subpopulations (von Freeden-Jeffry et al., 1995). However, in all cases T cells were detected in the periphery of these mice, suggesting that T cell development can occur, at least to some limited extent, in the absence of an IL-7-mediated phase of expansion. Conversely, IL-7 transgenic nude mice have increased numbers of peripheral cells expressing Thy-1, cell surface TcR, CD4, and CD8 (Rich et al., 1995). In addition, T cell specific activation and proliferative responses were also restored in these mice, suggesting that IL-7 plays a fundamental role in thymocyte maturation.

Although much of the data extant indicate that IL-7-mediated signals play a preeminent role in thymopoiesis, IL-7 is not the sole operative cytokine. Small numbers of mature T cells are found in the periphery of IL-7R$^{-/-}$ mice. The fact that many of the hematopoietic receptors use shared subunits raises the caveat that deletion of the IL-7Rα chain may also impinge on other receptor:ligand interactions. In this regard, a novel cytokine (thymic stromal-derived lymphopoietic—TSLP) whose activities strongly parallel those of IL-7 has recently been cloned (Williams et al., in preparation). Further the IL-7Rα chain appears to be a subunit of the TSLP receptor (Park, in preparation).

VI. IL-7 IN MYELOPOIESIS

IL-7 has been primarily regarded as a lymphoid-restricted cytokine due to its profound activities on mature T cells and B- and T cell precursors. However, receptors for IL-7 have also been detected on murine bone marrow cells and bone marrow macrophages (Park et al., 1990). Culture of human peripheral blood mononuclear cells (PBMC) in IL-7 induces the synthesis and expression of a wide range of inflammatory cytokines, including IL-6, TNF, IL-1α, IL-1β, and MIP-1β (Alderson et al., 1991; Ziegler et al., 1991). Although dose-response studies have shown that the induction of such inflammatory cytokines requires substantially higher concentrations of IL-7 (10 ng/ml) than are required to augment T cell responses (0.1 ng/ml), at optimal concentration (100 ng/ml) IL-7 induced PBMC to produce 10-fold greater levels of IL-6 than did optimal concentrations of IL-1β (10 ng/ml), and the kinetics of induction were as rapid as those induced by LPS.

The effects of IL-7 on the propagation and maturation of other hematologic cell lineages has also been investigated. Interestingly, IL-7 has been shown to enhance myeloid colony formation *in vitro* by various colony stimulating factors (CSFs) from murine Lin$^-$, Sca-1$^+$ bone marrow cells (Jacobsen et al., 1993). Further studies have demonstrated that although IL-7 has no direct myeloproliferative effects, it synergizes with CSF-1, GM-CSF, IL-3 and Steel factor (a.k.a. mast cell growth factor, stem cell factor, c-*kit* ligand) in the generation of Mac-1$^+$ mononuclear cells (Morrissey et al., 1991a; Jacobsen et al., 1993, 1994b; Fahlman et al., 1994). Similar results have been obtained using human CD34$^+$ progenitor cells (Jacobsen et al., 1994a). *In vivo* administration of IL-7 to mice has also been shown to mobilize myeloid progenitor cells from the bone marrow to the spleen, blood, and liver (Damia et al., 1992; Grzegorzewski et al., 1994). Recent data also indicate that IL-7 treatment mobilizes pluripotent stem cells with long-term reconstituting activity to the peripheral blood (Grzegorzewski et al., 1995). Finally, IL-7 has also been demonstrated to accelerate the repopulation of lymphoid cells and some myeloid cells in mice that had been immunocompromised by either treatment with cyclophosphamide or sub-lethal irradiation (Morrissey et al., 1991a; Faltynek et al., 1992).

VII. IL-7 ON NEURONAL CELLS

IL-7 has also been evaluated for its effects on cultures of hippocampal neuronal cells and found to significantly enhance their survival in a dose-dependent manner (Araujo and Cotman, 1993). The increases in neuronal survival were more pronounced with increasing time in culture with increased numbers of both astroglia and microglia compared to control cultures. Data have also been presented that indicate that IL-7, either alone or in combination with transforming growth factor-α (TGFα), acts to promote differentiation (as assessed by morphological and immunocytochemical criteria) of neuronal progenitor cells (Mehler et al., 1993). Culture of neuronal progenitor cells with IL-7 also induced expression of sodium channels with developmentally immature physiological properties (i.e., tetrodotoxin-insensitive electrophysiological responses).

VIII. EFFECTS OF IL-7 OVEREXPRESSION *IN VIVO*

The fact that IL-7 can be directly mitogenic to cells expressing IL-7R suggests the possibility that autocrine or paracrine stimulation of cells may be tumorigenic. Indeed, infection of an IL-7-dependent pre-B cell line with a retrovirus containing the IL-7 gene not only resulted in conversion of the cells to IL-7 independence, but also the tumorigenic transformation of those cells (Overell et al., 1991). The tumorigenic transformation of these cells by IL-7 expression in this case was clearly due to an autocrine loop because the proliferation of these cells was inhibited by neutralizing IL-7 specific mAb. Interestingly, although conversion of an IL-7 dependent pre-B cell line to IL-7 independence resulted in a dramatic change in tumorigenicity, the majority of human B cell lines tested do not appear to be maintained by IL-7/IL-7R autocrine loops. In one study IL-7 mRNA was detected in five of six B cell lines derived from patients with AIDS-associated Burkitt's lymphomas, three of three B cell lines derived from patients with American Burkitt's lymphoma, and five of six normal lymphoblastoid B cell lines, but only one of five B cell lines from patients with African Burkitt's lymphoma, and two of seven EBV^- B cell lines (Benjamin et al., 1994). Biologically active IL-7 was found to be secreted by five of six AIDS-associated Burkitt's lymphoma cell lines and six of seven normal lymphoblastoid cell lines, but was not produced by any of the seven EBV^- cell lines tested. Further, mRNA for IL-7R was only detected in eight of 26 of the B cell lines tested, and none of these cell lines produced IL-7.

Further efforts to evaluate the effects of overexpression of IL-7 have used transgenic mouse lines in which IL-7 was expressed in the bone marrow, spleen, and thymus, but not in the kidney, liver, brain, or heart (Samaridis et al., 1991). Consistent with the role of IL-7 in lymphopoiesis, the frequency and total number of B cell precursors and mature B cells is increased in both the bone marrow and spleen. The number of all major subsets of thymocytes and mature peripheral T cells is also increased. However, in contrast to predictions based on the proposed role of IL-7 in myelo- and monocytopoiesis the number of granulocytes and macrophages in the bone marrow of IL-7 transgenic mice is similar to that found in non-transgenic littermates.

Interestingly, a number of disorders have also been observed in IL-7 transgenic mice. For example, mice expressing IL-7 under the control of the SRα promoter (which results in IL-7 expression in the brain, thymus, lung, liver, spleen, kidney, lymph nodes, skin, and skeletal muscle) develop severe dermatitis with erythroderma and alopecia (Uehira et al., 1993). The lesions in these mice are characterized by severe mononuclear cell infiltrates comprised primarily of T cells expressing the TcR-γ/δ. Other studies have shown that mice carrying the IL-7 transgene under the control of the I-Eα chain promoter show a high incidence of lymphoproliferative disease (Fisher et al., 1993). The lesions in these mice are characterized by oligoclonal expansion of early B cells and bipotential B cell/macrophage stem cells. Mice carrying the IL-7 transgene under the control of an immunoglobulin heavy-chain promoter and enhancer (which express IL-7 in the bone marrow, lymph nodes, spleen, thymus, and skin) show a marked diminution of $CD4^+CD8^+$ thymocytes, the development of B and T cell lymphomas and a progressive skin disorder characterized by progressive alopecia, hyperkeratosis, and exfoliation (Rich and Leder, 1995). The skin disorder is associated with, and appears to be mediated by, infiltrating T-lineage cells since the phenotype can be adoptively transferred to syngeneic wild-type mice by transplantation of either skin or lymphoid cells from the IL-7 transgeneic mice.

The observations regarding skin disorders in IL-7 transgenic mice also appear to have a counterpart in Sezary syndrome in humans (Foss et al., 1994). *In vitro* studies of lymphoma cells from five of seven patients with Sezary syndrome showed synergistic proliferative responses when cultured in the presence of IL-7 and IL-2 compared with either cytokine alone. Constitutive IL-7R expression was detected by both RT-PCR and ^{125}I-IL-7 affinity cross-linking on cells from two of these patients, and coexpression of both IL-7R and IL-2R was induced on cells from all seven patients upon incubation in IL-7. In addition, lymphona cells from three of five of the patients tested expressed IL-7 mRNA. Skin from three of five patients with Sezary syndrome, as well as skin from a normal individual, also expressed IL-7 mRNA. Collectively, the data suggest that autocirine and paracrine mechanisms involving IL-7 may play a role in the growth regulation of some, but not all, B and T cell lymphomas, and perhaps cutaneous T cell lymphoma.

IX. IMMUNOMODULATORY ACTIVITIES OF IL-7

IL-7 is directly mitogenic to thymocytes (Conlon et al., 1989) but not particularly so for purified mature T cells unless a co-mitogen such as ConA is also added (Chazen et al., 1989; Morrissey et al., 1989; Armitage et al., 1990; Grabstein et al., 1990). In contrast, antigen-dependent cloned T cells and activated splenic T cells can be stimulated to proliferate in response to IL-7 (Grabstein et al., 1990). Lymphokine activated killer (LAK) cells have also been shown to be induced by culture of lymphocytes with IL-7, although the kinetics of their induction is somewhat slower than that induced by IL-2 (Alderson et al., 1990; Lynch and Miller, 1990). In humans, the LAK were phenotypically heterogeneous and included $CD8^+$ T cells, $CD56^+$ NK cells and $TcR\gamma\delta8^+$ cells (Alderson et al., 1990). In cultures of heterogeneous murine lymphoid cells the prinicipal LAK activity is mediated by $CD8^+$ cells with no apparent contribution by $NK1.1^+$ cells. However, $CD4^+$ LAK cells were found to be induced if they were cultured with IL-7 (Lynch and Miller, 1990). IL-7 has also been shown to promote secretion of IL-2, IL-4, IL-6, and IFNγ upon restimulation of human peripheral blood T cells with phorbol ester and calcium ionophore (Armitage et al., 1992) and induction of macrophage inflammatory protein-1 (MIP-1) β and nonspecific tumoricidal activity by human peripheral blood monocytes (Alderson et al., 1991; Ziegler et al., 1991).

One potentially important immunomodulatory effect of IL-7 is its ability to enhance the generation of cytolytic T lymphocytes (CTL) *in vitro* (Alderson et al., 1990; Jicha et al., 1992; Komschlies et al., 1994). Although the initial studies evaluated enhanced generation of allospecific CTL (Alderson et al., 1990), the generation of tumor-specific CTL was also augmented by inclusion of IL-7 into the culture medium (Lynch et al., 1991). Further, *in vivo* comparisons of the immunotherapeutic efficacy of tumor-specific CTL generated with IL-7 demonstrated them to be up to eightfold more effective than CTL generated in medium alone and two to threefold more effective than CTL generated in the presence of IL-2 alone (Lynch et al., 1991). In a second study, anti-tumor CTL generated in the presence of IL-7 was found to be equivalent to CTL generated with IL-2 in the elimination of pulmonary metastases (Jicha et al., 1991). In addition, inclusion of IL-7 into cultures of anti-tumor CTL has been shown to promote their long-term (>31 months) survival in the absence of repeated stimulation with either tumor cells or tumor antigen (Lynch and Miller, 1994).

IL-7 has also been tested for its influence on disease progression in a limited range of virally-mediated pathologic conditions. Analysis of the potential immunomodulatory effects of IL-7 in the generation of CTL responses to a short syntheric peptide (NPP) derived from influenza A nucleoprotein (147-158 R156$^-$) demonstrated that naive BALB/c splenocytes were able to generate H2-restricted influenza A-specific CTL when cultured with NPP in the presence of IL-7, but not in its absence (Kos et al., 1992). In addition, mice that have been intravenously injected with the polycythemia-inducing strain of the Friend virus complex (FVC) followed by treatment with recombinant IL-7 showed both prolonged survival time and an increased proportion of long-term survival compared to untreated mice (Lu et al., 1992). IL-7 has also been evaluated in preliminary studies for its ability to enhance HIV-specific immune responses. One study investigated the ability of recombinant IL-7 to augment immune responses of mice immunized with either alum-associated or liposome-associated recombinant HIV envelope protein env-2-35F2 (a non-glycosylated denatured gp120 of HIV strain ISF2; Bui et al., 1994). Antibody responses to both alum- and liposome-associated env-2-35F2 were found to be enhanced by IL-7. CTL responses were also enhanced in mice immunized with the HIV-env antigen in alum-, but not liposome-associated antigen. The addition of IL-7 to cultures of mitogen-stimulated peripheral blood lymphocytes from HIV-1 infected individuals results in up to a threefold enhancement of env-specific CTL activity (Carini and Essex, 1994). Augmentation of both proliferation and cytotoxic activity were time and dose-dependent, were IL-2 independent and primarily affected CD8$^+$ cells. Interestingly, evaluation of a group of five HIV-1$^+$ individuals who were negative for env- and gag-specific CTL activity showed that they not only had markedly decreased levels of cell-surface IL-7R, but also a significant increase in the "shedding" of soluble IL-7R into the culture supernatant (Carini et al., 1994).

IL-7 has also been found to affect immune responses to parasitic infections. *In vitro* IL-7 treatment of murine macrophages infected with

Leishmania major has been found to mediate a 45% reduction in both the percentage of infected cells and the parasite burden per cell (Gessner et al., 1993). Simultaneous treatment of infected macrophages with IL-7 and IFNγ has led to a nearly complete elimination. Surprisingly, IL-7 treatment of genetically susceptible BALB/c mice at the time of infection with *L. major* has led to enhanced lesion formation, increased parasite burden in spleens and lymph nodes, and decreased survival kinetics (Gessner et al., 1995). Conversely, IL-7 treatment of mice infected with *Toxoplasma gondii* appears to have a beneficial effect (Kasper et al., 1995). The data indicate that the effects of IL-7 administration are indirect. Mice treated with IL-7 beginning at the time of infection have shown increased numbers of NK cells and $CD8^+$ cells and increased survival rates (83%) compared to control mice (0%). Increases in antigen-specific CTL and serum IgM have also been noted. The data suggest an indirect effect of IL-7 in augmenting the immune response. Elevated levels of mRNA for both IFNγ and IL-2 have been detected in splenocytes from IL-7 treated mice infected with *T. gondii*, and the protective effects of IL-7 administration could be blocked by IFNγ-specific mAb.

X. CONCLUSION

Although initially identified as a cytokine that promotes pre-B cell (actually pro-B cell) growth, it is now clear that IL-7 mediates effects on a wide range of cell types. Data generated during the past several years have demonstrated the exciting immunomodulatory potentials of IL-7 in T cell mediated immune responses. However, the potential use of IL-7 to promote B and T cell lymphopoiesis in clinical settings has yet to be meaningfully evaluated. Similarly, the role for IL-7 in augmenting immune responses in clinical settings to antigens expressed on tumor cells, to parasitic infections, and to viral antigens also awaits investigation. It is hoped that the results of studies in these areas will soon be undertaken to determine whether the promise shown by this cytokine in pre-clinical studies will be borne out. Certainly, the addition of a cytokine with the substantial beneficial immunomodulatory activities attributed to IL-7 in both *in vitro* and *in vivo* studies in murine systems, and in vitro studies using human lymphoid cells, would be a welcome addition in the immunotherapeutic treatment of cancer and HIV.

ACKNOWLEGMENTS

I would like to thank Drs. Eugene Maraskovsky and Linda Park for their comments during the writing of this manuscirpt. I would also like to thank Anne C. Bannister for her excellent editorial assistance in the preparation of this manuscript.

REFERENCES

Alderson, M. R., Sassenfeld, H. M., & Widmer, M. B. (1990). Interleukin 7 enhances cytolytic T lymphocyte generation and induces lymphokine-activated killer cells from human peripheral blood. J. Exp. Med. 172, 577-587.

Alderson, M. R., Tough, T. W., Ziegler, S. F., & Grabstein, K. H. (1991). Interleukin 7 induces cytokine secretion and tumoricidal activity by human peripheral blood monocytes. J. Exp. Med. 173, 923-930.

Appasamy, P. M., Kenniston, T. W., Jr., Weng, Y., Holt, E. C., Kost, J., & Chambers, W. H. (1993). Interleukin 7-induced expression of specific T cell receptor γ variable region genes in murine fetal liver cultures. J. Exp. Med. 178, 2201-2206.

Araujo, D. M., & Cotman, C. W. (1993). Trophic effects of interleukin-4, -7 and -8 on hippocampal neuronal cultures: Potential involvement of glial-derived factors. Brain Res. 600, 49-55.

Armitage, R. J., Namen, A. E., Sassenfeld, H. M., & Grabstein, K. H. (1990). Regulation of human T cell proliferation by IL-7. J. Immunol. 144, 938-941.

Armitage, R. J., Macduff, B. M., Ziegler, S. F., & Grabstein, K. H. (1992). Multiple cytokine secretion by IL-7-stimulated human T cells. Cytokine 4, 461-469.

Benjamin, D., Sharma, V., Knobloch, T.J., Armitage, R. J., Dayton M. A. & Goodwin, R.G. (1994). B cell IL-7. Human B cell lines constitutively secrete IL-7 and express IL-7 receptors. J. Immunol. 152:4749-4757.

Bui, T., Dykers, T., Hu, S. L., Faltynek, C. R., & Ho, R. J. (1994). Effect of MTP-PE liposomes and interleukin-7 on induction of antibody and cell-mediated immune responses to a recombinant HIV-envelope protein. J. Acquir. Immune Defic. Syndr. 7, 799-806.

Carini, C., & Essex, M. (1994). Interleukin 2-independent interleukin 7 activity enhances cytotoxic immune response of HIV-1-infected individuals. AIDS Res. Hum. Retroviruses 10, 121-130.

Carini, C., McLane, M. F., Mayer, K. H., & Essex, M. (1994). Dysregulation of interleukin-7 receptor may generate loss of cytotoxic T cell response in human immunodeficiency virus type 1 infection. Eur. J. Immunol. 24, 2927-2934.

Chazen, G. D., Pereira, G. M. B., LeGros, G., Gillis, S., & Shevach, E. M. (1989). Interleukin 7 is a T-cell growth factor. Proc. Natl. Acad. Sci. USA 86, 5923-5927.

Conlon, P. J., Morrissey, P. J., Nordan, R. P., Grabstein, K. H., Prickett, K. S., Reed, S. G., Goodwin, R., Cosman, D., & Namen, A. E. (1989). Murine thymocytes proliferate in direct response to interleukin-7. Blood 74, 1368-1373.

Dadi, H., Ke, S., & Roifman, C. M. (1994). Activation of phosphatidylinositol-3 kinase by ligation of the interleukin-7 receptor is dependent on protein tyrosine kinase activity. Blood 84, 1579-1586.

Damia, G., Komschlies, K. L., Faltynek, C. R., Ruscetti, F. W., & Wiltrout, R. H. (1992). Administration of recombinant human interleukin-7 alters the frequency and number of myeloid progenitor cells in the bone marrow and spleen of mice. Blood 79, 1121-1129.

Dibirdik, I., Langlie, M.-C., Ledbetter, J. A., Tuel-Ahlgren, L., Obuz, V., Waddick, K. G., Gajl-Peczalska, K., Schieven, G. L., & Uckun, F. M. (1991). Engagement of interleukin-7 receptor stimulates tyrosine phosphorylation, phosphoinositide turnover, and clonal proliferation of human T-lineage acute lymphoblastic leukemia cells. Blood 78, 564-570.

Fahlman, C., Blomhoff, H. K., Veiby, O. P., McNiece, I. K., & Jacobsen, S. E. W. (1994). Stem cell factor and interleukin-7 synergize to enhance early myelopoiesis in vitro. Blood 84, 1450-1456.

Faltynek, C. R., Wang, S., Miller, D., Young, E., Tiberio, L., Kross, K., Kelley, M., & Kloszewski, E. (1992). Administration of human recombinant IL-7 to normal and irradiated mice increases the numbers of lymphocytes and some immature cells of the myeloid lineage. J. Immunol. 149, 1276-1282.

Fernández, E., Vicente, A., Zapata, A., Brera, B., Lozano, J. J., Martínez-A, C., & Toribio, M. L. (1994). Establishment and characterization of cloned human thymic epithelial cell lines. Analysis of adhesion molecule expression and cytokine production. Blood 83, 3245-3254.

Fisher, A.G., Burdet, C., LeMeur, M., Haasner, D., Gerber, P., & Ceredig, R. (1993). Lymphoproliferative disorders in an IL-7 transgenic mouse line. Leukemia (England) Suppl 2 pS66-68.

Foss, F.M., Koc, Y., Stetler-Stevenson, M.A., Nguyen, D.T., O'Brien, M.C., Turner, R., & Sausville, E.A. (1994). Costimulation of cutaneous T cell lymphoma cells by interleukin-7 and interleukin-2: Potential autocrine and paracrine effectors in the Sezary Syndrome. J. Clin. Oncol. 12, 326-335.

Gessner, A., Vieth, M., Will, A., Schröppel, K., & Röllinghoff, M. (1993). Interleukin-7 enhances antimicrobial activity against *Leishmania major* in murine macrophages. Infect. Immun. 61, 4008-4012.

Gessner, A., Will, A., Vieth, M., Schröppel, K., & Röllinghoff, M. (1995). Stimulation of B-cell lymphopoiesis by interleukin-7 leads to aggravation of murine leishmaniasis. Immunology 84, 416-422.

Giri, J. G., Ahdieh, M., Eisenman, J., Shanebeck, K., Grabstein, K., Kumaki, S., Namen, A., Park, L. S., Cosman, D., & Anderson, D. (1994). Utilization of the β and γ chains of the IL-2 receptor by the novel cytokine IL-15. EMBO J. 13, 2822-2830.

Goodwin, R. G., Friend, D., Ziegler, S. F., Jerzy, R., Falk, B. A., Gimpel, S., Cosman, D., Dower, S. K., March, C. J., Namen, A. E., & Park, L. S. (1990). Cloning of the human and murine interleukin-7 receptors: Demonstration of a soluble form and homology to a new receptor superfamily. Cell 60, 941-951.

Goodwin, R. G., Lupton, S., Schmierer, A., Hjerrild, K. J., Jerzy, R., Clevenger, W., Gillis, S., Cosman, D., & Namen, A. E. (1989). Human interleukin 7: Molecular cloning and growth factor activity on human and murine B-lineage cells. Proc. Natl. Acad. Sci. USA 86, 302-306.

Grabstein, K. H., Namen, A. E., Shanebeck, K., Voice, R. F., Reed, S. G., & Widmer, M. B. (1990). Regulation of T cell proliferation by IL-7. J. Immunol. 144, 3015-3020.

Grabstein, K. H., Waldschmidt, T. J., Finkelman, F. D., Hess, B. W., Alpert, A. R., Boiani, N. E., Namen, A. E., & Morrissey, P. J. (1993). Inhibition of murine B and T lymphopoiesis *in vivo* by an anti-interleukin 7 monoclonal antibody. J. Exp. Med. 178, 257-264.

Grzegorzewski, K., Komschlies, K. L., Mori, M., Kaneda, K., Usui, N., Faltynek, C. R., Keller, J. R., Ruscetti, F. W., & Wiltrout, R. H. (1994). Administration of recombinant human interleukin-7 to mice induces the exportation of myeloid progenitor cells from the bone marrow to peripheral sites. Blood 83, 377-385.

Grzegorzewski, K. J., Komschlies, K. L., Jacobsen, S. E. W., Ruscetti, F. W., Keller, J. R., & Wiltrout, R. H. (1995). Mobilization of long-term reconstituting hematopoietic stem cells in mice by recombinant human interleukin 7. J. Exp. Med. 181, 369-374.

Hardy, R. R., Carmack, C. E., Shinton, S. A., Kemp, J. D., & Hayakawa, K. (1991). Resolution and characterization of pro-B and pre-pro-B cell stages in normal mouse bone marrow. J. Exp. Med. 173, 1213-1225.

Hatakeyama, M., Kono, T., Kobayashi, N., Kawahara, A., Levin, S. D., Perlmutter, R. M., & Taniguchi, T. (1991). Interaction of the IL-2 receptor with the src-family kinase p56lck: identification of novel intermolecular association. Science 252, 1523-1528.

Jacobsen, F. W., Veiby, O. P., Skjønsberg, C., & Jacobsen, S. E. W. (1993). Novel role of interleukin 7 in myelopoiesis: Stimulation of primitive murine hematopoietic progenitor cells. J. Exp. Med. 178, 1777-1782.

Jacobsen, F. W., Rusten, L. S., & Jacobsen, S. E. W. (1994a). Direct synergistic effects of interleukin-7 on *in vitro* myelopoiesis of human $CD34^+$ bone marrow progenitors. Blood 84, 775-779.

Jacobsen, F. W., Veiby, O. P., & Jacobsen, S. E. W. (1994b). IL-7 stimulates CSF-induced proliferation of murine bone marrow macrophages and $Mac\text{-}1^+$ myeloid progenitors *in vitro*. J. Immunol. 153, 270-276.

Jicha, D. L., Mulé, J. J., & Rosenberg, S. A. (1991). Interleukin 7 generates antitumor cytotoxic T lymphocytes against murine sarcomas with efficacy in cellular adoptive immunotherapy. J. Exp. Med. 174, 1511-1515.

Jicha, D. L., Schwarz, S., Mulé, J. J., & Rosenberg, S. A. (1992). Interleukin-7 mediates the generation and expansion of murine allosensitized and antitumor CTL. Cell. Immunol. 141, 71-83.

Kasper, L.H., Matsuura, T., & Khan, I. A. (1995). IL-7 stimulates protective immunity in mice against the intracellular pathogen, Toxoplasma gondii. J. Immunol. 155, 4798-4804.

Kawahara, A., Minami, Y., & Taniguchi, T. (1994). Evidence for a critical role for the cytoplasmic region of the interleukin 2 (IL-2) receptor γ chain in IL-2, IL-4, and IL-7 signalling. Mol. Cell. Biol. 14, 5433-5440.

Komschlies, K. L., Gregorio, T. A., Gruys, M. E., Back, T. C., Faltynek, C. R., & Wiltrout, R. H. (1994). Administration of recombinant human IL-7 to mice alters the composition of B-lineage cells and T cell subsets, enhances T cell function, and induces regression of established metastases. J. Immunol. 152, 5776-5784.

Kondo, M., Takeshita, T., Ishii, N., Nakamura, M., Watanabe, S., Arai, K.-I., & Sugamura, K. (1993). Sharing of the interleukin-2 (IL-2) receptor γ chain between receptors for IL-2 and IL-4. Science 262, 1874-1877.

Kos, F.J., & Mullbacher, A. (1992). Induction of primary anti-viral cytotoxic T cells by in vitro stimulation with short synthetic peptide and interleukin-7. Eur. J. Immunol. 22:3183-3185.

Lee, G., Namen, A. E., Gillis, S., Ellingsworth, L. R., & Kincade, P. W. (1989). Normal B cell precursors responsive to recombinant murine IL-7 and inhibition of IL-7 activity by transforming growth factor-β. J. Immunol. 142, 3875-3883.

Lu, L., Zhou, Z., Wu, B., Xiao, M., Shen, R.-N., Williams, D. E., Kim, Y.-J., Kwon, B. S., Ruscetti, S., & Broxmeyer, H. E. (1992). Influence of recombinant human interleukin (IL)-7 on disease progression in mice infected with friend virus complex. Int. J. Cancer 52, 261-265.

Lynch, D. H., & Miller, R. E. (1990). Induction of murine lymphokine-activated killer cells by recombinant IL-7. J. Immunol. 145, 1983-1990.

Lynch, D. H., & Miller, R. E. (1994). Interleukin 7 promotes long-term *in vitro* growth of antitumor cytotoxic T lymphocytes with immunotherapeutic efficacy *in vivo*. J. Exp. Med. 179, 31-42.

Lynch, D. H., Namen, A. E., & Miller, R. E. (1991). *In vivo* evaluation of the effects of interleukins 2, 4 and 7 on enhancing the immunotherapeutic efficacy of anti-tumor cytotoxic T lymphocytes. Eur. J. Immunol. 21, 2977-2985.

Mehler, M. F., Rozental, R., Dougherty, M., Spray, D. C., & Kessler, J. A. (1993). Cytokine regulation of neuronal differentiation of hippocampal progenitor cells. Nature 362, 62-65.

Morrissey, P. J., Goodwin, R. G., Nordan, R. P., Anderson, D., Grabstein, K. H., Cosman, D., Sims, J., Lupton, S., Acres, B., Reed, S. G., Mochizuki, D., Eisenman, J., Conlon, P. J., & Namen, A. E. (1989). Recombinant interleukin-7, pre-B cell growth factor, has costimulatory activity on purified mature T cells. J. Exp. Med. 169, 707-716.

Morrissey, P. J., Conlon, P., Braddy, S., Williams, D. E., Namen, A. E., & Mochizuki, D. Y. (1991a). Administration of IL-7 to mice with cyclophosphamide-induced lymphopenia accelerates lymphocyte repopulation. J. Immunol. 146, 1547-1552.

Morrissey, P. J., Conlon, P., Charrier, K., Braddy, S., Alpert, A., Williams, D., Namen, A. E., & Mochizuki, D. (1991b). Administration of IL-7 to normal mice stimulates B-lymphopoiesis and peripheral lymphadenopathy. J. Immunol. 147, 561-568.

Morrow, M. A., Lee, G., Gillis, S., Yancopoulos, G. D., & Alt, F. W. (1992). Interleukin-7 induces N-myc and c-myc expression in normal precursor B lymphocytes. Genes Dev. 6, 61-70.

Muegge, K., Vila, M. P., & Durum, S. K. (1993). Interleukin-7: A cofactor for V(D)J rearrangement of the T cell receptor β gene. Science 261, 93-95.

Namen, A. E., Lupton, S., Hjerrild, K., Wignall, J., Mochizuki, D. Y., Schmierer, A., Mosley, B., March, C. J., Urdal, D., Gillis, S., Cosman, D., & Goodwin, R. G. (1988a). Stimulation of B-cell progenitors by cloned murine interleukin-7. Nature 333, 571-573.

Namen, A. E., Schmierer, A. E., March, C. J., Overell, R. W., Park, L. S., Urdal, D. L., & Mochizuki, D. Y. (1988b). B cell precursor growth-promoting activity. Purification and characterization of a growth factor active on lymphocyte precursors. J. Exp. Med. 167, 988-1002.

Noguchi, M., Nakamura, Y., Russell, S. M., Ziegler, S. F., Tsang, M., Cao, X., & Leonard, W. J. (1993). Interleukin-2 receptor γ chain: A functional component of the interleukin-7 receptor. Science 262, 1877-1880.

Overell, R.W., Clark, L., Lynch, D., Jerzy, R., Schmierer, A., Weisser, K.E., Namen, A.E., & Goodwin, R.G. (1991). Interleukin-7 retroviruses tranform pre-B cells by an autocrine mechanism not evident in Abelson murine leukemia virus. Mol. Cell. Biol. 11, 1590-1597.

Page, T. H., Willcocks, J. L., Taylor-Fishwick, D. A., & Foxwell, B. M. J. (1993). Characterization of a novel high affinity human IL-7 receptor. J. Immunol. 151, 4753-4763.

Park, L. S., Friend, D. J., Schmierer, A. E., Dower, S. K., & Namen, A. E. (1990). Murine interleukin 7 (IL-7) receptor. Characterization on an IL-7-dependent cell line. J. Exp. Med. 171, 1073-1089.

Peschon, J. J., Morrissey, P. J., Grabstein, K. H., Ramsdell, F. J., Maraskovsky, E., Gliniak, B. C., Park, L. S., Ziegler, S. F., Williams, D. E., Ware, C. B., Meyer, J. D., & Davison, B. L. (1994). Early lymphocyte expansion is severely impaired in interleukin 7 receptor-deficient mice. J. Exp. Med. 180, 1955-1960.

Rich, B.E., Campos-Torres, J., Tepper, R.I., Moreadith, R.W., & Leder, P. (1995). Cutaneous lymphoproliferation and lymphomas in interleukin-7 transgenic mice. J. Exp. Med. 177, 305-316.

Rich, B.E., & Leder, P. (1995) Transgenic expression of interleukin-7 restores T cell populations in nude mice. J. Exp. Med. 181, 1223-1228.

Russell, S. M., Keegan, A. D., Harada, N., Nakamura, Y., Noguchi, M., Leland, P., Friedmann, M. C., Miyajima, A., Puri, R. K., Paul, W. E., & Leonard, W. J. (1993). Interleukin-2 receptor γ chain: A functional component of the interleukin-4 receptor. Science 262, 1880-1883.

Russell, S. M., Johnston, J. A., Noguchi, M., Kawamura, M., Bacon, C. M., Friedmann, M., Berg, M., McVicar, D. W., Witthuhn, B. A., Silvennoinen, O., Goldman, A. S., Schmalstieg, F. C., Ihle, J. N., O'Shea, J. J., & Leonard, W. J. (1994). Interaction of IL-2Rβ and γc chains with Jak1 and Jak3: Implications for XSCID and XCID. Science 266, 1042-1045.

Samaridis, J., Casorati, G., Traunecker, A., Iglesias, A., Gutierrez, J.C., Muller, U., & Palacios, R. (1991). Development of lymphocytes in interleukin-7 transgenic mice. Eur. J. Immunol. 21:453-460.

Schlossman, S. F., Boumsell, L., Gilks, W., Harlan, J. M., Kishimoto, T., Morimoto, C., Ritz, J., Shaw, S., Silverstin, R. L., Springer, T. A., Tedder, T. F., & Todd, R. F. (1994). CD Antigen 1993. Blood 83, 879-880.

Seckinger, P., & Fogereau, M. (1994). Activation of *src* family kinases in human pre-B cells by IL-7. J. Immunol. 153, 97-109.

Seckinger, P., Milili, M., Schiff, C., & Fougereau, M. (1994). Interleukin-7 regulates c-myc expression in murine T cells and thymocytes: A role for tyrosine kinase(s) and calcium mobilization. Eur. J. Immunol. 24, 716-722.

Takeshita, T., Ohtani, K., Asao, H., Kumaki, S., Nakamura, M., & Sugamura, K. (1992). An associated molecule, p64, with IL-2 receptor β chain. Its possible involvement in the formation of the functional intermediate-affinity IL-2 receptor complex. J. Immunol. 148, 2154-2158.

Uckun, F. M., Dibirdik, I., Smith, R., Tuel-Ahlgren, L., Chandan-Langlie, M., Schieven, G. L., Waddick, K. G., Hanson, M., & Ledbetter, J. A. (1991a). Interleukin 7 receptor ligation stimulates tyrosine phosphorylation, inositol phospholipid turnover, and clonal proliferation of human B-cell precursors. Proc. Natl. Acad. Sci. USA 88, 3589-3593.

Uckun, F. M., Tuel-Ahlgren, L., Obuz, V., Smith, R., Dibirdik, I., Hanson, M., Langlie, M.-C., & Ledbetter, J. A. (1991b). Interleukin 7 receptor engagement stimulates tyrosine phosphorylation, inositol phospholipid turnover, proliferation, and selective differentiation to the CD4 lineage by human fetal thymocytes. Proc. Natl. Acad. Sci. USA 88, 6323-6327.

Uehira, M., Matsuda, H., Hikita, I., Sakata, T., Fujiwara, H., & Nishimoto, H. (1993). The development of dermatitis infiltrated by gamma delta T cells in IL-7 transgenic mice. Int. Immunol. 5:1619-1627.

Venkitaraman, A. R., & Cowling, R. J. (1992). Interleukin 7 receptor functions by recruiting the tyrosine kinase $p59^{fyn}$ through a segment of its cytoplasmic tail. Proc. Natl. Acad. Sci. USA 89, 12083-12087.

Venkitaraman, A. R., & Cowling, R. J. (1994). Interleukin-7 induces the association of phosphatidylinositol 3-kinase with the α chain of the interleukin-7 receptor. Eur. J. Immunol. 24, 2168-2174.

Vicari, A., de Moraes, M. do C., Gombert, J.-M., Dy, M., Penit, C., Papiernik, M., & Herbelin, A. (1994). Interleukin 7 induces preferential expansion of $V\beta 8.2^{+}CD4^{-}8^{-}$ and $V\beta 8.2^{+}CD4^{+}8^{-}$ murine thymocytes positively selected by class I molecules. J. Exp. Med. 180, 653-661.

Vissinga, C.S., Fatur-Saunders, D.J., & Takei, F. (1992). Dual role of IL-7 in the growth and differentiation of immature thymocytes. Exp. Heamatol. 20, 998-1003.

Vollger, L. W., & Uittenbogaart, C. H. (1993). Interleukin-7 promotes the generation of phenotypically mature CD45RA positive human thymocytes *in vitro*. Cytokine 5, 157-168.

von Freeden-Jeffry, U., Vieira, P., Lucian, L. A., McNeil, T., Burdach, S. E. G., & Murray, R. (1995). Lymphopenia in interleukin (IL)-7 gene-deleted mice identifies IL-7 as a nonredundant cytokine. J. Exp. Med. 181, 1519-1526.

Watson, J. D., Morrissey, P. J., Namen, A. E., Conlon, P. J., & Widmer, M. B. (1989). Effect of IL-7 on the growth of fetal thymocytes in culture. J. Immunol. 143, 1215-1222.

Whitlock, C. A., & Witte, O. N. (1982). Long-term culture of B lymphocytes and their precursors from murine bone marrow. Proc. Natl. Acad. Sci. USA 79, 3608-3612.

Wiles, M. V., Ruiz, P., & Imhof, B. A. (1992). Interleukin-7 expression during mouse thymus development. Eur. J. Immunol. 22, 1037-1042.

Zeng, Y. X., Takahashi, H., Shibata, M., & Hirokawa, K. (1994). JAK3 Janus kinase is involved in interleukin 7 signal pathway. FEBS Lett. 353, 289-293.

Ziegler, S. F., Tough, T. W., Franklin, T. L., Armitage, R. J., & Alderson, M. R. (1991). Induction of macrophage inflammatory protein-1β gene expression in human monocytes by lipopolysaccharide and IL-7. J. Immunol. 147, 2234-2239

THE BIOLOGICAL, BIOCHEMICAL, AND MOLECULAR PROFILE OF LEUKOCYTE CHEMOTACTIC AND ACTIVATING CYTOKINE, INTERLEUKIN-8

Akihisa Harada, Naofumi Mukaida, and Kouji Matsushima

Growth Factors and Cytokines in Health and Disease
Volume 2A, pages 277-317.

ISBN: 0-7623-0117-1

I. INTRODUCTION

The recruitment of cells into the inflammatory site is the most direct and crucial defense mechanism of the host against foreign stimuli. Migration of cells is presumed to be influenced by factors with chemotactic activities. The types of infiltrated cells vary depending largely upon the stimuli and the time intervals from the tissue injury. The existence of chemotactic cytokines against neutrophils was first suggested by Tono-oka et al. (1980) showing that monocyte culture supernatants contained chemotactic activities against neutrophils. In the early 1980s these chemotactic activities were attributed to partially purified IL-1 or TNFα (Luger et al., 1983; Sauder et al., 1984). However, subsequent studies using purified recombinant IL-1 or TNFα revealed that both lacked chemotactic activities against neutrophils, indicating that some other factor(s) distinct from IL-1 or TNFα was responsible for neutrophil chemotaxis.

Yoshimura, Matsushima, and coworkers (1987) purified this factor from lipopolysaccharide (LPS)-stimulated human monocyte culture supernatants. They also cloned the cDNA for this neutrophil chemotactic factor from the cDNA library of LPS-stimulated human monocytes (Matsushima et al., 1988). They named this novel molecule monocyte-derived neutrophil chemotactic factor (MDNCF). Several other groups subsequently purified the identical neutrophil chemotactic factor from phytohemmagglutinin (PHA)- or LPS-stimulated human peripheral blood mononuclear cells (PBMC). They called this molecule neutrophil activating factor (NAF; Walz et al., 1987), neutrophil activating protein (NAP; Schröder et al., 1987), or granulocyte chemotactic peptide (GCP; Van Damme et al., 1988). Moreover, they demonstrated the sequence identity with the predicted amino acid sequence of a "β-thromboglobulin-like protein" previously described by Schmidt and Weismann (1987). Further studies revealed that MDNCF could chemoattract T lymphocytes as well as neutrophils, and therefore was renamed as interleukin-8 (IL-8; Larsen et al., 1989a).

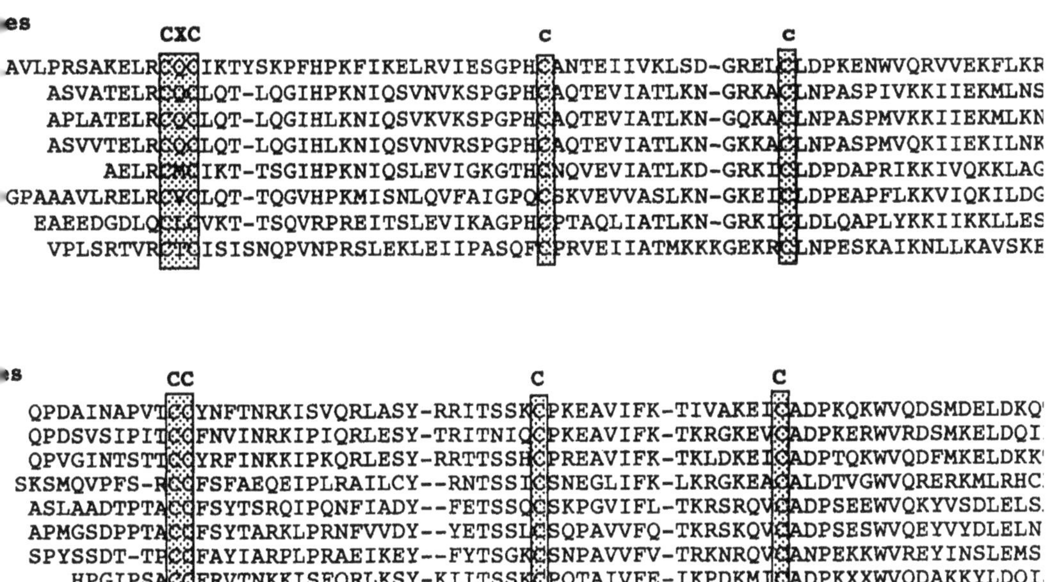

ison of amino acid sequences of C-X-C and C-C chemokines. Four well-conserved cysteines are aligned.

The discovery of IL-8 rapidly led to the extensive studies on related leukocyte chemotactic cytokines that exhibit cell type-specific leukocyte chemoattraction. More than 15 chemotactic cytokines have been identified so far. These cytokines possess four cysteine residues at the well-conserved positions and basic heparin-binding property with a molecular mass of 8-10 kDa. Moreover, they exhibit 20-50% identity with each other in amino acid sequence and structures of genomic DNAs are well conserved among these cytokines, indicating that they constitute a supergene family. Since most of these cytokines have been shown to induce the directional migration of particular type of leukocytes, they are now termed as "chemokines."

Chemokine family proteins can be divided into two subgroups, C-X-C and C-C chemokines, depending on the presence of one amino acid between the first two cysteine residues (Oppenheim et al., 1991). The C-X-C chemokines generally exhibit chemotactic activities against neutrophils and lymphocytes but not against monocytes, whereas C-C chemokines exhibit chemotactic activities against monocytes and T lymphocytes but not against neutrophils (Figure 1).

II. STRUCTURES OF IL-8

The analysis on IL-8 cDNA revealed that the precursor protein consisted of 99 amino acids with a highly hydrophobic N-terminal region that represented a putative signal peptide (Yoshimura et al., 1987; Matsushima et al., 1988).The cleavage of the N-terminus at different sites produces several forms of IL-8, 79-, 77-, 72-, and 69-amino acid forms (Yoshimura et al., 1989). These variants are produced by numerous types of cells. The most active form of IL-8 secreted by stimulated monocytes predominantly starts at serine residue and consists of 72 amino acids. The intermediate forms of 77- and 79-amino acid of IL-8 are produced mainly by endothelial cells and can be cleaved to 72-amino acid form by specific proteases such as thrombin and plasmin *in vitro* (Hébert et al., 1990; Nakagawa et al., 1991). This mature 72-amino acid form is 10 times more potent in biological activity than the 77 residue of IL-8 *in vitro*, although both forms of IL-8 show similar activity *in vivo*. This may due to the processing of 77 amino acid to 72 amino acid form *in vivo* (Nourshargh et al., 1992).

The mature 72 amino acid form has 14 basic residues that contribute to the basic characteristic of the protein (pI 10). Since purified natural

IL-8 has been completely sequenced by Edman degradation method, IL-8 is presumed not to be glycosylated. The most probable heparin binding site is the sequence of Lys64-Phe65-Leu66-Lys67-Arg68 at the carboxyl-terminal region, since affinity to heparin is decreased with progressive truncation of this portion. However, neutrophil chemotactic activity of IL-8, when bound to heparin or heparan-sulfate, was enhanced up to fourfold, and the stimulus-dependent increase in cytosolic free Ca^{2+} was also markedly enhanced(Webb et al., 1993), suggesting a functional role of heparan sulfate and heparin on endothelial cells and in basement membrane as an enhancer of IL-8 activity.

Structural studies of recombinant IL-8 by nuclear magnetic resonance (NMR; Clore et al. 1990) and X-ray crystallography (Baldwin et al., 1990, 1991) have revealed that IL-8 is a dimer at a high concentration with six interchain H-bond from residues 23-29 and four interchain salt bridges. The dimer interface is formed by three anti-parallel β-strands from each monomer and the α-helices are formed in the carboxyl-terminal region. Thus, the IL-8 dimer is stabilized by formation of a six-stranded anti-parallel β-sheet and by hydrophobic interactions with the helices. It is not known whether IL-8 can be present as a monomer or a dimer at a low concentration under physiological conditions. However, since the biological activity of IL-8 is maximal at 10 ng/ml, significant concentrations of monomeric IL-8 are presumed to exist. In order to prevent formation of hydrogen bonds between monomers, Rajarathnam et al. (1994) chemically synthesized the IL-8 analog whose amide nitrogen of leucine-25 was selectively methylated. The IL-8 analog, which was confirmed to be a monomer by analytical ultracentrifugation and NMR, showed an equipotent activity of neutrophil activation as native IL-8 (Rajarathnam et al., 1994). These results indicate that the monomer is also a functional form of IL-8.

Alanine scanning mutagenesis demonstrated that the sequence Glu4-Lue5-Arg6 was critically involved in the receptor binding (Hébert et al., 1991). N-terminal truncation studies suggested that this portion was essential for IL-8 activity (Clark-Lewis et al., 1991). C-X-C chemokines except platelet factor 4 (PF4) and γ-interferon inducible protein 10 (γIP 10) have the Glu4-Lue5-Arg6 motif in common and activate neutrophils. Both PF4 and IP10, of which this motif is absent, lack the binding and activating activity to neutrophils. PF4 gained the ability to bind to IL-8 receptors on neutrophils if its amino terminal portion was replaced to Glu-Lue-Arg (Baggiolini et al., 1994). On the other hand, when Glu-Leu-Arg sequence was introduced into γIP 10 and MCP-1/MCAF, neither of them activated neutrophils. These data suggest that, besides the Glu-Leu-Arg sequence, additional residues or regions are necessary for binding to IL-8 receptors (Clark-Lewis et al., 1993). Further studies of structure-functional analysis revealed that the turn of residues 30-35 was also a key structure to provide the correct geometry for the 7-34 disulfide bridge that controlled the conformation of the Glu-Leu-Arg motif (Clark-Lewis et al., 1994). In the view of the receptor binding and neutrophil activation, IL-8 antagonist was synthesized by modification of the Glu-Leu-Arg sequence at the N-terminus (Moser et al., 1993a). Twenty-six analogs of IL-8 were synthesized with deletions or amino acid replacements in the Glu-Leu-Arg region. The most potent antagonists were Arg6-IL-8 (7-72) and Ala4-Ala5-Arg6-IL-8 (7-72). These analogs inhibited IL-8 receptor binding, exocytosis, and the respiratory burst as well as chemotaxis.

III. THE IL-8 GENE AND REGULATION OF ITS EXPRESSION

A. Positive Regulation of IL-8 Transcription

The gene encoding for IL-8 protein locates in a gene cluster with several other members of C-X-C chemokines such as GROα, GROβ, GROγ, PF4, platelet basic protein (PBP), and IP 10 on human chromosome 4q12-21 (Modi et al., 1990; Richmond et al., 1988; Griffin et al., 1987; Tunnacliffe et al., 1992; Luster et al., 1987) while the genes for the C-C chemokines (MCAF/MCP-1, MIP-1α, MIP-1β, I-309, and RANTES) are clustered on chromosome 17 (Mehrabian et al., 1991; Irving et al., 1990; Donlon et al., 1990; Miller et al., 1990).

IL-8 gene consists of four exons and three introns with a single "CAT"- and "TATA"-like structure (Mukaida et al., 1989). The 5'-flanking region of the IL-8 gene contains potential binding sites for several transcriptional factors, such as AP-1, AP-2, octamer motif, NF-kB, IFN regulatory factor-1, glucocorticoid receptor, and hepatocyte nuclear factor-1. The structure of exon-intron junctions resembles each other among the members of chemokine family, suggesting that the members of this family have diverged from a common ancestral gene.

IL-1 and TNF as well as LPS are strong inducers of IL-8 production in a variety of cells. IL-8 production is regulated both at transcriptional and at post-transcriptional levels. Molecular biological analysis on the transcription of the IL-8 gene revealed that three cis-elements in the 5'-flanking region of the IL-8 gene, AP-1 binding, C/EBP-like, and kB-like sites were required for transcriptional activation of the IL-8 gene by IL-1 or TNF. Interestingly, the relative importance of these three sites differs among cell types. In fibrosarcoma 8387 cell line and glioblastoma cell line T98G, C/EBP-like and kB-like sites are necessary for IL-8 gene transcription (Mukaida et al., 1990), whereas in gastric cancer-derived MKN-45 cell line and human T cell leukemia cell-derived Jurkat cell line, AP-1 binding and kB-like sites are indispensable (Yasumoto et al., 1992). In any case, the IL-8 kB-like site is essentially involved in IL-8 gene activation.

Synergistic action between TNF and INF-γ was observed in several immune responses. Although INF–γ, by itself, could not induce the IL-8 production, it augmented TNF-induced IL-8 production in MKN-45 cells (Yasumoto et al., 1992). INF–γ did not change the number or affinity constant of TNF receptors on MKN-45 cells, indicating that the synergism is a post-receptor event. The expression of chloramphenicol acetyl transferase (CAT) gene linked to the IL-8 promoter was synergistically induced by TNF and INF–γ, indicating that the synergism occurred at transcriptional level. Moreover, mutation of either AP-1 binding or kB-like sites abolished the synergistic induction of CAT activity. Furthermore, gel retardation assays revealed that TNF and INF–γ synergistically induced the nuclear factors bound to the AP-1 binding and kB-like site of the IL-8 gene. These findings indicate that in MKN-45 cells, INF–γ and TNF synergistically activate essential nuclear factors for the IL-8 gene transcription, leading to the synergistic activation of the IL-8 gene (Yasumoto et al., 1992).

B. Negative Regulation of IL-8 Transcription

Many anti-inflammatory agents and cytokines are reported to suppress IL-8 mRNA expression and inhibit IL-8 production. Cytokines such as IL-4, IL-10, transforming growth factor-β, and INF-α, β, γ as well as immunomodulatory drugs such as glucocorticoid, vitamin D3, and cyclosporin A inhibit the induction of IL-8 mRNA in human PBMC. We observed that an immunosuppressive drug, FK506 also suppressed the IL-8 production in a human T cell line, Jurkat cells activated with phorbolester (PMA), and calcium ionophore (ionomycin). So far, extensive studies have been done to reveal the mechanism of IL-2 gene repression by FK506 (Schreiber and Crabtree, 1992). The calcium/calmodulin-dependent protein phosphatase, calcineurin, mediates the induction of NF-AT and NF-IL2A of the IL-2 gene. In T lymphocytes, FK506 binds to FK506 binding protein (FKBP) and this FK506/FKBP complex binds to calcineurin, thus inhibiting the phosphatase activity of calcineurin and the IL-2 gene repression. In contrast to the result on the IL-2 gene, the AP-1 binding and kB-like sites were indispensable for the IL-8 gene activation in Jurkat cells as well as in MKN-45 cells. The IL-8 AP-1 binding site or kB-like site-driven transcription was inhibited by FK506, indicating that the AP-1 binding and kB-like sites of IL-8 gene are FK506-sensitive (Okamoto et al., 1994). It remains to be investigated whether FK506 inhibits IL-8 production through the calcineurin-dependent mechanism. Hence, the elucidation of the suppressive effect on the AP-1 binding and kB-like sites of the IL-8 gene by FK506 may bring new insights into the understanding of immunosuppressive action of FK506.

A glucocorticoid, widely used as an anti-inflammatory agent, inhibits gene transcription of several cytokines including IL-8 and MCAF. Neither mutation nor deletion of AP-1 site in the IL-8 gene abolished the IL-1 induced transcription and the gene repression by dexamethasone (DEX), indicating that AP-1 site is not a main target of DEX (Mukaida et al., 1994) as opposed to the case of glucocorticoid-mediated collagenase gene repression. Subsequent studies revealed that a glucocorticoid interfered with the binding of the most essential transcription factor, NF-kB, to its cognate cis-element without inhibiting the nuclear translocation of the factor, thereby inhibiting IL-8 gene transcription.

INF-β and -γ also inhibited IL-8 mRNA accumulation after induction of TNF, IL-1, or poly(I)·poly(C) in normal fibroblasts (Oliveira et al., 1992). Further analysis defined that both NF-kB and NF-IL6 were involved for the inhibition of IL-8 gene expression by IFN (Oliveira et al., 1994).

Molecular analyses of IL-8 gene transcription suggest that IL-8 production seems to be controlled mainly at the activation step of the transcription factor(s) bound to the NF-kB site. Considering the potential role of IL-8 in inflammatory reaction, NF-kB activation pathway will be a main target of new drugs and agents to inhibit IL-8 production.

IV. IL-8 RECEPTOR

Several groups demonstrated the presence of specific binding sites of IL-8 with a high affinity to human neutrophils. A single high affinity receptor (Kd=8 x 10^{-10} M) on peripheral neutrophils was estimated to be 67kDa and 59 kDa based on the cross-linking analysis and the binding site was calculated to be about 20,000/cell (Samanta et al., 1989). An additional receptor with different affinity was also reported, but this finding was not confirmed by others (Basemen et al., 1989; Grob et al., 1990). The competitive binding analysis was performed using chemokines including GRO and NAP-2 (Walz and Baggiolini, 1989; Walz et al., 1989). NAP-2 is a chemotactic factor that has been characterized as a degradation product of PBP and has less potent chemotactic activity to interleukin-8/NAP-1. In these experiments, GRO and, to a lesser degree, NAP-2 could displace the binding of radiolabeled IL-8 to neutrophils (Moser et al., 1991). In desensitizing experiments, stimulation of IL-8 against neutrophils showed the transient rise of intracellular free calcium and desensitization against subsequent stimulation of GRO or NAP-2. A similar desensitization was observed when the first stimulation was NAP-2 or GRO. These observations indicated the existence of two different IL-8 receptors; one with high affinity (Kd = 0.1-0.4 nM) for all three ligands, and the other with high affinity for only IL-8 and low affinity (Kd = 100nM) for GRO and NAP-2 (Baggiolini et al., 1994).

cDNAs encoding two human IL-8 receptors were independently isolated and characterized (Holmes et al., 1991; Murphy and Tiffany, 1991). These two types of receptors are named as IL-8R type A (IL-8RA) and IL-8R type B (IL-8RB), respectively. The cDNA of IL-8RA encodes a mature protein with 350 amino acids and the COS cells transfected with this clone showed the specific binding to IL-8 with Kd = 3.6nM. On the other hand, Murphy reported that cDNA encoded 360 amino acid proteins of IL-8RB. They exhibit 77% amino acid sequence homology with each other and about 30% homology with the receptors for other chemotactic factors such as fMLP, C5a, and PAF. From the deduced amino acid sequence, both types of IL-8 receptors are presumed to have seven transmembrane domains and to be members of the G-protein coupled receptor family. At the beginning, the affinity of IL-8RB was presumed to be lower than that of IL-8RA. However, subsequent studies demonstrated that IL-8 bound to both types of receptors with a high affinity (Kd = 2 nM), whereas GRO bound to IL-8RA with low affinity (Kd = 450 nM), but to IL-8RB with high affinity (Lee et al., 1992a). The cells transfected with the IL-8RA exhibited the increase in intracellular calcium concentration only in reponse to IL-8, whereas IL-8RB transfectant exhibited the increase equally well in response to GRO and IL-8. The experiments on chimeric receptor revealed that the difference of

N-terminal domain of the IL-8 receptors was responsible for the binding affinities against IL-8 and GRO (LaRosa et al., 1992).

IL-8 receptors in other species have also been cloned and the binding properties of these receptors were examined. Rabbit F3R, previously claimed to be a rabbit fMLP receptor, has turned out to be the rabbit IL-8 receptor (Beckmann et al., 1991; Lee et al., 1992b). Rabbit neutrophils bound radiolabeled human IL-8 and human NAP-2 but neither radiolabeled human GRO nor human PF4. These results could explain why rabbit neutrophils mobilized intracellular calcium in response to human IL-8 and human NAP-2 but neither to human GRO nor human PF4. COS-7 cells transfected with rabbit F3R clone bound radiolabeled human IL-8 but not other ligands such as human NAP-2, GRO, or PF4 and the binding were displaced only by unlabeled IL-8 (Thomas et al., 1994). These data suggest that the receptor encoded by F3R cDNA specifically binds human IL-8 and the rabbit neutrophils possess an additional type of IL-8 receptor. This F3R is similar to the human IL-8RA that shows a high-affinity binding capacity to human IL-8 and a low affinity binding capacity to GRO. Recently, a novel rabbit IL-8 receptor, distinct from F3R cDNA, was isolated and moleculary cloned (Prado et al., 1994). This cDNA clone (5B1) encodes 358 amino acids and exhibits 80% amino acid homology with the human IL-8RB, 74% with the rabbit IL-8RA, and 73% with human IL-8RA. 5B1 is similar to human IL-8 RB since both receptors can bind NAP-2 with a similarly high affinity. However, 5B1 cannot bind GRO with a high affinity in contrast to human IL-8 RB. It remains to be investigated whether 5B1 can bind rabbit GRO with a high affinity.

A mouse homologue of IL-8 receptor was independently cloned by several groups including us (Cerretti et al., 1993; Suzuki et al., 1994; Harada et al., 1994). The cDNA encodes 359 amino acids with seven putative transmembrane portions similar to human IL-8RA, IL-8RB, and rabbit IL-8R. Its amino acid sequence shows 64 and 69% homology to those of human IL-8RA and IL-8RB, respectively (Figure 2). COS-7 cells transfected with this cDNA bind human IL-8, but do not bind MCAF/MCP-1 (Harada et al., 1994). Further studies demonstrated that transient expression of the isolated cDNA binds only to KC and MIP-2 with a high affinity (~5 nM), suggesting that these two chemokines are the ligands of this receptor (Bozic et al., 1994). The interspecific back-cross analysis localized that this gene to chromosome 1 in an adjacent position to the *Lsh-Ity-Bcg* gene that was presumed to be involved in host resistance to infections with intracellular pathogens (Morris et al.,

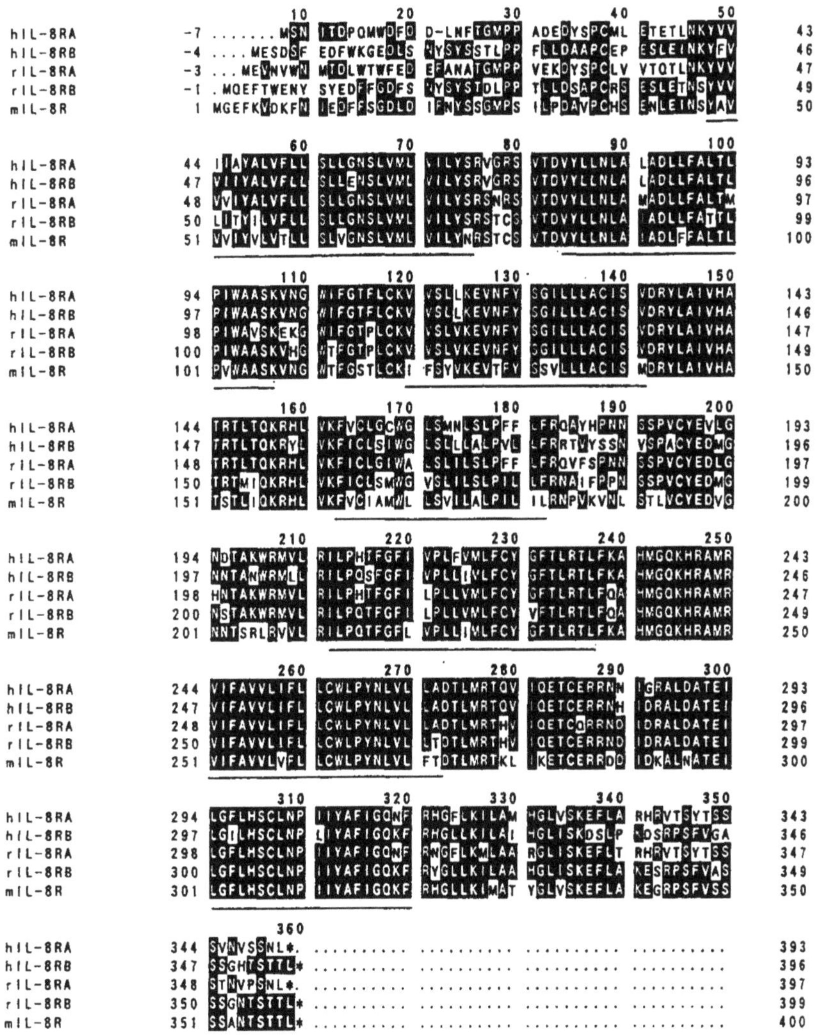

Figure 2. Comparison of amino acid sequences of human IL-8 receptor (hIL-8RA, hIL-8RB), rabbit IL-8 receptor (rIL-8RA, rIL-8RB), and mouse homolog of IL-8 receptor (mIL-8R). Identical resides are highlighted. The seven transmembrane domains are underlined.

1992). Although the product of the *Lsh-Ity-Bcg* locus is distinct from IL-8 receptor, it cannot be completely excluded that the IL-8 receptor may participate in host resistance to infections.

Genomic analysis of human IL-8 receptors identified the existence of pseudogene of IL-8RA (IL-8RAP), and all three genes, IL-8RA, IL-8RB, and IL-8RAP, form a cluster of human genes on chromosome 2q34-q35 (Ahuja et al., 1992).

Except at the amino terminal portion, human IL-8RA and IL-8RB show more than 80% sequence similarity with each other at the amino acid level. The polyclonal monospecific antibodies against the N-terminal portion of each type of IL-8 receptors generated by us failed to inhibit the binding of IL-8 to receptors, suggesting that the other extracellular portion of the receptor is also involved in ligand binding. The extensive mutation studies of IL-8RA by replacing extracellular acidic residues of the receptor with alanine revealed that the acidic residue, Asp11, in N-terminal portion and two other residues, Glu275 and Arg280, in the third extracellular portion of the receptor were critical for ligand binding (Hébert et al., 1993). These three residues come close to each other by the formation of a disulfide bridge between Cys30 and Cys277.

Immunofluorescence analysis of human peripheral blood leukocytes revealed that a majority of monocytes and CD16+ natural killer (NK) cells were stained with polyclonal antibodies that specifically recognized the NH2-terminal region of each type of IL-8 receptors (Morohashi et al., 1995). IL-8 receptor mRNA is expressed at high levels even in resting neutrophils but at low levels in monocytes (Baggiolini et al., 1994). Low levels of IL-8RA mRNA were detected in the myeloid cell lines, HL-60 and AML 193, but no significant levels of IL-8RA mRNA were detected in THP-1 cells, lymphocytes, or Jurkat cells. On the other hand, IL-8RB transcripts were found in all blood cells. IL-8RA was also detected on non-leukocytic cells such as melanoma, melanocyte, and cytomegalovirus(CMV)-infected human embryonic fibroblasts (Moser et al., 1993; Murayama et al, 1994). These data suggest that regulation of gene expression for the two receptors is different. Recent analysis of the genomic organization of both types of IL-8 receptor genes revealed that the open reading frame was transcribed from a single exon, but the organization of 5'-untranslated regions of IL-8RA and IL-8RB gene were completely different. IL-8RA gene has one additional exon in the 5'-flanking region (Ahuja et al., 1994; Sprenger et al., 1994a). Two types of mRNA, 2.0 and 2.4 kb in length, for IL-8RA are expressed due to different polyadenylation signals. This genomic structure of IL-8RA is similar to other members of the rhodopsin superfamily. The genes of C5a receptor (Gerard et al., 1993), the fMLP receptor (Murphy et al., 1993), and the dopamine D1A receptor (Minowa et al., 1992) have generally one or two additional exons in their 5'-untranslated regions. In contrast, the IL-8RB gene consists of 11 exons and seven distinct mRNA are expressed by alternative splicing of these exons (Sprenger et al., 1994b). The proximal region of exon 1 of IL-8RA gene and exon 3 of IL-8RB gene have limited sequence similarity but had strong constitutive promoter activity in CAT assay (Ahuja et al., 1994).

In addition to these IL-8 receptors on leukocytes, a promiscuous chemokine receptor on erythrocytes has been identified (Darbonne et al., 1991). This receptor has a broad spectrum of binding capacity for chemokines including IL-8, GRO, MCP-1/MCAF, and RANTES (Neote et al., 1993). The molecular mass of the erythrocyte chemokine receptor is a 39kDa glycosylated protein that does not couple to G protein (Horuk et al., 1993a). Furthermore, the molecule on erythrocytes has been identified as Duffy blood group antigen, which was the erythrocyte receptor for the malaria parasite, *Plasmodium vivax* (Horuk et al., 1993b). Recently, Duffy antigen glycoprotein D cDNA has been cloned. This clone composed of 338 amino acid residues in length and a molecular mass was about 36 kDa. The protein is highly hydrophobic and it appears to traverse the membrane seven times based on data derived from hydropathy plots (Horuk, 1994). Both Duffy-positive and negative individuals possess this gene. Duffy negative ones do not express this receptor on erythrocytes although the precise molecular defect has not been identified (Chaudhuri et al., 1993). Radiolabeled IL-8, GRO, MCP-1, and RANTES bound to human embryonic kidney cells transfected with this cDNA and displaced heterologous chemokines as observed previously on red blood cells. However, this failed to transmit an intracellular signal in response to chemokines on transfected cells (Neote et al., 1994). This receptor is expressed in a wide variety of organs such as the lung, muscle, spleen, heart, pancreas, and brain. Although the precise physiological role of this receptor in these organs is not known at this moment, it is presumed to be a "scavenger receptor" for chemokines in blood and other tissues.

V. SIGNAL TRANSDUCTION OF IL-8 RECEPTOR

Recent cloning of IL-8 receptors demonstrated that they belong to a G-protein coupled receptor family. Pretreatment of neutrophils with

Bordetella pertussis toxin abolished the increase of IL-8-induced cytosolic free Ca^{2+} while respiratory burst elicited by IL-8 was blocked by inhibitors of protein kinase C (Thelen et al., 1988), indicated that IL-8 receptor signals are mediated by a guanine nucleotide-binding protein (G-protein) and protein kinase C. In general, the heteromeric G proteins are composed of three subunits α, β, and γ. In a resting state, GDP binds the α-subunit of G proteins. Ligand-receptor interaction replaces a-subunit with GTP and dissociates α-GTP from β- and γ-subunits. The G protein α-subunit then activates the enzyme phospholipase C, which generates inositol triphosphate (IP_3) and diacylglycerol. IP3 induces the release of Ca^{2+} from intracellular stores while diacylglycerol leads to the activation of protein kinase C. IL-8 induces GTP binding and GTPase activity in human neutrophils (Kupper et al., 1992). Recent investigations of signal pathway in response to IL-8 and GRO show that IL-8 signals are mediated by the free βγ subunits of the heterotrimeric Gi protein and that free βγ subunits could specifically activate the β2 isoform of phospholipase C (Wu et al., 1993). Furthermore, IL-8 activates phosphatidylinositol-4-phosphate kinase that synthesizes phosphatidylinositol-4,5-biphosphate (Pike et al., 1992), thereby inducing formation of IP3. However, the precise molecular mechanism of IL-8 signal transmission remains to be investigated.

VI. BIOLOGICAL ACTIVITIES OF IL-8

A. *In Vitro* Activities of IL-8

Isolation and purification of IL-8 were based on its ability to cause chemotaxis and degranulation of neutrophils. Subsequent studies on the biological functions of IL-8 revealed the various effects against numerous types of cells other than neutrophils (Table 1).

IL-8 stimulates a number of function of human neutrophils including induction of shape change (Thelen et al., 1988), release of lysosomal enzymes such as myeloperoxidase, α-mannosidase and β-gluconidase (Peveri et al., 1988), induction of respiratory burst (Schröder et al., 1987), adhesion to endothelium (Detmers et al., 1990), generation of superoxide and hydrogen peroxide (Thelen et al., 1988), and generation of bioactive lipids (Schröder, 1989). IL-8 has been shown to increase the expression of adhesion molecules and complement receptor type 1 on neutrophils (Paccaud et al., 1990). IL-8 increases the adhesion of neutro-

Table 1. *In Vitro* Biological Functions of C-X-C Chemokines

Chemokines	*Target cells*	*Effect*
IL-8	Neutrophils	Chemotaxis
		Shape change
		Lysosomal enzyme release
		Respiratory burst
		Increased cytosolic free Ca2+
		Generation of superoxide and hydrogen peroxide
		Generation of bioactive lipids
		Up-Regulation of adhesion molecules (CD11a, CD11b, CD11c, and CD18)
		Increased adhesion to unstimulated endothelial cells
		Transendothelial migration
		Activation of 5-lipoxygenase
		Release of leukotriene B4 and 15-HETE
		Growth inhibition of Candida albicans
	T cells	Chemotaxis
	B cells	Inhibition of IL-4-induced Ig E production
	Basophils	Chemotaxis
		Inhibition of histamine release
		Increased leukotriene release
	Monocytes	Intracellular calcium influx
		Respiratory burst
	Keratinocytes	Proliferation
	Fibroblasts	Decrease of collagen mRNA expression
		Enhancement of cytomegalovirus replication
	Endothelial cells	Blood vessel proliferation
	Melanoma cells	Haptotactic response
	Smooth muscle	Chemotaxis
	Synovial fibroblats	Inhibition of collagen expression
	Huvec cells	Chemotaxis
	Stem cells	Suppression of colony formation of immature myeloid progenitors
GROα	Neutrophils	Chemotaxis
		Lysosomal enzyme release
	Monocytes	Intracellular calcium influx
		Respiratory burst
	Fibroblasts	Growth stimulatory activity
		Decrease of collagen mRNA expression
	Melanoma cells	Growth stimulatory activity
	Synovial fibroblats	Inhibition of collagen expression
GROβ, GROγ	Neutrophils	Chemotaxis
		Shape change
		Induction of cytosolic free Ca2+
		Degranulation
	Basophils	Chemotaxis

(continued)

Table 1. (Continued)

Chemokines	*Target cells*	*Effects*
	Fibroblasts	Growth stimulatory activity
	Stem cells	Suppression of colony formation of immature Myeloid progenitors
PF-4	Neutrophils	Chemotaxis
	Monocytes	Chemotaxis
	Fibroblasts	Chemotaxis
	Stem cell	Inhibition of megakaryocytopoiesis
	T cells	Immunostimulant
	Endothelial cells	Increased ICAM-1 expression
	Chorioallantoic membrane	Inhibition of angiogenesis
	Fetal rat bone	Inhibition of bone resorption
ENA 78	Neutrophils	Chemotaxis Shape change Induction of cytosolic free Ca2+ Degranulation
NAP-2	Neutrophils	Chemotaxis
	Basophils	Induction of cytosolic free Ca2+ Induction of histamine release
IP-10	T cells	Chemotaxis
	Monocytes	Chemotaxis
	NK cells	Chemotaxis

phils to unstimulated human umbilical cord vein-endothelial cells *in vitro*, whereas it inhibits the binding of neutrophils to endothelial cells pre-activated by IL-1, TNF, or LPS (Gimbrone et al., 1989). Furthermore, IL-8 enhances transendothelial migration by inducing rapid shedding of leukocyte adhesion molecule-1 (LECAM-1; Huber et al., 1991). IL-8 also enhances the growth inhibitory activity of neutrophils to *Candida albicans* (Djeu et al., 1990).

Besides the effects on neutrophils, IL-8 induces chemotaxis against a wide variety of cells. IL-8 induces the chemotaxis of $CD4^+$ or $CD8^+$ human peripheral T lymphocytes (Larsen et al., 1989a; Leonard et al., 1990; Taub et al., 1993a; Oppenheim 1993). These lymphocytes respond to 10 times lower concentrations of IL-8 than they do to neutrophils. In contrast to IL-8, another member of C-X-C chemokines, IP 10 induces chemotaxis of stimulated, but not unstimulated, human peripheral blood T lymphocytes as well as monocytes (Taub et al., 1993b). IL-8 also has chemotactic activities against basophils. Moreover, at high concentrations, IL-8 causes histamine and leukotriene release from human basophils when the cells are pretreated with IL-3 or GM-CSF whereas it inhibits histamine release from basophils at lower concentrations (White et al., 1989; Dahinden et al., 1989; Kuna et al., 1991).

Previously, IL-8 was reported to induce eosinophil chemotaxis pretreated with IL-3 or GM-CSF (Warringa et al., 1991). Recent observation has demonstrated that human eosinophils showed significant migratory responses toward IL-8 at concentration 1-100 nM only after being primed with IL-5. Unprimed eosinophils only showed a significant migratory response toward RANTES, but not toward IL-8 (Schweizer et al., 1994). Moreover, circulating eosinophils from patients with atopic dermatitis and allergic asthma but not from normal volunteers exhibits an increased migratory response toward IL-8 (Bruijnzeel et al., 1993; Warringa et al., 1993). These observations support the notion that eosinophils from allergic patients may be primed to obtain high responsiveness to locally produced chemokines such as IL-8 or RANTES, thus migrating into the allergic inflammatory locus.

Monocytes produce various cytokines including C-X-C and C-C chemokines and are highly responsive particularly to C-C chemokines such as MCAF/MCP-1, RANTES, I-309, MCP-2, and MCP-3. *In vitro* biological functions of these C-C chemokines are summarized in Table 2. Several studies demonstrated that monocytes did not exhibit chemotaxis in response to IL-8, but showed slight intracellular calcium influx and repiratory burst in response to this chemokine (Walz et al., 1991). However, IL-8 induces the chemokinesis of lymphokine-activated killer cells (Sebok et al., 1993) and inhibits the production of immunoglobulin E by human B cells stimulated with IL-4 (Kimata et al., 1992).

IL-8 also acts on non-leukocytic tissue cells such as melanoma cells, fibroblasts, and human umbilical vein endothelial cells *in vitro*. IL-8 is haptotactic and promotes adhesion of melanoma cells. Primary and metastasized melanoma cells constitutively produced IL-8 as well as GRO when they were cultured *in vitro* (Wang et al., 1990; Kuna et al., 1991). These observations suggest the possible involvement of IL-8 or IL-8-related molecules in the progression of melanoma. Recent studies demonstrated that human IL-8 could induce the migration of human umbilical vein endothelial cells *in vitro* and the angiogenesis in the rat cornea (Koch et al., 1992; Hu et al., 1993). These observations were confirmed by the fact that the neutralizing antisera to IL-8 could attenuate both *in vitro* endothelial cell chemotaxis and *in vivo* cornea neovascularization (Smith et al., 1994). IL-8 and GROa could inhibit collagen expression in synovial fibroblasts (Unemori et al., 1993). Moreover, IL-8

Table 2. *In Vitro* Biological Functions of C-C Chemokines

Chemokine	*Target cells*	*Effect*
MCAF/MCP-1	Monocytes	Chemotaxis Intracellular calcium influx respiratory burst Expression of b2-integrins (CD11b/CD18 and CD11c/CD18) Production of IL-1 and IL-6 Augmentation of cytostatic activity N-acetyl b-glucuronaminidase release
	T cells	Chemotaxis
	Basophils	Histamine release Leukotriene release Chemotaxis Intracellular calcium influx
	Stem cells	Suppression of colony formation of progenitor cells
MCP-2	Monocytes	Chemotaxis
	T cells	Chemotaxis
MCP-3	Monocytes	Chemotaxis
	T cells	Chemotaxis
	Basophils	Chemotaxis Release of histamine Release of leukotriene
	Eosinophils	Chemotaxis
MIP-1α	Neutrophils	Chemotaxis Oxidative burst Degranulation Lysosomal enzyme release
	Monocytes	Chemotaxis Intracellular calcium influx
	T cells	Chemotaxis for CD8+ T lymphocytes
	Basophils	Chemotaxis Release of histamine
	Eosinophils	Chemoaxis Release of cationic protein
	Mast cells	Chemotaxis
	Stem cells	Proliferation of granulocyte-macrophage progenitor cells Suppression of immature progenitor cells
MIP-1β	Monocytes	Chemotaxis Intracellular calcium influx
	T cells	Chemotaxis for CD4+ T lymphocytes Promotion of adhesion to VCAM-1
	Stem cells	Inhibition of the efect of MIP-1α
RANTES	Monocytes	Chemotaxis Intracellular calcium influx
	T cells	Chemotaxis for CD4+ memory T cells

(Continued)

Table 2. (Continued)

Chemokine	*Monocyutes*	*Effect*
	Basophils	Chemotaxis
		Release of histamine
	Eosinophils	Chemotaxis
		Respiratory burst
		Release of cationic protein

also enhances replication of cytomegalovirus in human fibroblasts *in vitro* (Murayama et al., 1994).

A number of cytokines can directly or indirectly have stimulating, enhancing, and/or suppressing effects on stem cells or progenitor cells. Several chemokines have been implicated in negative regulation of hematopoiesis. An inhibitor of stem cell proliferation turned out to be identical to macrophage inflammatory protein (MIP)-1α, a member of C-C chemokines (Graham et al., 1990). Further investigation revealed that MIP-1α, but not MIP-1β, inhibits the proliferation of murine day 12-CFU-S and CFU-A both *in vitro* and *in vivo* (Graham et al., 1990; Broxmeyer et al, 1991). In humans, MIP-1α reversibly inhibits the proliferation of CFU-granulocyte, erythrocyte, macrophage, megakaryocyte (GEMM), burst-forming unit-erythroid (BFU-E), and CFU-GM, whereas more mature progenitor cells such as CFU-E, CFU-G, and CFU-M are stimulated (Broxmeyer et al., 1989). Moreover, MIP-1α, MIP-2α, PF4, IL-8, and MCAF also suppress colony formation of immature subsets of myeloid progenitor cells stimulated by GM-CSF and steel factor (Broxmeyer et al., 1993). In contrast, MIP-1β, MIP-2β, GROα, NAP-2, and RANTES were not suppressive nor did they synergize with other chemokines. Furthermore, MIP-1α, in combination of IL-3, maintained long-term bone marrow cultures (LTBMC) for at least eight weeks, suggesting that MIP-1α may have an important role in maintaining immature hematopoietic progenitors in quiescent state (Verfaille et al., 1994).

B. *In Vivo* Activities of IL-8

An extensive effort has been made to elucidate the pathophysiological roles of IL-8 *in vivo* over the past several years. To make animal models for the study of IL-8, IL-8 homologs in other species have been moleculary cloned including guinea pigs, sheep (Yoshimura and Johnson, 1993), rabbits (Yoshimura and Yuhki , 1991), pigs (Goodman et al., 1992), and

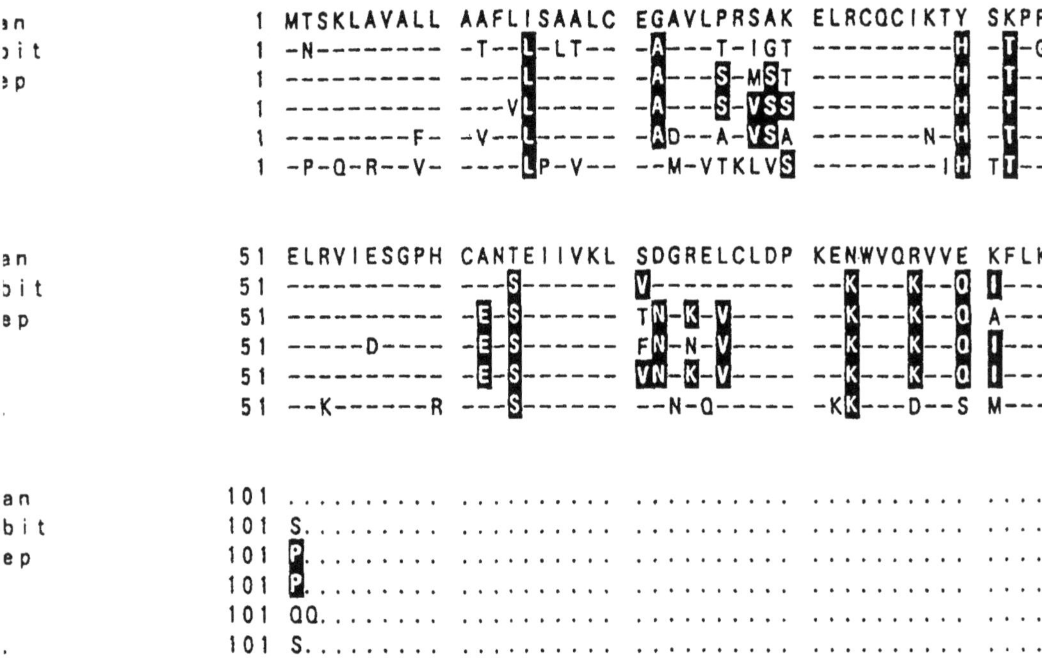

3. Comparison of amino acid sequences of human, rabbit, sheep, dog, pig, and guinea pig (G.P.) IL-8. Ide dashed lines.

dogs (Ishikawa et al., 1993; Figure 3). However, IL-8 homolog in rat or mouse has not been identified so far (Yoshimura and Johnson, 1993).

Intradermal injection of human IL-8 induces massive local neutrophil infiltration in mice (Furuta et al. 1989), rats (Larsen et al., 1989a, 1989b), rabbits (Van Damme and Opdenakker, 1990), dogs (Thomsen et al., 1991), sheep (Cloditz and Watson, 1992), and humans (Leonard et al., 1991; Swensson, et al., 1991). These data demonstrate that intradermal injection of human IL-8 induces a rapid and concentration-dependent neutrophil infiltration. In rats, both IL-8 and GROα induce similar neutrophil infiltration, whereas NAP-2 is significantly less active. However, in the rabbit skin, IL-8 and NAP-2 are of identical potency (Zwahlen et al., 1993). Subsequent studies show that intradermal injection of human IL-8 in rabbits causes neutrophil dependent plasma leakage at injected sites Foster et al. 1989; Rampart et al., 1989). Neutrophil migration into the injected site is rapid, reaching maximum levels within 30 min after injection and lasting for 8-10 (Colditz et al., 1990). The action of IL-8 lasted much longer than other chemotactic factors such as fMLP, C5a, PAF, or LTB4. Intracutaneous injections of fMLP showed the maximum neutrophil infiltration at 1 h after the injection and declined to baseline by 4 h. PAF and LTB4 induced transient neutrophil infiltration that diminished to background levels within 4-6 h after injection (Colditz and Movat, 1984; Movat et al., 1984). These data suggest that IL-8 is rather resistant to inactivaion by proteolytic enzyme and the clearance from the injected site is very slow. This may be due to the biochemical characteristics of IL-8 that has many positively charged amino acid side chains on the surface and can interact with acidic extracellular matrix components (Hébert and Baker, 1993).

Intradermal injection of human IL-8 into human subjects causes essentially similar histological changes observed in rabbits. The repeated intravenous injection of IL-8 in rabbits causes accumulation of activated neutrophils into the lung, where inflammatory and proliferative changes occur (Zwahlen et al., 1993). Furthermore, several reports demonstrated that other types of cells besides neutrophils were also induced to infiltrate by local administration of IL-8 *in vivo*. In rats, human IL-8 also induces lymphocyte infiltration at lower concentrations (Larsen et al., 1989a). The time course of neutrophil and lymphocyte infiltration induced by IL-8 is different (Kudo et al., 1991). The peak response in neutrophil recruitment was observed soon after subcutaneous injection of IL-8 whereas lymphocytes infiltrated slowly with the peak extravasation occurring several hours after IL-8 injection. In sheep, intradermal injection of human IL-8 causes the infiltration of a small number of lymphocytes (Colditz et al., 1992). Injection of porcine alveolar macrophage chemotactic factor-I (ACMF-I), pig IL-8, into porcine lung showed about 25-fold increase in lymphocytes in the bronchoalveolar lavage fluid (Goodman et al., 1992). Recent studies demonstrated that human IL-8 induced massive T lymphocytes infiltration in mice with severe combined immunodeficiency (SCID) engrafted with human peripheral blood lymphocytes (Taub and Oppenheim 1994). In guinea pigs, intraperitoneal injection of recombinant human IL-8 induced T cells and eosinophils to infiltrate into lungs. Although this may not be the direct proof of chemotactic activity of human IL-8 against these cells, IL-8 may play an important role in regulating eosinophil recruitment as well as neutrophils. Furthermore, also in guinea pigs, intradermal injection of human IL-8 induces dose-dependent accumulation of intravenously-administered [111In]eosinophils into the skin sites as well as neutrophils (Collins et al., 1993). The accumulation of eosinophils in response to IL-8 becomes apparent from 1-2 h after intradermal injection of IL-8. This delay may suggest that the action of IL-8 against eosinophils is indirect, due to the production of secondary eosinophil chemotactic factor(s).

Among the members of chemokine family, only members of C-C chemokines, RANTES, and MIP-1α have been described to have *in vitro* chemotactic activities against eosinophils, monocytes, and T lymphocytes (Kameyoshi et al., 1992; Rot et al., 1992; Taub et al., 1993a; Schall et al., 1990). Although human RANTES does not evoke any dermal response in rat and rabbit, intradermal injection of RANTES in dogs resulted in a dose- and time-dependent eosinophil infiltration at the injection site. In contrast, human MIP-1α elicits no response in canine dermis. This may be explained by the observation that the RANTES/MIP-1α receptors on human THP-1 cells bind both RANTES and MIP-1α and compete each other (Gao et al., 1993), whereas the canine RANTES receptor has no measurable binding affinity for human MIP-1α. These observations, further, suggested that *in vivo* studies using exogenous cytokines of other species should be performed only after the target receptors against the exogenously added ligand were characterized functionally (Meurer et al., 1993). Table 3 summarized *in vivo* activities described so far for both C-X-C and C-C chemokines.

Systemic administration of IL-8 in baboons, either as a bolus or as a continuous intravenous infusion, caused a rapid, transient granulocytepenia followed by granulocytosis that persists as long as IL-8 levels

Table 3. *In Vivo* Biological Functions of C-X-C and C-C Chemokines

Chemokine	*Application site*	*Effect*
IL-8	Intradermal injection	Neutrophil recruitment in mice, rats, rabbits, dogs, sheep, and humans
		Plasma exudation in rabbits
		T cell infiltration in rats
		Neutrophils and i.v. administered [111In] eosinophil
		Recruitment in guinea pigs
		Human T cell infiltration in SCID mice
	Repeated i.v. injection	Lung and tissue damage in rabbits
	I.v. administration	Inhibition of neutrophil accumulation in dermis
		Neutrohilia in baboons
	Lung instillation	Lymphocytes infiltration in BALF in pigs
	Intraarticular injection	Neutrophil infiltration in rabbits
	Transgenic mice	Neutrophil accumulation in the lung, liver, and spleen without tissue damage
		Decreased L-selectin expression on peripheral neutrophils
		Normal L-selectin expession on bone marrow neutrophils
		No-up-regulation of β2-integrin on peripheral neutrophils
	Corneal bioassay	Angiogenesis in rat cornea
GROα	Intradermal injection	Neutrophil accumulation in rat and rabbit
GCP-2	Intradermal injection	Granulocyte infiltation in rabbits
PF-4		Neutrophil and mononuclear cel infiltration in mice
		Induction of fibrosis
		Inhibits angiogenesis
IP-10		Human T cell infiltration in SCID mice
		Anti-tumor effect
MCAF/MCP-1	Subcutaneous injection	Mmonocyte infiltration in mice, rabbits, and rats
		Human T cell infiltration in SCID mice
MCP-2	Intradermal injection	Monocyte infiltration in rabbits
MCP-3	Intradermal injection	Monocyte infiltration in rabbits
MIP-1α	Subcutaneous injection	Monocyte infiltration in rabbits
		Non-prostaglandin mediated fever induction
		Human T cell infiltation in SCID mice
MIP-1β	Subcutaneous injection	Monocyte infiltration in rabbit
		Non-prostaglandin mediated fever induction
		Human T cell infiltation in SCID mice
RANTES	Intradermal injection	Eosinophil and monocyte infiltration in mice
		Eosinophil infiltration in dog
		Human T cell infiltation in SCID mice

were detectable in the circulation (Van Zee et al., 1992). IL-8, however, failed to induce TNFα, IL-1β, or IL-6 production, hemodynamic change, or acute organ damage.

IL-8 transgenic mice also exhibit accumulation of neutrophils in the microcirculation of the lung, liver, and spleen without neutrophil infiltration into any tissues, plasma exudation, and apparent tissue damage (Simonet et al., 1994). Interestingly, L-selectin expression is decreased on the surface of circulating neutrophils in these transgenic mice, whereas that on bone marrow neutrophils or neutrophil precursors was normal. Furthermore, no significant up-regulation of the β2-integrins, Mac-1, and LFA-1 was observed on peripheral blood neutrophils in these transgenic mice. These results indicate directly that IL-8 modulates the surface expression of L-selectin without showing any effects on β2-integrin levels. Neutrophil migration into inflamed peritoneal cavity was also inhibited in transgenic mice, which is consistent with the previous observations that intravenous administration of IL-8 inhibited neutrophil accumulation to intradermal sites of inflammation (Hechmann et al., 1991). More detailed studies using these mice will clarify the putative role of IL-8 in tissue damage in various inflammatory responses.

VII. PATHOPHYSIOLOGY OF IL-8

Since the discovery of IL-8, there has been a remarkable number of studies undertaken concerning the pathophysiological role of IL-8 *in vivo*. Neutralizing antibody against the ligand, receptors, or targeting of these genes by homologous recombination may be an essential experimental approach for the understanding of the pathophysiological roles of the ligand *in vivo*.

We generated mouse anti-human IL-8 monoclonal antibodies and confirmed that one of the antibodies inhibited the binding of recombinant rabbit IL-8 to rabbit peritoneal neutrophils with 50% inhibition at a molar ratio of nearly 1:1. The anti-IL-8 antibody also inhibited the chemotactic activity of recombinant rabbit IL-8 to rabbit peritoneal neutrophils. Using this specific antibody against rabbit IL-8, we explored the critical role of endogenously produced IL-8 in inflammatory reactions in rabbits.

Skin redness is an important sign of cutaneous inflammation. Intradermal injection of LPS induced local skin redness with massive neutrophil infiltration (Harada et al., 1993). Administration of the anti-IL-8 antibody significantly prevented the neutrophil infiltration until 6 h after LPS-injection. Moreover, it inhibited skin redness at the LPS injected sites at 24 h after LPS injection when neutrophil infiltration was not suppressed. These observations suggest that IL-8 is essentially involved in the activation of neutrophils as well as neutrophil infiltration.

Neutrophil infiltration into joint cavities has been observed during acute flares in rheumatoid arthritis (RA) and the infiltrated cells amplify the inflammatory response, thus enhancing tissue damage. Intra-articular injection of LPS or IL-1 in rabbits causes rapid and remarkable neutrophil infiltration into the knee joints followed by tissue destruction of synovium, thickening of synovial lining, and edematous change of subsynovial tissue. Intra-articular injection of IL-8 in rabbits also causes synovial lining cells become ovoid, pleomorphic, and multilayered with neutrophil infiltration, exhibiting similar pathological changes of rheumatoid arthritis (Endo et al., 1991). Simultaneous intra-articular injection of LPS or IL-1 together with anti-IL-8 antibody significantly reduces neutrophil infiltration, joint redness, limpness, and prevents damage of synovial membrane (Akahoshi et al., 1993), suggesting that IL-8 produced by LPS or IL-1 is essentially involved in neutrophil-dependent tissue damage in this model.

Accumulation of immune complexes in renal glomerulus, "electron-dense deposits," is the hallmark of immune complex-type glomerulonephritis. Neutrophils are the major infiltrated cells at the acute phase of this type of nephritis. Repeated injection of crystallized bovine serum albumin (BSA) with complete Freund's adjuvant into rabbits induces a significant number of neutrophils infiltrated into glomeruli followed by fusion of epithelial foot process, resulting in marked urinary excretion of protein and albumin with concomitant increase of urinary IL-8 levels. An anti-IL-8 antibody, though partially inhibited neutrophil infiltration into glomeruli, completely prevented the fusion of epithelial foot process as well as excretion of protein and albumin into urine (Wada et al., 1994a), suggesting that infiltrated neutrophils are the main contributors to the renal dysfunction and that IL-8 is essentially involved in the activation as well as infiltration of these neutrophils into glomeruli.

Reperfusion of blood flow into ischemic organ causes severe tissue injury with massive neutrophil sequestration. This ischemia-reperfusion injury is observed in several clinical conditions such as myocardial infarction, cerebral infarction, and multiple organ failure. In the lung reperfusion injury model in rabbits, increased numbers of neutrophils in bronchoalveolar lavage fluid (BALF), massive neutrophil infiltration in interstitium, and destruction of pulmonary architecture were observed as well as severe pulmonary edema. IL-8 was produced by bronchiolar ciliary cells and alveolar macrophages as revealed by immunohistochemical analyses. An anti-IL-8 antibody inhibited neutrophil infiltration into lung tissue and the increase of neutrophils in BALF. Moreover, anti-IL-8 treatment inhibited the destruction of pulmonary architecture (Sekido et al., 1993). Thus, IL-8 is a major and essential mediator responsible for neutrophil recruitment and neutrophil-dependent tissue damage observed in this reperfusion injury model.

All of these results provide the direct evidence on the essential involvement of IL-8 in establishing various inflammatory reaction *in vivo*. This assumption is further supported by the experiments of neutralization of IL-8 in a rabbit model of endotoxin-induced pleurisy (Broaddus et al., 1994).

Several additional research groups reported on experiments using a neutralizing antibody to human IL-8 in order to investigate the role of IL-8 in other species. Mulligan et al. (1993) reported that the injection of an anti-human IL-8 antibody (DM/C7) into rats prevented glycogen-induced accumulation of neutrophils and protected lung and dermal vascular injury after deposition of immunoglobulin G immune complexes. However, they failed to identify the molecule(s) in rats which this monoclonal antibody recognized. On the contrary, only anti-CINC antibody, but not DM/C7, could diminish glomerular inflammation or proteinuria in immune complex-mediated inflammation using the rat model of anti-glomerular basement membrane (GBM)-induced nephritis (Wu et al., 1994). Identification of the specific antigen(s) recognized by DM/C7 in rats is inevitable for the accurate evaluation of the experimental results in rats or mice, in which IL-8 homolog has not been discovered.

Target disruption of the murine homolog of IL-8 receptor by homologous recombination was recently reported (Cacalano et al., 1994). Gene disruption of IL-8 receptor caused massive megakaryocyte poiesis, increase of plasma cells in bone marrow and spleen with enlargement of these organs, elevation of neutrophil number in peripheral blood, and elevation of serum IgE levels as well as dysfunction of neutrophil infiltration. However, neutrophils possess normal killing activities against bacteria. These observations suggest that IL-8 or its related molecule(s) have a wide variety of effects on physiological hematopoiesis as observed *in vitro* (Broxymyer et al., 1993; Laterveer et al., 1993) as well as inflammatory responses.

VIII. IL-8 AND HUMAN DISEASES

Most acute inflammatory diseases are associated with neutrophil infiltration in the lesions. The number of reports concerning the production of IL-8 in these diseases has been rapidly increasing since the cloning of this cytokine. IL-8 was first clinically identified in skin lesions of psoriasis (Schröder and Christopher 1986; Camp et al., 1986; Fincham et al., 1988). IL-8 is also considered to be involved in the recruitment of neutrophils into the synovial fluids of rheumatoid arthritis (Brennan et al., 1990; Peichl et al., 1991; Koch et al., 1991), osteoarthritis (Symon et al., 1992), and gout (Terkeltaub et al., 1991). The serum level of IL-8 correlates with the titers of the rheumatoid factor but not with other clinical parameters such as C-reactive protein (CRP) or erythrocyte sedimentation rate (ESR; Rampart et al., 1992). Although serum IL-8 levels also increased in acute myocardial infarction (Abe et al., 1993), there was no apparent relevance to the clinical severity of the disease.

Several independent research groups reported increased IL-8 levels in bronchoalveolar lavage fluids (BALF) from patients with adult respiratory distress syndrome (ARDS; Miller et al., 1992), idiopathic pulmonary fibrosis (Carré et al., 1991; Lynch et al., 1992), and in pleural fluid of patients with empyema (Broaddus et al., 1992). Further study raised the possibility that elevated IL-8 levels in BALF at admission could predict the subsequent development of ARDS (Donnely et al., 1993). The patients with chronic airway disease by persistent infection of *Pseudomonas aeruginosa* show enhanced IL-8 production in BALF and IL-8-derived neutrophil chemotactic activity, causing neutrophil accumulation in the airway and progressive lung injuries (Oishi et al., 1994). Increased IL-8 gene expression was also observed in bronchial epithelial cells from patients with asthma (Marini et al., 1992) and cystic fibrosis (Nakamura et al., 1992). However, pathophysiological significance of IL-8 in these diseases remains to be investigated.

Increased urinary IL-8 levels were observed in patients with several types of glomerulonephritis, Ig A nephropathy, acute glomerulonephritis (AGN), purpura nephritis, membranous proliferative glomerulonephritis (MPGN), and lupus nephritis, but not in patients with focal glomerulosclerosis and membranous nephropathy (Wada et al., 1994b). The former groups are characterized pathologically by the infiltration of polymorphonuclear cells and/or mononuclear cells and proliferation of mesangial cells, whereas the latter are not. Besides these renal diseases, increased urinary levels of IL-8 were observed in most cases of urinary tract infections (Ko et al., 1993), suggesting that the increase in the number of neutrophils in urine is ascribed to local production of IL-8 in the urinary tract. From these clinical data of IL-8, detection of IL-8 in biofluids from various diseases will be one of the important indicators of the acute inflammatory reaction.

IX. CONCLUDING REMARKS

During the past several years, research on IL-8 and related chemokines *in vitro* and *in vivo* have been extensively carried out and have revealed their pivotal involvement not only in leukocyte recruitment and activation but also in hematopoiesis, angiogenesis, and immuneresponses under various pathophysiological conditions.

Based on the above-mentioned pathophysiological roles of IL-8, intervention of IL-8 becomes clinically important. Accumulating evidence indicates that IL-8 gene expression is tightly controlled to regulate the neutrophil migration and activation. IL-8 gene repression is a new way to control the leukocyte infiltration, thereby alleviating inflammation. IL-8 gene repression can be accomplished by already available immunosuppressants such as glucocorticoid, cyclosporin A, and FK506. Furthermore, since NF-kB is inevitable for the activation of IL-8 gene expression in a variety of cells, it is reasonable to target the NF-kB activation pathway to inhibit IL-8 production. We have recently developed a ligand-dependent cell-free activation system of NF-kB (LPS; Ishikawa et al., 1995), which should be useful for the screening of natural and synthetic compounds that block the activation of NF-kB without considering the cell permeability or cell toxicity of the compounds.

Direct inhibition of the action of IL-8 is also important in regulating the activation of neutrophils. Mouse monoclonal antibody to human IL-8 is one of the prototypes needed to design specific antagonists to IL-8. Humanization of this monoclonal antibody has already been achieved, and efficacy of this antibody will soon be tested in various animal inflammation models. Further investigation of the action of this factor and the subsequent development of its antagonistic agents and synthetic compounds which intervene its production, the activities, the binding to the receptors, or the receptor signaling pathway will also be another useful therapeutic approach for acute inflammatory diseases.

REFERENCES

Abe, Y., Kawakami, M., Kuroki, M., Yamamoto, T., Fujii, M., Kobayashi, H., Yaginuma, T., Kashii, A., Saito, M., & Matsushima, K. (1993). Transient rise in serum interleukin-8 concentration during acute myocardial infarction. Brit. Heart J. 70, 132-134.

Ahuja, S.K., Özçelik, T., Milatovitch, A., Francke, U., & Murphy, P.M. (1992). Molecular evolution of the human interleukin-8 receptor gene cluster. Nature Genet. 2, 31-36.

Ahuja, S.K., Shetty, A., Tiffany, H.L., & Murphy, P.M. (1994). Comparison of the genomic organiation and promoter function for human interleukin-8 receptors A and B. J. Biol. Chem. 269, 26381-26389.

Akahoshi, Y., Endo, H., Kondo, H., Kashiwazaki, S., Kasahara, T., Mukaida, N., Harada, A., & Matsushima, K. (1993). Essential involvement of interleukin 8 in neutrophil recruitment in rabbit with acute experimental arthritis induced by lipopolysaccharide and interleukin 1. Lymphokine Cytokine Res. 13, 113-116.

Baggiolini, M., Dewald, B., & Moser, B. (1994). Interleukin-8 and related chemotactic cytokines-CXC and CC chemokines. Adv. Immunol. 55, 97-179.

Baldwin, E.T., Franklin, K.A., Appella, E., Yamada, M., Matsushima, K., Woldawer, A. & Weber, I.T. (1990). Crystalization of human interleukin-8. A protein chemotactic for neutrophils and T lymphocytes. J. Biol. Che. 265, 6851-6853.

Baldwin, E.T., Weber, I.T., St. Charles, R., Xuan, J.C., Appella, E., Yamada, M., Matsushima, K., Edwards, B.F.P., Clore, G.M., Gronenborn, A.M., & Woldawer, A. (1991). Crystal structure of interleukin 8: Symbiosis of NMR and crystallography. Proc. Natl. Acad. Sci. USA 88, 502-506.

Basemer, J., Hujber, A., & Kuhn, B. (1989), Specific binding, internalization and degradation of human neutrophil activating factor by human polymorphonuclear leukocytes. J. Biol. Chem. 264, 17409-17415.

Beckmann, M.P., Munger, W.E., Kozlosky, C., VandenBos, T., Price, V., Lyman, S., Gerard, N.P., Gerard, C. & Cerreti, D.P.(1991). Molecular cloning of the interleukin-8 receptor. Biochem. Biophys. Res. Commun. 179, 784-789.

Bozic, C.R., Gerard, N.P., von Uexkull-Guldenband, C., Kolalowski, L.F., Jr.,Conklyn, M.j., Breslow, R., Showell, H.J., & Gerard, C. (1994). The murine interleukin 8 type B receptor homologue and its ligand. J. Biol. Chem. 269, 29355-29358.

Brennan, F., Zachariae, C., Chantry, D., Larsen, C.G., Turner, M., Maini, R.N., Matsushima, K., & Feldman, M (1990). Detection of interleukin- 8 (IL 8) biological activity in synovial fluid from patients with rheumatoid arthritis and production of IL 8 mRNA by isolated synovial cells. Eur. J. Immunol. 20, 2141-2144.

Broaddus, V.C., Hébert, C.A., Vitangeol, R.V., Hoeffel, J.M., Bernstein, M.S., & Boylan, A.M. (1992). Interleukin-8 is a major neutrophil chemotactic factor in pleural liquid of patients with empyema. Am. Rev. Respir.Dis. 146, 825-830.

Broaddus, V.C., Boylan, A.M., Hoeffel, J.M., Kim, K.J., Sadick, M., Chuntharapai, C., & Hébert, C.A. (1994). Neutralization of IL-8 inhibits neutrophil influx in a rabbit model of endotoxin-induced pleurisy. J. Immunol. 152, 2960-2967.

Broxmeyer, H.E., Sherry, B., Lu, L., Cooper, S., Carow, S.D., Wolpe, S.D., & Cerami, A. (1989). Myelopoietic enhancing effects of murine macrophage inflammatory proteins 1 and 2 on colony formation *in vitro* by murine and human bone marrow granulocyte/macrophage progenitor cells. J. Exp. Med. 170, 1583-1594.

Broxmeyer, H.E., Sherry, B., Lu, L., Cooper, S., Ruscetti, F.W., Williams, D.E., Arosio, P., Kwon, B.S., & Cerami., A. (1991). Macrophage inflammatory protein (MIP)-1β abrogates the capacity of MIP-1a to suppress myeloid progenitor cell growth. J. Immunol. 147, 2586-

Broxmeyer, H.E., Sherry, B., Cooper, S., Lu, L., Maze, R., Beckmann, M.P., Cerami, A., & Ralph, P. (1993). Comparative analysis of the human macrophage inflammatory protein family of cytokines (chemokines) on proliferation of human myeloid progenitor cells. Interacting effects involving suppression, synergistic suppression, and blocking of suppression. J. Immunol. 150, 3448-3458.

Bruijnzeel, P.L.B., Kuijper, P.H.M., Rihs, S., Betz, S., Warringa, R.A.J., & Koenderman, A. (1993). Eosinophil migration in atopic dematitis. I: Increased migratory responses to N-formyl-methionyl-leucyl-phenylalanine, neutrophil-activating factor, platelet-activating factor, and platelet factor 4. J. Invest. Dermatol. 100. 137-142.

Cacalano, G., Lee, J., Kikly, K., Ryan, A.M., Pitts-Meek, S., Hultgren, B., Wood, W.I., & Moore, M.W. (1994). Neutrophil and B cell expansion in mice that lack the murine IL-8 receptor homolog. Science 265, 682-684.

Camp, R.D.R., Fincham, N.J., Cunningham, F.M., Greaves, M.W., Morris, J., & Chu, A. (1986). Psoriatic skin lesions contain biologically active amounts of an interleukin 1-like compound. J. Immunol. 137, 3469-3474.

Carré, P., Mortenson, R., King, T., Jr., Noble, P.W., Sable, C.L., & Riches, D.W. (1991). Increased expression of the interleukin-8 gene by alveolar macrophages in idiopathic pulmonary fibrosis. J. Clin. Invest.88, 1802-1810.

Cerreti, D.P., Nelson, N., Kozlosky, C.J., Morrissey, P.J., Copeland, N.G., Gilbert, D.J., Jenkins, N.A., Dosik, J.K., & Mock, B.A. (1993). The murine homologue of the human interleukin-8 receptor type B maps near the Ity-Lsh-Bcg disease resistance locus. Genomics 18, 410-413.

Chaudhuri, A., Polyakova, J., Zbrzezna, V., Williams, K., Gulati, S., & Pogo, A.O. (1993). Cloning of glycoprotein D cDNA, whcih encodes the major subunit of the Duffy blood group system and the receptor for the Plasmodium vivax malaria parasite. Proc. Natl. Acad. Sci. USA 90, 10793-10797.

Clark-Lewis, I., Schumacher, C., Baggiolini. M., & Moser, B. (1991). Structure-activity relationships of interleukin-8 determined using chemically synthesized analogs. Critical role of NH2-terminal residues and evidence for uncoupling of neutrophil chemotaxis, exocytosis, and receptor binding activities. J. Biol. Chem. 266, 23128-23134.

Clark-Lewis, I., Dewald, B., Geiser, T., Moser, B., & Baggiolini, M. (1993). Platelet factor 4 binds to interleukin 8 receptors and activates neutrophils when its N terminus is modified with Glu-Leu-Arg. Proc. Natl. Acad. Sci. USA 90, 3574-3577.

Clark-Lewis, I., Dewald, B., Loetscher, M., & Baggiolini, M. (1994). Structure requirement for interleukin-8 function identified by design of analogs and CXC chemokine family. J. Biol. Chem. 269, 16075-16081.

Clore, G.M., Appela, E., Yamada M., Matsushima, K., & Gronenborn, A.M. (1990). Three-dimensional structure of interleukin 8 in solution. Biochemistry 29, 1689-1696.

Colditz, I.G., & Movat, H.Z. (1984). Kinetics of neutrophil accumulation in acute inflammatory lesions induced by chemotaxins and chemotaxinigens. J. Immunol. 133, 2169-2173.

Colditz, I.G., & Watson, D.L. (1992). The effect of cytokines and chemotactic agonists on the migration of T lymphocytes into skin. Immunology. 76, 272-278.

Colditz, I.G., Zwahlen, R.D., & Baggiolini, M. (1990). Neutrophil accumulation and plasma leakage induced by neutrophil-activating peptide-1. J. Leukocyte Biol. 48, 129-137.

Collins, P.D., Weg, V.B., Faccioli, L.H., Watson, M.L., Moqbel, R., & williams, T.J. (1993). Eosinophil accumulation induced by human interleukin-8 in the guinea-pig *in vivo*. Immunology 79, 312-318.

Dahinden, C.A., Kurimoto, Y., De Weck, A.L., Lindley, I., Dewald, B., & Baggiolini, M. (1989). The neutrophil-activating peptide NAF/NAP-1 induces histamine and leukotriene release by interleukin-3 primed basophils J. Exp. Med. 170, 1787-1792.

Darbonne, W.C., Rice, G.C., & Mohler, M.A., (1991). Red blood cells are a sink for interleukin-8, a leukocyte chemotaxin. J. Clin. Invest. 88. 1362-1369.

Detmers, P.A., Lo, S.K., Olsen-Edbert, E., Walz, A., Baggiolini, M., & Cohn, Z.A. (1990). Netrophil-activating protein 1/interleukin-8 stimulates the binding activity of the leukocyte adhesion receptor CD11b/CD18 on human neutrophils. J. Exp. Med. 171, 1155-1162.

Djeu, JY., Matsuhsima, K., Oppenheim, J.J., Shiotsuki, K., & Blanchard, D.K. (1990). Functional activation of human neutrophils by recombinant monocyte-derived neutrophil chemotactic factor/interleukin 8. J. Immunol. 144, 2205-2210.

Donnely, S.C., Strieter, R.M., Kunkel, S.L., Walz, A., Robertson, C.R., & Carter, D.C. (1993). Interleukin-8 and development of adult respiratory distress syndrome in at-risk patient groups. Lancet 341, 643-647.

Donlon, T.A., Krensky, A.M., Wallace, M.R., Collins, F.S., Lovett, M., & Clayberger, C. (1990). Localization of a human T-cell-specific gene, RANTES (D17S136E), to chromosome 17q11.2-q12. Genomics. 6, 548-553.

Endo, H., Akahoshi, T., Takagishi, K., Kashiwazaki, S.,& Matsushima, K.(1991). Elevation of interleukin-8 (IL-8) levels in joint fluids of patients with rheumatoid arthritis and the induction by IL-8 of leukocyte infiltration and synovitis in rabbit joints. Lymphokine Cytokine Res. 10, 245-252.

Fincham, N.J., Camp, R.D.R., Gearing, A.J.H., Bird, C.R., & Cunningham, F.M. (1988). Neutrophil chemoattractant and IL-1-like activity in samples from psoriatic skin lesions. Further characterization. J. Immunol. 140, 4294-4299.

Foster, S.J., Aked, D.M., Schröder, J.M., & Christophers, E. (1989).Acute inflammatory efects of a monocyte-derived neutrophil-activating peptide in rabbit skin. Immunology 67, 181-183.

Furuta, R., Yamagishi, J., Kotani, H., Sakamoto, F., Fukui, T., Matsui, Y., Sohmura, Y yamada, M., Yoshimura, T., Larsen, C.G., Oppenheim, J.J., & Matsushima, K. (1989). Production and characterization of recombinant human neutrophil chemotactic factor. 106, 436-441.

Gao, J., Kuhns, D., Tiffany, H., McDermott, D., Li,X., Franke, U., & Murphy, P.(1993). Structure and function of the human MIP-1a /RANTES receptor. J. Exp. Med. 177, 1421-1427.

Gerard, N.P., Bao, L., Xiao-Ping, H., Eddy, R.L., Jr., Shows, T.B., & Gerard, C. (1993). Human chemotaxis receptor genes cluster at 29q13.3-13.4. Characterization of the human C5a receptor gene. Biochemistry 32, 1243-1250.

Gimbrone, M.A.Jr., Obin, M.S., Brock, A.F., Luis, E.A., Hass, P.E., Hébert, C.A., Yip, Y.K., Leung, D.W., lowe, D.G., kohr, W.J., Darbonne, W.C., Bechtol, K.B.,& Baker, J.B. (1989). Endothelial interleukin-8: A novel inhibitor of leukocyte-endothelial interactions. Science 246, 1601-1603.

Goodman, R.B., Foster, D.C., Mathewes, S.L., Osborn, S.G., kuijper, J.L., Forstrom, J.W., & Martin, T.R. (1992). Molecular cloning of porcine alveolar macrophage-derived neutrophil chemotactic factor I and II: Identification of porcine IL-8 and another intercrine-alpha protein. Biochemistry 31, 10483-10490.

Graham, G.J., Wright, E.G., Howick, R., Wolpe, S.D., Wilkie, N.M., Donaldson, D., Lorimore, S., & Pragnall, I.B. (1990). Identification and characterization of an inhibitor of hematopoietic stem cell proliferation. Nature 34, 442-444.

Griffin, C.A., Emanuel, B.S., LaRocco, O., Schwartz, E., & Poncz, M. (1987). Human platelet factor 4 gene is mapped to 4q12-q21. Cytogenet. Cell. Genet. 45, 67-69.

Grob, P.M., David, E., Warren, T.C., Deleon, R.P., Farina, P.R., & Homon, C.A. (1990). Characterization of a specific receptor for monocyte-derived neutrophil chemotactic factor/interleukin-8. J. Biol. Chem. 265, 831-8316.

Harada, A., Sekido, N., Kuno, K., Akiyama, M., Kasahara, T., Nakanishi, I., Mukaida, N., & Matsushima, K. (1993). Expression of recombinant rabbit IL-8 in Escherichia coli and establishment of the essential involvement of IL-8 in recruiting neutrophils into lipopolysaccharide-induced inflammatory site of rabbit skin. Int. Immunol. 6,681-690.

Harada, A., Kuno, K., Nomura, H., Mukaida, N., & Matsushima, K. (1994). Cloning of a cDNA encoding a mouse homolog of the interleukin-8 receptor. Gene. 142, 297-300.

Hébert, C.A., Luscinskas, F.W., Kieley, J.M., Luis, E.A., Darbonne, W.C., Bennet, G.L., Liu, C.C., Orbin, M.S., Gimbrone, M.R.Jr., & Baker, J.B. (1990). Endothelial and leukocyte forms of interleukin-8. Conversion by thrombin and interractions with neutrophils. J. Immunol. 145, 3033-3040.

Hébert, C.A., & Baker, J.B. (1993). Interleukin-8: A review. Cancer Invest. 11, 743-750.

Hébert, C.A., Vitangcol, R.V., & Baker, I.T. (1991). Scanning mutagenesis of interleukin-8 identifies a cluster of residues required for receptor binding. J. Biol. Chem. 266. 18989-18994.

Hébert, C.A., Chuntharapao, A., Smith, M., Colby, T., Kim, J., & Horuk, R. (1993). Partial functional mapping of the human intereukin-8 type A receptor. Identification of a major ligand binding domain. J. Biol. Chem. 268, 18549-18553.

Hechman, D., Cybalsky, M., & Fuchs, H. (1991). Intravascular IL-8: Inhibitor of polymorphonuclear leukocyte accumulation at sites of acute inflammation. J. Immunol. 147, 883-892.

Holmes, W.E., Lee, J., Kuang, W.J., Rice, C.G., & Wood, W.I. (1991). Structure and function expression of a human interleukin-8 receptor. Science. 253, 1278-1280.

Horuk, R. (1994). The interleukin-8-receptor family: From chemokine to malaria. Immunol. Today 15, 169-174.

Horuk, R., Colby, T.J., Darbonne, W.C., Schall, T.J., & Neote, K. (1993a). The human erythrocyte inflammatory peptide (chemokine) receptor. Biochemical characterization, solubilization, and development of a binding assay for the soluble receptor. Biochemistry. 32, 5733-5738.

Horuk, R., Chitnis, C.E., Darbonne, W.C., Colby, T.J., Rybicki, A., Hadley, T.J., & Miller, L.H. (1993b). A receptor for the malaria parasite Plasmodium vivax: The erythrocyte chemokine receptor. Science 261, 1182-1184.

Hu, D.E., Hori, Y., & Fan, T.P.D. (1993). Interleukin-8 stimulates angigenesis in rats. Inflammation 17, 135-143.

Huber, A.R., Kunkel, S.L., Todd, R.T., & Weiss, S.J. (1991). Regulation of transendothelial neutrophil migration by endogeneous interleukin-8. Science 254, 99-102.

Irving, S.G., Zipfel,P.F., Balke, J.,McBride, O.W., Mortom, C.C., Burd, P.R., Sienbenlist, U., & Kelly, K. (1990). Two inflammatory mediator cytokine genes are closelylinked and variably amplified on chromosome 17q. Nucleic Acid Res. 18, 3261-3270.

Ishikawa, J., Suzuki, S., Hotta, K., Hirota, Y., Mizuno, S., & Suzuki, K. (1993). Cloning of a canine gene homologous to the human interleukin-8-encoding gene. Gene 131, 305-306.

Ishikawa, Y., Mukaida, N., Kuno, K., Rice, N., Okamoto, S., & Matsushima, K. (1995). Establishment of lipopolysaccharide-dependent nuclear factor kb activation in a cell-free system. J. Biol. Chem. 270, 4158-4164.

Kameyoshi, Y., Dorschner, A., Mallet, A., Christopher, E., & Schröder, J.M. (1992). cytokine RANTES released by thrombin-stimuated platelets is a potent attractant for human eosinophils. J. Exp. Med. 176, 587-592.

Kimata, H., Yoshida, A., Ishioka, C., Lindley, I., & Mikawa, H. (1992). Interleukin-8 (IL-8) selectively inhibits immunoglobulin E production by IL-4 in human B cells. J. Exp. Med. 176, 1227-1231.

Ko, Y.-C., Mukaida, N., Ishiyama, S., Tokue, A., Kawai, T., Matsushiam, K, & Kasahara, T. (1993). Elevated interleukin-8 levels in the urine of patients with urinary tract infections. Infect. Immun. 61, 1307-1314.

Koch, A.E, Kunkel, S.L, Burrows, J.C, Evanoff, H.L., haines, G.K., Pope, R.M., & Strieter, R.M. (1991). Synovial tissue macrophage as a source of the chemotactic cytokine IL-8. J. Immunol. 147, 2187-2195.

Koch, A.E., Polverini, P.J., Kunkel, S.L., Harlow, L.A., DiPietro, L.A., Elner, V.M., Elner, S.G., & Strieter, R.M. (1992). Interleukin-8 as a macrophage-derived mediator of angiogenesis. Science 258, 1798-1801.

Kudo, C., Araki, A., Matsushima, K., & Sendo, F. (1991). Inhibition of IL-8-induced W3/25$^+$ (CD4$^+$) T lymphocyte recruitment into subcutaneous tissues of rats by selective depletion of in vivo neutrophils with a monoclonal abtibody. J. Immunol. 147, 2196-2201.

Kuna, P., Reddigari, S.R., Kornfeld, D., & Kaplan, A.P. (1991). IL-8 inhibits histamine release from human basophils induced by histamine-releasing factors, connective tissue activating peptide III, and IL-3. J. Immunol. 147, 1920-1924.

Kupper, R.W., Dewald, B., Jacobs, K.H., Baggiolini, M., & Gierschick, P. (1992). G-protein activation by interleukin-8 and related cytokines in human neutrophil plasma membranes. Biochem J. 282, 429-434.

LaRosa, G.J., Thoma, K.M., Kaufmann, M.E., Mark, R., White, M., Taylor, L.,Gray, G., Witt, D., & Navarro, J. (1992). Amino terminus of the interleukin-8 receptor is a major determinant of receptor subtype secificity. J. Biol. Chem. 267, 25402-25406.

Larsen, C.G., Anderson, A.O., Appella, E., Oppenheim, J.J., & Matsushima, K. (1989a). Neutrophil activating protein (NAP-1) is also chemotactic for T lymphocytes. Science 243, 1464-1466.

Larsen C.G., Anderson, A.O., Oppenheim, J.J., & Matsushima, K. (1989b). Production of interleukin-8 by human dermal fibrablasts and keratinocytes in response to interleukin-1 or tumor necrosis factor. Immunology 68-31-36.

Laterveer, L., Hamilton, M., Lindley, I., Willemze, R., & Fibbe, W.E. (1993). A single dose of interleukin-8 induces mobolization of myeloid progenitor cells with radioprotective capacity. Blood, 82 (Suppl.), 293a.

Lee, J., Horuk, R., Rice, G.C., Bennett,G.L., Camerato, T. & Wood, W.I. (1992a). Characterization of two high affinity human interleukin-8 receptors. J. Biol. Chem. 267, 16283-16287.

Lee, J., Kuang, W.J., Rice, G.C., & Wood, W.I. (1992b). Characterization of complementary DNA clones encoding the rabbit IL-8 receptor. J. Immunol. 148, 1261-1264.

Leonard, E.J., Skeel, A., Yoshimura, T., Noer, K., Kutbirt, S., & Van Epps, D. (1990). Leukocyte specificity and binding of human neutrophil attractant/activating protein-1. J. Immunol. 144, 1323-1330.

Leonard, E.J., Yoshimura, T., Tanaka, S., & Raffled, M.(1991). Neutrophil recruitment by intradermally injected neutrophil attractant/activation protein-1. J. Invest. Dermatol. 96,690-694.

Luger, T.A., Charon, J.A., Colot, M., Micksche, M., & Oppenheim, J.J. (1983). chemotactic properties of partially purified human epidermal cell-derived thymocyte-activating factor (ETAF) for polymorphonuclear and mononuclear cells. J. Immunol. 131, 816-820.

Luster A.D., Unkeless, J.C., Changanti R.S.K., Kersey, J.H., & Ravetch, J.V. (1987). Interferon inducible gene maps to a chromosomal band associated with a (4:11) translocation in acute leukemia cells. Proc. Natl. Acad. Sci. USA 84, 2868-2871.

Lynch, J.P., III, Standiford, T.J., Rolfe, M.W., Kunkel, S.L., & Strieter, R.M. (1992). Neutrophilic alveolitis in idiopathic pulmonary fibrosis: The role of interleukin-8. Am. Rev. Respir. Dis. 145, 1433- 1439.

Marini, M., Vittori, E., Hollemborg, J., & Mattoli, S. (1992). Expression of the potent inflammatory cytokines, granulocyte-macrophage-colony-stimulating factor and interleukin-6 and interleukin-8, in bronchia epethelial cells of patients with asthma. J. Allergy Cli. Invest. 89, 1257-1267.

Matsuhsima, K., Yoshimura, T., Lavu, S., Kobaysahi, Y., Lew, W., Appella, E., Kung, H.F., Leonard, E.J., & Oppenheim, J.J. (1988). Molecular cloning of a human monocyte-derived neutrophil chemoatctic factor (MDNCF) and the induction of MDNCF mRNA by interleukin 1 and tumor necrosis factor. J. Exp. Med. 167,1883-1893.

Mehrabian, M., Sparkes, R.S., Mohandas, T., Fogelman, A.M., & Lusis, A. (1991). Localization of monocyte chemotactic protein-1 gene (SCYA2) to human chromosome 17q11.2-q21.1. Genomics 9, 200-203.

Meurer, R., Van Rper, G., Feeney, W., Cunningham, P., Hora, D.Jr., Springer, M.S., MacIntyre, D.E., & Rosen, H. (1993). Formation of eosinophilic and monocytic intradermal inflammatory site in the dog by injection of human RANTES but not human monocyte chemoattractant protein 1, human macrophage inflammatory protein 1 a, or human interleukin 8. J. Exp. Med. 178, 1913-1921.

Miller, M.D., Wilson, S.D., Dorf, M.E., Seuanez, H.N., O'Brien, S.J., & Krangel, M.S. (1990). Sequence and chromosomal localization of the I309 gene: Relation ship to genes encoding a family of inflammatory cytokines. J. Immunol. 145, 2737-2744.

Miller, E.J., Cohen, A.B., Nagao, S., Griffith, D., Maunder, R.J., Martin, T.R., Weiner-Kronish, J.P., Sticherling, M., Christopher, E., & Matthay, M.A. (1992). Elevated levels of NAP-1/interleukin-8 are present in the airspace of patients with the adult respiratory distress syndrome and are associated with increased mortality. Am. Rev. Resir. Dis. 146, 427-432.

Minowa, M.T., Minowa, T., Monsma, F.J., Sibley, D.R., & Mouradien, M.M. (1992). Characterization of the 5'-flanking region of the human D1A dopamine receptor. Proc. Natl. Acad. Sci. USA 89, 3045-3049.

Modi, W.S., Dean, M., Seuanez, H.N., Mukaida, N., Matsushima, K., & O'Brien, S.J. (1990). Monocyte-derived neutrophil chemotactic factor (MDNCF/IL-8) resides in a gene cluster along with several other members of the platelet factor 4 gene superfamily. Hum Genet. 84, 185-187.

Morohashi, H., Miyawaki, T., Nomura, H., Kuno, K., Murakami, S., Matsushima, K., & Mukaida, N. (1995). Expression of both types of human interleukin-8 receptors on mature neutrophils, monocytes, and natural killer cells. J. Leukocyte Biol. 57, 180-187.

Morris, S.W., Nelson, N., Valentine, M.B., Shapiro, D.N., Look, A.T., Kozlosky, C.J., Beckmann, M.P., & Cerretti, D.P. (1992). Assignment of the genes encoding human interleukin-8 receptor types 1 and 2 and an interleukin-8 receptor pseudogene to chromosome 2q35. Genomics 14, 685-691.

Moser, B., Schumacher, C., von Tscharner, V., Clark-Lewis, I., & Baggiolini, M. (1991). Neutrophil-activating peptide 2 and gro/melanoma growth-stimulatory activity interact with neutrophil-activating peptide 1/interleukin 8 receptors on human neutrophils. J. Biol. Chem. 266, 10666-10671.

Moser, B., Dewald, B., Barella, L., Schumacher, C., Baggiolini, M., & Clark-Lewis, I. (1993a). Interleukin-8 antagonists generated by N-termeinal modification. J. Biol. Chem. 268, 7125-7128.

Moser, B., Barella, L., Mattei, S., Schumacher, C. Boulay, F., Colombo, M.P., & Baggiolini, M. (1993b). Expression of transcripts for two interleukin 8 receptors in human phagocytes, lymphocytes and melanoma cells. Biochem. J. 294, 285-292.

Movat, H.Z., Rettl, C., Burrowes, C.E., & Johnston, M.G. (1984). The *in vivo* effect of leukotriene B4 on polymorphonuclear leukocytes and the microcirculation: Comparison with activated complement (C5a des Arg) and enhancement by prostaglandin E2. 115, 233-244.

Mukaida, N., Shiroo, M., & Matsushima, K. (1989). Genomic structure of the human monocyte-derived neutrophil chemotatic factor IL-8. J. Immunol. 143, 1366-1371.

Mukaida, N., Mahé, Y., & Matsushima, K. (1990). Cooperative interaction of nuclear factor-kB- and cis-regulatory enhancer binding protein-like factor binding elements in activating the interleukin-8 gene by pro-inflammatory cytokines. J. Biol.Chem. 265, 21128-21133.

Mukaida, N., Morita, M., Ishikawa, Y., Rice, N., Okamoto, S-I., Kasahara, T., & Matsushima, K. (1994). Novel mechanism of glucocorticoid-mediated gene repression. Nuclear factor-kB is target for glucocorticoid-mediated interleukin 8 gene repression. J. Biol.Chem. 265, 21128-21133.

Mulligan, M.S., Jones, M.L., Bolanowski, M.A., Baganoff, M.P., Dppeler, C.L., Meyers, D.M., Ryan, U.S., & Ward, P.A. (1993). Inhibition of lung inflammatory reactions in rats by an anti-human IL-8 antibody. J. Immunol. 150, 5585-5595.

Murayama, T., Kuno, K., Jisaki, F., Sakamoto, D., Obuchi, M., Furukawa, T., Mukaida, N., & Matsushima, K. (1994). Enhancement of human cytomegalovirus replication in a human lung fibroblast cell line by interleukin-8. J. Virol. 68, 7582-7585.

Murphy, P.H., & Tiffany, H.E. (1991). Cloning of complementary DNA encoding a functional human interleukin-8 receptor. Science 253, 1280-1283.

Murphy, P.M., Tiffany, H.L., McDermott, D., & Ahuja, S.K. (1993). Sequence and organization of the human N-formyl peptide receptor-encoding gene. Gene 133, 285-290.

Nakagawa, H., Hatakeyama, S., Ikesue, A., & Miyai, H. (1991). Generation of interleukin-8 by plasmin from AVLPR-interleukin-8, the human fibroblast-derived neutrophil chemotactic factor. FEBS Lett. 282, 412-414.

Nakamura, H., Yoshimura, K., McElvaney, N.G., & Crystal, R.G. (1992). Neutrophil elastase in respiratory epithelial lining fluid of individuals with cystic fibrosis induces interleukin-8 gene expression in a human bronchial epithelial cells line. J. Clin. Invest. 89, 1478-1484.

Neotoe, K., Darbonne, W., Ogez, J., Horuk, R., & Schall, T.J. (1993). identification of a promiscuous inflammatory peptide receptor on the surface of red blood cells. J. Biol. Chem. 268, 12247-12249.

Neote, K., Mak, J.Y., Kolakowski, L.F.,Jr., & Schall, Y.J. (1994). Functional and biochemical analysis of the cloned Duffy antigen: Identity with the red blood cell chemokine receptor. Blood 84, 44-52.

Nourshargh, S., Perkins, J.A., Showell, H.J., Matsuhsima, K., Williams, T.J., & Collins, P.D. (1992). A comparative study of the neutrophil stimulatory activity *in vitro* and proinflammatory properties *in vivo* of 72 amino acid and 77 amino acid IL-8. J. Immunol. 148, 106-111.

Oishi, K., Sonoda, F., Kobayashi, S., Iwagaki, A., Nagatake, T., Matsushima, K., & Matsumoto, K. (1994). Role of interleukin-8 (IL-8) and an inhibitory effect of erythromycin on IL-8 release in the airway of patients with chronic airway diseases. Infect. Immun. 62, 4145-4152.

Okamoto, S-I., Mukaida, N., Yasumoto, K., Rice, N., Ishikawa, Y., Horiguchi, H., Murakami, S., & Matushima, K. (1994). The interleukin-8 AP-1 and kB-like sites are genetic end targets of FK506-sensitive pathway accompanied by calcium mobilization. J. Biol. Chem. 269, 8582-8589.

Oliveira, I.C., Sciavolino, P.J., Lee, T.H., & Vilcek, J. (1992). Downregulation of interleukin-8 gene expression in human fibroblast: unique mechanism of transcriptional inhibition by interferon. Proc. Natl. Acad. Sci. USA 89, 9049-9053.

Oliveira, I.C., Mukaida, N., Matsushima, K., & Vilcek, J. (1994). Transcriptional inhibition of the intereukin-8 gene by interferon is mediated by the NF-kB site. Mol. Cell. Biol. 14, 5300-5308.

Oppenheim, J.J. (1993). *In vitro* properties of attractive chemokines and *in vivo* veritas. At Combined Meeting of the 8th Intl. Lymphokine Workshop and the 4th Intl. Workshop of cytokines. Osaka, Oct. 21.

Oppenheim, J.J., Zachariae, C.O.C, Mukaida, N., & Matsushima, K. (1991). Properties of the novel proinflammatory supergene "intercrine" cytokine family. Annu. Rev. Immunol. 9, 617-648.

Paccaud, J.P., Shifferli, J.A., & Baggiolini, M. (1990). NAP-1/IL 8 induces upregulation of CR1 receptors in human neutrophil leukocytes. Biochem. Biophys Res. Comm. 166, 187-192.

Peichl, P., Ceska, M., Effenberger, F.,Haberhauer, G., Broell, H., & Lindley, I.J.D.(1991). Presence of NAP-1/IL-8 in synovial fluids indicates a possible pathogenic role in rheumatoid arthritis. Scand. J. Immunol. 34, 333-339.

Peveri, P., Walz, A., Dewald, B., & Baggiolini, M. (1988). A novel neutrophil-activating factor produced by human mononuclear phagocytes. J. Exp. Med. 167, 1547-1559.

Pike, M.C., Costello, K.M., & Lamb, K.A. (1992). IL-8 stimulates phosphatidylinositol-4-phosphate kinase in human polymorphonuclear leukocytes. J. Immunol. 148, 3158-3164.

Prado, G.N., Thomas, K.M., Suzuki, H., LaRosa, G.J., Wilkinson, N., Folco, E., & Navarro, J. (1994). Molecular characterization of a novel rabbit interleukin-8 receptor isotype. J. Biol. Chem. 269, 12391-12394.

Rajarathnam, K., Sykes, B.D., Kay, C.M., Dewald, B., Geiser, T., Baggiolini, M., & Clak-Lewis, I. (1994). Neutrophil activation of monomeric interleukin-8. Science 264. 90-92.

Rampart, M., Van Damme, J., Zonnkeyn, L.,& Herman, A.C.(1989). Granulocyte chemotactic protein/interleukin-8 induces plasma leakage and neutrophil accumulation in rabbit skin. Am. J. Pathol. 135, 21-25

Rampart, M., Herman, A.G., Grillet, B., Opdenakker, G., & Van Damme, J. (1992). Development and application of a radioimmunoassay for interleukin-8: Detection of interleukin-8 in synovial fluids from patients with inflammatory joint disease. Lab. Invest. 66, 512-518.

Richmond, A., Balentein, E., & Thomas, H.G. (1988). Molecular characterization and chromosomal mapping of melanoma growth stimulatory activity, a growth factor structurally related to b-thromboglobulin. EMBO J. 7, 2025-2033

Rot, A., Krieger, M., Brunner, T., Bishoff, S., Schall, T., & Dahinden, C.(1992). RANTES and MIP-1a induce the migration and activation of normal human eosinophil granulocytes. J. Exp. Med. 176, 1489-1495.

Samanta, A.K., Oppenehim, J.J., & Matsushima, K. (1989). Identification of a specific receptor for monocyte-derived neutrophi chemotactic factor (MDNCF) on human neutrophils. J. Exp. Med. 169, 1185-1189.

Sauder, D.N., Mounessa, N.L., Katz, S.I., Dinarello, C.A., & Gallin, J.I.(1984).Chemotactic cytokines: The role of leukocytic pyrogen and epidermal cell thymocyte-activating factor in neutrophil chemotaxis. J. Immunol. 132, 828-832.

Schall, T., Bacon, K.,Toy, K., & Goeddel, D. (1990). Selective attraction of monocytes and T lymphocytes of the memory phenotype by cytokine RANTES. Nature 347, 669-671.

Schmidt, J., & Weissmann, C. (1987). Induction of mRNA for a serine protease and a b-thromboglobuin-like protein in mitogen-stimulated human leukocytes. J. Immunol. 139, 250-256.

Schreiber, S.L., & Crabtree, G.R. (1992). The mechanism of action of cyclosporin A and FK506. Immunol. Today 13, 136-142.

Schröder, J.M. (1989). The monocyte-derived neutrophil activating peptide (NAP/interleukin 8) stimulates human neutrophil arachidonate-5-lipoxygenase, but not the release of cellular arachidonate. J. Exp. Med. 170, 847-863.

Schröder, J.M.,& Christopher, E. (1986). Identification of C5ades arg and an anionic neutrophil-activating peptide (ANAP) in psoriatic scales. J. Invest. Dermatol. 87, 53-58.

Schröder, J.M., Mrowietz, U., Morita, E., & Christophers, E. (1987). Purification and partial biochemical characterization of a human monocyte-derived, neutrophil-activating peptide that lacks iterleukin 1 activity. J. Immunol. 139, 3474-3483.

Schweizer, R.C., Welmers, B.A., Raaijmakers, J.A., Zanen, P., Lammes, J.W., & Koenderman, L.(1994). RANTES-and interleukin-8-induced responses in normal human eosinophils: Effects priming with interleukin-5. Blood 83, 3697-3704.

Sebok, K., Woodside, D., Al-Aoukatty, A., Ho, A.D., Gluck, S., & Maghazachi, A.A. (1993). IL-8 induces the locomotion of human IL-2-activated natural killer cells. Involvement of a guanine nucleotide binding (Go) protein. J. Immunol. 150, 1524-1534.

Sekido, N., Mukaida, N., Harada, A., Nakanishi, I., Watanabe, Y., & Matsushiama, K. (1993). Prevention of lung reperfusion injury in rabbits by a monoclonal antibody against interleukin-8. Nature 365, 654-657.

Simonet, W.S., Hughes, T.M., Nguyen, H.Q., Trebasky, L.D., Danilenko, D.M., & Medlock, E.S. (1994). Long-term impaired neutrophil migration in mice overexpressing human interleukin-8. J. Clin. Invest. 94, 1310-1319.

Smith, D.R., Polverini, P.J., Kunke, S.L., Orringer, M.B., Whyte, R.I., Burdick, M.D., Wilke, C.A., & Strieter, R.M. (1994). Inhibition of interleukin 8 attenuates angiogenesis in bronchogenic carcinoma. J. Exp. Med. 179, 1409-1415.

Sprenger, H., Llyod, A.R., Meyer, R.G., Johnston, J.A., & Kelvin, D.J. (1994a). Genomic structure, characterization, and identification of the promoter of the human IL-8 receptor A gene. J. Immunol. 153, 2524-2532.

Sprenger, H., Lloyd, A.R., Lautens, L.L., Bonner, T.I., & Kelvin, D.J. (1994b). Structure, genomic organization, and expression of the human interleukin-8 receptor B gene. J. Biol. Chem. 269, 11065-11072.

Suzuki, H., Prado, G.N., Wilkinson, N., & Navarro, J. (1994). the N terminus of interleukin-8 (IL-8) receptor confers high affinity binding to human IL-8. J. Biol. Chem. 269, 18263-18266.

Swensson, O., Schubert, C., Christopher, E.,& Schröder, J.M. (1991). Inflammatory properties of neutrophil-activatingprotein-1/interleukin 8 (NAP-1/IL-8) in human skin: A light- and electronmicroscopic study. J. Invest. Dermatol. 96, 682-689.

Symon, J.A., Wong, W.L., Palladino, M.A., & Duff, G.W. (1992). Interleukin 8 in rheumatoid arthritis and osteoarthritis. Scand. J. Rheumatol. 21, 92-94.

Taub, D.D., & Oppenheim, J.J. (1994). Chemokines, inflammation and the immune system. Therapeutic Immunol. 1, 229-246.

Taub, D.D., Conlon, K., Lloyd, A., oppenheim, J., & Klein, D.(1993a). Preferetial migration of activated CD4+ and CD8+ cells in response to MIP-1α and MIP-1β. Science 260, 355-358.

Taub, D.D., Lloyd, A.R., Conlon, K., Wang, J.M., Ortaldo, J.R., Harada, A., Matsushima, K., Kelvin, D.J., & Oppenheim, J.J. (1993b). Recombinant human interferon-inducible protein 10 is a chemoattractant for human monovytes and T lymphocytes and promotes T cell adhesion to endothelial cells. J. Exp. Med. 177, 1809-1814.

Terkeltaub, R., Zachariae, C., Santoro, D., Martin, J., Peveri, P., & Matsushima, K. (1991). Monocyte-derived neutrophil chemotactic factor/interleukin-8 is a potential mediator of crystal-induced inflammation. Arthritis Rheum. 34, 894-903.

Thelen, M., Peveri, P., Kernen, P., von Tschrarner, V., Walz, A., & Baggiolini, M. (1988). Mechanism of neutrophil activation by NAF, a novel monocyte-derived peptide agonist. FASEB J. 2, 2702-2706.

Thomas, K.M., Taylor, L., Prado, G., Romeo, J., Moser, B., Car, B., Walz., A., Baggiolini, M., & Navarro, J. (1994). Functional and ligand binding specificity of the rabbit neutrophil IL-8 receptor. J. Immunol. 152, 2496-2500.

Thomsen, M.K., Lasen, C.G., Thomsen, H.K., Kirstein, D., Skak-Nielsen, T., Ahnfelt-Ronne, I., & Thestrup- Pederson, K. (1991). Recombinant human interleukin-8 is a potent activator of canine neutrophil aggregation, migration, and leukotriene B4 biosynthesis. J. Invest. Dermatol. 96, 260-266.

Tono-oka, T., Nakayama, M., & Matsumoto, S. (1980). Human monocyte-derived chemotactic factor for granulocytes. Immunology 39, 607-613.

Tunnacliffe, A.,Majumdar, S., Yan, B., & Poncz, M. (1992). Genes for b-thromboglobulin and pllatelet factor 4 are closely linked and form part of a cluster of related genes on chromosome 4. Blood 79, 2896-2900.

Unemori, E.N., Amento, E.P., Bauer, E.A., & Horuk, R. (1993). Melanoma growth-stimulatory activity/GRO decreases collagen expression by human fibroblasts. Regulation by C-X-C but not C-C cytokines. J. Biol. Chem. 268, 1338-1342.

Van Damme, J., & Opdenakker, G.(1990). Interaction of interferons with skin reactive cytokines: From interleukin-1 to interleukin-8. J. Invest. Dermatol. 95, 90S-93S.

Van Damme, J., Van Beeumen, J., Opdenakker, G., & Billiau, A. (1988). A novel, NH2-termina sequence-characterized human monokine possessing neutrophil chemotactic, skin-reactive, and granulocytosis-promoting activity. J. Exp. Med. 167, 1364-1376.

Van Zee, K., Fischer, E., Hawes, A.S., Hébert, C.A., Terrell, T.G., Banker, J.B., Lowry, S.F., & Moldawer, L.L. (1992). Effects of intavenous IL-8 administration in nonhuman primates. J. Immunol. 148, 1746-1752.

Verfaillie, C.M., Catanzarro, P.M., & Li, W. (1994). Macrophage inflammatory protein 1a, interleukin 3, and diffusible marrow stromal factors maintain human hematopoietic stem cells for at least eight weeks *in vitro*. J. Immunol. 179, 643-649.

Wada, T., Tomosugi, N., Naito, T., Yokoyama, H., Kobayashi, K., Harada, A., Mukaida, N., & Matsushima, K. (1994a). Prevention of proteinuria by the administration of anti-interleukin 8 antibody in experimental acute immune complex-induced glomerulonephritis. J. Exp. Med. 180, 1135-1140

Wada, T., Yokoyama, H., Tomosugi, N., Hisada, Y., Ohta, S., Naito, T., Kobayashi, H., Mukaida, N., & Matsushima, K. (1994b). Detection of urinary interleukin-8 in glomerular diseases. Kidney Int. 46, 455-460.

Walz, A., & Baggiolini, M. (1989). A novel cleavage product of b-thromboglobulin formed in cultures of stimulated mononuclear cells activates human neutrophils. Biochem. Biophys. Res. Commun. 159, 969-975.

Walz, A., Peveri, P., Aschauer, H., & Baggiolini, M. (1987). Purification and amin acid sequencing of NAF, a novel neutrophil-activating factor produced by monocytes. Biochem. Biophys. Res. commun. 149, 755-761.

Walz, A., Dewald, B., von Tscharner, V., & Baggiolini, M. (1989). Effects of the neutrophil activating peptide NAP-2, platelet basic protein, connective tissue-activating peptide III amd platelet factor 4 on human neutrophils. J. Exp. Med. 170, 1745-1750.

Walz, A., Melni, F., Clark-Lewis, I., von Tscharner, V., & Baggiolini, M. (1991). [Ca2+]i changes and respiratory burst in human neutrophils and monocytes induced by NAP-1/interleukin-8, NAP-2, and gro/MGSA. J.Leukocyte Biol. 50, 279-286.

Wang. J.M., Taraboletti, G., Matsuhsima, K., Van Damme, J., & Mantovani, A. (1990). Induction of haptotactic migration of melanoma cells by neutrophil activating protein/interleukin-8. Biochem. Biophys. Res. Commun. 169, 165-170.

Warringa, R.A., Loenderman, L., Kok, P.T.M., Kreukniet, J., & Bruijnzeel, P.L.B. (1991). Modulation and induction of eosinophil chemotaxis by granulocyte-macrophage colony stimulating factor and interleukin-3. Blood 77, 2694-2700.

Warringa, R.A.J., Mengelers, H.J.J., Raaijmakers. J.A.M., Bruijnzeel, P.L.B, & Koenderman, L. (1993). Upregulation of formyl-peptide and interleukin-8-induced eosinophil chemotaxis in patients with allergic asthma. J. Allergy Clin. Immunol. 91, 1198-1205.

Webb, L.M.C., Ehrengruber, M.U., Clark-Lewis, I., Baggiolini, M., & Rot, A. (1993). Binding to heparan sulfate or heparin enhances neutrophil responses to interleukin 8. Proc. Natl. Acad. Sci. USA 90, 7158-7162.

White, M.V., Yoshimura, T., Hook, W., Kaliner, M.A., & Leonard, E.J. (1989). Neutrophil attractant/activation protein-1 (NAP-1) causes human basophil histamine release. Immunol. Lett. 22, 151-154.

Wu, D., LaRosa, G.J., & Simon, M.I. (1993). G protein-coupled signal transduction pathways for interleukin-8. Science 261, 101-103.

Wu, X., Wittwer, A.J., Carr, L.S., Crippes, B.A., DeLarco, J.E., & Lefkowith, J.B. (1994). cytokine-induced neutrophil chemoattractant mediates neutrophil influx in immune complex glomerulonephritis in rat. J. Clin. Invest. 94, 337-344.

Yasumoto, K., Okamoto, S-I., Mukaida, N., Murakami, S., Mai, M., & Matsushima, K. (1992). Tumor necrosis factor a and interferon g synergistically induce interleukin 8 production in a human gastric cancer cell line through acting concurrently on AP-1 and NF-kB-like binding sites of the interleukin 8 gene. J. Biol. Chem. 266, 19611-19617.

Yoshimura, T., & Johnson, D.G. (1993). cDNA cloning and expression of guinea pig neutrophil attractant protein-1 (NAP-1). NAP-1 is highly conserved in guinea pig. J. Immunol. 151, 6225-6236.

Yoshimura, T., & Yuhki, N. (1991). Neutrophil attractant/activation protein-1 and monocyte chemoattractant protein-1 in rabbit. cDNA cloning and their expression in spleen cells. J. Immunol. 146, 3483-3488.

Yoshimura, T., Matsushima, K., Tanaka, S., Robinson, E.A., Apella, E., Oppenheim, J.J., & Leonard, E.J. (1987). Purification of a human monocyte-derived neutrophil chemotactic factor that has peptide sequence similarity to other host defense cytokines. Proc. Natl. Acad. Sci. USA 84, 9233-9237.

Yoshimura, T., Robinson, E.A., Appella, E., Matsuhsima, K., Showalter, S.D., Skeel, A., & Leonard, E.J. (1989). Three forms of monocyte-derived neutrophil chemotactic factor (MDNCF) distinguished by different lengths of the amino-terminal sequence. Mol. Immunol. 26, 87-93.

Zwahlen, R., Walz, A., & Rot, A. (1993). *In vitro* and *in vivo* activity and pathophysiology of human interleukin-8 and related peptides. Int. Rev. Exp. Pathol. 34 Pt B, 27-42.

INTERLEUKIN-10

Alexander H. Enk

Growth Factors and Cytokines in Health and Disease
Volume 2A, pages 319-353.

ISBN: 0-7623-0117-1

I. INTRODUCTION

Interleukin-10 (IL-10) is the term for a pleiotropic cytokine with a variety of stimulatory and inhibitory effects on the immune system. Interleukin-10 was originally described in 1990 by Moore, Mosmann, and their coworkers as "cytokine synthesis inhibitory factor," a name that already emphasizes one of the effects that finally led to the detection of IL-10 as a novel cytokine. Just like other cytokines, however, IL-10 has more than one biologic activity; the most important activity usually depends on the local context in which the cytokine is produced.

For IL-10 there were three observations that finally led to its discovery as a distinct molecule. The first and probably most important was the observation by Fiorentino, Mosmann, and coworkers (1989) that T cells of the T helper 2 (Th2)-subtype secreted factors or a factor that were capable of inhibiting the cytokine production of Th1-cells. Indeed it was demonstrated in these early publications that CSIF-mediated inhibition of Th1 cytokine production was an indirect effect, possibly needing the presence of functional APC. Furthermore, it was shown that the effect on cytokine synthesis occurred at the mRNA level and needed several hours to come into effect. The effect was shown to be distinct from any other cytokine characterized until then (Fiorentino et al., 1989).

In addition to this inhibitory effect on Th1 cell cytokine synthesis other investigators had described an activity in the supernatants of mouse B cell lymphomas that strongly enhanced the proliferation of thymocytes in response to IL-2 and IL-4 (Suda et al., 1991). Even more surprising was the finding that the same lymphoma supernatants contained a factor that sustained the viability of mouse mast cell lines in culture and augmented their proliferative response toward IL-3 or IL-4 (O'Garra et al., 1990).

These biological activities led investigators at DNAX to screen cDNA libraries of Th2 clones for the IFN-γ-suppressing activity and finally isolate a cDNA clone for murine IL-10 (Moore et al., 1990). Cross-hybridization experiments with the mouse cDNA finally identified clones coding for human IL-10 (Vieira et al., 1991). Antibodies raised to this cytokine then demonstrated that the activities on thymocytes and mast cells were also due to the effects of IL-10 (Mosmann et al., 1990).

This chapter is designed to describe the now discovered numerous activities of IL-10 on different cell systems under different physiological or pathological conditions. In the first part a general introduction is given about the known characteristics of the IL-10 gene and protein, as well as its homology to an open reading frame of the Epstein-Barr virus (EBV) BCRF1, as well as the IL-10 receptor. Thereafter, the *in vitro* effects of IL-10 on various cell systems are described, whereby a certain redundancy is unavoidable and in fact intended. The last part is then devoted to IL-10 action in vivo under physiological, as well as certain pathological conditions in various disease models.

II. THE IL-10 GENE

A. Chromosomal Localization

The murine (mIL-10) and human (hIL-10) Interleukin-10 genes are located on mouse and human chromosome 1 (Kim et al., 1992). The mIL-10 gene comprises five exons arrayed over approximately 5.1kb of DNA and it is reported that the genomic clone itself encodes an expressible mIL-10 protein (Moore et al., 1993). The structure of the mIL-10 gene generally resembles that of other cytokine genes. Intron-exon junctions do not disrupt amino acid codons, and therefore it is possible for alternative splicing events to occur without disruption of the open reading frame (ORF). However, so far there is no evidence for altered forms of the IL-10 protein or mRNA, which would suggest alternative splicing (Kim et al., 1992).

The noncoding regions of the mIL-10 gene contain a number of sequence elements that have been associated with regulation of transcription of cytokine genes (e.g., TATA-box, AP-1 binding sites). The combination of motifs found in the IL-10 gene is similar to the 5' flanking region of the mouse and human IL-6 genes, although the sequences found in the mIL-10 gene are stretched over a much larger area of the gene, than those of IL-6 (Kim et al., 1992). As for IL-6, production of IL-10 is not inhibited by cyclosporine A or FK506. Furthermore, the late expression of IL-10 after activation of monocytes and keratinocytes is suggestive of an autoregulatory, negative feedback mechanism for IL-10 at least in these cells (de Waal Malefyt et al., 1991; Enk and Katz, 1992).

The mIL-10 is located closely linked to the En-1, Ren-2, and Ly-5 loci of the chromosome 1. This region is syntenic with the corresponding region of human chromosome 1, and the hIL-10 has also been mapped to chromosome 1. As has been shown for mIL-10, hIL-10 will likely reside in the 1q region as well. This finding might be of importance, as several translocation breakpoints and DNA rearrangements associated with human neoplasia have been assigned to 1q (Kim et al., 1992).

B. IL-10 cDNA Clones

The primary structure of mIL-10 cDNA was determined by screening expression libraries from the Th2 clone D10.G2 for IL-10 bioactivity on HDK.1 cells (Th1 cell clone; Moore et al., 1990). In this assay, the IFN-γ-suppressing activity of IL-10 was assayed for and used as a marker for IL-10 production. Thousands of clones were assayed until single clones expressing CSIF-activity were identified and a cDNA encoding IL-10 was cloned. Subsequently, cDNA clones encoding human IL-10 were identified by cross-hybridization with the mouse cDNA (Vieira et al., 1991). A high degree of nucleotide sequence homology (> 80%) is exhibited by mIL-10 and hIL-10 cDNA clones, with the only significant difference being the insertion of a human Alu repetitive sequence element in the 3'-untranslated region of the hIL-10 cDNA clone.

C. The IL-10 Protein

Very similar ORF are encoded by the mouse and human IL-10 cDNAs (73% amino acid homology) of 178 amino acids, including the hydrophobic leader sequences. Whereas hIL-10 is active on mouse cells, no such activity has been found for mIL-10 on human cells.

IL-10 is a member of the four α-helix bundle family of cytokines based on its primary structure. Whereas hIL-10 is an 18kD polypeptide lacking detectable carbohydrate, mIL-10 is N-glycosylated at a site near its N-terminus missing from hIL-10. This glycosylation seems to be heterogeneous, resulting in a mixture of 17, 19, and 21 kD species (Mosmann et al., 1990). Glycosylation is not required for biological activity as a mutant lacking the N-linked site is active, and recombinant mIL-10 (rmIL-10) expressed in *Escherichia coli* (and therefore lacking glycosylation) retains all known biological activities (Moore et al., 1993).

Early immunoprecipitation studies indicated that mIL-10 is expressed as a noncovalent homodimer, an observation that seems to hold true for hIL-10 as well. It is not certain at the moment whether mIL-10 or hIL-10 monomers are biologically active. mIL-10 and hIL-10 are rapidly inactivated by exposure to acid (pH < 5.5), but appear to be stable in basic conditions up to pH 11. The mature N-terminus of mIL-10 was determined by radiochemical sequencing of rmIL-10 to be Gln22, whereas the N-terminus of rhIL-10 is Ser19 (Moore et al., 1993). In the meantime recombinant mIL-10 and hIL-10 have been expressed in COS7 cells, mouse myeloma cells, Chinese hamster ovary cells (CHO), *E. coli*, and a baculovirus expression system. So far the biological activities have been described as indistinguishable. Elongated mIL-10 and hIL-10 with peptide "tags" of at least eight amino acids at the N-terminus and 21 amino acids at the C-terminus have been prepared by the DNAX group without any detectable loss of biological activity.

D. IL-10 and its Homology to the EBV Gene BCRF1

hIL-10 and mIL-10 exhibit strong DNA and amino acid sequence homology to an ORF in the EBV genome, BCRF1 (Hsu et al., 1990). The original reports revealed that mIL-10 had a sequence identity of approximately 70% on the amino acid level, when the mature polypeptides were compared. It was found that BCRF1 shared most, but not all of the activities of mIL-10, so that it was assumed that BCRF1 might represent a processed viral homolog of the cellular IL-10 gene, that was therefore called viral IL-10 (vIL-10) from then on (Moore et al., 1990; Hsu et al., 1990; Vieira et al., 1991).

In the meantime it has become clear that the homology of hIL-10 and mIL-10 is confined to the mature protein coding sequence and is not detected in the signal sequence or 5' or 3'-untranslated sequences. Of the three sequences, mIL-10 and hIL-10 are most closely related with a homology at the DNA level of about 81%. DNA sequences for mature hIL-10 and BCRF1 proteins have only 71% homology. Additionally hIL-10 and BCRF1 are more closely related in amino acid sequence (84%; Moore et al., 1991; Vieira et al., 1991). Using a panel of rat anti-hIL-10 antibodies it has been demonstrated that there is extensive cross-reactivity with vIL-10, but not with mIL-10 (Moore et al., 1993).

Unlike hIL-10 or mIL-10, the BCRF1 gene contains no introns. In aggregate, the findings suggest that mIL-10 and hIL-10 genes evolved from a common ancestor, vIL-10 may represent an ancestral processed cellular cytokine gene, and that vIL-10 has been largely constrained to resemble the human cytokine. In support of this, BCRF1 ORF encodes a 17kDa secreted polypeptide which, like hIL-10, contains little or no carbohydrate (Hsu et al., 1990). In assays, where vIL-10 activity can be detected, its specific activity appears to be three to10-fold lower than hIL-10.

vIL-10 is expressed only during the productive phase of EBV replication, namely in the late phase of the lytic cycle of EBV (Hsu et al., 1990). Thus, it is predicted that vIL-10 is not involved during latency periods in vivo, but only during periods of viral replication and spread. As mentioned above, vIL-10 displays some of the activities of IL-10 that will be discussed in more detail later.

E. The IL-10 Receptor

The presence of an IL-10-binding structure has been demonstrated on human and mouse cells by binding studies using iodinated IL-10 (Tan et al., 1993). It was shown that high affinity binding structures (Kd ~50pM) for hIL-10 are present on B-lymphoma lines, T cell lines, and monocytic lines. For mIL-10 binding to mouse mast cell lines has been observed (Tan et al., 1993). Whereas the binding of mIL-10 to its receptor seemed to be species specific, human IL-10 receptor binding was not. Thus, hIL-10 was capable of competing for binding to murine IL-10 receptors with mIL-10, whereas mIL-10 could not compete for binding to human IL-10 receptors. The density of receptors per cell for hIL-10 and mIL-10 was shown to be between 100 and 300 receptors/cell with only a single class of binding site identified thus far (Tan et al., 1993).

This observation was confirmed by Weber Nordt and coworkers who demonstrated the presence of IL-10R on murine B cells, Th1 cells, and LPS-stimulated fibroblasts (Weber Nordt et al., 1994).

F. Cellular Sources for IL-10

In the murine system IL-10 is expressed by CD4+ Th2 clones, as well as some Th0 cells and some $CD8^+$ T cell clones (Fiorentino et al., 1989, 1991; MacNeil et al., 1990; Mosmann., 1991; Chen and Zlotnik., 1991;

Hisatsune et al., 1992). In addition, B cells (especially Ly-1 B cells; O'Garra et al., 1990; Kantor, 1991; Benjamin et al., 1992), activated mast cells, activated macrophages, and keratinocytes have been shown to produce IL-10.

In the human system IL-10 is expressed by T cells (and Th1, Th2, Th0 clones), by CD8+ T cells and T cell clones, monocytes and macrophages, keratinocytes, activated B cells, as well as B cell lymphomas (de Vries et al., 1991; de Waal Malefyt et al., 1991; Vieira et al., 1991; Spits and de Waal Malefyt., 1992; Emilie et al., 1992; Benjamin et al., 1992; Yssel et al., 1992). Although IL-10 is usually thought of as a Th2-type cytokine, its expression pattern is rather diverse, again resembling IL-6 more than IL-4.

It should be mentioned again at this point that IL-10 is induced rather late in all cell types mentioned following activation of the cell (de Waal Malefyt et al., 1991; Kennedy et al., 1992; Enk and Katz, 1992). This finding points to an important counterregulatory function of IL-10 in the induction of proinflammatory immune processes and complements its role as "cytokine synthesis inhibitory factor."

III. *IN VITRO* FUNCTIONS OF IL-10

A. Characterization of IL-10 Based on Its Inhibitory Functions on APC

The distinction of $CD4^+$ T helper cell clones in Th1 and Th2 cells based on their ability to secrete certain cytokines, was first made in the murine system by Mosmann and colleagues (Fiorentino et al., 1989). It was demonstrated that whereas Th1 cells produce IL-2, IFN-γ, and lymphotoxin (LT), and mediate delayed-type hypersensitivity reactions (DTH) functionally, Th2 cells release IL-4, IL-5, IL-6, and so forth, and provide help for B cell immunoglobulin production (Mosmann et al., 1986). These findings were later extended into the human system by Romagnani and colleagues (Del Prete et al., 1993). Very early on, however, it was observed that there was a strong crossregulation between Th1- and Th2-types of cells. Whereas IFN-γ (released by Th1 clones) was shown to inhibit proliferation of Th2 cells, there was an unknown product released by Th2 cell clones that inhibited proliferation of Th1 cell clones as well as cytokine production (Mosmann, 1991a, 1991b, 1991c, 1991d; Mosmann and Moore, 1991; Street and Mosmann, 1991). This factor was then termed "cytokine synthesis inhibitory factor" or CSIF.

It was observed in the beginning that CSIF inhibited IFN-γ synthesis in response to antigen plus APC or anti-CD3 plus APC, but not IFN-γ synthesis stimulated by anti-CD3 bound to a polystyrene surface, or by ConA (Fiorentino et al., 1989). Additionally, it was shown that CSIF had no measurable effect on the ability of Th1 cells to proliferate in response to IL-2. Therefore, an indirect effect of CSIF (IL-10) involving the APC was assumed.

This hypothesis was tested when purified IL-10 was available after identification of the coding sequences. When purified populations of murine macrophages or B cells were used as APC together with a Th1 cell clone, IL-10 only inhibited proliferation when macrophages, but not B cells were used as APC (Fiorentino et al., 1991). Preincubation of macrophages for at least seven hours resulted in a marked suppression of Th1 cell proliferation even if IL-10 was removed from the proliferation assays afterwards, whereas shorter preincubation periods seemed to be less efficient. Proliferation of Th2 cell clones remained unaffected by IL-10 (Fiorentino et al., 1991).

mIL-10 also inhibited the production of cytokines at both the mRNA and protein level by mouse Th1 cell clones in the presence of macrophage-containing APC and either antigen or soluble anti-CD3 (Fiorentino et al., 1991). In contrast, cytokine production of Th2 cell clones was not significantly affected. The inhibition of cytokine synthesis of Th1 cells by IL-10 was most pronounced for IFN-γ and IL-3, cytokines that are synthesized late after activation, but could also be observed to an extent for IL-2, LT, and GM-CSF.

IL-10 also inhibited cytokine synthesis in the human system. Inhibition of IFN-γ, TNF-α, GM-CSF, and LT production by human PBMC stimulated by PHA or anti-CD3 mAb by hIL-10 was observed at the protein and RNA levels (Hsu et al., 1990; Vieira et al., 1991). Again, IL-10 only inhibited cytokine production by $CD4^+$ human T cell clones when monocytes, but not B cells were used as APC. However, in the human system it has been demonstrated that IL-10 inhibits not only the APC-dependent cytokine production of Th1 cells, but also Th0 or Th2 like cells. Although the stimulatory cells used in mouse and man are somewhat different (splenic vs. peripheral blood APC), it is clear that IL-10 exerts slightly more pronounced effects in the human system than in mice (Fiorentino et al., 1989; Hsu et al., 1990; Vieira et al., 1991; Emilie et al., 1992).

In addition to inhibiting the macrophage-induced cytokine synthesis of T cell clones, IL-10 also inhibited the cytokine synthesis of human natural killer (NK) cells (Hsu et al., 1990, 1992). Thus, hIL-10 suppressed the production of IFN-γ and TNF-α in a macrophage-dependent, IL-2 driven stimulation, but had no effect on the cytokine production of NK cells induced by IL-2 alone (Hsu et al., 1990, 1992). This was also true for the IL-2-driven proliferation and generation of lymphokine-activated killer cells (LAK) that was not inhibited by IL-10 (Kaye and Bancroft, 1992).

As has already been mentioned, IL-10 does not only inhibit the production of cytokines by T cell clones, but also inhibits T cell proliferation (Ding and Shevach, 1992). Again the effect is indirect and dependent on the APC used (Macatonia et al., 1993). There is no effect on B cells, whereas the proliferation induced by monocyte/macrophage APC is inhibited in human T cells and T cell clones. The inhibition can be overcome, though apparently not completely, by addition of IL-2 (Ding and Shevach, 1992).

Similarly, mIL-10 inhibited the proliferation of mitogen-stimulated murine $CD4^+$ and $CD8^+$ T cells, as well as antigen-stimulated Th1 clones. In this case, however, the effects were more pronounced in the antigen-specific proliferation and the effect could be largely overcome by addition of IL-2 (Ding and Shevach, 1992). Furthermore, a marked decrease of expression of IL-2R p55 was observed, when IL-10 was added to the cocultures, suggesting that the effect of IL-10 was due, in part, to inhibition of APC-dependent stimulation of IL-2 production (Ding and Shevach, 1992).

The situation was more complex when dendritic cells (DC) were used as APC. Using splenic DC in the murine system, Macatonia et al. (1993) did not observe any effect of IL-10 on the proliferative response of Th1 cell clones. In addition, mIL-10 did not have any effect on the proliferation of primary $CD4^+$ or $CD8^+$ T cells. In contrast, when cytokines from supernatants were assessed, there was a marked inhibition of IFN-γ production, when $CD4^+$ T cells were used as responders. Production of IL-2 by the T cells, however, was not affected by IL-10-treatment. The contrasting effects of IL-10 on DC-induced IFN-γ production and proliferation of T cells suggest different pathways for stimulation of IFN-γ synthesis and proliferation (Macatonia et al., 1993).

The situation was quite different, when murine Langerhans cells (LC) as immature DC of the epidermis were used as APC in primary MLR or clonal responses (Enk et al., 1993). Enk et al. (1993) reported that while mIL-10 did not exert an inhibitory effect on the LC-induced proliferative response of naive $CD4^+$ T cells in allogeneic or anti-CD3 proliferation assays, production of IFN-g, and IL-2 were significantly reduced in the supernatants of the cocultures. These effects could be observed on the mRNA and protein level. In contrast, IL-10 inhibited cytokine production and proliferation of Th1 clones, when LC were used as APC. Proliferation of Th2 cell clones was not affected by IL-10 treatment. As in the macrophage system, preincubation of LC for at least 8 h in IL-10 significantly enhanced the antiproliferative effect in this culture system. The suppressive effect of IL-10 could only be observed, however, if freshly prepared LC were used as APC. LC that were cultured for two or more days prior to exposure to IL-10 were resistent to the inhibitory effect of IL-10. In further studies it was demonstrated that IL-10-treated freshly prepared LC not only were bad stimulators of clonal Th1 cell responses, but also induced T cell anergy in the Th1 clone (Enk et al., 1993). Thus, LC were converted from potent accessory cells to tolerance-inducing cells in vitro. These data might also explain why Macatonia et al. (1993) did not observe an effect of IL-10 on splenic DC, as these cells are considered to be fully mature DC, whereas freshly prepared LC are considered to be immediate precursors of mature DC. LC that are cultured in vitro for several days have been shown to mature into fully potent DC and are considered to be the *in vitro* correlates of splenic or lymph node DC.

In the human system, Peguet-Navarro et al. (1994) using human LC as APC also observed an inhibitory effect of hIL-10 on the proliferation of T cells. In contrast to the murine system, however, hIL-10 also inhibited the immunostimulatory functions of LC in primary allogeneic proliferation assays (Peguet Navarro et al., 1994). Again the effect of IL-10 was most pronounced when the cytokine was added to cocultures early on suggesting an effect of IL-10 on the early steps of T cell activation. The inhibitory effect on LC-induced T cell proliferation could be observed for both $CD4^+$ and $CD8^+$ T cells. Also, again in contrast to the murine system, even the accessory functions of cultured LC were inhibited by IL-10, indicating a more pronounced effect of IL-10 in the human system (Peguet Navarro et al., 1994).

Using human DC prepared from cord blood, Rousset et al. confirmed an inhibitory effect of IL-10 on DC-APC function. It was demonstrated that IL-10 suppressed the immunostimulatory effects of cord blood DC for $CD4^+$ and $CD8^+$ T cells by inhibiting their proliferation. Again the effect observed was on early T cell activation and could be observed for immature and mature DC.

B. Mechanism of Inhibition of APC Function

For macrophages, several mechanisms have been proposed to explain the inhibitory effect of IL-10 on accessory functions. In the human system it was observed by de Waal Malefyt that IL-10 suppressed the expression of MHC class II molecules on monocytes, an effect that could not be observed for B cells or DC (Go et al., 1990; de Waal Malefyt et al., 1991; Enk et al., 1993). The suppressive effect of hIL-10 can be seen for the constitutive, as well as the IFN-γ-induced expression of MHC class II molecules. Because APC-induced Ca^{2+} fluxes in human T cell clones were diminished when monocytes preincubated with IL-10 were used as APC, it was thought that MHC class II suppression by IL-10 prevented rather than suppressed T cell activation (de Waal Malefyt et al., 1991).

However, several points of evidence suggested that the suppressive effect of IL-10 on monocyte/macrophage MHC class II expression was insufficient to explain the inhibitory effect of IL-10 on accessory functions of these cells. First of all IL-10 did not suppress MHC class II expression of macrophages in the murine system, although it also inhibited macrophage APC functions (Moore et al., 1993). Second, the effect of IL-10 on MHC class II expression of human monocytes could be reversed by IL-4, but the suppressive effect on accessory functions persisted (de Waal Malefyt et al., 1991), and third, the inhibitory effects of IL-10 were observed in stimulations that were independent of MHC, but did require monocyte APC function such as APC-dependent NK cell cytokine synthesis (Hsu et al., 1992).

Together with the observation that freshly prepared LC (with an immature phenotype) were susceptible to the inhibitory effects of IL-10, while cultured LC (with a mature phenotype and a lot of costimulatory molecules) were resistant toward IL-10 effects these findings indicated that some costimulatory molecule was likely to be the target of IL-10 on monocytes and other APC. This was further confirmed by the data reporting tolerance induction in Th1 cell clones by IL-10-pretreated LC (Enk et al., 1993). Indeed, Ding and coworkers reported that the up-regulation of the costimulatory molecule B7 was inhibited by IL-10-treatment of murine macrophages, whereas expression of MHC class II molecules or ICAM-1 expression were not affected (Ding and Shevach, 1992; Enk et al., 1993; Ding et al., 1993). The suppressive effect of IL-10 on T cell proliferation in this system could be overcome by addition of B7-transfected L cells demonstrating that indeed the effect of IL-10 on

B7 was responsible for the anti-proliferative influence of this cytokine.

In humans similar results were obtained for IL-10-treated monocytes by Willems et al. (1994). These investigators showed that IL-10 inhibited the up-regulation of B7, as well as ICAM-1 on human monocytes (Willems et al., 1994). This was observed for the basal expression of ICAM-1, as well as for the IFN-γ-induced up-regulation of ICAM-1 and B7 in a dose-dependent fashion.

In the DC-system the effect of IL-10 has not been elucidated in detail thus far. Enk et al. (1993) reported for murine LC that IL-10 renders these cells from potent inducers of a T cell response to tolerizing APC *in vitro*. As this effect could not be observed if cultured cells were used in the assay, it was assumed that the up-regulation of a costimulatory molecule on freshly prepared LC was inhibited by IL-10. This was confirmed in a costimulator assay where allogeneic LC were cultured for three days in the absence or presence of IL-10 and then added to cultures of freshly prepared antigen-pulsed LC plus a Th1 cell clone in the presence of IL-10. Whereas proliferation was usually inhibited in these cultures by IL-10, proliferation could be restored by addition of allogeneic cultured LC that were cultured in the absence of IL-10, but not by LC cultured in the presence of IL-10 (Enk et al., 1993). Thus, it was demonstrated that indeed the up-regulation of a costimulatory molecule was inhibited by IL-10. As antigen-presentation in the absence of costimulation leads to T cell anergy, anergy experiments were performed. These experiments showed that IL-10-treated LC were capable of inducing T cell tolerance (Enk et al., 1993). Unfortunately, the nature of the costimulatory molecule involved in anergy induction of LC has not been elucidated thus far. Staining for B7-1, B7-2, ICAM-1, or HSA (heat stable antigen) did not show an inhibitory effect of IL-10 on these molecules on LC. Apart from membrane-bound molecules, cytokines or other soluble mediators could also be affected by IL-10. Thus far there is evidence that IL-10 inhibits production of IL-1β and IL-12 by LC/DC (Enk et al., unpublished). Both molecules are important mediators of a successful T cell sensitization. Further studies are necessary to determine whether suppression of these molecules is responsible for the inhibitory effects of IL-10 on LC/DC-APC function. For human LC, Peguet-Navarro described a suppressive effect of IL-10 on MHC class II expression. In their system expression of MHC class II molecules was downregulated following an 18h incubation of LC in IL-10 (Peguet Navarro et al., 1994). Although these authors did not investigate the effects of IL-10 on hLC B7-expression, they were unable to detect an effect of IL-10 on other costimulatory molecules such as ICAM-1 and LFA-3 (Peguet Navarro et al., 1994). Interestingly, addition of IL-1b to IL-10-treated LC/T cell cultures partially reversed the inhibitory effect of IL-10 on T cell proliferation indicating that indeed suppression of IL-1b production by IL-10 might be a mechanism of action of IL-10 in this system (Peguet Navarro et al., 1994).

C. Effects of IL-10 on T Cells

Many of the effects of IL-10 on APC-function use T cells as the final effector cells. However, besides these indirect effects, IL-10 also has some well-documented direct effects on T cells. Nevertheless, it should be emphasized that IL-10 is not a cytokine primarily important for T cell growth or development as can be seen in the IL-10 knockout mouse that does not seem to have a primary T cell defect (Kuhn et al., 1993).

For murine thymocytes mIL-10 has been shown to have a function as growth cofactor. By itself, mIL-10 does not induce proliferation of fetal thymocytes, but in the presence of IL-2 and IL-4 proliferation of fetal (day 15) thymocytes is enhanced (MacNeil et al., 1990; Bendelac and Schwartz, 1991). In addition, mIL-10 serves as an enhancer of proliferation of adult thymocytes in the presence of combinations of IL-2, IL-4, or IL-7 alone (Bendelac and Schwartz, 1991). Furthermore, except for double positive thymocytes, all thymocyte subpopulations were capable of producing IL-10 following polyclonal activation. The presence of IL-10 in fetal and adult thymocytes and its effect as a growth cofactor might indicate a role for IL-10 as a T cell differentiation factor.

For human thymocytes, analogous activities of hIL-10 have yet to be described, as proliferation or phenotype of total thymocytes or thymocyte subsets were not affected by IL-10. It has been reported though that the viability of triple negative thymocytes is enhanced by IL-10 (Moore et al., 1993).

For $CD8^+$ T cells Zlotnik and coworkers have reported that IL-10 enhances the proliferation, cytotoxic precursor frequency, and clone size following activation of the cells in the presence of ConA and IL-2 (Chen and Zlotnik, 1991). Additionally IL-10 also enhanced the cytotoxic activity of $CD8^+$ effector cells in a lectin-dependent cytotoxicity assay. The same authors also reported that murine $CD8^+$ cells that were grown in the presence of IL-2, IL-10, and several other cytokines acquired a potential to produce more IFN-γ upon subsequent stimulation, compared to cells grown in the same cytokine cocktail without addition of IL-10

(Chen and Zlotnik, 1991). It thus seems that IL-10 enhances T cell proliferation in some APC-independent systems, while being inhibitory when APCs are present.

In human T cells the situation seems to be somewhat different. Here IL-10 seems to have a direct inhibitory effect on the proliferation of peripheral blood T cells and $CD4^+$ T cell clones (Bejarano et al., 1992; te Velde et al., 1992; de Waal Malefyt et al., 1993; Taga and Tosato, 1992; Taga et al., 1993). Thus, it has been reported by de Waal Malefyt et al. (1993) that the antigen-specific proliferative response of human T cell clones was inhibited by hIL-10 when L cells that were transfected with the relevant MHC class II molecules were used as APCs. The MHC class II expression was not affected by IL-10 in this system, since the constitutive expression of the transfected MHC class II genes was under the control of an SV40 promoter. Proliferative responses of human Th0, Th1, and Th2 cells were likewise inhibited by IL-10 when L cells transfected with the FcγRII (CD32) were used in combination with either anti-CD3 mAb, or a mitogenic pair of anti-CD2 mAbs (de Waal Malefyt et al., 1993). As a control, species-specific mIL-10 did not inhibit T cell proliferation, indicating that hIL-10 acts directly on the responder T cells. When supernatants from proliferation assays were tested for cytokine production, inhibition of proliferation could be correlated to inhibition of IL-2 production. In contrast, production of IL-4 was unaffected, indicating that IL-10 has a differential effect on cytokine production by T cells. The mechanism by which IL-10 inhibits human T cell proliferation has remained unclear thus far.

In addition to inhibiting proliferation of human T cells IL-10 has been shown by Schandene et al. (1993, 1994) to inhibit B7-dependent production of IL-5 by human T cells in the absence of accessory cells. IL-10 inhibited IL-5 production on the protein and mRNA level. It was demonstrated that after T cell activation endogenously produced IL-10 inhibited IL-5 production (Schandene et al., 1994). These data indicate that IL-10, apart from being a potent downregulator of Th1 cytokines via functional APCs, can also directly control production of other Th2-type cytokines.

As another property of hIL-10 Taga et al. have described an effect of IL-10 on T cell apoptosis (Taga et al., 1993). When human T cell clones were deprived of IL-2, apoptosis occurred. If IL-10 was added to IL-2 deprived T cell cultures, cell viability was improved significantly from 35-70%. The effect of IL-10 was dose-dependent and could be reversed by addition of an anti-IL-10 mAb (Taga et al., 1993).

Yet another effect of IL-10 was recently described by Takanaski et al. (1994) for human peripheral blood eosinophils. In this system IL-10 inhibited survival of human eosinophils when LPS-stimulation was used to induce survival, but not when GM-CSF was used. Additionally IL-10 also suppressed LPS-induced cytokine secretion by eosinophils, as demonstrated for TNF-α, IL-8, and GM-CSF. As a mechanism of action suppression of GM-CSF production by eosinophils by IL-10 is identified (Takanaski et al., 1994).

In addition to influencing growth, differentiation, and proliferation of mouse and human T cells, IL-10 has also been shown to act as a selective chemoattractant for $CD8^+$ T cells (Jinquan et al., 1993). No such effect could be observed for IL-10 and $CD4^+$ T cells. In a checkerboard analysis hIL-10 was shown to be chemotactic, but not chemokinetic with a selective chemotactic activity for $CD8^+$ T cells, not other cells such as monoctes or granulocytes. Furthermore, hIL-10 was shown to inhibit the chemotactic response of $CD4^+$ T cells, but not $CD8^+$ T cells toward IL-8 (Jinquan et al., 1993).

D. Effects of IL-10 on B Cells

The production of IL-10 by B cells and B cell lymphomas has been known for some time (O'Garra 1990; Go et al., 1990). As a major source of IL-10, O'Garra and coworkers identified $CD5^+$ (Ly-1) B cells (Kantor, 1991; Suda et al., 1991; O'Garra and Howard., 1992). In contrast to normal splenic B cells, Ly-1 B cells are a self-replenishing population which is relatively abundant in the peritoneal cavity and is minimally represented in spleen and lymph nodes of normal adult mice. In older mice, clonal expansion of Ly-1 B cells has been observed giving rise to speculation about a possible bias of immune reactivity in older mice toward a Th2 response. Ishida et al. (1992a, 1992b, 1993) furthermore demonstrated later that IL-10 acts as an autocrine growth factor for Ly-1 B cells. Mice treated continuously from birth to eight weeks of age with a neutralizing anti-IL-10 mAb lacked peritoneal-resident Ly-1 B-lymphocytes, exhibited greatly reduced IgM levels, and were unable to generate significant *in vivo* antibody responses to bacterial antigens α1,3-dextran or phosphorylcholine, antigens known to activate the Ly-1 subset of B cells (Ishida et al., 1992). In contrast, conventional splenic B cells of anti-IL-10-treated mice were normal with respect to total numbers, phenotype, and function. The mechanism of Ly-1 B cell depletion appeared to be related to elevation of endogenous IFN-α levels in anti-IL-10-treated mice, since coadministration of neutralizing anti-IFN-γ antibodies substantially restored the number of peritoneal Ly-1 B cells in the peritoneal cavity. Thus, IL-10 seems to influence lymphocyte development of certain B cells and therefore might be an important modulator of humoral immune responses.

In addition to its effect on these specialized B cells, IL-10 exerted various effects on conventional B cells. Highly purified, small, dense splenic B cells from unstimulated mice show increased expression of MHC class II antigens and enhanced viability when cultured with rmIL-10 compared with B cells cultured in medium alone (Go et al., 1990). Unlike IL-4, IL-10 did not up-regulate the expression of Fcε receptor II (CD23) or induce proliferation of small resting B cells, indicating differential mechanisms of action of the two cytokines. Interestingly, IL-10 did not upregulate MHC class II antigens on B cells from X-chromosome-linked immunodeficiency (XID) mice, while the same cells showed normal up-regulation of MHC class II antigens in response to IL-4 (Go et al., 1990). However, IL-10 still sustained viability of XID B cells in culture suggesting that IL-10 may act through at least two distinct receptor-signal transduction systems in mouse B cells, one of which transduces the signal for upregulation of class II MHC expression. This is supported by studies of vIL-10.

In humans, hIL-10 is a potent growth and differentiation factor for activated B cells (Rousset et al., 1992). hIL-10 enhanced the proliferation of Staphylococcus aureus strain Cowan (SAC)-activated B cells, but had no effect on unstimulated cells. Even at optimal concentrations, the effect of IL-10 in this system was less pronounced than when B cells were costimulated with IL-2 or IL-4 (Rousset et al., 1992). When B lymphocytes were stimulated with anti-CD40 mAb, IL-4 and IL-10 were equally active, whereas IL-2 was inactive.

IL-10 also serves as a B cell differentiation factor. IL-10 induced SAC-preactivated B cells to produce high levels of IgM, IgG, and IgA more effectively than IL-2 or IL-4 (Rousset et al., 1992). When B cells were activated via the CD40 molecule, addition of IL-10 augmented production of IgM, IgG, and IgA. Again, IL-10 was much more efficient than IL-2 or IL-4. Unlike IL-4, neither vIL-10 nor hIL-10 induced anti-CD40-activated B cells to secrete IgE. In fact, IL-10 inhibited IgE-synthesis by PBMC in a different culture system. All four isotypes were detected when IL-4 and IL-10 were added simultaneously. Whereas the combination of IL-4 and IL-10 had additive effects on B cell proliferation, the levels of secreted IgM, IgG, and

IgA in response to both cytokines were less than those observed with IL-10 alone.

Defrance et al. assessed the effects of hIL-10 and TGF-β on naive, surface IgD^+ ($sIgD^+$) B cells as compared with more differentiated, sIgD- B lymphocytes (Briere et al., 1993, 1994). hIL-10 dramatically increased the Ig secretion by both types of B cells preactivated by SAC or anti-CD40 mAb. Whereas $sIgD^+$ B cells mainly produced IgM, IgG constituted the predominant Ig isotype produced by sIgD- B cells in response to IL-10. Furthermore, naive B cells produced IgA when cultured with TGF-β and IL-10 in the presence of anti-CD40 mAb and CDw32-transfected L cells. In contrast, TGF-β suppressed the IL-10-mediated IgG, IgM, and IgA secretions by sIgD- B cells. Naive B cells could not be induced to secrete IgA by TGF-β and IL-10 after cross-liniking of sIg, suggesting that ligation of CD40 was an obligatory signal required for commitment of sIgD+ B cells to IgA secretion. IL-10 and TGF-β, therefore, seem to cooperate as switch factors for naive human B cells to secrete IgA. The data suggest an important role for IL-10 in regulating B cell proliferation and differentiation. The nature of this effect seems to be dependent upon the way the B cells are activated, on the surrounding cytokine milieu, and the state of differentiation of the cell (Defrance et al., 1992; Briere et al., 1993).

Using a different system, Pecanha et al. (1992, 1993) demonstrated that IL-10 ihibited Ig secretion by dextran-conjugated, anti-δ-antibody-activated B cells as well as the Ag-specific responses of B cells to TNP-Ficoll in the presence of IL-5. IL-10 had no inhibitory effect on IL-2-stimulated responses, indicating a selective effect on a specific mode of B cell activation. IL-10 also did not inhibit anti-δ-dextran-stimulated B cell proliferation, and the IL-10-induced inhibition of Ig secretion did not result from suppression of IFN-γ production. These data suggest that IL-10 exerts different effects on B cells depending on their maturation stage. IL-10 seems to promote early phases in B cell differentiation and suppress the late steps in the activation of Ag-stimulated B cell-activation.

E. IL-10 Effects on Mast Cells

Both hIL-10 and mIL-10 supported the viability of mouse mast cell lines *in vitro* and synergized with IL-3 and IL-4 to stimulate proliferation of these cell lines in culture (Thompson Snipes et al., 1991). The effect was most pronounced if all three cytokines were added simultaneously.

IL-10 alone had no effect on mast cell proliferation. In addition, IL-10 augmented both the number and size of mast cell colonies from mesenteric lymph node precursors isolated form nematode-infected mice in the presence of IL-3 and IL-4 in vitro. Mast cells in these cultures have the characteristics of mucosal-like mast cells (Thompson Snipes et al., 1991).

Ghilyad et al. (1992a, 1992b, 1993) also recently showed that IL-10 enhanced the expression of the mucosal MC protease MMCP-2 gene in IL-3-pretreated bone marrow-derived mast cells. The induction was strictly dependent on the presence of IL-10 and could be reversed by withdrawing IL-10 from the culture medium. These data qualify IL-10 as a growth and differentiation factor for murine mast cells (Ghildyal et al., 1992a, 1992b, 1993).

F. IL-10 Effect on Macrophages/Monocytes

Apart from its effects on the accessory functions of monocytes and macrophages, IL-10 also has pronounced effects on the morphology, phenotype, cytokine production, and function of mouse and human cells.

Morphologically IL-10 induced rounding up of cells in cultures of purified murine peritoneal macrophages and reduced adherence to plastic (Moore et al., 1993). Both hIL-10 and mIL-10 strongly inhibit production of monokines by human and mouse macrophages following activation (Bogdan et al., 1991; de Waal Malefyt et al., 1991; Fiorentino et al., 1991). hIL-10 inhibited synthesis of IL-1α, IL-1β, IL-6, IL-8, TNF-α, GM-CSF, and G-CSF by human monocytes following activation by LPS or LPS plus IFN-γ at both mRNA and protein levels. In an analogous fashion, mIL-10 inhibited production and mRNA levels of IL-1, IL-6, and TNFα by murine macrophage cell lines and peritoneal macrophages activated by LPS or IFN-γ (Bogdan et al., 1991; de Waal Malefyt et al., 1991; Fiorentino et al., 1991).

Human and mouse monocytes activated by LPS or IFN-γ also produce high levels of IL-10. IL-10 was first detectable 7 h after activation of the monocytes reaching peak strength after 24-48 h, indicating that the production of IL-10 was relatively late compared with that of IL-1α, IL-1β, IL-6, IL-8, TNFα, GM-CSF, and G-CSF. Production of IL-10 was inhibited by IL-4.

Human IL-10 also had autoregulatory effects (de Waal Malefyt et al., 1991). It strongly inhibited IL-10 mRNA synthesis in LPS-activated monocytes. In addition, endogenously produced hIL-10 strongly down-regulates class II MHC expression on LPS-activated monocytes via an autoregulatory mechanism. Thus, endogenously produced IL-10 has important autoregulatory and anti-inflammatory functions.

IL-10 also induced changes in the expression of cell surface molecules on monocytes. As discussed, IL-10 inhibited class II MHC expression on human monocytes. IL-10 did not affect the expression of CD11abc, CD14, CD54, or CD58 on monocytes, and even up-regulated expression of FCγRI correlating with an enhanced capacity of these cells to lyse Rh+ erythrocytes (ADCC function; te Velde et al., 1992).

mIL-10 was shown to inhibit production of reactive oxygen intermediates (H_2O_2) and reactive nitrogen intermediates (NO) by macrophages following activation of the cells by IFN-γ (Gazzinelli et al., 1992; Cunha et al., 1992). IL-10 inhibited NO-production if it was present before or during the first 12h of IFN-γ activation, that is, the kinetics were similar to those observed for the CSIF activity of IL-10. The inhibition was reversible by a later incubation with IFN-γ in the absence of IL-10. Production of NO by IFN-γ-activated macrophages is dependent on endogenous TNFα, and inhibition by IL-10 of NO production by IFN-γ-activated monocytes was substantially due to inhibition of TNFα production (Oswald et al., 1992).

G. Effect of IL-10 on Dendritic Cells

Except for affecting the APC functions of DC, IL-10 also has some direct effects on this cell type. Peguett-Navarro et al. (1994) have reported that hIL-10 inhibits class II MHC expression of human Langerhans cells (LC), an effect that could not be observed in the murine system (Oswald et al.,1992).

Additionally, IL-10 also seems to suppress cytokine production of DC/LC. It was noted that IL-10 inhibits IL-1β_ production of murine LC (Enk et al., unpublished). Indirect evidence for a similar effect was obtained by Peguet-Navarro in human LC (Peguet Navarro et al., 1994). Here the suppressive effect of IL-10 on LC APC function could be reversed by addition of IL-1β, which is usually produced by LC. Furthermore, there is evidence that IL-10 suppresses production of IL-12 by human and murine DC obtained from proliferating precursors (Enk et al., unpublished).

H. Viral IL-10 (BCRF1)

As already mentioned vIL-10 shares many of the activities of regular IL-10 and exhibits CSIF activity on both human and mouse cells, although with a much lower activity as hIL-10. Like regular IL-10, vIL-10 deactivates macrophages and inhibits the production of superoxide anions following activation of the cells (Niiro et al., 1992, 1994).

However, there are also differences in the spectrum of action of vIL-10 and regular IL-10. Thus, vIL-10 did not induce MHC class II expression on mouse or human B cells, but did sustain the viability of the B cells in culture (Go et al., 1990). Furthermore, the abilities of vIL-10 to costimulate the proliferation of mouse thymocytes or mouse mast cells were substantially reduced or absent (Rousset et al., 1992; Defrance et al., 1992). As mentioned previously, these data might indicate the existence of at least two types of receptor/signal transduction systems for mIL-10, one of which reacts poorly with vIL-10.

For human cells vIL-10 has been reported to enhance the reactivation of virus-specific cytotoxic T cells and the HLA-unrestricted killer cell responses *in vitro*. It was speculated that since vIL-10 is expressed during the lytic cycle of EBV, the expression of vIL-10 may enhance immune responses to EBV-infected cells during periods of virus replication *in vivo*. In this way, the virus might limit its own replication and maintain the apathogenic virus carrier state characteristic of EBV (Stewart and Rooney, 1992).

In addition, it was shown by Taga et al. (1994) that vIL-10 inhibited the apoptotic cell death of T lymphocytes from patients with infectious mononucleosis (EBV-infection). T cells from patients with EBV-infection are known to undergo rapid apoptosis when incubated *in vitro*. When these cells were cultured in cellular or vIL-10, survival was increased from 35-70%. The effect was reversible by addition of anti-IL-10 mAb. Thus, IL-10 promoted the survival of infectious mononucleosis T cells and might be critical for the establishment of immunologic memory after resolution of the disease (Taga et al., 1994).

IV. EFFECT OF IL-10 *IN VIVO*

A. IL-10 as a Regulator of Th2 Cell Development

The question of whether IL-10 serves as a critical mediator for the development of Th2 cells has not been fully addressed, yet. Although the original data on the effect of IL-10 on dendritic cells did not find an effect of the cytokine on the proliferation induced by DC (the cell type thought to be most important for the induction of a primary immune response;), this was not confirmed in later reports (Enk et al., 1993). Here several investigators did see an effect of IL-10 on the proliferation induced by certain immature LC or by human cord blood DC. Furthermore, even the original experiments showed an inhibition of DC-induced IFN-β production by T cells in primary stimulation cultures.

Probably the most revealing experiments to identify cytokines critical for directing differentiation of naive T cells were done by Hsieh and coworkers (1992). Using an αβ-TCR transgenic mouse line as a source for large numbers of antigen-specific, naive T cells, these investigators demonstrated that addition of IL-4, but not IL-10 to primary stimulation cultures was sufficient to drive T cell development toward a Th2 phenotype. However, with no cytokines added and neutralization of endogenous IL-10 by addition of anti-IL-10 mAb to primary cultures, a Th1 phenotype ensued (Hsieh et al., 1992). Furthermore, whereas effects of IL-4 on T cell differentiation were independent of the APC used, neutralization of IL-10 only showed an effect on T cell differentiation when splenocytes, but not when B cell hybridomas were used as APC. These data would suggest, that IL-4 directly promotes differentiation of Th2 cells, while IL-10 inhibits Th1 differentiation by skewing APC function.

B. IL-10 in Infectious Diseases

The most extensive data on IL-10 and its role in infectious diseases have been collected in various epxerimental parasite infections. In most of these models it was already known that a Th1 immune response implicated protective immunity, while a Th2 response was considered to be ineffective and harmful.

One of the best studied infectious disease models in the mouse is the infection with Schistosoma mansoni (Kullberg et al., 1992; Murphy et al., 1993). The infection of mice with this parasite results in a strong Th2 response with a concomitant suppression of a "healthy" Th1 response. Thus, it was demonstrated by several investigators that infection of mice with Schistosoma mansoni resulted in significant suppression of IFN-γ which is supposed to be essential for killing the parasite by macrophages (Gazzinelli et al., 1992). As a mechanism of killing, nitric oxide production by macrophages is essential. *In vitro* suppression of IFN-γ by T cells from infected mice could be restored by the addition of anti-IL-10 mAb suggesting a crucial function for the cytokine in the development of disease. As mentioned above it was shown before that nitric oxide-dependent killing of parasites was inhibited by IL-10. Inhibition of NO production and parasite elimination by IL-10 appears to be due to inhibition of TNFα production by IL-10, since exogenous TNFα reportedly abrogated the effect of IL-10 (Oswald et al., 1992). In aggregate, IL-10 not only inhibited T cell production of the macrophage activating cytokine IFN-γ_, but directly down-regulated macrophage activation as well.

Similar results have been obtained by other investigators for *Trypanosoma cruzi,* murine leishmaniasis, and mycobacterial infections (Silva et al., 1992, 1993; Reed et al., 1994). In trypanosomal infections, susceptible and resistant strains of mice have been described. Whereas spleen cells of resistant mice showed only little production of IL-10 during acute infection *in vitro*, a lot of IL-10 could be detected in susceptible animals (Silva et al., 1992). In addition, the addition of anti-IL-10 mAb to the cultures of spleen cells from susceptible mice restored the ability of these cells to produce IFN-γ.

In infectious leishmaniasis cells isolated from draining lymph nodes of resistant strains of mice expressed the Th1 cytokines IL-2 and IFN-γ, whereas cells from susceptible mice contained mRNA for IL-10 and IL-4 (Heinzel et al., 1991). Interestingly, elimination of Th2-producing cells *in vivo* by anti-CD4 mAb treatment resulted in healing of infected susceptible mice. Similar findings were obtained in patients with Leishmania donovani infections. Whereas during acute infection PBMC from patients produced high levels of IL-10, healing was associated with coexpression of IFN-γ.

In mice experimentally infected with Mycobacteria, neutralization of IL-10 by mAb *in vivo* resulted in resistance to systemic infection (Denis and Ghadirian., 1993; Bermudez and Champsi., 1993). A similar immunosuppressive function was described for IL-10 in mycobacterial infection in man (Yamamura et al., 1991; Modlin and Nutman, 1993; Sieling et al., 1993; Barnes et al., 1993).

On the other hand it was demonstrated in a murine model of Herpes virus keratitis that IL-10 can prevent an ocular inflammation leading to permanent blindness by suppressing proinflammatory cytokines without compromising clearance of the infecting virus from the eye (Tumpey et al., 1994).

In a similar fashion IL-10 has also been shown to suppress the release of TNF and prevent lethality in experimental endotoxemia (Howard et al., 1993; Gerard et al., 1993; van der Poll et al., 1994). When IL-10 was injected into mice prior to LPS-challenge, amounts of LPS-induced

TNFα_ were significantly reduced. Also IL-10 efficiently prevented the hypothermia generated by LPS and prevented mortality. The protective effect could even be observed if IL-10 was applied 30 min before LPS-challenge and could be reversed by anti-IL-10 mAB. Similar results were obtained for *Staphylococcal enterotoxin* B-induced lethal shock (Bean et al., 1993). As an autoregulatory feedback loop it was shown in man and chimpanzees that the endotoxin-induced rise in TNFα also induced a protective rise in IL-10 levels simultaneously functioning to inhibit ongoing TNF synthesis.

In the mouse model of murine acquired immunodeficiency syndrome (MAIDS) there is also a strong, dominant Th2 immune response and a suppressed Th1 immunity. In this model, virus infection is accompanied by prompt polyclonal T cell activation, most likely by a retrovirus-associated superantigen, followed by $CD4^+$ cell-dependent B cell proliferation and differentiation, and immune deficiency. In the first four weeks of infection, cytokine expression is shifted from a mixed (Th0) pattern to a Th2 pattern, characterized by high expression of IL-4, IL-6, and IL-10, and little expression of IL-2 and IFN-γ (Gazzinelli et al., 1992). In humans the shift of a Th1 to a Th2 phenotype during progression of HIV-infection has been discussed rather controversially (Graziosi et al., 1994; Clerici et al., 1994). Whereas Clerici et al. obtained clear evidence for a dominant Th2 phenotype at later stages of disease progression, subsequent work by others has failed to confirm these data. However, it has been shown that lymphomas from HIV-infected patients showed IL-10 gene expression in about 50% of all cases investigated, whereras expression was rarely detected in non-HIV cases. It is speculated that IL-10 might stimulate AIDS lymphoma growth and proliferation in an autocrine way (Emilie et al., 1992, 1993).

C. IL-10 in Autoimmunity

It has been shown by Howard et al. (1993) in the NZB x NZW F_1 hybrid mouse (New Zealand black x New Zealand white), a mouse model for human systemic lupus erythematosus who develop fatal immune-complex-mediated glomerulonephritis around 6-9 months after birth (in females), that continuous administration of anti-IL-10 mAb from birth delays the onset of autoimmunity (Ishida et al., 1994). The effect of IL-10-neutralization seemed to be mediated via elevation of TNFα levels since coadministration of anti-TNFα mAb neutralized the protective effect. On the contrary, injection of IL-10 enhanced the onset of autoimmune-glomerulonephritis. It seems, therefore, possible that IL-10 might also play a role in the induction of human SLE.

In a transgenic mouse system expressing IL-10 in the A cell area of the pancreas in nonobese-diabetic mice (NOD), IL-10 expression accelerated the outbreak of diabetes significantly. On the other hand, however, Wogensen et al. (1993, 1994) showed in a different transgenic mouse model overexpressing IL-10 in the islet of Langerhans β cells that IL-10 expression only led to pronounced inflammation in the pancreas, without inflammation of the islets of Langerhans and without diabetes (Wogensen et al., 1993, 1994; Lee et al., 1994). As a mechanism, recruitment of inflammatory cells by enhanced expression of MHC class I and II molecules and adhesion molecules induced by IL-10 is discussed. It should be noted that here normal BALB/c mice were used as founder animals, whereas NOD mice that are prone to develop diabetes were founders in the first report possibly explaining the different results.

In the Th1 type autoimmune disease model of experimental autoimmune encephalomyelitis analysis of cytokine mRNA expression in the central nervous system during the course of the disease showed that expression of IL-10 in the CNS is assosiated with recovery from disease (Kennedy et al., 1992). Whereas the acute phase of the disease was characterized by rapid increases in the levels of proinflammatory cytokines such as IL-2, IL-4, IL-6, IFN-γ, and IL-1α, IL-10 was only expressed modestly in the acute phase of the disease. However, stabilization of clinical symptoms was not only characterized by a rapid decline in mRNA levels for IL-2, IL-4, IL-6, and IFN-γ, but also by a concomitant rise in IL-10 mRNA levels, suggesting a role for IL-10 in the natural recovery from this disease.

In patient with severe combined immunodeficiency (SCID) that received an HLA mismatched transplantation of hematopoietic stem cells, high levels of IL-10 production *in vivo* were associated with tolerance (Bacchetta et al., 1994). It was shown that $CD4^+$ autoreactive T cells from these patients after stimulation produced high amounts of IL-10 and very low amount of IL-2, inhibiting their proliferative response. Additionally, high levels of IL-10 could be detected in the supernatants of patient-derived monocytes that proved to be major sources of IL-10. High endogenous production of IL-10, therefore, seems to be a general phenomenon in SCID patients with allogeneic stem cell transplantation and might be of importance for tolerance maintenance.

The expression of hIL-10 mRNA and protein could also be demonstrated in synovial tissue from patients with rheumatoid arthritis and osteoarthritis (Katsikis et al., 1994). IL-10 was spontaneously produced in synovial membrane cultures from all patients tested without further stimulation, whereas none of the controls showed IL-10 production. As cellular sources for IL-10 macrophages and T cells in the synovial membrane lining layer were identified. Neutralization of IL-10 by anti-IL-10 mAb resulted in an upregulation of proinflammatory cytokines TNFα and IL-1β. The findings indicate that IL-10 serves as an important immunoregulatory component in the cytokine network of these rheumatic diseases.

Finally the generation of IL-10 knock-out mice has lead to further evidence that IL-10 is an important regulator of inflammatory and autoimmune diseases. In IL-10-deficient mice, generated by gene targeting, lymphocyte development and antibody responses were normal, but animals were growth retarded and anemic (Kuhn et al., 1993). They also suffered from chronic enterocolitis. The alterations in intestinal tissue included extensive mucosal hyperplasia, inflammatory reactions, and aberrant expression of MHC class II molecules on epithelia. However, mutants kept under specific pathogen-free conditions develop only a local inflammation limited to the proximal colon. These results indicate that bowel inflammation in the mutants originates from uncontrolled immune responses stimulated by enteric antigens and that IL-10 is an essential immunoregulator in the intestinal tract.

D. IL-10 in Skin

During the past several years it has been demonstrated that the epidermis and especially the activated keratinocyte is a rich source of cytokines and growth factors. Epidermal keratinocyte-derived-IL-10 has been described by Enk and coworkers originally in the induction phase of contact sensitivity reactions in mice (Enk and Katz, 1992; Enk et al., 1993, 1994). It was demonstrated that up-regulation of IL-10 mRNA occurred rather later after allergen-application compared with the other proinflammatory cytokines (Enk and Katz, 1992). IL-10 mRNA was found to be induced only after 4 h following hapten application, reaching peak strength after 12 h and being down-regulated again after 24 h. This has been confirmed by Ferguson et al. (1994) in the meantime. The up-regulation of IL-10 was found to be allergen-specific and could not be induced by irritants or tolerogens. The late induction of the cytokine and its probable down-modulatory effect on IFN-γ and IP-10 (interferon-induced protein 10) in the epidermis indicated a counter-regulatory function of IL-10 in skin immune responses. To further investigate the effect of IL-10 on immune cells in the epidermis, epidermal LC were exposed to rmIL-10 and the effect of IL-10 on LC-MHC class II-expression was analyzed (Enk et al., 1993). In our experiments, however, no effect of IL-10 on MHC class II expression of LC could be detected. In addition, when IL-10-treated LC were used as APC for naive $CD4^+$ T cells in allogeneic and anti-CD3-induced proliferation assays, no effect of IL-10 on T cell proliferation could be observed. However, if supernatants from these proliferation assays were assayed for IFN-γ and IL-2-production, both cytokines were found to be suppressed. When Th1 or Th2 cell clones were used as responder cells for IL-10-treated LC, the proliferative response of Th2 cell clones remained unaffected, while Th1 cell clones were markedly inhibited. As could be shown in costimulator assays, the expression of some costimulatory molecule on LC was inhibited by IL-10-treatment. Furthermore, anergy experiments showed that LC was converted from potent inducers of T cell proliferation and priming to tolerizing APC *in vitro*. Similar results have been obtained in a tumor-antigen model by Granstein and colleagues (unpublished) in the meantime.

In order to investigate whether IL-10 also subserves tolerizing function *in vivo*, mice were injected with rmIL-10 prior to application of a contact allergen to the overlying epidermis (Enk et al., 1994). This procedure inhibited reproducible sensitization in all animals tested, whereas controls responded normally. When IL-10-treated animals were subjected to a second sensitization protocol 14 days later in the absence of IL-10 treatment, inhibition of sensitization persisted, indicating that the mice had become tolerant. When those mice were exposed to a different allergen again 14 days later, they responded normally, indicating that they were immunologically fully competent. Additionally, draining lymph node cells from mice treated with IL-10 did not proliferate upon stimulation with the allergen they had received together with IL-10 *in vitro*, but did proliferate normally upon stimulation with allergen number two. When the epidermal cytokine pattern of mice treated with IL-10 was analyzed following allergen stimulation and compared with controls, a marked inhibition of cytokine signals for TNFa, IL-1b, and IL-1a was noted in IL-10-treated animals. Thus, IL-10 is capable of rendering mice tolerant toward a distinct contact allergen *in vivo* and it appears that tolerance is mediated by disturbance of the epidermal cytokine milieu by IL-10.

A similar regulatory role for the induction process of contact sensitivity was also described by Ferguson et al. (1994) for IL-10 *in vivo*.

These authors confirmed the previously mentioned time course analysis and demonstrated that during peak IL-10 expression in the epidermis no transfer of contact sensitivity by lympoid cells was possible, indicating that IL-10 is an important regulator of contact sensitivity reactions. Furthermore, these authors showed that IL-10 injected into the skin of sensitized mice prior to antigenic challenge inhibited the elicitation of contact sensitivity. This was also confirmed in further detail by Schwarz et al. (1994). In addition, studies on the induction of contact sensitivity in IL-10 knock-out mice indicate that IL-10 is an important counter-regulatory factor in this immune response. It has been shown that IL-10-deficient mice develop ear necrosis when treated epicutaneously with a contact allergen, indicating that the antiinflammatory effects of IL-10 are essential in this disease model and supporting the experimental data reported above (unpublished). Production of IL-10 by activated keratinocytes has been described for human keratinocytes as well in the meantime (Grewe et al., 1995). As an additional stimulator of IL-10 expression in mouse and man UV-light has been described (Rivas and Ullrich, 1992). Rivas and Ullrich provided experimental evidence that the well-known immunosuppressive effects of UV-light were largely due to IL-10 production following insolation. Injection of anti-IL-10 mAb prior to or together with UV-exposure was able to almost completely block the immunosuppressive effect of UV-light. This was confirmed for the human system by Grewe et al. (1995) as well. Thus, expression of IL-10 following UV-exposure might be an important factor in the induction of tumors. It might be that IL-10-mediated suppression of immune surveillance leads to a deficit in the detection of newly induced tumor cells by the immune system, or, what is worse, to the induction of specific tolerance by IL-10-tolerized APC toward tumor tissue.

V. CONCLUSIONS

As outlined previously the biologic activities of IL-10 are multiple and diverse. Although IL-10 was originally identified as "cytokine synthesis inhibitory factor" and although a lot of knowledge has certainly been accumulated about this function of IL-10 in the meantime, this is at most one (important) effect of IL-10 among many others. It has been delineated previously that IL-10 is also a potent T cell growth and differentiation factor, that it directly suppresses T cells in humans; and that it is involved in numerous infectious or autoimmune diseases. Other functions, such as IL-10's role in the induction of tumor-tolerance or immunosuppression induced by tumors, have only received moderate attention thus far and certainly need to be explored in more detail. Also the molecular basics of IL-10 action, as well as a better understanding of the IL-10 receptor are studies that need to be done in the future.

ACKNOWLEDGMENTS

The author thanks Drs. Jürgen Knop and Stephen I. Katz for their continued support and wonderful mentorship over numerous years. This work was supported, in part, by the Deutsche Forschungsgemeinschaft.

REFERENCES

Bacchetta, R., Bigler, M., Touraine, J.L., Parkman, R., Tovo, P.A., Abrams, J., de Waal Malefyt, R., de Vries, J.E., & Roncarolo, M.G. (1994). High levels of interleukin 10 production in vivo are associated with tolerance in SCID patients transplanted with HLA mismatched hematopoietic stem cells. J. Exp. Med. 179, 493-502.

Barnes, P.F., Abrams, J.S., Lu, S., Sieling, P.A., Rea, T.H., & Modlin, R.L. (1993). Patterns of cytokine production by mycobacterium-reactive human T-cell clones. Infect. Immun. 61, 197-203.

Bean, A.G., Freiberg, R.A., Andrade, S., Menon, S., & Zlotnik, A. (1993). Interleukin 10 protects mice against staphylococcal enterotoxin B-induced lethal shock. Infect. Immun. 61, 4937-4939.

Bejarano, M.T., de Waal Malefyt, R., Abrams, J.S., Bigler, M., Bacchetta, R., de Vries, J.E., & Roncarolo, M.G. (1992). Interleukin 10 inhibits allogeneic proliferative and cytotoxic T cell responses generated in primary mixed lymphocyte cultures. Int. Immunol. 4, 1389-1397.

Bendelac, A., & Schwartz, R.H. (1991). CD4+ and CD8+ T cells acquire specific lymphokine secretion potentials during thymic maturation. Nature 353, 68-71.

Benjamin, D., Knobloch, T.J., & Dayton, M.A. (1992). Human B-cell interleukin-10: B-cell lines derived from patients with acquired immunodeficiency syndrome and Burkitt's lymphoma constitutively secrete large quantities of interleukin-10 [see comments]. Blood 80, 1289-1298.

Bermudez, L.E., & Champsi, J. (1993). Infection with Mycobacterium avium induces production of interleukin-10 (IL-10), and administration of anti-IL-10 antibody is associated with enhanced resistance to infection in mice. Infect. Immun. 61, 3093-3097.

Bogdan, C., Vodovotz, Y., & Nathan, C. (1991). Macrophage deactivation by interleukin 10. J. Exp. Med. 174, 1549-1555.

Briere, F., Bridon, J.M., Servet, C., Rousset, F., Zurawski, G., & Banchereau, J. (1993). IL-10 and IL-13 as B cell growth and differentiation factors. Nouv. Rev. Fr. Hematol. 35, 233-235.

Briere, F., Servet Delprat, C., Bridon, J.M., Saint Remy, J.M., & Banchereau, J. (1994). Human interleukin 10 induces naive surface immunoglobulin D+ (sIgD+) B cells to secrete IgG1 and IgG3. J. Exp. Med. 179, 757-762.

Chen, W.F., & Zlotnik, A. (1991). IL-10: A novel cytotoxic T cell differentiation factor. J. Immunol. 147, 528-534.

Clerici, M., Wynn, T.A., Berzofsky, J.A., Blatt, S.P., Hendrix, C.W., Sher, A., Coffman, R.L., & Shearer, G.M. (1994). Role of interleukin-10 in T helper cell dysfunction in asymptomatic individuals infected with the human immunodeficiency virus. J. Clin. Invest. 93, 768-775.

Cunha, F.Q., Moncada, S., & Liew, F.Y. (1992). Interleukin-10 (IL-10) inhibits the induction of nitric oxide synthase by interferon-gamma in murine macrophages. Biochem. Biophys. Res. Commun. 182, 1155-1159.

de Vries, J.E., de Waal Malefyt, R., Yssel, H., Roncarolo, M.G., & Spits, H. (1991). Do human TH1 and TH2 CD4+ clones exist? Res. Immunol. 142, 59-63.

de Waal Malefyt, R., Abrams, J., Bennett, B., Figdor, C.G., & de Vries, J.E. (1991). Interleukin 10(IL-10) inhibits cytokine synthesis by human monocytes: An autoregulatory role of IL-10 produced by monocytes. J. Exp. Med. 174, 1209-1220.

de Waal Malefyt, R., Haanen, J., Spits, H., Roncarolo, M.G., te Velde, A., Figdor, C., Johnson, K., Kastelein, R., Yssel, H., & de Vries, J.E. (1991). Interleukin 10 (IL-10) and viral IL-10 strongly reduce antigen-specific human T cell proliferation by diminishing the antigen-presenting capacity of monocytes via downregulation of class II major histocompatibility complex expression. J. Exp. Med. 174, 915-924.

de Waal Malefyt, R., Yssel, H., & de Vries, J.E. (1993). Direct effects of IL-10 on subsets of human CD4+ T cell clones and resting T cells. Specific inhibition of IL-2 production and proliferation. J. Immunol. 150, 4754-4765.

Defrance, T., Vanbervliet, B., Briere, F., Durand, I., Rousset, F., & Banchereau, J. (1992). Interleukin 10 and transforming growth factor beta cooperate to induce anti-CD40-activated naive human B cells to secrete immunoglobulin A. J. Exp. Med. 175, 671-682.

Del Prete, G., De Carli, M., Almerigogna, F., Giudizi, M.G., Biagiotti, R., & Romagnani, S. (1993). Human IL-10 is produced by both type 1 helper (Th1) and type 2 helper (Th2) T cell clones and inhibits their antigen-specific proliferation and cytokine production. J. Immunol. 150, 353-360.

Denis, M., & Ghadirian, E. (1993). IL-10 neutralization augments mouse resistance to systemic Mycobacterium avium infections. J. Immunol. 151, 5425-5430.

Ding, L., Linsley, P.S., Huang, L.Y., Germain, R.N., & Shevach, E.M. (1993). IL-10 inhibits macrophage costimulatory activity by selectively inhibiting the up-regulation of B7 expression. J. Immunol. 151, 1224-1234.

Ding, L., & Shevach, E.M. (1992). IL-10 inhibits mitogen-induced T cell proliferation by selectively inhibiting macrophage costimulatory function. J. Immunol. 148, 3133-3139.

Emilie, D., Galanaud, P., Raphael, M., & Joab, I. (1993). Interleukin-10 and acquired immunodeficiency syndrome lymphomas [letter; comment]. Blood 81, 1106-1107.

Emilie, D., Touitou, R., Raphael, M., Peuchmaur, M., Devergnee, O., Rea, D., Coumbraras, J., Crevon, M.C., Edelman, L., Joab, I., et al. (1992). *In vivo* production of interleukin-10 by malignant cells in AIDS lymphomas. Eur. J. Immunol. 22, 2937-2942.

Enk, A.H., Angeloni, V.L., Udey, M.C., & Katz, S.I. (1993). Inhibition of Langerhans cell antigen-presenting function by IL-10. A role for IL-10 in induction of tolerance. J. Immunol. 151, 2390-2398.

Enk, A.H., & Katz, S.I. (1992). Identification and induction of keratinocyte-derived IL-10. J. Immunol. 149, 92-95.

Enk, A.H., Saloga, J., Becker, D., Bmadzadeh, M., & Knop, J. (1994). Induction of hapten-specific tolerance by interleukin 10 *in vivo*. J. Exp. Med. 179, 1397-1402.

Ferguson, T.A., Dube, P., & Griffith, T.S. (1994). Regulation of contact hypersensitivity by interleukin 10. J. Exp. Med. 179, 1597-1604.

Fiorentino, D.F., Bond, M.W., & Mosmann, T.R. (1989). Two types of mouse T helper cell. IV. Th2 clones secrete a factor that inhibits cytokine production by Th1 clones. J. Exp. Med. 170, 2081-2095.

Fiorentino, D.F., Zlotnik, A., Mosmann, T.R., Howard, M., & O'Garra, A. (1991). IL-10 inhibits cytokine production by activated macrophages. J. Immunol. 147, 3815-3822.

Fiorentino, D.F., Zlotnik, A., Vieira, P., Mosmann, T.R., Howard, M., Moore, K.W., & O'Garra, A. (1991). IL-10 acts on the antigen-presenting cell to inhibit cytokine production by Th1 cells. J. Immunol. 146, 3444-3451.

Gazzinelli, R.T., Makino, M., Chattopadhyay, S.K., Snapper, C.M., Sher, A., Hugin, A.W., & Morse, H.C. (1992). CD4+ subset regulation in viral infection. Preferential activation of Th2 cells during progression of retrovirus-induced immunodeficiency in mice. J. Immunol. 148, 182-188.

Gazzinelli, R.T., Oswald, I.P., James, S.L., & Sher, A. (1992). IL-10 inhibits parasite killing and nitrogen oxide production by IFN-gamma-activated macrophages. J. Immunol. 148, 1792-1796.

Gerard, C., Bruyns, C., Marchant, A., Abramowicz, D., Vandenabeele, P., Delvaux, A., Fiers, W., Goldman, M., & Velu, T. (1993). Interleukin 10 reduces the release of tumor necrosis factor and prevents lethality in experimental endotoxemia. J. Exp. Med. 177, 547-550.

Ghildyal, N., Friend, D.S., Nicodemus, C.F., Austen, K.F., & Stevens, R.L. (1993). Reversible expression of mouse mast cell protease 2 mRNA and protein in cultured mast cells exposed to IL-10. J. Immunol. 151, 3206-3214.

Ghildyal, N., McNeil, H.P., Gurish, M.F., Austen, K.F., & Stevens, R.L. (1992a). Transcriptional regulation of the mucosal mast cell-specific protease gene, MMCP-2, by interleukin 10 and interleukin 3. J. Biol. Chem. 267, 8473-8477.

Ghildyal, N., McNeil, H.P., Stechschulte, S., Austen, K.F., Silberstein, D., Gurish, M.F., Somerville, L.L., & Stevens, R.L. (1992b). IL-10 induces transcription of the gene for mouse mast cell protease-1, a serine protease preferentially expressed in mucosal mast cells of Trichinella spiralis-infected mice. J. Immunol. 149, 2123-2129.

Go, N.F., Castle, B.E., Barrett, R., Kastelein, R., Dang, W., Mosmann, T.R., Moore, K.W., & Howard, M. (1990). Interleukin 10, a novel B cell stimulatory factor: unresponsiveness of X chromosome-linked immunodeficiency B cells. J. Exp. Med. 172, 1625-1631.

Graziosi, C., Pantaleo, G., Gantt, K.R., Fortin, J.P., Demarest, J.F., Cohen, O.J., Sekaly, R.P., & Fauci, A.S. (1994). Lack of evidence for the dichotomy of TH1 and TH2 predominance in HIV-infected individuals [see comments]. Science 265, 248-252.

Grewe, M., Gryfko, K., & Krutmann, J. (1995). Interleukin-10 production by human keratinocytes: Regulation by Ultaviolet B and Ultraviolet A1 Radiation. J. Invest. Dermatol. 104, 3-6.

Heinzel, F.P., Sadick, M.D., Mutha, S.S., & Locksley, R.M. (1991). Production of interferon gamma, interleukin 2, interleukin 4, and interleukin 10 by CD4+ lymphocytes *in vivo* during healing and progressive murine leishmaniasis. Proc. Natl. Acad. Sci. USA 88, 7011-7015.

Hisatsune, T., Minai, Y., Nishisima, K., Enomoto, A., Moore, K.W., Yokota, T., Arai, K., & Kaminogawa, S. (1992). A suppressive lymphokine derived from Ts clone 13G2 is IL-10. Lymphokine Cytokine. Res. 11, 87-93.

Howard, M., Muchamuel, T., Andrade, S., & Menon, S. (1993). Interleukin 10 protects mice from lethal endotoxemia. J. Exp. Med. 177, 1205-1208.

Hsieh, C.S., Heimberger, A.B., Gold, J.S., O'Garra, A., & Murphy, K.M. (1992). Differential regulation of T helper phenotype development by interleukins 4 and 10 in an alpha beta T-cell-receptor transgenic system. Proc. Natl. Acad. Sci. U S A 89, 6065-6069.

Hsu, D.H., de Waal Malefyt, R., Fiorentino, D.F., Dang, M.N., Vieira, P., de Vries, J., Spits, H., Mosmann, T.R., & Moore, K.W. (1990). Expression of interleukin-10 activity by Epstein-Barr virus protein BCRF1. Science 250, 830-832.

Hsu, D.H., Moore, K.W., & Spits, H. (1992). Differential effects of IL-4 and IL-10 on IL-2-induced IFN-gamma synthesis and lymphokine-activated killer activity. Int. Immunol. 4, 563-569.

Ishida, H., Hastings, R., & Howard, M. (1992a). Multiple IL-10 antibody treatment blocks the development of Ly-1 lineage B cells. Ann. N. Y. Acad. Sci. 651, 264-265.

Ishida, H., Hastings, R., Kearney, J., & Howard, M. (1992b). Continuous anti-interleukin 10 antibody administration depletes mice of Ly-1 B cells but not conventional B cells. J. Exp. Med. 175, 1213-1220.

Ishida, H., Hastings, R., Thompson Snipes, L., & Howard, M. (1993). Modified immunological status of anti-IL-10 treated mice. Cell Immunol. 148, 371-384.

Ishida, H., Muchamuel, T., Sakaguchi, S., Andrade, S., Menon, S., & Howard, M. (1994). Continuous administration of anti-interleukin 10 antibodies delays onset of autoimmunity in NZB/W F1 mice. J. Exp. Med. 179, 305-310.

Jinquan, T., Larsen, C.G., Gesser, B., Matsushima, K., & Thestrup Pedersen, K. (1993). Human IL-10 is a chemoattractant for CD8+ T lymphocytes and an inhibitor of IL-8-induced CD4+ T lymphocyte migration. J. Immunol 151, 4545-4551.

Kantor, A.B. (1991). The development and repertoire of B-1 cells (CD5 B cells). Immunol. Today 12, 389-391.

Katsikis, P.D., Chu, C.Q., Brennan, F.M., Maini, R.N., & Feldmann, M. (1994). Immunoregulatory role of interleukin 10 in rheumatoid arthritis. J. Exp. Med. 179, 1517-1527.

Kaye, P.M., & Bancroft, G.J. (1992). Leishmania donovani infection in scid mice: Lack of tissue response and *in vivo* macrophage activation correlates with failure to trigger natural killer cell-derived gamma interferon production *in vitro*. Infect. Immun. 60, 4335-4342.

Kennedy, M.K., Torrance, D.S., Picha, K.S., & Mohler, K.M. (1992). Analysis of cytokine mRNA expression in the central nervous system of mice with experimental autoimmune encephalomyelitis reveals that IL-10 mRNA expression correlates with recovery. J. Immunol. 149, 2496-2505.
Kim, J.M., Brannan, C.I., Copeland, N.G., Jenkins, N.A., Khan, T.A., & Moore, K.W. (1992). Structure of the mouse IL-10 gene and chromosomal localization of the mouse and human genes. J. Immunol. 148, 3618-3623.
Kuhn, R., Lohler, J., Rennick, D., Rajewsky, K., & Muller, W. (1993). Interleukin-10-deficient mice develop chronic enterocolitis [see comments]. Cell 75, 263-274.
Kullberg, M.C., Pearce, E.J., Hieny, S.E., Sher, A., & Berzofsky, J.A. (1992). Infection with Schistosoma mansoni alters Th1/Th2 cytokine responses to a non-parasite antigen. J. Immunol. 148, 3264-3270.
Lee, M.S., Wogensen, L., Shizuru, J., Oldstone, M.B., & Sarvetnick, N. (1994). Pancreatic islet production of murine interleukin-10 does not inhibit immune-mediated tissue destruction. J. Clin. Invest. 93, 1332-1338.
Macatonia, S.E., Doherty, T.M., Knight, S.C., & O'Garra, A. (1993). Differential effect of IL-10 on dendritic cell-induced T cell proliferation and IFN-gamma production. J. Immunol. 150, 3755-3765.
MacNeil, I.A., Suda, T., Moore, K.W., Mosmann, T.R., & Zlotnik, A. (1990). IL-10, a novel growth cofactor for mature and immature T cells. J. Immunol. 145, 4167-4173.
Modlin, R.L., & Nutman, T.B. (1993). Type 2 cytokines and negative immune regulation in human infections. Curr. Opin. Immunol. 5, 511-517.
Moore, K.W., O'Garra, A., de Waal Malefyt, R., Vieira, P., & Mosmann, T.R. (1993). Interleukin-10. Annu. Rev. Immunol. 11, 165-190.
Moore, K.W., Rousset, F., & Banchereau, J. (1991). Evolving principles in immunopathology: Interleukin 10 and its relationship to Epstein-Barr virus protein BCRF1. Springer Semin. Immunopathol. 13, 157-166.
Moore, K.W., Vieira, P., Fiorentino, D.F., Trounstine, M.L., Khan, T.A., & Mosmann, T.R. (1990). Homology of cytokine synthesis inhibitory factor (IL-10) to the Epstein-Barr virus gene BCRFI [published erratum appears in Science 1990 Oct 26;250(4980):494]. Science 248, 1230-1234.
Mosmann, T.R., Cherwinski, H., Bond, M.W., Giedlin, M.A., & Coffman, R.L. (1986). Two types of murine helper T cell clone. I. Definition according to profiles of lymphokine activities and secreted proteins. J. Immunol. 136, 2348-2357.
Mosmann, T.R. (1991a). Regulation of immune responses by T cells with different cytokine secretion phenotypes: Role of a new cytokine, cytokine synthesis inhibitory factor (IL10). Int. Arch. Allergy Appl. Immunol. 94, 110-115.
Mosmann, T.R. (1991b). Cytokine secretion patterns and cross-regulation of T cell subsets. Immunol. Res. 10, 183-188.
Mosmann, T.R. (1991c). Role of a new cytokine, interleukin-10, in the cross-regulation of T helper cells. Ann. N. Y. Acad. Sci. 628, 337-344.
Mosmann, T.R. (1991d). Cytokine secretion phenotypes of TH cells: How many subsets, how much regulation? Res. Immunol. 142, 9-13.
Mosmann, T.R., & Moore, K.W. (1991). The role of IL-10 in crossregulation of TH1 and TH2 responses. Immunol. Today 12, A49-A53.

Mosmann, T.R., Schumacher, J.H., Fiorentino, D.F., Leverah, J., Moore, K.W., & Bond, M.W. (1990). Isolation of monoclonal antibodies specific for IL-4, IL-5, IL-6, and a new Th2-specific cytokine (IL-10), cytokine synthesis inhibitory factor, by using a solid phase radioimmunoadsorbent assay. J. Immunol. 145, 2938-2945.

Murphy, E., Hieny, S., Sher, A., & O'Garra, A. (1993). Detection of *in vivo* expression of interleukin-10 using a semi-quantitative polymerase chain reaction method in Schistosoma mansoni infected mice. J. Immunol. Methods 162, 211-223.

Niiro, H., Otsuka, T., Abe, M., Satoh, H., Ogo, T., Nakano, T., Furukawa, Y., & Niho, Y. (1992). Epstein-Barr virus BCRF1 gene product (viral interleukin 10) inhibits superoxide anion production by human monocytes. Lymphokine Cytokine. Res. 11, 209-214.

Niiro, H., Otsuka, T., Kuga, S., Nemoto, Y., Abe, M., Hara, N., Nakano, T., Ogo, T., & Niho, Y. (1994). IL-10 inhibits prostaglandin E2 production by lipopolysaccharide-stimulated monocytes. Int. Immunol. 6, 661-664.

O'Garra, A., & Howard, M. (1992). Cytokines and Ly-1 (B1) B cells. Int. Rev. Immunol. 8, 219-234.

O'Garra, A., Stapleton, G., Dhar, V., Pearce, M., Schumacher, J., Rugo, H., Barbis, D., Stall, A., Cupp, J., Moore, K., et al. (1990). Production of cytokines by mouse B cells: B lymphomas and normal B cells produce interleukin 10. Int. Immunol. 2, 821-832.

Oswald, I.P., Gazzinelli, R.T., Sher, A., & James, S.L. (1992). IL-10 synergizes with IL-4 and transforming growth factor-beta to inhibit macrophage cytotoxic activity. J. Immunol. 148, 3578-3582.

Oswald, I.P., Wynn, T.A., Sher, A., & James, S.L. (1992). Interleukin 10 inhibits macrophage microbicidal activity by blocking the endogenous production of tumor necrosis factor alpha required as a costimulatory factor for interferon gamma-induced activation. Proc. Natl. Acad. Sci. USA 89, 8676-8680.

Pecanha, L.M., Snapper, C.M., Lees, A., & Mond, J.J. (1992). Lymphokine control of type 2 antigen response. IL-10 inhibits IL-5- but not IL-2-induced Ig secretion by T cell-independent antigens. J. Immunol. 148, 3427-3432.

Pecanha, L.M., Snapper, C.M., Lees, A., Yamaguchi, H., & Mond, J.J. (1993). IL-10 inhibits T cell-independent but not T cell-dependent responses *in vitro*. J. Immunol. 150, 3215-3223.

Peguet Navarro, J., Moulon, C., Caux, C., Dalbiez Gauthier, C., Banchereau, J., & Schmitt, D. (1994). Interleukin-10 inhibits the primary allogeneic T cell response to human epidermal Langerhans cells. Eur. J. Immunol. 24, 884-891.

Reed, S.G., Brownell, C.E., Russo, D.M., Silva, J.S., Grabstein, K.H., & Morrissey, P.J. (1994). IL-10 mediates susceptibility to Trypanosoma cruzi infection. J. Immunol. 153, 3135-3140.

Rivas, J.M., & Ullrich, S.E. (1992). Systemic suppression of delayed-type hypersensitivity by supernatants from UV-irradiated keratinocytes. An essential role for keratinocyte-derived IL-10. J. Immunol. 149, 3865-3871.

Rousset, F., Garcia, E., Defrance, T., Peronne, C., Vezzio, N., Hsu, D.H., Kastelein, R., Moore, K.W., & Banchereau, J. (1992). Interleukin 10 is a potent growth and differentiation factor for activated human B lymphocytes. Proc. Natl. Acad. Sci. USA 89, 1890-1893.

Schandene, L., Alonso Vega, C., Willems, F., Gerard, C., Delvaux, A., Velu, T., Devos, R., de Boer, M., & Goldman, M. (1994). B7/CD28-dependent IL-5 production by human resting T cells is inhibited by IL-10. J. Immunol. 152, 4368-4374.

Schandene, L., Gerard, C., Crusiaux, A., Abramowicz, D., Velu, T., & Goldman, M. (1993). Interleukin-10 inhibits OKT3-induced cytokine release: *In vitro* comparison with pentoxifylline. Transplant. Proc. 25, 55-56.

Schwarz, A., Grabbe, S., Riemann, H., Aragane, Y., Simon, M., Manon, S., Andrade, S., Luger, T.A., Zlotnik, A., & Schwarz, T. (1994). *In vivo* effects of interleukin-10 on contact hypersensitivity and delayed-type hypersensitivity reactions. J. Invest. Dermatol. 103, 211-216.

Sieling, P.A., Abrams, J.S., Yamamura, M., Salgame, P., Bloom, B.R., Rea, T.H., & Modlin, R.L. (1993). Immunosuppressive roles for IL-10 and IL-4 in human infection. *In vitro* modulation of T cell responses in leprosy. J. Immunol. 150, 5501-5510.

Silva, J.S., Barral Netto, M., & Reed, S.G. (1993). Aggravation of both Trypanosoma cruzi and murine leukemia virus by concomitant infections. Am. J. Trop. Med. Hyg. 49, 589-597.

Silva, J.S., Morrissey, P.J., Grabstein, K.H., Mohler, K.M., Anderson, D., & Reed, S.G. (1992). Interleukin 10 and interferon gamma regulation of experimental *Trypanosoma cruzi* infection. J. Exp. Med. 175, 169-174.

Spits, H., & de Waal Malefyt, R. (1992). Functional characterization of human IL-10. Int. Arch. Allergy Immunol. 99, 8-15.

Stewart, J.P., & Rooney, C.M. (1992). The interleukin-10 homolog encoded by Epstein-Barr virus enhances the reactivation of virus-specific cytotoxic T cell and HLA-unrestricted killer cell responses. Virology 191, 773-782.

Street, N.E. and Mosmann, T.R. (1991). Functional diversity of T lymphocytes due to secretion of different cytokine patterns. FASEB J. 5, 171-177.

Suda, T., MacNeil, I., Fischer, M., Moore, K.W., & Zlotnik, A. (1991). Identification of a novel thymocyte growth factor derived from B cell lymphomas. Adv. Exp. Med. Biol. 292, 115-120.

Taga, K., Cherney, B., & Tosato, G. (1993). IL-10 inhibits apoptotic cell death in human T cells starved of IL-2. Int. Immunol. 5, 1599-1608.

Taga, K., Chretien, J., Cherney, B., Diaz, L., Brown, M., & Tosato, G. (1994). Interleukin-10 inhibits apoptotic cell death in infectious mononucleosis T cells. J. Clin. Invest. 94, 251-260.

Taga, K., Mostowski, H., & Tosato, G. (1993). Human interleukin-10 can directly inhibit T-cell growth. Blood 81, 2964-2971.

Taga, K., & Tosato, G. (1992). IL-10 inhibits human T cell proliferation and IL-2 production. J. Immunol. 148, 1143-1148.

Takanaski, S., Nonaka, R., Xing, Z., O'Byrne, P., Dolovich, J., & Jordana, M. (1994). Interleukin 10 inhibits lipopolysaccharide-induced survival and cytokine production by human peripheral blood eosinophils. J. Exp. Med. 180, 711-715.

Tan, J.C., Indelicato, S.R., Narula, S.K., Zavodny, P.J., & Chou, C.C. (1993). Characterization of interleukin-10 receptors on human and mouse cells. J. Biol. Chem. 268, 21053-21059.

te Velde, A.A., de Waal Malefijt, R., Huijbens, R.J., de Vries, J.E., & Figdor, C.G. (1992). IL-10 stimulates monocyte Fc gamma R surface expression and cytotoxic activity.

Distinct regulation of antibody-dependent cellular cytotoxicity by IFN-gamma, IL-4, and IL-10. J. Immunol. 149, 4048-4052.

Thompson Snipes, L., Dhar, V., Bond, M.W., Mosmann, T.R., Moore, K.W., & Rennick, D.M. (1991). Interleukin 10: A novel stimulatory factor for mast cells and their progenitors. J. Exp. Med. 173, 507-510.

Tumpey, T.M., Elner, V.M., Chen, S.H., Oakes, J.E., & Lausch, R.N. (1994). Interleukin-10 treatment can suppress stromal keratitis induced by herpes simplex virus type 1. J. Immunol. 153, 2258-2265.

van der Poll, T., Jansen, J., Levi, M., ten Cate, H., ten Cate, J.W., & van Deventer, S.J. (1994). Regulation of interleukin 10 release by tumor necrosis factor in humans and chimpanzees. J. Exp. Med. 180, 1985-1988.

Vieira, P., de Waal Malefyt, R., Dang, M.N., Johnson, K.E., Kastelein, R., Fiorentino, D.F., deVries, J.E., Roncarolo, M.G., Mosmann, T.R., & Moore, K.W. (1991). Isolation and expression of human cytokine synthesis inhibitory factor cDNA clones: Homology to Epstein-Barr virus open reading frame BCRFI. Proc. Natl. Acad. Sci. USA 88, 1172-1176.

Weber Nordt, R.M., Meraz, M.A., & Schreiber, R.D. (1994). Lipopolysaccharide-dependent induction of IL-10 receptor expression on murine fibroblasts. J. Immunol. 153, 3734-3744.

Willems, F., Marchant, A., Delville, J.P., Gerard, C., Delvaux, A., Velu, T., de Boer, M., & Goldman, M. (1994). Interleukin-10 inhibits B7 and intercellular adhesion molecule-1 expression on human monocytes. Eur. J. Immunol. 24, 1007-1009.

Wogensen, L., Huang, X., & Sarvetnick, N. (1993). Leukocyte extravasation into the pancreatic tissue in transgenic mice expressing interleukin 10 in the islets of Langerhans. J. Exp. Med. 178, 175-185.

Wogensen, L., Lee, M.S., & Sarvetnick, N. (1994). Production of interleukin 10 by islet cells accelerates immune-mediated destruction of beta cells in nonobese diabetic mice. J. Exp. Med. 179, 1379-1384.

Yamamura, M., Uyemura, K., Deans, R.J., Weinberg, K., Rea, T.H., Bloom, B.R., & Modlin, R.L. (1991). Defining protective responses to pathogens: Cytokine profiles in leprosy lesions [published erratum appears in Science 1992 Jan 3; 255(5040):12]. Science 254, 277-279.

Yssel, H., de Waal Malefyt, R., Roncarolo, M.G., Abrams, J.S., Lahesmaa, R., Spits, H., & de Vries, J.E. (1992). IL-10 is produced by subsets of human CD4+ T cell clones and peripheral blood T cells. J. Immunol. 149, 2378-2384.

THE INTERLEUKIN-2 RECEPTOR

Mark A. Goldsmith and Warner C. Greene

Growth Factors and Cytokines in Health and Disease
Volume 2A, pages 355-402

ISBN: 0-7623-0117-1

I. THE MULTIMERIC HUMAN IL-2 RECEPTOR

A. Analysis of Receptor Subunits

Resting human T cells are physiologically activated by antigen engagement of the clonotypic T cell receptors that are constitutively displayed at their cell surfaces (Chan et al., 1994). With many CD4 T lymphocytes, this ligand-receptor interplay triggers an intracellular signaling cascade leading to the production of the T cell growth factor, interleukin-2 (IL-2), and the concomitant display of functional high affinity membrane receptors for IL-2 (Goldsmith and Greene, 1994; Smith, 1988). IL-2 binding to these high affinity IL-2 receptors generates a second series of intracellular signals culminating in T cell division and clonal expansion of the antigen-reactive cell population. Subsequent decline in IL-2 receptor display and diminished production of IL-2 combine to terminate the T cell immune response.

Current evidence suggests that the fully functional high affinity human IL-2 receptor is composed of three distinct transmembrane subunits termed IL-2Rα, IL-2Rβ, and γc (Table 1). Each of these receptor chains serves a distinct role in the assembly and function of the high

Table 1. Characteristic of the IL-2 Receptor Subunits

	Gene		*mRNA*		*Mature Protein*		
Receptor Subunit	*Chromosome*	*Exons*	*Size (kb)*	*Predominant Distribution*	*Size (kD)*	*Length (amino acids)*	*CR Super-family*
IL-2Rα	10p14-15	8	1.5/3.5	activated T, B	50-55	251	No
IL-2Rβ	22q11.2-12	NR	4.0	resting T, B NK	70-75	525	Yes
γc	Xq13.1	8	1.8/3.6	resting T, B, NK, monocytes	64	347	Yes

Note: Shown is the chromosomal localization and genomic structure of the genes encoding each subunit. The sizes of the predominant mRNA species are also indicated, along with the major cell types expressing these transcripts. Also shown ae the lengths and masses of each mature protein and membership in the cytokine receptors (CR) super family.

affinity IL-2 receptor complex. Salient features of each of the known receptor subunits are discussed below.

IL-2Rα

The IL-2Rα gene is located on chromosome 10p at band 14-15 (Leonard et al., 1985b) and organized within eight exons (Leonard et al., 1985a). Transcription of the IL-2Rα gene occurs in a highly inducible manner, a property that distinguishes it from the constitutively expressed IL-2Rβ and γc genes (Leonard et al., 1984; Nikaido et al., 1984; Cosman et al., 1984). The induced transcription of the IL-2Rα gene is known to involve members of the NF-κB/Rel family of enhancer binding proteins as well as CArG and Sp1 binding proteins (Ballard et al., 1989). IL-2Rα transcription proceeds from at least two different initiation sites and terminates at one of three different polyadenylation signals (Leonard et al., 1984). Variable splicing of the fourth exon has also been detected (Leonard et al., 1984). However, IL-2Rα transcripts lacking this exon encode a nonfunctional IL-2Rα polypeptide (Cullen et al., 1988). Translation of the functional IL-2Rα transcripts results in a 272 amino acid primary translation product whose 21 residue signal sequence is co-translationally cleaved to yield a 33 kD primary translation product (Leonard et al., 1984; Nikaido et al., 1984; Cosman et al., 1984). N-linked glycosylation subsequently occurs within the endoplasmic reticulum involving sugar addition at two distinct sites to produce 35 kD and 37 kD precursors. These polypeptides are then exported to the Golgi apparatus where they undergo additional O-linked glycosylation.

While O-glycosylation appears important for effective surface display of IL-2Rα, N-glycosylation is required for neither membrane expression nor ligand binding (Lowenthal et al., 1985). Additional post-translational modifications of IL-2Rα include constitutive sulfation and serine-based phosphorylation. The mature 50-55 kD IL-2Ra protein contains a 219 amino acid extracellular domain, a 19 residue membrane spanning region, and a short 13 amino acid cytoplasmic tail. The unexpectedly short size of this IL-2Rα cytoplasmic tail first suggested the possible existence of additional receptor subunits. It should be noted that initial identification of the IL-2Rα subunit was markedly facilitated by the isolation of the anti-Tac monoclonal antibody (Uchiyama et al., 1981; Leonard et al., 1982). More recently, this antibody has been used as a therapeutic agent to eliminate various IL-2R-expressing tumor cells or activated T cells involved in graft rejection. Additionally, the detection of soluble circulating forms of IL-2Ra reflecting alternative splicing and/or proteolytic cleavage of membrane IL-2Ra has proven valuable in the diagnosis of various tumors and heightened states of T cell activation.

IL-2Rβ

The IL-2Rβ gene is located on chromosome 22, and in contrast to the IL-2Rα gene, is constitutively expressed on the surface of T cells, B, cells and NK cells (Tsudo et al., 1989a; Hatakeyama et al., 1989a; Gnarra et al., 1990; Dukovich et al., 1987). Although various immune stimuli may modestly augment IL-2Rβ mRNA and protein levels, these effects reflect post-transcriptional changes in IL-2Rβ stability rather than augmented IL-2Rβ transcription. Translation of IL-2Rβ mRNA produces a 551 residue primary translation product that includes a 26 residue signal peptide (Hata Keyama). The mature IL-2Rβ polypeptide contains four potential sites for N-glycosylation, but it is unknown which of these sites are utilized. This receptor chain is also distinguished by the presence of eight extracellular cysteine residues. The spacing of two pairs of these conserved cysteines and the presence of a canonical WSXWS motif near the plasma membrane have revealed membership by IL-2Rβ in the cytokine receptor superfamily (Bazan, 1980a). This superfamily collection of genes encodes the receptors for a number of interleukins, including IL-3, IL-4, IL-5, IL-6, IL-7, IL-9, IL-13, and IL-15.

The IL-2Rβ polypeptide is divided into a 214 extracellular domain similar in length to IL-2Rα, a 25 residue transmembrane domain, and a substantial 286 amino acid cytoplasmic domain. Of note, this cytoplasmic domain does not contain consensus sequences encoding a tyrosine kinase or other known enzymatic activities. However, this cytoplasmic tail does contain Box 1 and Box 2 elements similar to those found in many other cytokine receptors (Murakami et al., 1991). It is now clear that these conserved motifs play an important role in the generation of intracellular signals (Goldsmith et al., 1994). The cytoplasmic tail of IL-2Rβ is also remarkable for the presence of six tyrosine residues, one or more of which undergoes ligand induced phosphorylation (see following and Sharon et al., 1989; Mills et al., 1990; Asao et al., 1990).

γc

The gene encoding γc (*IL2RG*) is located on the X chromosome (Noguchi et al., 1993b) and is constitutively expressed in nearly all hematopoietic cells. Transcripts from the *IL2RG* gene encode a 369 residue polypeptide (γc) that includes a 22 amino acid signal peptide, a 232 residues extracellular domain, a 29 amino acid transmembrane segment, and an 86 amino acid C-terminal cytoplasmic tail (Takeshita et al., 1992a). Like IL-2Rβ, γc is a member of the cytokine receptor superfamily as indicated by the presence of the conserved cysteine tetrad and the WSXWS pentad motif. In addition, the extracellular domain of the γc subunit contains a heptamerically spaced repeat of four leucine residues that resembles a leucine zipper motif. It is possible that this domain is involved in subunit-subunit interactions within the IL-2R complex. The cytoplasmic portion of γc contains versions of the Box 1 and Box 2 elements that are separated by a sequence that resembles an SH2 domain. However, several signature amino acids characteristic of functional SH2 domains are missing in this γc element, raising doubt about its ability to bind to phosphotyrosine residues like classical SH2 elements.

B. Multiple Affinity Forms of the Human IL-2 Receptor

The existence of three distinct IL-2 receptor subunits has helped provide a molecular basis for the multiple affinity forms of the human IL-2 receptor displayed on the surface of various lymphoid cells (Figure 1). It is now recognized that expression of the IL-2Rα subunit alone gives rise to the classical low affinity form of the receptor (Kd of 10^{-8}M; Sabe et al., 1984; Green et al., 1985). While these receptors represent the most prevalent form of receptor found on activated T cells, they are not believed to mediate intracellular signaling given the short cytoplasmic tail of IL-2Rα. Expression of the β or γc subunits alone leads to the appearance of binding sites with either very low (Kd of 10^{-7}M; Ringheim et al., 1991) or undetectable binding affinity, respectively. Furthermore, no functional response can be ascribed to engagement of these isolated subunits. In contrast, heterodimeric combinations of these chains leads to the appearance of different affinity classes of receptors, one of which is functional. Specifically, combinations of the α and β chains give rise to pseudo-high affinity receptors (Kd of 10^{-10}M; Arima et al., 1992b). The β-γc chain combination yields intermediate affinity receptors (Kd of 10^{-9}M; Takeshita et al., 1990); thus far, the third potential heterodimer pair α-γc has not been identified. The intermediate affinity β-γc dimers, which are detectable on some resting T cells and NK cells (Dukovich et al., 1987), are capable of effective signal transduction in the presence of high concentrations of IL-2. In contrast, no function has yet been ascribed to ligand engagement of the pseudo-high affinity α-β heterodimers. The trimolecular α-β-γc complex present on activated T and B cells corresponds to the classic high affinity receptor complex (K_d of 10^{-11}M). Due to its high affinity, this receptor functions effectively at the low ligand concentrations likely encountered *in vivo*.

The distinct binding affinities manifested by these different affinity forms of the human IL-2 receptor derive from the specific binding properties of each of the three receptor subunits (Lowenthal and Greene, 1987; Wang and Smith, 1987). For example, IL-2 binding to the α chain is characterized by rapid rates of both ligand association and dissociation ($t_{1/2}$ of 4 and 6 sec, respectively). In contrast, IL-2 binding to the β-γc complex occurs with a much slower on-rate and a very slow off-rate ($t_{1/2}$ of 40 min and 4 h, respectively) . Although the γc chain alone has not been shown to bind IL-2 in the current assay systems, recent studies suggest that this receptor subunit further retards the dissociation of IL-2 from either the β-γ or α-β-γ receptor complexes (Voss et al., 1993). In general, the various IL-2 receptor complexes appear to exert the most "favorable" binding properties of each of the individual subunits, such as the rapid on-rate of the a chain and the slow off-rate of the β-γcomplex.

The inter-subunit interactions of these different chains are also quite different. For example, the α and β chains appear to associate with each other in the absence of IL-2 in the murine system as evidenced by their coimmunoprecipitation (Saragovi and Malek, 1988). In addition, the α chain, which alone normally does not undergo receptor-mediated endocytosis via clathrin-coated pits, is effectively internalized in the presence of β, γc, and IL-2. Indeed, the α chain is even internalized in the presence of mutants of IL-2 that fail to engage the α subunit (Kuziel et al., 1993). Specifically, when Ala is substituted for Phe at position 42 in IL-2, an IL-2 mutant is produced that is unable to interact with α. However, in the presence of the β-γ chains and this IL-2 analogue, the unliganded α chain promotes high affinity receptor binding and function. These results suggest that the α chain interacts with β in the absence of IL-2, and that this interaction leads to a conformational change in one or both subunits that in turn alters the structure and function of the ligand binding site. Similar conformation effects might be exerted by the γc subunit, although studies in this regard have not yet been described.

In contrast to the α-β interaction, IL-2 binding appears to be required for recruitment of the gc subunit into the high affinity receptor complex (Takeshita et al., 1990). The γc chain was initially identified by

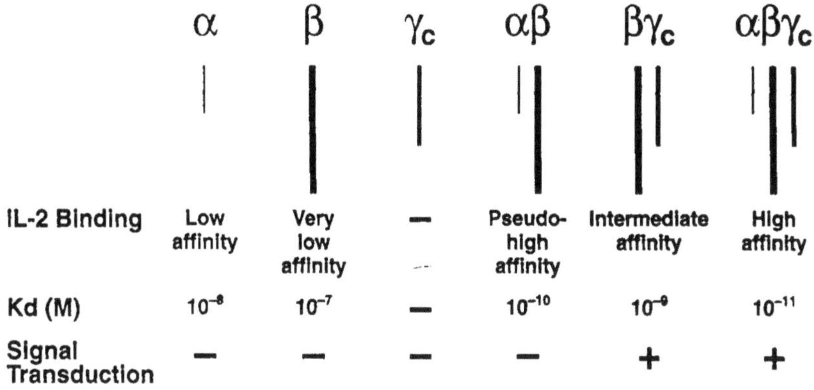

Figure 1. Multiple binding affinities of the IL-2 receptor. The binding characteristics of each IL-2 receptor subunit for IL-2 are indicated, as well as the affinities displayed by various combinations of these receptor subunits.

coimmunoprecipitation with anti IL-2Rβ antibodies in the presence of IL-2. Indeed, as discussed later in this chapter, ligand-induced heterodimerization of the β and γc chains plays a central role in the initiation of growth signal transduction.

C. The Cytokine Receptor Superfamily

As noted above, both the IL-2Rβ and γc subunits of the high affinity IL-2 receptor complex are now recognized as members of a superfamily of homologous cytokine receptors (Bazan, 1990a). Specifically, these receptors are distinguished by the presence of extracellular fibronectin-like domains containing a tetrad of cysteines whose spacing is highly conserved. A second signature motif is the WSXWS sequence positioned near the external plasma membrane. It seems likely that the cysteines contribute to the production of the ligand binding site, although the function of WSXWS motif remains undefined. Many cytokine receptor members also display conserved elements within their cytoplasmic tails termed Box 1 and Box 2 (Murakam et al., 1991). Both the IL-2Rβ and γc chains contain such elements, and both of these conserved elements have been shown by deletion to be important for effective growth signaling (Goldsmith et al., 1994). None of the cytokine receptor superfamily members contain intrinsic tyrosine kinase enzymatic activity within their cytoplasmic tails. However, it is now clear that various tyrosine kinases including members of the Janus family of kinases (JAK1, -2, -3, Tyk-2) associate with these tails, and that JAK activation is triggered by receptor subunit homodimerization or heterodimerization induced by ligand binding (Darnell et al., 1994).

An important step in defining the structural basis of subunit cooperation in the cytokine receptor superfamily was recently taken when the three-dimensional x-ray structure was solved for the growth hormone receptor (GHR) bound to its physiological ligand (de Vos et al., 1992). These studies demonstrated a homodimeric structure for the growth hormone receptor that intriguingly bound growth hormone in an asymmetric manner through various N-terminal residues. It seems quite likely that the IL-2Rβ and γ subunits form an analogous, albeit heterodimeric, structure the assembly of which is driven by concurrent engagement of the two receptor subunits (Nakamura et al., 1994; Nelson et al., 1994).

D. Structural Analysis of IL-2 Binding

Over the past several years, considerable information has been generated regarding properties of IL-2 required for its binding. Like many of the cytokines, IL-2 is arranged in four helices arrayed in an up-up-down-down bundle configuration (Brandhuber et al., 1987; Bazan, 1992). Deletion mutagenesis has indicated that the N-terminal 20 amino acids residues are essential for ligand binding to IL-2Rβ (Ju et al., 1987). Further, simple replacement of Asp-20 with Lys leads to an IL-2 analogue that no longer functions due to its inability to interact with IL-2Rβ subunit (Collins et al., 1988; Zurawski and Zurawski, 1989). These results suggest that one or more residues in the N-terminal A α-helix are likely involved in contacting the IL-2Rβ chain. In contrast, replacement of Phe-42 with Ala or of Arg-38 with Glu or Ala results in markedly diminished binding to IL-2Rα (Kuziel et al., 1993; Grant et al., 1992; Sauve et al., 1991; Weigel et al., 1989). These results suggest that the A-B interloop region may mediate interactions between IL-2 and the α subunit. Finally, mutations in the C-terminus of murine IL-2 at Gln-141 lead to analogues that fail to interact normally with β-γc dimers but bind well to α-β dimers (Zurawski et al., 1990). These findings suggest that the C-terminal D α-helix of IL-2 interacts with the γc subunit. Thus, although a higher resolution molecular dissection is required, a simple structural model of the quaternary complex of IL-2 bound to the heterotrimeric IL-2 receptor complex would place the β and γ chains contacting opposite faces of the bound IL-2 molecule, with the α chain supporting the ligand via interactions at one end. In essence, each of the three chains forms a portion of a "cytokine cradle" for IL-2.

E. Structural Analysis of Ligand Binding by IL-2Rα, β, and γ

Regarding the individual IL-2 receptor subunits, much less is known about which residues are required for ligand binding. In the case of the α chain, both internal deletion and truncation studies have revealed that the N-terminal 163 amino acid residues (derived from exons 1-4) contain all of the information needed for effective IL-2 binding (Rusk et al., 1988; Robb et al., 1988). Furthermore, residues 1-6 and 35-43 from exon 2 appear to be essential for ligand binding as well as recognition of this subunit by several anti-receptor antibodies. His-120 from exon 4 also appears to be critical for IL-2 binding, although antibodies interacting with other parts of exon 4 selectively block high affinity binding without impairing low affinity binding (Robb et al., 1988). These latter results raise the possibility that a domain within exon 4 of IL-2Rα might be involved in interactions with other receptor subunits needed for assembly of the high affinity binding site rather than for direct contact with IL-2. Virtually no information exists regarding specific residues in IL-2Rβ that are needed for IL-2 binding, although the N-terminal 212 residues of the extracellular domain of IL-2Rβ appear sufficient for binding, at very low affinity. Similarly, little is known about γc. Further structure-function studies are clearly needed to delineate precisely those residues from each receptor subunit that either mediate ligand binding or stabilize inter-chain interactions.

F. Sharing of Subunits in Different Cytokine Receptors

Recent studies of several cytokine receptor superfamily members have revealed that select receptor subunits participate in the formation of different receptors. For example, the β subunit of the IL-3 receptor also is a part of the GM-CSF receptor while the large 130 kD subunit of the IL-6 receptor also participates in the formation of the leukemia inhibitory factor (LIF), ciliary neurotrophic factor (CNTF), and oncostatin M receptors (Miyajima et al., 1992). This biological parsimony also extends to the human IL-2 receptor family where the γc chain has been shown to be a part of the assembly of the IL-4, IL-7, IL-9, and IL-15 receptors (Giri et al., 1994; Russel et al., 1993, 1994; Noguchi et al., 1993a; Mondo et al., 1994). Indeed, the IL-15 receptor also employs the

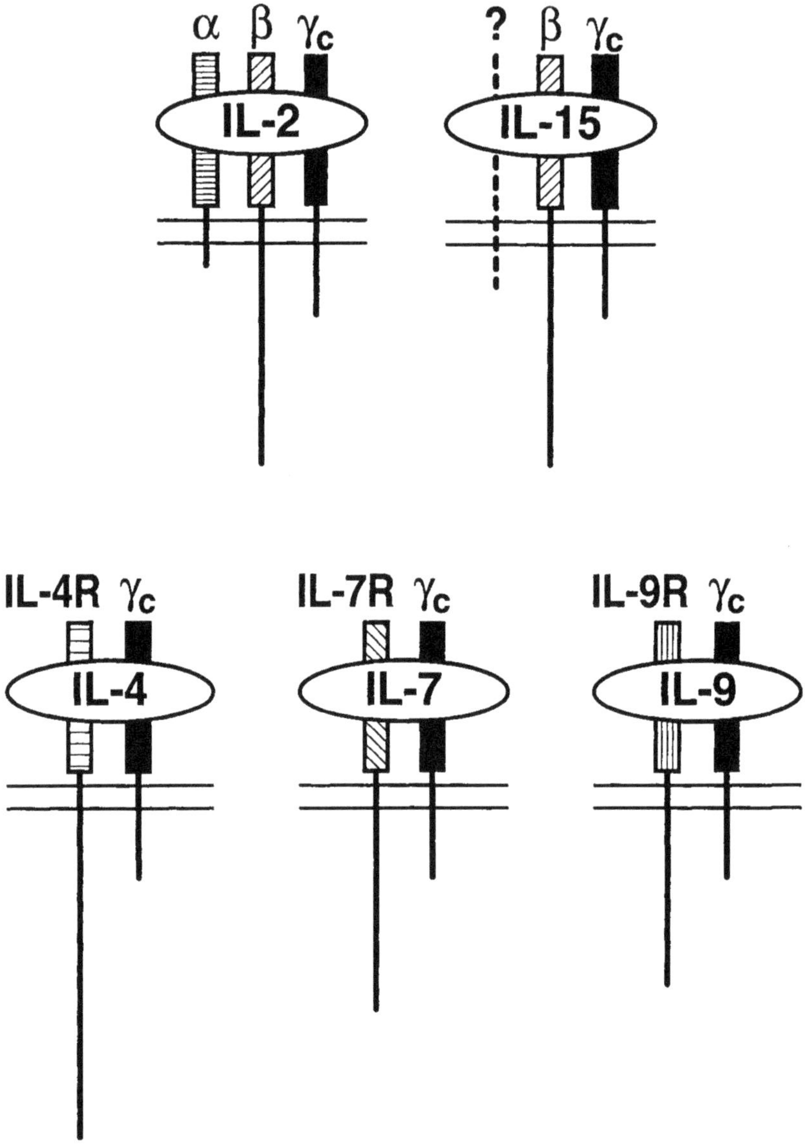

Figure 2. The γc-containing cytokine receptor family. The cytokine receptors depicted schematically here utilize the γc subunit as well one of more specific subunits, as indicated.

IL-2Rβ chain, while the α chain is unique. Such wide sharing of the γc subunit has provided a compelling explanation for a previously puzzling biological observation. Specifically, it rapidly became clear that the immunodeficiency observed in X-linked severe combined immunodeficiency reflecting defects in the γc chain gene were far more pronounced than the immunodeficiency observed in IL-2 deficient mice prepared by homologous recombination (Schorle et al., 1991). This finding led to the hypothesis that the γc chain might participate in the formation of additional cytokine receptors that might account for the more severe functional deficiencies encountered in the absence of γc (Noguchi et al., 1993b). The recent findings that γc participates in the IL-4, 7, 9, and 15 receptors (Figure 2) has now substantiated the original hypothesis and suggests an important functional role for these receptors *in vivo*.

II. THE BIOLOGIC ROLE OF THE IL-2/IL-2R SYSTEM

A. Effects on T Lymphocytes

Although the IL-2/IL-2R system is now recognized to regulate the growth and differentiation of a wide variety of hematopoietic cell types, IL-2 was initially known as T cell growth factor. Indeed, IL-2 was first identified as an activity derived from mitogen-activated peripheral blood mononuclear cells that could support the long-term growth of human T cells in culture (Morgan et al., 1976). Subsequently, mature T cells were shown to be the specific cellular source of this lymphokine (Gillis et al., 1978; Baker et al., 1978; Crabtree, 1989). T cells elaborate IL-2 during an immune response upon triggering of the T cell antigen receptor (TCR) complex by antigen, mitogens, or anti-TCR monoclonal antibodies; these agents, when provided in the context of appropriate co-stimuli, initiate a program of cellular activation that culiminates in the *de novo* synthesis and secretion of IL-2 (reviewed in Crabtree, 1989). The production of IL-2 is considered a hallmark of the T helper-1 (Th1) subset of T lymphocytes (reviewed in Seder and Paul, 1994).

The biological effects of IL-2 on T lymphocytes are pleiotropic. Its primary role in normal immune responses is thought to be the enhancement of clonal expansion following exposure to a specific antigen. Resting T cells, which display only intermediate affinity (IL-2Rβ and γc) receptors for IL-2, acquire the ability to bind IL-2 with high affinity through the expression of IL-2Rα chains during antigen-mediated T cell activation (see Regulation of Receptor Expression). The concurrent induction of both IL-2 itself as well as high affinity receptors for IL-2 provides a means for the expansion of antigen-specific T cell populations. Both paracrine and autocrine mechanisms may be utilized in this process (Depper et al., 1984; Meuer et al., 1984 and reviewed in Smith, 1988), although recent experiments involving the conditional ablation of IL-2-producing cells in transgenic mice implicate an autocrine pathway as the dominant mode of clonal expansion (Minas et al., 1993). The duration of a given immune response appears to be regulated by subsequent declines in both IL-2 synthesis and high affinity receptor display (Depper et al., 1984; Smith and Cantrell, 1985), which together terminate cellular growth and other events driven by IL-2. Evidence has also been reported that IL-2 may, under certain conditions, participate in a regulatory loop that promotes apoptosis by mature T lymphocytes (Lenardo, 1991). Finally, it has recently been reported that signal transduction through the γc chain specifically prevents induction of T cell anergy (Boussiotis et al., 1994). The IL-2R, therefore, plays a pivotal yet complex role in determining the fate (i.e., expansion, apoptosis or anergy) of a given T cell clone.

In addition to its effects on cell growth, IL-2 serves as a differentiation signal for T cells. It has been recognized for some time that IL-2 stimulates the production of interferon-γ (IFN-γ) (Farrar et al., 1982) and interleukin-4 (Howard et al., 1983). Engagement of the IL-2R by IL-2 may directly trigger expression of lymphokine genes, as has been observed with the GM-CSF gene in an established helper T cell line (Ruegemer et al., 1990). Alternatively, IL-2 may act through the IL-2R to promote acquisition of a lymphokine-producing phenotype. For example, *in vitro* data have emerged suggesting that IL-2 is required to support the differentiation of naive precursors of helper T cells into either Th1 cells (secreting IL-2, IFN-γ and TNF-β) or Th2 cells (secreting IL-4, IL-5, IL-6, IL-10 and IL-13; Le Gros et al., 1990; Ben-Sasson et al., 1990). In addition, IL-2 has been shown in certain circumstances to promote the differentiation of cytolytic T cells (Kern et al., 1981) and the resumption of cytolytic activity by deactivated cytolytic T cells (Orosz et al., 1985). Thus, intracellular signals emanating from the IL-2R may affect cellular programming as well as growth.

An area of significant controversy has been the role of the IL-2/IL-2R system in T cell development in the thymus. High levels of IL-2Rα have been described on immature thymocytes within the fetal and adult murine thymus (Ceredig et al., 1985; Hubu et al., 1985; Nakano et al.,

1987; Raulet, 1985; von Boehmer et al., 1985), as have IL-2Rβ chains (Takeuchi et al., 1992; Falk et al., 1993). Both low- and high-affinity IL-2 receptors have also been detected by IL-2-binding studies of fetal thymocytes (Zuniga-Pflucker et al., 1990). Moreover, although murine thymocytes in suspension have not been noted to grow in response to IL-2, significant proliferative responses were observed in cultures of intact murine fetal thymic lobes in the presence of low concentrations of IL-2 (Takeuchi et al., 1992; Zuniga-Pflucker et al., 1990). A specific role for the IL-2R in thymic maturation was further suggested by the finding that anti-IL-2Rα monoclonal antibodies arrested thymocyte development in thymic organ cultures (Jenkinson et al., 1987) and in newborn mice (Tentori et al., 1988). Despite these observations, however, the biologic significance of IL-2/IL-2R interactions in thymic development was challanged by the recent report that mice genetically deficient for IL-2 display normal thymocyte and peripheral T cell development (Schorle et al., 1991). Presumably, a complex interplay among several cytokine/cytokine receptor systems modulates these developmental processes, an interplay that may be obscured by functional redundancy among some of these factors.

B. Effects on Other Cells

Other cells of the immune system are also regulated by interactions between IL-2 and the IL-2R. For example, resting B cells constitutively express IL-2Rα, IL-2Rβ, and γc (Nakanishi et al., 1992; Nakarai et al., 1994), and the display of IL-2R subunits is modulated both by agents triggering the B-cell antigen receptor (Waldmann et al., 1984) and by the cytokines IL-2 and IL-4 (Nakanishi et al. 1992). As with T cells, IL-2 promotes the growth of mature B lymphocytes *in vitro* (Nakanishi et al., 1992; Jelinek and Lipsky, 1987). In addition, it regulates the differentiation state of B cells by augmenting J-chain synthesis (Blackman et al., 1986; Tigges et al., 1989) and the assembly and secretion of immunoglobulin (Jelinek and Lipsky, 1987; Blackman et al., 1986; Tigges et al., 1989). Intriguingly, IL-2Rα chains have recently been found on pre-B and immature B cells in the bone marrow (Chen et al., forthcoming), suggesting a possible role for IL-2 and its receptor in B cell development.

The IL-2/IL-2R system also regulates the expression of certain effector functions, including cytolysis, by non-T cells. For example, natural killer (NK) cells not only proliferate in response to IL-2 but also produce IFN-γ and exhibit enhanced cytolytic activity in the presence of IL-2

(Ortaldo et al., 1984; Trinchieri et al., 1984; Lanier et al., 1985; Talmadge et al., 1986). Interestingly, in transgenic mice constitutively expressing both human IL-2Rα chains as well as human IL-2 there was a specific and marked expansion of the large granular lymphocyte population (LGL) exhibiting elevated NK-like cytolytic activity, a finding that further supports a role for IL-2 in regulating the growth and function of this cell compartment (Ishida et al., 1989). High concentrations of IL-2 also promote the expansion of a population of human lymphoid cells (termed lymphokine-activated killer or LAK cells) that exhibit potent cytolytic activity against tumor target cells, a process that has been turned to clinical utility (reviewed in Rosenberg and Lotze, 1986). It is thought that intermediate-affinity IL-2Rβ/γc heterodimers mediate the proliferation and activation of such NK and LAK cells in response to IL-2 (Siegel et al., 1987; Kehrl et al., 1988; Voss et al., 1992).

Non-lymphoid cells of several hematopoietic lineages also display IL-2R that are responsive to IL-2. Macrophage precursors derived from bone marrow (Baccarini et al., 1989) as well as primary peripheral blood monocytes (Ohashi et al., 1989; Bosco et al., 1994) express intermediate-affinity cell surface IL-2R, and these receptors can be converted to high-affinity receptors through the induction of IL-2Rα chains by treatment with IFN-γ or lipopolysaccharide (Herrmann et al., 1985; Holter et al., 1986, 1987; Hancock et al., 1987). That such receptors may have a functional role *in vivo* is suggested by the observation that alveolar macrophages obtained from patients with active pulmonary sarcoidosis display IL-2Rα chains (Hancock et al., 1987). Furthermore, IL-2 has been reported to enhance the cytolytic activity of peripheral blood monocytes (Malkovsky et al., 1987), promote antibody-dependent tumoricidal activity by macrophages (Ralph et al., 1988), induce the expression of cytokines by macrophages and monocytes (Strieter et al., 1989; Kovacs et al., 1989) and augment microbicidal activity by monocytes (Whal et al., 1987); enhancement of antifungal activity by neutrophils has also been reported (Djev et al., 1993). Thus, IL-2 influences the expression of effector functions by many distinct hematopoietic cell types.

Finally, IL-2R has also been detected in certain non-hematopoietic cell types. For example, isolated oligodendrocytes appear to display IL-2Rα chains (Saneto et al., 1986). Additionally, IL-2 has been reported to modulate the growth (Saneto et al., 1986; Benvisk et al., 1986) and differentiation (Benviste et al., 1986) of oligodendrocyte precursors. IL-2Rα and IL-2Rβ chains have also been detected on human embryonic fibroblasts (Plasisance et al., 1992). The biologic significance of IL-2R expression in such varied cell types nonetheless remains to be determined.

III. SIGNAL TRANSDUCTION BY THE IL-2 RECEPTOR COMPLEX

A. Receptor Subunit Dimerization

The mechanism of transmembrane signaling by the IL-2R has been the subject of intensive investigation. It has long been evident that the first recognized subunit of this receptor, IL-2Rα, contained a cytoplasmic tail that is too short to encompass the complete signal transduction apparatus. The subsequent identification of two other subunits, IL-2Rβ and γc, offered additional components of the receptor complex with more extensive cytoplasmic regions that were expected to participate substantively in the signaling process. These subunits have consequently become the focus of investigation in this area.

Two basic models could explain the initiation of signal transduction upon ligand binding. First, the engagement of ligand might elicit conformational changes in the receptor that are transmitted from the extracellular domain across the plasma membrane, thereby altering the structure of the cytoplasmic region of the receptor. Such transmembrane conformational changes have been proposed in many other receptor systems, and a mechanistic model has been described for the bacterial aspartate receptor that hinges upon changes in vertical positioning of the receptor relative to the plane of the membrane (Milligan and Koshland, 1991). Although no direct evidence for conformational changes in the IL-2R have been described, circumstantial evidence has emerged from studies employing an IL-2 analogue with altered receptor binding properties that are consistent with conformational changes in the extracellular regions of the receptor complex during engagement of IL-2 (Kuziel et al., 1993; Grant et al., 1992). In addition, we have recently reported evidence supporting the concept that conformational changes in receptor structure may be transmitted from the intracellular region to the extracellular domains (Goldsmith et al., 1995a). Such changes may involve modulation of the vertical positioning of individual receptor subunits relative to one another or to the plane of the membrane, and such adjustments may well be transmitted in either direction across the cell surface. Therefore, a conformational mechanism may contribute to the transmission of signals by the IL-2R during the binding of IL-2.

A second model to account for the initiation of transmembrane signaling depends upon receptor subunit oligomerization during the engagement of ligand. The delineation of the three-dimensional structure of the growth hormone receptor bound to growth hormone provided a compelling visual representation of the dimerization of receptor subunits in the presence of the cognate ligand (Devos et al., 1992). These studies capped a striking series of biophysical experiments that demonstrated a step-wise ligand-binding process in which one molecule of growth hormone facilitates the approximation of two growth hormone receptor subunits (GHR; Cunningham et al., 1991); this juxtaposition of the GHR extracellular domains during ligand binding presumably results in a similar ordering of the intracellular regions of the receptor subunits, which activates associated signaling components. These reports serve as a powerful paradigm for considering other members of the cytokine receptor superfamily. Compelling data also emerged from studies of mutants of the erythropoietin receptor (EPOR) in which introduction of cysteine residues into the predicted regions of extracellular domain approximation resulted in the formation of intersubunit disulfide bonds (Yashimura et al., 1990; Watowich et al., 1992, 1994); these covalent linkages permitted ligand-independent EPOR subunit homodimerization and, hence, constitutive signal transduction. The recently reported three-dimensional structure of the EPOR is consistent with this concept (Livnah et al., 1996). These findings, together with the GHR observations, strongly suggest that subunit dimerization among members of the cytokine receptor superfamily contributes to the initiation of signal transmission.

Data have recently emerged supporting a similar role for subunit oligomerization in IL-2R function. A first hint that IL-2R subunits are approximated during the binding of ligand derives from the recognition that individual IL-2R subunits bind IL-2 with only low or even undetectable affinities, while receptor heterodimers (IL-2Rβ/γc, IL-2Rα/β) and heterotrimers (IL-2Rα/β/γc) bind IL-2 with significantly higher affinities (see Section IB for further discussion). Moreover, IL-2 markedly increases the coimmunoprecipitation of IL-2Rβ and γc chains (Takeshita et al., 1990, 1992b). Thus, IL-2 appears to provide nucleation for the assembly of the IL-2R complex. That dimerization of IL-2R subunits is crucial for eliciting signal transduction has been demonstrated by two types of studies employing chimeric receptors. In one line of investigation, the extracellular domains of the IL-2Rβ and γc subunits were replaced by that of the IL-2Rα chain (Nakamura et al., 1994). These hybrid receptors could be induced to transmit signals by antibodies that crosslinked the β and γc chimeras. In a second line of study, hybrid receptors were prepared that linked the IL-2R intracellular domains with extracellular domains derived from other receptors that could be engaged by natural ligands (Nelson et al., 1994). In these experiments, engagement of individual hybrid receptor subunits failed to elicit biologic responses, while concurrent engagement of the β and γc chimeras elicited full biologic responses. Indeed, even the earliest detectable events such as activation of receptor-associated Janus kinases occurrs only upon dimerization of these two subunits and not upon triggering either subunit alone (Goldsmith et al., 1995b). These findings collectively implicate IL-2R subunit oligomerization during ligand binding as a critical incipient event in signal transduction.

B. Receptor Phosphorylation

A dominant theme in receptor-mediated signal transduction research is the activation of intracellular kinases. For receptors containing intrinsic kinases activity, such as the protein tyrosine kinase receptors (reviewed in Ullrich and Schlessinger, 1990), receptor cross-linking promotes either *trans-* or autophosphorylation of the receptor itself. Individual phosphotyrosine residues of the receptor subsequently serve as loading sites for specific signaling intermediates that participate in signal propagation (Valins and Kazlauskas, 1993; Kazlauskas and Cooper, 1989). For receptors lacking intrinsic kinase catalytic activity, a similar mechanism is employed that relies upon recruitment of cytoplasmic kinases to phosphorylate tyrosine residues within a common peptide motif (reviewed in Weiss, 1993).

The role of receptor phosphorylation in cytokine receptor function has been controversial. In the related interleukin-4 (IL-4) system, evidence has been reported that tyrosine-497 of the IL-4 receptor (IL-4R) regulates both growth signal transduction and activation of the signaling intermediate IRS-1 (Keegan et al., 1994). In addition, the IL-4R itself has been shown undergo tyrosine phosphorylation upon exposure to IL-4 (Wang et al., 1992; Kzuhara and Harada, 1993). Recently, cytoplasmic phosphotyrosines of the IL-4R were also implicated in regulating activation of a transcription factor associated with the IL-4R, IL-4 Stat (Hou et al., 1994). However, another group has reported that concurrent substitution of phenylalanine for all cytoplasmic tyrosines of the IL-4R had no effect on the signaling function of the receptor (Seldin and Leder, 1994). Resolution of these conflicting findings awaits further investigation.

Signal transduction by the IL-2R may also be regulated by receptor phosphorylation. The IL-2Rβ chain contains six cytoplasmic tyrosine residues, five of which lie in positions that are conserved between the human and mouse chains. Several groups have reported that this chain undergoes rapid tyrosine phosphorylation in the presence of IL-2 (Mills et al., 1990; Asao et al., 1990; Shackelford and Trowbridge, 1991). Similarly, the γc chain contains four conserved tyrosines, and tyrosine phosphorylation of this subunit upon activation by IL-2 has been described (Asao, 1992). Therefore, both IL-2R subunits involved in signal transduction are substrates for receptor-associated tyrosine kinase(s).

Little information has been available regarding the biological significance of these phosphorylation events. However, recent findings implicate certain of these tyrosine residues in receptor function. We have found that concurrent substitution of phelylalanine for all six cytoplasmic tyrosines of IL-2Rβ virtually abrogates its growth signal transduction competence upon expression in either B or T lymphocytes (Goldsmith et al., 1995b). Moreover, the C-terminal two tyrosines of IL-2Rβ appear to be functionally redundant for one another in a pro-B cell line, and at least one of these (tyrosine-392) is clearly a target for tyrosine phosphorylation. Interestingly, data derived from *in vitro* phosphopeptide experiments suggests that this amino acid mediates the association of IL-2Rβ with phosphatidylinositol-3 kinase (PI3K; Truitt et al., 1994). It is also possible that STAT factors involved in transcription regulation of IL-2-responsive genes utilize such phosphotyrosines in a manner analogous to the IL-4 Stat/IL-4R interaction (see the subsection on Janus kinases). The complete profile of molecules assembling onto phosphotyrosine-containing domains remains undetermined. In addition, the kinase responsible for IL-2R phosphorylation has not been defined, although preliminary *in vitro* data suggest that the Janus kinase JAK1 constitutively bound to IL-2Rβ may mediate this event (Liu, Greene, and Goldsmith, unpublished observations).

A comparable function for γc phosphorylation has not been evident from parallel studies with this chain. Unlike the IL-2Rβ chain, substilatyrosine residues of γc had no effect on growth signal transduction function in T cells (Goldsmith et al., 1995b). It, therefore, remains uncertain why these amino acids are conserved across species and why they serve as targets for phosphorylation during receptor activation. It is conceivable that these residues are "innocent bystanders" that receive phosphate groups merely because of their proximity to activated, receptor-associated kinases. Alternatively, they permit receptor interactions with other signaling components that mediate pathways of signal transduction not fully measured in the investigations completed thus far (e.g., factors regulating differentiation). Further elucidation of the precise role(s) of these tyrosine phosphorylation events in the full IL-2R signaling program thus awaits additional experimentation.

It is also possible that phosphorylation of serine and/or threonine residues of the IL-2R influences signal transduction. Indeed, phorbol esters promote serine-phosphorylation of the IL-2Rα cytoplasmic tail (Shackelford and Trowbridge, 1991), but studies employing site-directed mutagenesis demonstrated that IL-2Rα phosphorylation sites are dispensable for either receptor downregulation or proliferation signaling (Hatakeyama et al., 1986). No systematic analysis of serines and threonines within the IL-2Rβ or γc chains has been reported. In this regard, it is of potential interest that the cytoplasmic tail of IL-2Rβ contains a region that is particularly rich in serine residues, and that this segment plays a crucial role in many signaling processes (see Hatakeyama et al., 1986b and Figure 3 for further discussion). It is tempting to speculate that phosphorylation of serine or threonine residues within this region contribute to receptor function. Our recent observation that substitution of alanines for a stretch of three adjacent serines within this segment abolishes growth signal transduction (Liu et al., 1995b) is consistent with such an hypothesis.

C. Receptor-activated Kinases

Overview

It is evident from the deduced primary sequences of the IL-2R subunits that the receptor lacks the cardinal features of tyrosine kinase catalytic domains, including the canonical Gly-X-Gly-X-X-Gly ATP-binding motif. However, that the IL-2Rβ and γc chains are targets for tyrosine phosphorylation during receptor activation implies a role for one or more tyrosine kinases in receptor-mediated signal transduction. Several studies have demonstrated that pharmacologic inhibitors of such kinases disrupt the usual biologic consequences of receptor triggering. For example, the tyrosine kinase inhibitors herbimycin A and genistein were found to abrogate the induction of both proliferation as well as IL-2Rα expression by IL-2 in various cells expressing IL-2R (Merida et al., 1991; Otani et al., 1992). Similarly, genistein was reported to inhibit

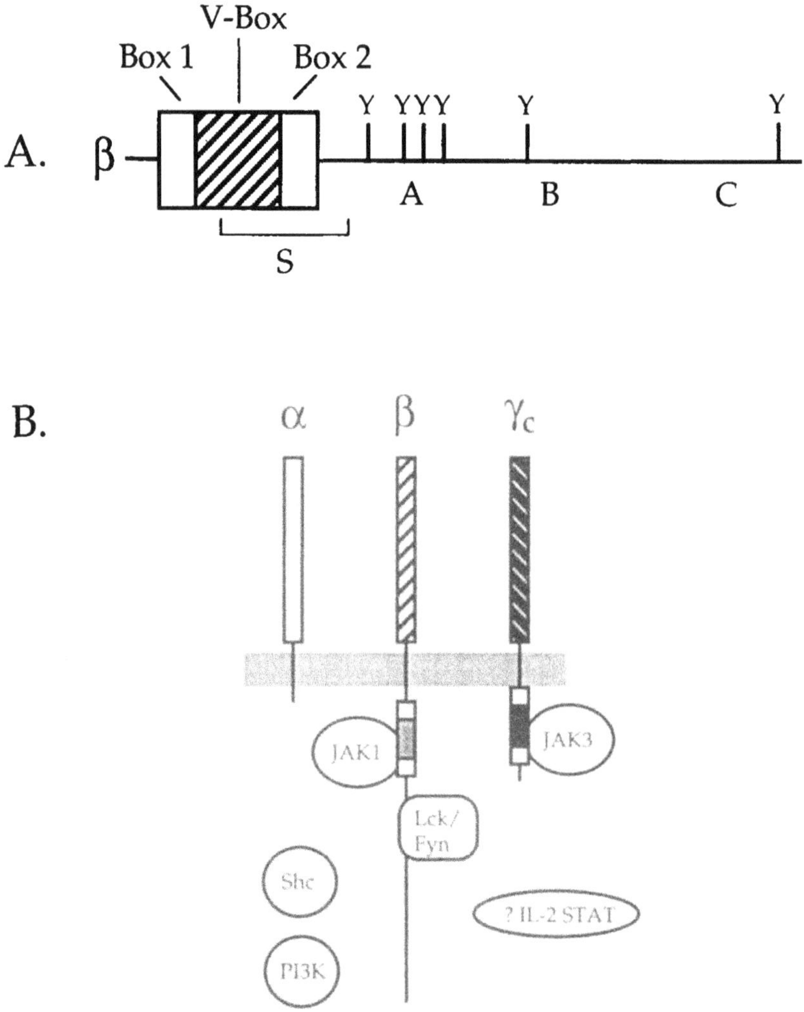

Figure 3. Functional architecture of the IL-2 receptor complex. (**A**) Features of the cytoplasmic tail of the IL-2Rβ chain are shown, with the membrane-proximal motifs at the left and the C-terminal structures at the right. The six cytoplasmic tyrosines are indicated (Y). See text for other designations. (**B**) Signal transduction intermediates associated with the IL-2R complex. The JAK1, JAK3, and Lck kinases have been shown to associate physically with the IL-2R, as indicated. The specific molecular basis of association of Shc or PI3K with the IL-2 receptor complex has not been described. Similarly, a putative IL-2R-specific STAT factor and its association with the IL-2R have been hypothesized but not yet reported.

IL-2-increased NK-mediated cytotoxicity, and this effect was attributed to an impaired physical association between $p56^{lck}$ and IL-2Rβ (see the next subsection and Nishio et al., 1994). Finally, the tyrosine kinase inhibitor leflunomide has also been shown to disrupt both antigen- and IL-2-driven proliferation of T cell clones (Nikcevich et al., 1994). Collectively these studies provide strong evidence that one or more tyrosine kinase is critically involved in IL-2R signal transduction function.

Since the receptor lacks intrinsic kinase function, it is reasonable to assume that one or more cytoplasmic enzyme containing such an activity is functionally, and perhaps physically, linked to the receptor complex. Evidence for such direct contact between an intracellular kinase and the receptor complex has derived from several independent immunoprecipitation studies. First, IL-2R complexes immunoprecipitated from the murine cytotoxic T cell line CTLL-2 following cross-linking to IL-2 were shown to be capable of tyrosine phosphorylating an exogenous substrate, histone 2B (Merida and Gaulton, 1990). Likewise, a tyrosine kinase activity capable of phosphorylating the IL-2Rβ chain itself was co-immunoprecipitated with the IL-2Rβ chains derived from both activated human lymphoblasts and the murine pro-B cell line BA/F3 (Fung et al., 1991). Finally, a 97-kDa protein with apparent autophosphorylation activity was reported to co-immunoprecipitate with the IL-2R, and this activity was induced by IL-2 *in vitro* (Michiel et al., 1991; Garcia et al., 1992). Together these observations implicate receptor-associated tyrosine kinase(s) in IL-2R signaling function. The phosphorylation state of cellular constituents may also be regulated by protein phosphatases, as IL-2 has been found to cause transient decreases in protein phosphatase PP1 activity (Matsuzawa et al., 1993) as well as enhanced expression of the leukocyte tyrosine phosphatase (LC-PTP) gene (Adachi et al., 1994).

Src-*Family Kinases*

The identity of the kinase(s) responsible for IL-2R signal transduction continues to be the subject of intensive investigation. Much attention has focused on enzymes recognized to influence signaling in other receptor systems. Several years ago it was reported that IL-2 induces the serine-/threonine-phosphorylation and concomitant activation of the T cell-specific kinase $p56^{lck}$ in IL-2-dependent human T cell clones (Horak et al., 1991). Subsequently, activation of this enzyme was found to correlate with induction of proliferation (Kim et al., 1993). That $p56^{lck}$ may indeed be a crucial receptor-associated signaling component was strongly suggested by the finding that it physically associates with the IL-2Rβ chain in both lymphocytes and co-transfected heterologous cells (Hatakeyama et al., 1991), and that this physical association is essential for IL-2-induced activation of $p56^{lck}$ (Minami et al., 1993).

The functional significance of the IL-2Rb/$p56^{lck}$ association has been challenged by several reports that $p56^{lck}$ is not essential for IL-2-mediated signaling. For example, the growth of certain HTLV-I-infected human T cell lines (Mills et al., 1992) and the myeloid line 32D (Otani et al., 1992) is modulated by IL-2 despite the complete absence of detectable $p56^{lck}$. In addition, a variant of the CTLL-2 line that lacks $p56^{lck}$ retains dependence upon IL-2 for viability and growth, although a reduction in IL-2-induced cytolytic activity was noted that could be reversed upon restoration of $p56^{lck}$ by transfection (Karnitz et al., 1992). Likewise, over-expression of $p56^{lck}$ in CTLL-2 cells by transfection failed to augment proliferative signaling to IL-2 (Taichman et al., 1992). Although these findings tend to undermine a biologic role of $p56^{lck}$ in IL-2R function, *src*-family kinases are thought to demonstrate some degree of functional promiscuity that would allow more than one such kinase to operate in a given receptor system. The determination that the related family members $p59^{fyn}$ and $p53/56^{lyn}$ also couple with the IL-2R in cell types lacking $p56^{lck}$ supports the possibility that such functional redundancy applies in the IL-2R system (Kobayashi et al., 1993). Nonetheless, the full role of *src* kinases in IL-2R function remains to be elucidated. Intriguingly, a novel family T cell-specific tyrosine kinases bearing some relationship to the *src*-kinases was recently identified and shown to be inducible by IL-2 (Siliciano et al., 1992; Tanaka et al., 1993). The possible role of these kinases in IL-2R signaling function also remains to be defined.

Janus Kinases

Elegant genetic studies in the interferon receptor system (reviewed in Hunter, 1993) have led to the identification of an important new class of cytoplasmic kinases (Janus or JAK kinases) regulating cytokine receptor function. These enzymes, which range in size from approximately 115 kDa to 140 kDa, are related to one another by first, the presence of both a classical kinase domain, and second, adjacent domain with features suggestive of a catalytic function. As described initially in the interferon system and subsequently in other related systems, specific JAK family members bind to the cytoplasmic tails of individual cytokine receptor subunits where, upon receptor dimerization, they initiate phosphorylation of cellular substrates. Among the best characterized of these substrates are the so-called STAT factors (signal transducer and activator of transcription), which transiently associate with the receptor complex; phosphorylation by a JAK kinase compels STAT factor dimerization and to transit to the nucleus in order to initiate transcription of certain target genes (reviewed in Darnell et al., 1994). The so-called JAK-STAT pathway is now recognized to participate in signal transduction by several cytokine receptor superfamily members.

The first suggestion that Janus kinases may be involved in IL-2R function came from the discovery of a 116-kDa-protein with probable tyrosine kinase activity associated with the IL-2Rβ chain (Kirken et al., 1993). Shortly thereafter the specific Janus family members JAK1 and JAK2 were reported to be physically associated with the IL-2R and to undergo tyrosine phosphorylation upon stimulation of the receptor (Tanaka et al., 1994). Subsequently, a Janus kinase that was specific for lymphocytes was identified and molecularly cloned (now known as JAK3; Kawamura et al., 1994; Witthuhn et al., 1994), which provided a possible basis for signal transduction specificity by the IL-2R. Indeed, the availability of specific antibodies allowed a determination that both JAK1 and JAK3 (but not JAK2) are activated by IL-2 (Witthuhn et al., 1994; Johnston et al., 1994; Asao et al., 1994). Specific JAK induction by IL-2 was later shown to require selective interactions between the cytoplasmic tails of individual IL-2R subunits and the JAKs (IL-2Rβ with JAK1; γc with JAK3; Russell et al., 1994; Miyazaki et al., 1994). Since the induction of both of these kinases depends upon concurrent engagement of both IL-2Rβ and γc by ligand (Goldsmith et al., 1995b), it appears as if assembly of the dimeric receptor complex in the presence of IL-2 concurrently activates these two receptor-associated kinases. It remains uncertain whether this process depends upon *trans*-phosphorylation among the Janus kinases, or instead upon as yet unidentified elements acting earlier in the pathway. The specific substrates of these kinases have not been identified, although both the receptor itself as well as members of the STAT family may be among these substrates.

Other Kinase-dependent Pathways

A wide variety of additional signaling intermediates have been reported to be activated following IL-2R engagement by IL-2. For example, the ribosomal p70 S6 kinase becomes activated following stimulation of cells by IL-2, and this event is inhibited by rapamycin, a macrolide that also inhibits IL-2-induced progression through the cell cycle (Calvo et al., 1992; Kuo et al., 1992; Tereda et al., 1992). In addition, the expression and catalytic activity of the serine/threonine kinase Raf-1 are regulated by IL-2 (Turner et al., 1991; Zmuidzinas, 1991). Similarly, IL-2 promotes assembly of the active, GTP-bound form of p21ras (Grabstein et al., 1994; Graves et al., 1992), a process that is inhibited by herbimycin A (Izquierdo et al., 1992; Izquierdo and Cantrell, 1993). The mitogen-activated protein (MAP) kinase member Erk2 has also been shown to be induced by IL-2 in some cell types (Perkins et al., 1993). One possible scenario, therefore, is that IL-2R-mediated induction of an early tyrosine kinase leads to activation of $p21^{ras}$, which in turn promotes activation of Erk2, Raf-1, and S6 kinase with consequent progression through the cell cycle.

Phosphoinositide kinases have also been implicated in IL-2R signaling function. Specifically, phosphatidylinositol 3-kinase (PI3K) has been shown to be activated by IL-2 in lymphocytes through a mechanism that is disrupted by tyrosine kinase inhibitors (Merida et al., 1991, 1993; Augustine et al., 1991). Moreover, IL-2 apparently triggers the non-covalent binding of PI3K to the IL-2R (Truitt et al., 1994; Remillard et al., 1991), and this physical association may depend upon phosphorylation of tyrosine-392 of IL-2Rβ (Truitt et al., 1994). Interestingly, recent evidence implicates $p56^{lck}$ in the IL-2-mediated regulation of PI3K activity (Taichman et al., 1993; Karnitz et al., 1994), perhaps providing insight into the relationship between early receptor-activated tyrosine kinases and PI3K function.

Finally, several cellular proteins thought to serve as adaptors or modulators of signaling intermediates have been shown to be regulated by the IL-2R. For example, IL-2 promotes the tyrosine phosphorylation of the SH2-containing adaptor protein $p52^{shc}$ (Burns et al., 1993), and this phosphorylation event correlates with IL-2-induced proliferation (Zhu et al., 1994). Furthermore, phosphorylation of this protein promotes its association with Grb2 (Zhu et al., 1994; Ravichandran and Burakoff, 1994) and its associated Ras GTP/GDP exchange factor mSOS (Ravichandran and Burakoff, 1994). Moreover, $p52^{shc}$ has been found to associate inducibly with the IL-2R (Ravichandran and Burakoff, 1994). The hematopoietic GTP/GDP exchange factor $p95^{vav}$ has also been reported to undergo tyrosine phosphorylation in response to stimulation of lymphocytes by IL-2 (Evans et al., 1993). Thus, these data together suggest that the IL-2R regulates multiple pathways of signal transduction. Further investigation is required to assemble a complete picture of the connections among these various pathways that account for induction of downstream biologic events by IL-2.

D. Regulation of Gene Transcription and Progression Through the Cell Cycle

At least some of the early signal transduction events described above culminate in the induction of transcription of a certain genes involved in differentiation, effector functions, or proliferation. The mechanisms by which these early events are translated into gene expression are poorly defined. It appears likely that one or more STAT factors are activated by the JAK1 and JAK3 kinases, and these factors would be expected to have specific gene targets within the nucleus. One such target, the interferon-regulatory factor 1 (IRF-1) gene, is regulated by multiple cytokines including IL-2 (Gilmour et al., 1994). It appears as if a DNA-binding factor (NAF) is activated rapidly by IL-2 to bind to a GAAA inverted repeat in the IRF-1 promoter; this DNA motif is also recognized by a prolactin-induced factor (PRLIF) that contains a protein constituent bearing some relationship to p91/STAT1 (Gilmour et al., 1994). Several groups have identified STAT5 (Kujii et al., 1995; Hou et al., 1995; Lin et al., 1995; Gaffen et al., 1995) and specifically the isoforms STAT5A and STAT5B (Gaffen et al., 1996) as the predominant factors activated via the IL-2R.

A second mechanism that may be involved in mediating some of these connections is the NFκB system of transcription factors (Blank et al., 1992). IL-2 has been shown to induce NFκB activation in some circumstances (Arima et al., 1992a), and a T cell line that proliferates via an IL-2-dependent autocrine pathway contains constitutively active NFκB (Hémar et al., 1991). Indeed, the IL-2Rα gene, which contains κB enhancer elements, is up-regulated by IL-2 (Smith and Cantrell, 1985; Kehrl et al., 1988; Depper et al., 1985; Le Gros et al., 1987). Therefore, NFκB factors may be involved in regulating certain genes in response to IL-2.

A variety of genes associated with cell cycle regulation are also activated by triggering of the IL-2R by IL-2. Transcripts for several proto-oncogenes, including *c-myc*, *c-myb* and the AP-1 factors (*c-fos/c-jun*) are induced by IL-2 in various cell systems (Granelli-Piperno et al., 1986; Stern and Smith, 1986; Reed et al., 1987; Shibuya et al., 1992). Similarly, expression of the hematopoietic kinase *pim-1* that has been associated with proliferation and leukemogenesis is markedly augmented by IL-2 and several other cytokines (Dautry et al., 1988). The specific relationships among these various genes and progression through the cell cycle await further clarification.

While many genes regulated by IL-2 have been identified, the full panoply of such IL-2-responsive genes has not yet been defined. One approach that has been used by two groups depends upon differential hybridization of a cDNA library prepared from IL-2-treated cells (Sabath et al., 1990; Beadling et al., 1993). This strategy yielded an array of cDNA clones, among which were many novel sequences as well as *pim-1* and other transcripts previously recognized to be involved in cell cycle regulation. Further studies are needed to clarify the specific roles of each of these genes in regulating cell cycle progression and other aspects of cell function. These libraries should also be useful in determining the mechanisms in receptor-mediated signal transduction that permit specificity in the responses to IL-2 and other related cytokines.

E. Structure/Function Relationships within the IL-2R Complex

IL-2Rβ

In addition to the many signaling intermediates that have been determined to act in *trans* during IL-2R signal transduction, a detailed picture of the functional domains and amino acids within the receptor itself has begun to emerge (Figure 3). Mutational analyses have focused largely upon the IL-2Rα and γc cytoplasmic tails, because an earlier study demonstrated that the cytoplasmic tail of IL-2Rα appeared to be dispensable for IL-2R signaling function (Kondo et al., 1987). Extensive work has been performed specifically with the IL-2Rβ chain, since several cell systems have been identified that fail to express this chain and in which signaling responses can be restored upon introduction of exogenous IL-2Rβ chains by gene transfer (Hatakeyama et al., 1986b; Otani et al., 1992; Tanaka et al., 1991).

Initial studies identified a serine-rich segment of the membrane-proximal region of the IL-2Rβ cytoplasmic tail (the S region, including amino acids 267-322 of the mature β protein) that is essential for proliferative signaling through the IL-2R but dispensable for the ligand-binding and internalization properties of the receptor (Hatakeyama et al., 1989a, 1989b). This segment also appeared to be critical for induction of the proto-oncogenes *c-fos* (Hatakeyama et al., 1992) and *c-myc* (Merida et al., 1993; Satoh et al., 1992), induction of NFκB (Arima et al., 1992a) activation of $p21^{ras}$ (Satoh et al., 1992), and functional coupling to receptor-associated tyrosine kinases (Fung et al., 1991; Merida et al., 1993) including $p56^{lck}$ (Minami et al., 1993). It was recently noted that the S region encompasses one of two membrane-proximal motifs ("Box 2"; Goldsmith et al., 1994) found among many members of the cytokine receptor superfamily (Murakami et al., 1991). Detailed mutagenesis studies revealed that this 14-residue segment (amino acids 298-309) as well as the more proximal "Box 1" motif (amino acids 248-260; Goldsmith et al., 1994) and the non-conserved segment connecting Box 1 with Box 2 (the "V-Box"; Liu et al., 1995) are essential for growth signaling.

The functionality of these domains seems to be accounted for by only a few individual amino acids, including aspartic acid-258 within Box 1 (Goldsmith et al., 1994) and leucine-299 within Box 2 (Goldsmith et al., 1994; Mori et al., 1991). We have observed that certain serines and non-serine residues within the V-Box are crucial not only for proliferative signaling, but also for the activation of early signaling events by the IL-2R, including receptor phosphorylation and Janus kinase induction (Liu et al., 1995). Recently, it has been shown that the distal portion of this region is essential for the binding of JAK1 to the IL-2Rβ chain, which likely accounts for some of the impaired signaling properties of such mutants (Miyazaki et al., 1994). It remains unclear, however, how leucine-299 within Box 2 regulates activation of $p56^{lck}$ since $p56^{lck}$ binds to IL-2Rβ independently of Box 2 (see Minami et al., 1993 and following). It is possible that the induction of $p56^{lck}$ is a secondary event that results from activation of JAK1 and JAK3 or of kinases operating even earlier in the pathway. The molecular function of the Box 1 and V-Box segments are also unclear, since many mutations within these segments that abrogate signaling function do not alter associations with JAK1 (Goldsmith and Greene, unpublished observations). Conceivably these regions of the receptor are involved in contact with other receptor-associated structures or in receptor subunit interactions.

A second region of the cytoplasmic domain of IL-2Rβ that has been implicated in signaling by the IL-2R is the 46-amino acid "A" or "acid" region (residues 313-382 of the mature protein) lying immediately downstream from Box 2 (Hatakeyama et al., 1989a, 1989b). Although deletion of this segment does not substantially alter growth signaling in a pro-B cell line (Hatakeyama et al., 1989b), this region may have divergent functions in various cell types (Minami et al., 1994; Greene, unpublished observations). This segment regulates the assocation with

$p56^{lck}$ (Hatakeyama et al., 1991) as well as the induction of *c-fos* transcription (Hatakeyama et al., 1992) and activation of $p21^{ras}$ (Satoh et al., 1992); the binding sites on IL-2Rbbb for other *src*-kinases have not been described. Interestingly, four of the six cytoplasmic tyrosines of IL-2Rβb lie within this region. Although these tyrosine residues are dispensable in the pro-B cell line (see Section IIIB), their functional significance in other cell types has not been determined. It may be noteworthy that the interaction between $p56^{lck}$ and IL-2Rβ is mediated by the catalytic domain of $p56^{lck}$ (Hatakeyama et al., 1991), suggesting that tyrosine residues within this segment may be important for such interactions in cell types expressing $p56^{lck}$.

Mutational analyses have also revealed crucial signaling elements within the C-terminal region of IL-2Rβ (designated the B and C segments; Goldsmith et al., 1994). These regions appear to be functionally redundant for mediating growth signals in BA/F3 cells. Since each of these segments contains a single tyrosine residue thought to be important for full growth signaling, it is unknown whether or not other amino acids within these segments are also biologically active. Recent studies have further mapped specific signaling pathways to various typosine residues of the IL-2Rβ (Gaffen et al., 1996).

Finally, it should be noted that at least one mutation of the extracellular domain of IL-2Rβ has been described that interferes with signaling function. Specifically, substitution of the conserved serine residues within the canonical Trp-Ser-X-Trp-Ser motif in the extracellular domain moderately affected growth signaling without impairing IL-2 binding (Miyazaki et al., 1991). This finding raises the intriguing possibility that this membrane-proximal region of the extracellular domain of IL-2Rβ may be involved in associations with other, perhaps unidentified components of the receptor complex. Alternatively, this extracellular motif may influence the transmission of conformational changes into the cell during receptor engagement by IL-2.

γc

A critical role for the γc chain in cytokine receptor function was revealed by the discovery that the molecular basis of X-linked severe combined immunodeficiency (X-SCID) lies in either impaired expression or truncations of γc (see Naguchi et al., 1993b and Section IIIF). However, relatively little is known about the specific contributions of γc to signaling by the IL-2R. Recent studies employing severely truncated γc chains demonstrated that mutants lacking the cytoplasmic tail exhibit *trans*-dominant negative behavior in relation to wild type receptors, strongly implying a critical role for the γc chain in normal IL-2R function (Kawahara et al., 1994). Similarly, our studies of receptor dimerization with chimeric receptors also revealed a crucial contribution of the γc cytoplasmic tail to IL-2R signaling (see Section IIIA).

The molecular architecture of the γc cytoplasmic tail recently has been the subject of intensive investigation. Like IL-2Rβ and other members of the cytokine receptor superfamily, the membrane-proximal portion of the γc cytoplasmic tail contains a partially conserved Box 1 motif and a segment bearing some similarity to a Box 2 motif, as well as a divergent peptide segment (V-Box) connecting these regions (Goldsmith et al., 1995b). A region overlapping the Box 1 and V-Box segments that is reminiscent of a partial SH2 domain has also been described (Takeshita et al., 1992b), but it is lacking in several features that typically define functional SH2 elements.

Initial IL-2R reconstitution studies using murine fibroblasts indicated that the C-terminal portion of γc (lying beyond the vestigial Box 2 motif) are essential for induction of AP-1 factors, but dispensable for induction of tyrosine kinase and *c-myc* expression (Asoa et al., 1993). We have determined that this distal portion of the molecule is also dispensable for growth signaling, but that the proximal Box 1, Box 2, and V-Box domains of γc contain essential elements for mediating proliferative signals (Goldsmith et al., 1995b). Other work has demonstrated that the binding of the JAK3 kinase to γc is regulated by sequences within the V-Box and Box 2 elements (Russell et al., 1994). These findings together suggest that a primary and critical signaling function of the proximal region of the γc cytoplasmic tail is the attachment of JAK3, which allows the γc chain to convey JAK3 into the receptor complex upon ligand binding. Indeed, very recent studies have demonstrated that the γc cytoplasmic tail is fully replaced by the tail of a heterologous receptor that binds JAK2 instead of JAK3 (Lai et al., 1996). These and related findings led to a new model of IL-2 receptor function in which the γc chains serve predominantly as a trigger for the initiation of signal transduction, which serves to drive specific signaling events (Lai et al., 1996).

IV. PATHOGENESIS OF DISEASES ASSOCIATED WITH THE IL-2 RECEPTOR

A. X-linked Severe Combined Immunodeficiency

Several years ago the crucial opportunity that immunodeficiency states provide as windows into normal immunobiology was noted:

"SCID (severe combined immunodeficiency) continues to point the way" (Gelfand, 1990). SCID is a heterogeneous collection of disorders involving deficiencies in both T and B cell function, among which the X-linked form (X-SCID) affecting predominantly males is the most common in North America. Typically X-SCID patients—who succumb to fatal infections during early childhood—have very few T lymphocytes but normal numbers of functionally impaired B lymphocytes. Both T and B cells in otherwise healthy, male obligate carriers exhibit a non-random pattern of X chromosome inactivation, which suggests that there is an *in vivo* selection pressure for expression of the normal allele of the culprit gene in both cell types even in heterozygotes (Conley et al., 1988). The definitive treatment for X-SCID in males carrying homozygous mutations is bone marrow transplantation at an early age. Interestingly, immune reconstitution is effective even with engraftment exclusively of T cells, implying that endogenous B cells carrying two mutant alleles can exhibit sufficient function to permit immune competence in the presence of normal T cells.

The gene for responsible for X-SCID has been mapped by linkage analysis to the long arm of the X chromosome at Xq13.1-q21.1 (De Saint Basile et al., 1987; Puck et al., 1989). A startling observation that linked this disorder to cytokine receptor function was made when the gene encoding γc was mapped to the same chromosomal locus (Noguchi et al., 1993b; Puck et al., 1993). Indeed, a variety of mutations have been identified in tissues derived from X-SCID patients that cause impaired expression of γc chains, expression of severely truncated chains, or expression of γc subunits with alterations in crucial extracellular features of the γc chain that determine IL-2-binding activity (Noguchi et al. 1993b; Puck et al., 1993; Di Santo et al., 1994). In addition, mice deleted of γc chains through homologous recombination exhibit profound deficiencies in immune development and function (Di Santo et al., 1995). Thus, there is little doubt that alterations of γc expression or function are responsible for the immune impairment of X-SCID.

The detailed manifestations of altered gc chains in developmental and effector aspects of the immune of system remain somewhat obscure. The initial expectation that γc dysfunction may be expressed entirely through the IL-2R was undermined by the relatively mild phenotypes of mice genetically lacking IL-2 expression (see Section II). This paradox was resolved by the subsequent demonstration that the γc chain also participates in the receptors for IL-4, IL-7, IL-9, and IL-15 (Russell et al., 1993; Noguchi et al., 1993a; Dondo et al., 1993, 1994). Therefore, it is now anticipated that at least some of the deficits resulting from mutations of γc may be linked to other cytokine systems. Certain mutations described thus far cause a loss of association between γc and JAK3, which provides a molecular basis for disrupted signal transduction by the IL-2R as well as by other cytokine receptors employing the γc subunit (Russell et al., 1994). One intriguing mutation that was discovered in a family with a moderate form of X-SCID lies in the V-Box segment of γc and causes a diminution, but not complete abrogation, of interaction between JAK3 and γc (Russell et al., 1994). Further investigation is underway to determine the specific functional contributions of the γc subunit to each of the various receptor systems and, hence, to the pathology of X-SCID and possibly other forms of immunodeficiency.

B. HTLV-I

Adult T Cell Leukemia

The type I human T cell leukemia virus (HTLV-I) corresponds to an oncogenic type C retrovirus that has been etiologically linked with the adult T cell leukemia (ATL; Poiesz et al., 1980; Yoshida et al., 1982). This disease is geographically clustered in areas of the world where HTLV-I infection is endemic including Japan, sub-Saharan Africa, the Caribbean basin, and parts of the southeastern United States. ATL is a highly aggressive and frequently fatal neoplasm of activated CD4 T cells. These leukemic cells are often distinguished by the deregulated expression of IL-2Rα chains at their cell surfaces (Kronke et al., 1985). Although the precise mechanism by which T cells are transformed by HTLV-I remains unknown, increasing evidence points to a central role for the HTLV *tax* gene product. The Tax protein corresponds to a 40 kD *trans*-activating factor that not only augments activity of the HTLV-I long terminal repeat, but also induces the expression of several cellular genes involved in T cell activation and growth (Smith and Greene, 1991). These cellular genes include IL-2Rα, IL-2, GM-CSF, Fos, TNF-β, vimentin, and others. Tax-induced activation of the IL-2Rα gene appears mediated through Tax induction of nuclear NF-κB/Rel expression. As noted previously, transcription of the IL-2Rα gene is critically regulated by the NF-κB/Rel family of transcription factors. A greater role for NF-kB in HTLV-I transformation may exist, since anti-sense NF-κB oligonucleotides have been shown to block HTLV-I Tax induced tumor cell growth in a mouse transgenic model (Kitajima et al., 1992). More recent studies also suggest that Tax may modestly increase IL-2Rβ gene expression, although the underlying mechanism remains unknown (Greene, unpublished results). It seems possible that the display of high affinity IL-2 receptors coupled with the autocrine or paracrine production of IL-2 might play a role in the early polyclonal phase of growth that is characteristic of HTLV-I mediated transformation *in vitro*. This receptor-ligand interaction might promote cellular proliferation that facilitates the occurrence of additional second stage intracellular events required to complete the transformation process. Alternatively, it is possible that other cytokines, such as IL-15, which function through the IL-2Rβ and γc subunits also play a role in HTLV-I pathogenesis.

TSP/HAM

HTLV-I infection has also been associated with a chronic demyelinating process termed tropical spastic paraparesis (TSP) or HTLV-I associated myelopathy (HAM) (Gessain et al., 1985; Osame et al., 1986). This syndrome is characterized by progressive demyelination of long motor neuron tracts in the spinal cord which leads to spasticity, paraparesis, and diminished muscle strength. Multiple sclerosis, a far more prevalent demyelinating disease, shares certain clinical features in common with TSP/HAM, however, no convincing link has been made between HTLV-I infection and multiple sclerosis.

Inspection of the peripheral blood lymphocytes of patients with TSP/HAM has revealed a high level of spontaneous proliferation coupled with heightened HTLV-I gene expression. Other studies have shown that the proliferation of these cells can be inhibited by the addition of anti-IL-2 or anti-IL-2 receptor antibodies mediated through an autocrine or paracrine mechanism involving IL-2 and its receptor. In this regard, the TSP/HAM findings mimic the early polyclonal proliferative phase of HTLV-I transformation observed in tissue culture.

REFERENCES

Adachi, M., Sekiya, M., Ishino, M., Sasaki, H., Hinoda, Y., Imai, K., & Yachi, A. (1994). Induction of protein-tyrosine phosphatase LC-PTP by IL-2 in human T cells. FEBS Lett. 338, 43-46.

Arima, N., Kuziel, W.A., Gardina, T.A., & Green, N.C. (1992a). IL-2-induced signal transduction involves the activation of nuclear NF-kappa B expression. J. Immunol. 149, 83-91.

Arima, N., Kamio, M., Imada, K., Hori, T., Hattori, T., Tsudo, M., Okuma, M., & Uchiyama, T. (1992b). Pseudo-high affinity interleukin 2 (IL-2) receptor lacks the third component that is essential for functional IL-2 binding and signaling. J. Exp. Med. 176, 1265-1272.

Asao, H. (1992). IL-2 dependent *in vivo* and *in vitro* tyrosine phoshorylation of IL-2 receptor g chain. FEBS 304, 141-145.

Asao, H., Takeshita, T., Nakamura, M., Nagata, K., & Sugamura, K. (1990). Interleukin 2 (IL-2)-induced tyrosine phosphorylation of IL-2 receptor p75 [published erratum appears in J. Exp. Med. 1990 Jun 1;171(6):2183]. J. Exp. Med. 171, 637-644.

Asao, H., Takeshita, T., Ishii, N., Kumaki, S., Nakamura, M., & Sugamura, K. (1993). Reconstitution of functional interleukin 2 receptor complexes on fibroblastoid cells: Involvement of the cytoplasmic domain of the g chain in two distinct signaling pathways. Proc. Natl. Acad. Sci. USA 90, 4127-4131.

Asao, H., Tanaka, N., Ishii, N., Higuchi, M., Takeshita, T., Nakamura, M., Shirasawa, T., & Sugamura, K. (1994). Interleukin 2-induced activation of JAK3: Possible involvement in signal transduction for c-myc induction and cell proliferation. FEBS Letters 351, 201-206.

Augustine, J.A., Sutor, S.L., & Abraham, R.T. (1991). Interleukin 2- and polyomavirus middle T antigen-induced modification of phosphatidylinositol 3-kinase activity in activated T lymphocytes. Mol. Cell Biol. 11, 4431-4440.

Baccarini, M., Schwinzer, R., & Lohmann-Matthes, M.L. (1989). Effect of human recombinant IL-2 on murine macrophage precursors. Involvement of a receptor distinct from the p55 (Tac) protein. J. Immunol. 142, 118-125.

Baker, P.E., Gillis, S., Ferm, M.M., & Smith, K.A. (1978). The effect of T cell growth factor on the generation of cytolytic T cells. J. Immunol. 121, 2168-2173.

Ballard, D.W., Bohnlein, E., Hoffman, J.A., Bogerd, H.P., Dixon, E.P., Franza, B.R., & Greene, W.C. (1989). Activation of the Interleukin-2 receptor a gene: Regulatory role for DNA-protein interactions flanking the κB enhancer. The New Biologist 1, 83-92.

Bazan, J.-F. (1990a). Structural design and molecular evolution of a cytokine receptor superfamily. Proc. Natl. Acad. Sci. USA 87, 6934-8.

Bazan, J.F. (1992). Unraveling the structure of IL-2 [letter]. Science 257:, 410-413.

Beadling, C., Johnson, K.W., & Smith, K.A. (1993). Isolation of interleukin 2-induced immediate-early genes. Proc. Natl. Acad. Sci. USA 90, 2719-2723.

Ben-Sasson, S.Z., Le Gros, G., Conrad, D.H., Finkelman, F.D., & Paul, W.E. (1990). IL-4 production by T cells from naive donors. IL-2 is required for IL-4 production. J. Immunol. 145, 1127-1136.

Benveniste, E.N., & Merrill, J.E. (1986). Stimulation of oligodendroglial proliferation and maturation by interleukin-2. Nature 321, 610-613.

Blackman, M.A., Tigges, M.A., Minie, M.E., & Koshland, M.E. (1986). A model system for peptide hormone action in differentiation: Interleukin-2 induces a B lymphoma to transcribe the J chain gene. Cell 47, 609-617.

Blank, V., Kourilsky, P., & Israel, A. (1992). NF-kappa B and related proteins: Rel/dorsal homologies meet ankyrin-like repeats. Trends in Biochem. Sci. 17, 135-140.

Bosco, M.C., Espinoza-Delgado, I., Schwabe, M., Russell, S.M., Leonard, W.J., Longo, D.L., & Varesio, L. (1994). The γ subunit of the interleukin-2 receptor is expressed in human moncytes and modulated by interleukin-2, interferon γ, and transforming growth factor β. Blood 83, 3462-3467.

Boussiotis, V.A., Barber, D..L., Nakarai, T., Freeman, G.J., Gribben, J.G., Bernstein, G.M., D'Andrea, A.D., Ritz, J., & Nadler, L.M. (1994). Prevention of T cell anergy by signaling through the γc chain of the IL-2 receptor. Science 266, 1039-1042.

Brandhuber, B.J., Boone, T., Kenney, W.C., & McKay, D.B. (1987). Three-dimensional structure of interleukin-2. Science 238, 1707-1709.

Burns, L.A., Karnitz, L.M., Sutor, S.L., & Abraham, R.T. (1993). Interleukin-2-induced tyrosine phosphorylation of p52*shc* in T lymphocytes. J. Biol. Chem. 268, 17659-17661.

Calvo, V., Crews, C.M., Vik, T.A., & Bierer, B.E. (1992). Interleukin 2 stimulation of p70 S6 kinase activity is inhibited by the immunosuppressant rapamycin. Proc. Natl. Acad. Sci. USA 89, 7571-7575.

Ceredig, R., Lowenthal, J.W., Nabholz, M., & MacDonald, H.R. (1985). Expression of interleukin-2 receptors as a differentiation marker on intrathymic stem cells. Nature 314, 98-100.

Chan, A.C., Desai, D.M., & Weiss, A. (1994). The role of protein tyrosine kinases and protein tyrosine phosphatases in T cell antigen receptor signal transduction. Annual Review of Immunology 12, 555-592.

Chen, J., Ma, A., Young, F., & Alt, F.W. (1994). Interleukin-2 receptor a chain (IL-2Rα) expression during early B lymphocyte differentiation. Intl. Immunol. 6, 1265-1268.

Collins, L., Tsien, W.H., Seals, C., Hakimi, J., Weber, D., Bailon, P., Hoskings, J., Greene, W.C., Toome, V., & Ju, G. (1988). Identification of specific residues of human interleukin 2 that affect binding to the 70-kDa subunit (p70) of the interleukin 2 receptor. Proc. Natl. Acad. Sci. USA 85, 7709-7713.

Conley, M.E., Lavoie, A., Briggs, C., Brown, P., Guerra, C., & Puck, J.M. (1988). Nonrandom X chromosome inactivation in B cells from carriers of x chromosome-linked severe combined immunodeficiency. Proc. Natl. Acad. Sci. USA 85, 3090-3094.

Cosman, D., Cerretti, D.P., Larsen, A., Park, L., March, C., Dower, S., Gillis, S., & Urdal, D. (1984). Cloning, sequence and expression of human interleukin-2 receptor. Nature 312, 768-771.

Crabtree, G.R. (1989). Contingent genetic regulatory events in T lymphocyte activation. Science 243, 355-361.

Cullen, B.R., Podlaski, F.J., Peffer, N.J., Hosking, J.B., & Greene, W.C. (1988). Sequence requirements for ligand binding and cell surface expression of the Tac antigen, a human interleukin-2 receptor. J Biol Chem. 263, 4900-4906.

Cunningham, B.C., Ultsch, M., Devos, A.M., Mulkerrin, M.G., Clauser, K.R., & Wells, J.A. (1991). Dimerization of the extracellular domain of the human growth hormone receptor by a single hormone molecule. Science 254, 821-825.

Darnell, J.E., Jr., Kerr, I.M., & Stark, G.R. (1994). Jak-STAT pathways and transcriptional activation in response to IFNs and other extracellular signaling proteins. Science 264, 1415-1421.

Dautry, F., Weil, D., Yu, J., & Dautry-Varsat, A. (1988). Regulation of pim and myb mRNA accumulation by interleukin 2 and interleukin 3 in murine hematopoietic cell lines. J. Biol. Chem. 263, 17615-17620.

De Saint Basile, G., Arveiler, B., Oberle, I., Malcolm, S., Levinsky, R.J., Lau, Y.L., Hofker, M., Debre, M., Fischer, A., & Griscelli, C. (1987). Close linkage of the locus for X chromosome-linked severe combined immunodeficiency to polymorphic DNA markers in Xq11-q13. Proc. Natl. Acad. Sci. USA 84, 7576-7579.

de Vos, A., Ultsch, M., & Kossiakoff, A.A. (1992). Human growth hormone and extracellular domain of its receptor: Crystal structure of the complex. Science 255, 306-312.

Depper, J.M., Leonard, W.J., Kronke, M., Noguchi, P.D., Cunningham, R.E., Waldman, T.A., & Green, W.C. (1984). Regulation of interleukin 2 receptor expression: effects of phorbol diester, phospholipase C, and reexposure to lectin or antigen. J. Immunol. 133, 3054-3061.

Depper, J.M., Leonard, W.J., Drogula, C., & Kronke, M. (1985). Interleukin 2 (IL-2) augments transcription of the IL-2 receptor gene. Proc. Natl. Acad. Sci. USA 82, 4230-4234.

DiSanto, J.P., Dautry-Varsat, A., Certain, S., Fischer, A., & De Saint Basile, G. (1994). Interleukin-2 (IL-2) receptor gamma chain mutations in X-linked severe combined immunodeficiency disease result in the loss of high-affinity IL-2 receptor binding. Eur. J. Immunol. 24, 475-479.

DiSanto, J.P., Muller, W., Guy-Grand, D., Fisher, A., & Rajewsky, K. (1995). Lymphoid development in mice with a targeted deletion of the interleukin 2 receptor γ chain. Proc. Natl. Acad. Sci. USA 92, 377-381.

Djeu, J.Y., Liu, J.H., Wei, S., Rui, H., Pearson, G.A., Leonard, W.J., & Blanchard, D.K. (1993). Function associated with IL-2 receptor-b on human neutrophils. J. Immunol. 150, 960-970.

Dukovich, M., Wano, Y., Le, T., Bich, T., Katz, P., Cullen, B.R., Kehrl, J.H., & Greene, W.C. (1987). A second human interleukin-2 binding protein that may be a component of high-affinity interleukin-2 receptors. Nature 327, 518-522.

Evans, G.A., Howard, O.M.Z., Erwin, Z., & Farrar, W.L. (1993). Interleukin-2 induces tyrosine phosphorylation of the *vav* proto-oncogene product in human T cells: Lack of requirement for the tyrosine kinase lck. Biochem. J. 294, 339-342.

Falk, I., Levelt, C.N., & Eichmann, K. (1993). Lineage relationships of the fetal thymocyte subset that expresses the beta chain of the interleukin-2 receptor. Eur. J. Immunol. 23, 3373-3376.

Farrar, J.J., Benjamin, W.R., Hilfiker, M.L., Howard, M., Farrar, W.L., & Fuller Farrar, J. (1982). The biochemistry, biology, and role of interleukin 2 in the induction of cytotoxic T cell and antibody-forming B cell responses. Immunol. Rev., 129-166.

Fung, M.R., Scearce, R.M., Hoffman, J.A., Peffer, N.J., Hammes, S.R., Hosking, J.B., Schmandt, R., Kuziel, W.A., Haynes, B.F., Mills, G.B., & Greene, W.C. (1991). A tyrosine kinase physically associates with the beta-subunit of the human IL-2 receptor. J. Immunol. 147, 1253-1260.

Gaffen, S.L., Lai, S.Y., Ha, M., Liu, X., Hennighausen, L., Greene, W.C., & Goldsmith, M.A. (1996). Distinct tyrosine residues within the IL-2 receptor β chain drive signal transduction specificity, redundancy and diversity. J. Biol. Chem. 271, 21381-21390.

Gaffen, S.L., Lai, S.Y., Xu, W., Gouillex, F., Groner, B., Goldsmith, M.A., & Greene, W.C. (1995). Signaling through the IL-2Rβ chain activates a STAT-5-like DNA binding activity. Proc. Natl. Acad. Sci. USA 92, 7192-7196.

Garcia, G.G., Evans, G.A., Michiel, D.F., & Farrar, W.L. (1992). Characterization of a tyrosine kinase activity associated with the high-affinity interleukin 2 receptor compelx. Biochem. J. 285, 851-856.

Gelfand, E.W. (1990). SCID continues to point the way. New Eng. J. Med. 322, 1741-1743.

Gessain, A., Barin, F., Vernant, J.C., Gout, O., Maurs, L., Calendar, A., & de The, G. (1985). Antibodies to the human T lymphotropic virus type-1 in patients with tropical spastic paraparesis. Lancet 2, 407-410.

Gillis, S., Ferm, M.M., Ou, W., & Smith, K.A. (1978). T cell growth factor: Parameters of production and a quantitative microassay for activity. J. Immunol. 120, 2027-2032.

Gilmour, K.C., & Reich, N.C. (1994). Receptor to nucleus signaling by prolactin and interleukin 2 via activation of latent DNA-binding factors. Proc. Natl. Acad. Sci.USA 91, 6850-6854.

Giri, J.G., Ahdieh, M., Eisenman, J., Shanebeck, K., Grabstein, K., Kumaki, S., Namen, A., Park, L.S., Cosman, D., & Anderson, D. (1994). Utilization of the beta and gamma chains of the IL-2 receptor by the novel cytokine IL-15. EMBO J. 13, 2822-2830.

Gnarra, J.R., Otani, H., Wang, M.G., McBride, O.W., Sharon, M., & Leonard, W.J. (1990). Human interleukin 2 receptor beta-chain gene: chromosomal localization and identification of 5' regulatory sequences. Proc. Natl.Acad. Sci. USA 87, 3440-3444.

Goldsmith, M.A., & Greene, W.C. (1994). Interleukin-2 and the interleukin-2 receptor. In: *The Cytokine Handbook* (Thomson, A., Ed.). Academic Press, Ltd, London, pp. 57-80.

Goldsmith, M.A., Xu, W., Amaral, M.C., Kuczek, E.S., & Greene, W.C. (1994). The cytoplasmic domain of the IL-2 receptor β chain contains both unique and functionally redundant signal transduction elements. J. Biol. Chem. 269, 14698-14704.

Goldsmith, M.A., Amaral, M.C., & Greene, W.C. (1995a). Ligand binding by the IL-2 receptor is modulated by intracellular determinants of the IL-2R receptor β-chain. J. Immunol. 154, 2033-2040.

Goldsmith, M.A., Lai, S.Y., Xu, W., Amaral, M.C., Kuczek, E.S., Parent, L.J., Mills, G.B., Tarr, K.L., Longmore, G.D., & Greeene, W.C. (1995b). Growth signal transduction by the human IL-2 receptor requires cytoplasmic tyrosines of the β chain and non-tyrosine residues of the γ_c chain. J. Biol. Chem. 270, 21729-21737.

Grabstein, K.H., Eisenman, J., Shanebeck, K., Rauch, C., Srinivasan, S., Fung, V., Beers, C., Richardson, J., Schoenborn, M.A., & Ahdich, M. (1994). Cloning of a T cell growth factor that interacts with the beta chain of the interleukin-2 receptor. Science 264, 965-968.

Granelli-Piperno, A., Andrus, L., & Steinman, R.M. (1986). Lymphokine and nonlymphokine mRNA levels in stimulated human T cells. J. Exp. Med. 163, 922-937.

Grant, A.J., Roessler, E., Ju, G., Tsudo, M., Sugamura, K., & Waldmann, T.A. (1992). The interleukin 2 receptor (IL-2R): The IL-2R alpha subunit alters the function of the IL-2R beta subunit to enhance IL-2 binding and signaling by mechanisms that do not require binding of IL-2 to IL-2R alpha subunit. Proc. Natl. Acad. Sci. USA 89, 2165-2169.

Graves, J.D., Downward, J., Izquierdo, P.M., Rayter, S., Warne, P.H., & Cantrell, D.A. (1992). The growth factor IL-2 activates p21ras proteins in normal human T lymphocytes. J. Immunol. 148, 2417-2422.

Greene, W.C., Robb, R.J., Svetlik, P.B., Rusk, C.M., Depper, J.M., & Leonard, W.J. (1985). Stable expression of cDNA encoding the human interleukin 2 receptor in eukaryotic cells. J. Exp. Med. 162, 363-368.

Habu, S., Okumura, K., Diamantstein, T., & Shevach, E.M. (1985). Expression of interleukin 2 receptor on murine fetal thymocytes. Eur. J. Immunol. 15, 456-460.

Hancock, W.W., Muller, W.A., & Cotran, R.S. (1987). Interleukin 2 receptors are expressed by alveolar macrophages during pulmonary sarcoidosis and are inducible by lymphokine treatment of normal human lung macrophages, blood monocytes, and monocyte cell lines. J. Immunol. 138, 185-191.

Hatakeyama, M., Minamoto, S., & Taniguchi, T. (1986). Intracytoplasmic phosphorylation sites of Tac antigen (p55) are not essential for the conformation, function, and regulation of the human interleukin 2 receptor. Proc. Natl. Acad. Sci. USA 83, 9650-9654.

Hatakeyama, M., Tsudo, M., Minamoto, S., Kono, T., Doi, T., Miyata, T., Miyasaka, M., & Taniguchi, T. (1989a). Interleukin-2 receptor beta chain gene: Generation of three receptor forms by cloned human alpha and beta chain cDNA's. Science 244, 551-556.

Hatakeyama, M., Mori, H., Doi, T., & Taniguchi, T. (1989b). A restricted cytoplasmic region of IL-2 receptor beta chain is essential for growth signal transduction but not for ligand binding and internalization. Cell 59, 837-845.

Hatakeyama, M., Kono, T., Kobayashi, N., Kawahara, A., Levin, S.D., Perlmutter, R.M., & Taniguchi, T. (1991). Interaction of the IL-2 receptor with the src-family kinase p56lck: Identification of novel intermolecular association. Science 252, 1523-1528.

Hatakeyama, M., Kawahara, A., Mori, H., Shibuya, H., & Taniguchi, T. (1992). c-fos gene induction by interleukin 2: Identification of the critical cytoplasmic regions within the interleukin 2 receptor beta chain. Proc. Natl. Acad. Sci. USA 89, 2022-2026.

Hémar, A., Cereghini, S., Cornet, V., Blank, V., Israël, A., Greene, W.C., & Dautry-Varsat, A. (1991). κ B binding proteins are constitutively expressed in an IL-2 autocrine human T cell line. J. Immunol. 146, 2409-2416.

Herrmann, F., Cannistra, S.A., Levine, H., & Griffin, J.D. (1985). Expression of interleukin 2 receptors and binding of interleukin 2 by gamma interferon-induced human leukemic and normal monocytic cells. J. Exp. Med. 162, 1111-1116.

Holter, W., Grunow, R., Stockinger, H., & Knapp, W. (1986). Recombinant interferon-gamma induces interleukin 2 receptors on human peripheral blood monocytes. J. Immunol. 136, 2171-2175.

Holter, W., Goldman, C.K., Casabo, L., Nelson, D.L., Greene, W.C., & Waldmann, T.A. (1987). Expression of functional IL 2 receptors by lipopolysaccharide and interferon-gamma stimulated human monocytes. J. Immunol. 138, 2917-2922.

Horak, I.D., Gress, R.E., Lucas, P.J., Horak, E.M., Waldmann, T.A., & Bolen, J.B. (1991). T-lymphocyte interleukin 2-dependent tyrosine protein kinase signal transduction involves the activation of p56lck. Proc. Natl. Acad. Sci. USA 88, 1996-2000.

Hou, J., Schindler, U., Henzel, W.J., Ho, T.C., Brasseur, M., & McKnight, S.L. (1994). An interleukin-4-induced transcription factor: IL-4 stat. Science 265, 1701-1706.

Hou, J., Schindler, U., Henzel, W.J., Wong, S.C., & McKnight, S.L. (1995). Identification and purfication of human stat proteins activated in response to interleukin-2. Immunity 2, 321-329.

Howard, M., Matis, L., Malek, T.R., Shevach, E., Kell, W., Cohen, D., Nakanishi, K., & Paul, W.E. (1983). Interleukin 2 induces antigen-reactive T cell lines to secrete BCGF-I. J. Exp. Med. 158, 2024-2039.

Hunter, T. (1993). Cytokine connections. Nature 366, 114-116.

Ishida, Y., Nishi, M., Taguchi, O., Inaba, K., Hattori, M., Minato, N., Kawaichi, M., & Honjo, T. (1989). Expansion of natural killer cells but not T cells in human interleukin 2/interleukin 2 receptor (Tac) transgenic mice. J. Exp. Med. 170, 1130-1115.

Izquierdo, M., & Cantrell, D.A. 1993. Protein tyrosine kinases couple the interleukin-2 receptor to p21ras. Eur. J. Immunol. 23, 131-135.

Izquierdo, M., Downward, J., Graves, J.D., & Cantrell, D.A. (1992). Role of protein kinase C in T-cell antigen receptor regulation of p21ras: Evidence that two p21ras regulatory pathways coexist in T cells. Mol. Cell. Biol. 12, 3305-3312.

Izuhara, K., & Harada, N. (1993). Interleukin-4 (IL-4) induces protein tyrosine phosphorylation of the IL-4 receptor and association of phoshatidylinositol 3-kinase to the IL-4 receptor in a mouse T cell line, HT2. J. Biol. Chem. 268, 13097-13102.

Jelinek, D.F., & Lipsky, P.E. (1987). Regulation of human B lymphocyte activation, proliferation, and differentiation. Adv .Immunol. 40, 1-59.

Jenkinson, E.J., Kingston, R., & Owen, J.J. (1987). Importance of IL-2 receptors in intra-thymic generation of cells expressing T-cell receptors. Nature 329, 160-162.

Johnston, J.A., Kawamura, M., Kirken, R.A., Chen, V., Blake, T.B., Shibuya, K., Ortaido, J.R., McVicar, D.W., & O'Shea, J.J. (1994). Phophorylation and activation of the Jak-3 Janus kinase in response to interleukin-2. Nature 370, 151-153.

Ju, G., Collins, L., Kaffka, K.L., Tsien, W.H., Chizzonite, R., Crowl, R., Bhatt, R., & Kilian, P.L. (1987). Structure-function analysis of human interleukin-2. Identification of amino acid residues required for biological activity. J. Biol, Chem. 262, 5723-5731.

Karnitz, L., Sutor, S.L., Torigoe, T., Reed, J.C., Bell, M.P., McKean, D.J., Leibson, P.J., & Abraham, R.T. (1992). Effects of $p56^{lck}$ deficiency on the growth and cytolytic effector function of an interleukin-2-dependent cytotoxic T-cell line. Mol. Cell. Biol, 12, 4521-4530.

Karnitz, L.M., Sutor, S.L., & Abraham, R.T. (1994). The src-family kinase, Fyn, regulates the activation of phosphatidylinositol 3-kinase in an interleukin 2-responsive T cell line. J. Exp. Med. 179, 1799-1808.

Kawahara, A., Minami, Y., & Taniguchi, T. (1994). Evidence for a critical role for the cytoplasmic region of the interleukin 2 (IL-2) receptor gamma chain in IL-2, IL-4, and IL-7 signaling. Mol. Cell Biol. 14, 5433-5440.

Kawamura, M., McVicar, D.W., Johnston, J.A., Blake, T.B., Chen, Y., Lal, B.K., Lloyd, A.R., Kelvin, D.J., Staples, J.E., Otaldo, J.R., & O'Shea, J.J. (1994). Molecular cloning of L-JAK, a Janus family protein-tyrosine kinase expressed in natural killer cells and activated leukocytes. Proc. Natl. Acad. Sci. USA 91, 6374-6378.

Kazlauskas, A., & Cooper, J.A. (1989). Autophosphorylation of the PDGF receptor in the kinase insert region regulates interactions with cell proteins. Cell 58, 1121-1133.

Keegan, A.D., Nelms, K., White, M., Wang, L.-M., Pierce, J.H., & Paul, W.E. (1994). An IL-4 receptor region containing an insulin receptor motif is important for IL-4-mediated IRS-1 phosphorylation and cell growth. Cell 76, 811-820.

Kehrl, J.H., Dukovich, M., Whelan, G., Katz, P., Fauci, A.S., & Greene, W.C. (1988). Novel interleukin 2 (IL-2) receptor appears to mediate IL-2-induced activation of natural killer cells. J. Clin. Invest. 81, 200-205.

Kern, D.E., Gillis, S., Okada, M., & Henney, C.S. (1981). The role of interleukin-2 (IL-2) in the differentiation of cytotoxic T cells: the effect of monoclonal anti-IL-2 antibody and absorption with IL-2 dependent T cell lines. J. Immunol. 127, 1323-1328.

Kim, Y.H., Buchholz, M.J., & Nordin, A.A. (1993). Murine T-lymphocyte proliferation induced by interleukin 2 correlates with a transient increase in $p56^{lck}$ kinase activity and the tyrosine phosphorylation of a 97-kDa protein. Immunology 90, 3187-3191.

Kirken, R.A., Rui, H., Evans, G.A., & Farrar, W.L. (1993). Characterization of an interleukin-2 (IL-2) induced tyrosine phosphorylated 116-kDa protein associated with the IL-2 receptor β-subunit. J. Biol. Chem 268, 22765-22770.

Kitajima, I., Shinohara, T., Bilakovics, J., Brown, D.A., Xu, X., & Nerenberg, M. (1992). Ablation of transplanted HTLV-I tax-transformed tumors in mice by antisense inhibition of NF-kappaB target sequences. Science 258, 1792-1795.

Kobayashi, N., Kono, T., Hata Keyama, M., Minami, Y., Miyazaki, T., Perlmutter, R.M., & Taniguchi, T. (1993). Functional coupling of the src-family protein tyrosine kinases p59fyn and p53/56lyn with the interleukin 2 receptor: Implications for redundancy and pleiotropism in cytokine signal transduction. Proc. Natl. Acad. Sci. USA 90, 4201-4205.

Kondo, M., Takeshita, T., Ishii, N., Nakamura, M., Watanabe, S., Arai, K-I., & Sugamura, K. (1993). Sharing of the interleukin-2 (IL-2) receptor g chain between receptors for IL-2 and IL-4. Science 262, 1874-1877.

Kondo, M., Takeshita, T., Higuchi, M., Nakamura, M., Sudo, T., Nishi Kawa, S., & Sugamura, K. (1994). Functional participation of the IL-2 receptor γ chain in IL-7 receptor complexes. Science 263, 1453-1454.

Kondo, S., Kinoshita, M., Shimizu, A., Saito, Y., Konishi, M., Sabe, H., & Honjo, T. (1987). Expression and functional characterization of artificial mutants of interleukin-2 receptor. Nature 327, 64-67.

Kovacs, E.J., Brock, B., Varesio, L., & Young, H.A. (1989). Il-2 induction of IL-1 beta mRNA expression in monocytes. J. Immunol. 143, 3532.

Kronke, M., Leonard, W.J., Depper, J.M., & Greene, W.C. (1985). Deregulation of interleukin-2 receptor gene expression in HTLV-1 induced adult T-cell leukemia. Science 228, 1215-1217.

Kuo, C.J., Chung, J., Fiorentino, D.F., Flanagan, W.M., Blenis, J., & Crabtree, G.R. (1992). Rapamycin selectively inhibits interleukin-2 activation of p70 S6 kinase. Nature 358, 70-73.

Kuziel, W.A., Ju, G., Grdina, T.A., & Greene, W.C. (1993). Unexpected effects of the IL-2 receptor alpha subunit on high affinity IL-2 receptor assembly and function detected with a mutant IL-2 analog. J. Immunol., 150, 3357-3365.

Lai, S.Y., Xu, W., Gaffen, S.L., Longmore, G.D., Greene, W.C., & Goldsmith, M.A., (1996). The molecular role of the common γ_c subunit in signal transduction reveals functional asymmetry within multimeric cytokine receptor complexes. Proc. Natl. Acad. Sci. USA 93, 231-235.

Lanier, L.L., Benike, C.J., Phillips, J.H., & Englemann, E.G. (1985). Recombinant interleukin 2 enhanced natural killer cell-mediated cytotoxicity in human lymphocyte subpopulations expressing the Leu 7 and Leu 11 antigens. J. Immunol. 134, 794-801.

Le Gros, G., Shackell, P.S., Le, G.S., & Watson, J.D. (1987). Interleukin 2 regulates the expression of IL 2 receptors on interleukin 3-dependent bone marrow-derived cell lines. J. Immunol. 138, 478-483.

Le Gros, G., Ben-Sasson, S.Z., Seder, R., Finkelman. F.D., & Paul, W.E. (1990). Generation of interleukin 4 (IL-4)-producing cells *in vivo* and *in vitro*: IL-2 and IL-4 are required for *in vitro* generation of IL-4-producing cells. J. Exp. Med. 172, 921-929.

Lenardo, M.J. (1991). Interleukin-2 programs mouse αβ T lymphocytes for apoptosis. Nature 353, 858-861.

Leonard, W.J., Depper, J.M., Uchiyama, T., Smith, K.A., Waldmann, T.A., & Greene, W.C. (1982). A monoclonal antibody that appears to recognize the receptor for human T-cell growth factor; partial characterization of the receptor. Nature 300, 267-269.

Leonard, W.J., Depper, J.M., Crabtree, G.R., Rudikoff, S., Pumphrey, J.. Robb, R.J., Kronke, M., Svetlik, P.B., Peffer, N.J., & Waldmann, T.A. (1984). Molecular cloning and expression of cDNAs for the human interleukin-2 receptor. Nature 311, 626-631.

Leonard, W.J., Depper, J.M., Kanehisa, M., Kronke, M., Peffer, N.J., Svetlik, P.B., Sullivan, M., & Greene, W.C. (1985a). Structure of the human interleukin-2 receptor gene. Science 230, 633-639.

Leonard, W.J., Dunlon, T.A., Lebo, R.V., & Greene, W.C. (1985b). Localization of the gene encoding the human interleukin-2 receptor on chromosome 10. Science 228, 1547-1549.

Lowenthal, J.W., & Greene, W.C. (1987). Contrasting interleukin 2 binding properties of the alpha (p55) and beta (p70) protein subunits of the human high-affinity interleukin 2 receptor. J. Exp. Med. 166, 1156-61.

Lowenthal, J.W., Tougne, C., MacDonald, H.R., Smith, K.A., & Nabholz, M. (1985). Antigenic stimulation regulates the expression of IL 2 receptors in a cytolytic T lymphocyte clone. J. Immunol. 134, 931-939.

Malkovsky, M., Loveland, B., North, M., Asherson, G.L., Gao, L., Ward, P., & Fiers, W. (1987).Recombinant interleukin-2 directly augments the cytotoxicity of human monocytes. Nature, 325, 262-265.

Matsuzawa, S., Matsuda, A., Mizuno, Y., & Kikuchi, K. (1993). IL-2 induces transient and specific decrease in cytosolic protein phosphatase PP1 activity in murine T cell lines. Biochem. Biophys. Res. Commun. 195, 989-996.

Merida, I., & Gaulton, G.N. (1990). Protein tyrosine phosphorylation associated with activation of the interleukin 2 receptor. J. Biol. Chem. 265, 5690-5694.

Merida, I., Diez, E., & Gaulton, G.N. (1991). IL-2 binding activates a tyrosine-phosphorylated phosphatidylinositol-3-kinase. J. Immunol. 147, 2202-2207.

Merida, I., Williamson, P., Kuziel, W.A., Greene, W.C., & Gaulton, G.N. (1993). The serine-rich cytoplasmic domain of the interleukin-2 receptor b chain is essential for interleukin-2-dependent tyrosine protein kinase and phosphatidylinositol-3-kinase activation. J. Biol. Chem. 268, 6765-6770.

Meuer, S.C., Hussey, R.E., Cantrell, D.A., Hodgdon, J.C., Scholssman, S.F., Smith, K.A., & Reinherz, E.L. (1984). Triggering of the T3-Ti antigen-receptor complex results in clonal T-cell proliferation through an interleukin 2-dependent autocrine pathway. Proc. Natl. Acad. Sci. USA 81, 1509-1513.

Michiel, D.F., Garcia, G.G., Evans, G.A., & Farrar, W.L. (1991). Regulation of the interleukin 2 receptor complex tyrosine kinase activity in vitro. Cytokine 3, 428-438.

Milligan, D.L., & Koshland D.E., Jr. (1991). Intrasubunit signal transduction by the aspartate chemoreceptor. Science 254, 1651-1654.

Mills, G.B., May, C., McGill, M., Fung, M., Baker, M., Sutherland, R., & Greene, W.C. (1990). Interleukin 2-induced tyrosine phosphorylation. Interleukin 2 receptor beta is tyrosine phosphorylated. J. Biol. Chem. 265, 3561-3567.

Mills, G.B., Arima, N., May, C., Hill, M., Schmandt, R., Li, J., Miyamoto, N.G., & Greene, W.C. (1992). Neither the LCK nor the FYN kinases are obligatory for IL-2-mediated signal transduction in HTLV-I-infected human T cells. Int. Immunol. 4, 1233-1243.

Minami, Y., Kono, T., Yamada, K., Kobayashi, N., Kawahara, A., Perlmutter, R.M., & Taniguchi, T. (1993). Association of p56lck with IL-2 receptor beta chain is critical for the IL-2-induced activation of p56lck. EMBO J. 12, 759-768.

Minami, Y., Oishi, I., Liu, Z-J., Nakagawa, S., Miyazaki, T., & Taniguchi, T. (1994). Signal transduction mediated by the reconstituted IL-2 receptor. J. Immunol. 152, 5680-5690.

Minasi, L.E., Kamogawa, Y., Carding, S., Bottomly, K., & Flavell, R.A. (1993). The selective ablation of interleukin 2-producing cells isolated from transgenic mice. J. Exp. Med. 177, 1451-1459.

Miyajima, A., Hara, T., & Kitamura, T. (1992). Common subunits of cytokine receptors and the functional redundancy of cytokines. Trends Biochem. Sci. 17, 378-382.

Miyazaki, T., Barsoumian, E.L., Hatakeyama, M., & Taniguchi, T. (1991). The integrity of the conserved 'WS motif' common to IL-2 and other cytokine receptors is essential for ligand binding and signal transduction. EMBO J. 10, 3191-3197.

Miyazaki, T., Kawahara, A., Fujii, H., Nakagawa, Y., Minami, Y., Liu, Z-J., Oishi, I., Silvennoinen, O., Witthuhn, B.A., Ihle, J.N., & Taniguchi, T. (1994). Functional activation of Jak1 and Jak3 by selective association with IL-2 receptor subunits. Science,266, 1045-1047.

Morgan, D.A., Ruscetti, F.W., & Gallo, R. (1976). Selective in vitro growth of T lymphocytes from normal human bone marrows. Science 193, 1007-1008.

Mori, H., Barsoumian, E.L., Hatakeyama, M., & Taniguchi, T. (1991). Signal transduction by interleukin 2 receptor beta chain: importance of the structural integrity as revealed by site-directed mutagenesis and generation of chimeric receptors. Int. Immunol. 3, 149-156.

Murakami, M., Narazaki, M., Hibi, M., Yawata, H., Yasukawa, K., Hamaguchi, M., Taga, T., & Kishimoto, T. (1991). Critical cytoplasmic region of the interleukin

6 signal transducer gp130 is conserved in the cytokine receptor family. Proc. Natl. Acad. Sci. USA 88, 11349-11353.

Nakamura, Y., Russell, S.M., Mess, S.A., Friedman, M., Erdos, M., Francois, C., Jacques, Y., Adelstein, S., & Leonard, W.J. (1994). Heterodimerization of the IL-2 receptor beta- and γ-chain cytoplasmic domains is required for signaling. Nature 369, 330-333.

Nakanishi, K., Hirose, S., Yoshimoto, T., Ishizashi, H., Hiroisha, K., Tanaka, T., Kono, T., Miyasaka, M., Taniguchi, T., & Higashino, K. (1992). Role and regulation of interleukin (IL)-2 receptor alpha and beta chains in IL-2-driven B-cell growth. Proc. Natl. Acad. Sci. USA 89, 3551-3555.

Nakano, N., Hardy, R.R., & Kishimoto, T. (1987). Identification of intrathymic T progenitor cells by expression of Thy-1, IL2 receptor and CD3. Eur. J. Immunol. 17, 1567-1571.

Nakarai, T., Robertson, M.J., Streuli, M., Wu, Z., Ciardelli, T.L., Smith, K.A., & Ritz, J. (1994). Interleukin 2 receptor g chain expression on resting and activated lymphoid cells. J. Exp. Med. 180, 241-251.

Nelson, B.H., Lord, J.D., & Greenberg, P.D. (1994). Cytoplasmic domains of the interleukin-2 receptor beta and gamma chains mediate the signal for T-cell proliferation. Nature, 369, 333-336.

Nikaido, T., Shimizu, A., Ishida, N., Sabe, H., Teshigawara, K., Maeda, M., Vchiyama, T., Yodoi, J., & Honjo, T. (1984). Molecular cloning of cDNA encoding human interleukin-2 receptor. Nature 311, 631-635.

Nikcevich, D.A., Finnegan, A., Chong, A.S.F., Williams, J.W., & Bremer, E. (1994). Inhibition of interleukin 2 (IL-2)-stimulated tyrosine kinase activity by leflunomide. Agents Actions 41, C279-C282.

Nishio, K., Miura, K., Ohira, T., Heike, Y., & Saijo, N. (1994). Genistein, a tyrosine kinase inhibitor, decreased the affinity of $p56^{lck}$ to β-chain of interleukin-2 receptor in human natural killer (NK)-rich cells and decreased NK-mediated cytotoxicity. P.S.E.B.M. 207, 227-233.

Noguchi, M., Nakamura, Y., Russell, S.M., Ziegler, S.F., Tsang, M., Cao, X., & Leonard, W.J. (1993a). Interleukin-2 receptor γ chain: A functional component of the interleukin-7 receptor. Science 262, 1877-1880.

Noguchi, M., Yi, H., Rosenblatt, H.M., Filiporich, A.H., Adelstein, S., Modi, W.S., McBride, O.W., & Leonard, W.J. (1993b). Interleukin-2 receptor γ chain mutation results in X-linked severe combined immunodeficiency in humans. Cell 73, 147-157.

Ohashi, Y., Takeshita, T., Nagata, K., Mori, S., & Sugamura, K. (1989). Differential expression of the IL-2 receptor subunits, p55 and p75 on various populations of primary peripheral blood mononuclear cells. J. Immunol. 143, 3548-3555.

Orosz, C.G., Scott, J.W., Gillis, S., & Finke, J.H. (1985). Reversal of phorbol ester-mediated reduction of cloned T lymphocyte cytolysis by interleukin 2. J. Immunol. 134, 324-329.

Ortaldo, J.R., Mason, A.T., Gerard, J.P., Henderson, L.E., Farrar, N., Hopkins, R.F., III, Heberman, R.B., & Rabin, H. (1984). Effects of natural and recombinant IL 2 on regulation of IFN gamma production and natural killer activity: Lack of involvement of the Tac antigen for these immunoregulatory effects. J. Immunol. 133, 779-783.

Osame, M., Usuku, K., Ijichi, N., Amitani, H., Igata, A., Matsumoto, M., & Tara, H. (1986). HTLV-I associated myelopathy, a new clinical entity. Lancet 1, 1031-1032.

Otani, H., Siegel, J.P., Erdos, M., Gnarra, J.R., Toledano, M.B., Sharon, M., Mostowski, H., Feinberg, MB., Pierce, J.H., & Leonard, W.J. (1992). Interleukin (IL)-2 and IL-3 induce distinct but overlapping responses in murine IL-3-dependent 32D cells transduced with human IL-2 receptor beta chain: Involvement of tyrosine kinase(s) other than p56lck. Proc. Natl. Acad. Sci. USA, 89, 2789-2793.

Perkins, G.R., Marvel, J., & Collins, M.K.L. 1993. Interleukin 2 activates extracellular signal-regulated protein kinase 2. J. Exp. Med. 170, 1429-1434.

Plasisance, S., Rubinstein, E., Alileche, A., Sahraoui, Y., Krief, P., Augery-Bourget, Y., Jasmin, C., Suarez, H., & Azzarone, B. (1992). Expression of the interleukin-2 receptor on human fibroblasts and its biological significance. Intl. Immunol. 4, 739-746.

Poiesz, B.F., Ruscetti, F.W., Gazdar, A.F., Bunn, P.A., Minna, J.D., & Galls, R.C. (1980). Detection and isolation of type C retrovirus particles from fresh cultured lymphocytes of a patient with cutaneous T-cell lymphoma. Proc. Natl. Acad. Sci. USA 77, 7415-7419.

Puck, J.M., Nussbaum, R.L., Smead, D.L., & Conley, M.E. (1989). X-linked severe combined immunodeficiency: Localization within the region Xq13.1-q21.1 by linkage and deletion analysis. Am. J. Human Genet. 44, 724-730.

Puck, J.M., Deschenes, S.M., Porter, J.C., Dutra, A.S., Brown, C.J., Willard, H.F., & Henthorn, P.S. (1993). The interleukin-2 receptor g chain maps to Xq13.1 and is mutated in X-linked severe combined immunodeficiency, SCIDX1. Human Mol. Gen. 2, 1099-1104.

Ralph, P., Nakoinz, I., & Rennick, D. (1988). Role of interleukin 2, interleukin 4, and alpha, beta, and gamma interferon in stimulating macrophage antibody-dependent tumoricidal activity. J. Exp. Med. 167, 712-717.

Raulet, D.H. (1985). Expression and function of interleukin-2 receptors on immature thymocytes. Nature 314, 101-103.

Ravichandran, K.S., & Burakoff, S.J. (1994). The adapter protein Shc interacts with the interleukin-2 (IL-2) receptor upon IL-2 stimulation. J. Biol. Chem. 269, 1599-1602.

Reed, J.C., Alpers, J.D., Scherle, P.A., Hoover, R.G., Nowell, P.C., & Prystowsky, M.B. (1987). Proto-oncogene expression in cloned T lymphocytes: mitogens and growth factors induce different patterns of expression. Oncogene 1, 223-228.

Remillard, B., Petrillo, R., Maslinski, W., Tsudo, M., Strom, T.B., Cantley, L., & Varticovski, L. (1991). Interleukin-2 receptor regulates activation of phosphatidylinositol 3-kinase. J. Biol. Chem. 266, 14167-14170.

Ringheim, G.E., Freimark, B.D., & Robb, R.J. (1991). Quantitative characterization of the intrinsic ligand-binding affinity of the interleukin 2 receptor beta chain and its modulation by the alpha chain and a second affinity-modulating element. Lymphokine Cytokine Res. 10, 219-224.

Robb, R.J., Rusk, C.M., & Neeper, M.P. (1988). Structure-function relationships for the interleukin 2 receptor: Location of ligand and antibody binding sites on the Tac receptor chain by mutational analysis [published erratum appears in Proc. Nat. Acad. Sci. USA 1988 Nov;85(21):8226]. Proc. Natl. Acad. Sci.USA 85, 5654-5658.

Rosenberg, S.A., & Lotze, M.T. (1986). Structure-function relationships for the IL-2 receptor system. V. Structure-activity analysis of modified and truncated forms of the Tac receptor protein: Site-specific mutagenesis of cysteine residues. Annu. Rev. Immunol. 4, 681-709.

Ruegemer, J.J., Ho, S.N., Augustine, J.A., Schlager, J.W., Bell, M.P., McKean, D.J., & Abraham, R.T. (1990). Regulatory effects of transformic growth factor-b on IL-2- and IL-4-dependent T cell-cycle progression. J. Immunol. 144:, 1767-1776.

Rusk, C.M., Neeper, M.P., Kuo, L.M., Kutny, R.M., & Robb, R.J. (1988). A larger number of L chains (Tac) enhance the association rate of interleukin 2 to the high affinity site of the interleukin 2 receptor. J. Immunol. 140, 2249-2259.

Russell, S.M., Keegan, A.D., Harado, N., Nakamura, Y., Noguchi, M., Leland, P., Friedmann, M.C., Miyajima, A., Puri, R.K., Paul, W.E., & Leonard, W.J. (1993). Interleukin-2 receptor γ chain: A functional component of the interleukin-4 receptor. Science 262, 1880-1883.

Russell, S.M., Johnston, J.A., Noguchi, M., Kawamura, M., Bacon, C.M., Friedmann, M., Berg, M., McVicar, D.W., Witthuhn, B.A., Silvennoinen, O., Goldman, A.S., Schmalstieg, F.C., Ihle, J.N. O'Shea, J.J., & Leonard, W.J. (1994). Interaction of IL-2Rβ and γc chains with Jak1 and Jak3: Implications for XSCID and XCID. Science 266, 1042-1045.

Sabath, D.E., Podolin, P.L., Comber, P.G., & Prystowsky, M.B. (1990). cDNA cloning and characterization of interleukin 2-induced genes in a cloned T helper lymphocyte. J. Biol. Chem. 265, 12671-12678.

Sabe, H., Kondo, S., Shimizu, A., Tagaya, Y., Yodoi, J., Kobayashi, N., Hatanaka, M., Matsunami, N., Maeda, M., & Noma, T. (1984). Properties of human interleukin-2 receptors expressed on non-lymphoid cells by cDNA transfection. Mol. Biol. Med. 2, 379-396.

Saneto, R.P., Altman, A., Knobler, R.L., Johnson, H.M., & de Vellis, J. (1986). Interleukin 2 mediates the inhibition of oligodendrocyte progenitor cell proliferation *in vitro*. Proc. Natl. Acad. Sci. USA 83, 9221-9225.

Saragovi, H., & Malek, T.R. (1988). Novel interleukin-2 receptor subunit detected by cross-linking under high-affinity conditions. J. Immunol. 141, 476-82.

Satoh, T., Minami, Y., Kono, T., Yamada, K., Kawahara, A., Taniguchi, T., & Kaziro, Y. (1992). Interleukin 2-induced activation of Ras requires two domains of interleukin 2 receptor beta subunit, the essential region for growth stimulation and Lck-binding domain. J. Biol. Chem. 267, 25423-25427.

Sauve, K., Nachman, M., Spence, C., Bailon, P., Campbell, E., Tsien, W.H., Kondos, J.A., Hakimi, J., & Ju, G. (1991). Localization in human interleukin 2 of the binding site to the alpha chain (p55) of the interleukin 2 receptor. Proc. Natl. Acad. Sci. USA 88, 4636-4640.

Schorle, H., Holtschke, T., Hunig, T., Schimpl, A., & Horak, I. (1991). Development and function of T cells in mice rendered interleukin-2 deficient by gene targeting. Nature 352, 621-624.

Seder, R.A., & Paul, W.E. (1994). Acquisition of lymphokine-producing phenotype by CD4+ cells. Ann. Rev. Immunol. 12, 635-673.

Seldin, D.C., & Leder, P. (1994). Mutational analysis of a critical signaling domain of the human interleukin 4 receptor. Proc. Natl. Acad. Sci. USA 91, 2140-2144.

Shackelford, D.A., & Trowbridge, I.S. (1991). Ligand-stimulated tyrosine phosphorylation of the IL-2 receptor beta chain and receptor-associated proteins. Cell Regul 2, 73-85.

Sharon, M., Gnarra, J.R., & Leonard, W.J. (1989). The beta-chain of the IL-2 receptor (p70) is tyrosine-phosphorylated on YT and HUT-102B2 cells. J. Immunol. 143, 2530-2533.

Shibuya, H., Yoneyama, M., Ninomiya, T.J., Matsumoto, K., & Taniguchi T. (1992). IL-2 and EGF receptors stimulate the hematopoietic cell cycle via different signaling pathways: demonstration of a novel role for c-myc. Cell 70, 57-67.

Siegel, J.P., Sharon, M., Smith, P.L., & Leonard, W.J. (1987). The IL-2 receptor β chain (p70): Role in mediating signals for LAK, NK, and proliferative activities. Science 238, 75-78.

Siliciano, J.D., Morrow, T.A., & Desiderio, S.V. (1992). itk, a T-cell-specific tyrosine kinase gene inducible by interleukin 2. Proc. Natl. Acad. Sci. USA 89, 11194-11198.

Smith, K.A. (1988). Interleukin-2: Inception, impact, and implications. Science 240, 1169-1176.

Smith, K.A., & Cantrell, D.A. (1985). Interleukin 2 regulates its own receptors. Proc. Natl. Acad. Sci. USA 82, 864-868.

Smith, M.R., & Greene, W.C. (1991). Molecular biology of the type I human T cell leukemia virus (HTLV-I) and adult T-cell leukemia. J. Clin. Inv. 87, 761-766.

Stern, J.B., & Smith, K.A. (1986). Interleukin-2 induction of T-cell G_1 progression and c-myb expression. Science 233, 203-206.

Strieter, R.M., Remick, D.G., Lynch, J.P., III, Spengler, R.B., & Kunkel, S.L. (1989). Interleukin-2 induced tumor necrosis factor alpha gene expression in human alveolar macrophages and blood monocytes. Am. Rev. Respir. Dis. 139, 335.

Taichman, R., Merida, I., Torigoe, T., Gaulton, G.N., & Reed, J.C. (1993). Evidence that protein tyrosine kinase p56-lck regulates the activity of phosphatidylinositol-3'-kinase in interleukin-2-dependent T-cells. J. Biol. Chem. 268, 20031-20036.

Taichman, R.S., Torigoe, T., Tanaka, S., Miyashita, T., & Reed, J.C. (1992). Gene transfer investigations of p56-LCK function in IL-2-dependent T-cell lines: Implications for mechanisms of IL-2-signal transduction. Cytokine 4, 441-453.

Takeshita, T., Asao, H., Suzuki, J., & Sugamura, K. (1990). An associated molecule, p64, with high-affinity interleukin 2 receptor. Intl. Immunol. 2, 477-480.

Takeshita, T., Asao, H, Ohtani, K., Ishii, N., Kumaki, S., Tanaka, N., Munakata, H., Nakamura, M., & Sugamura, K. (1992a). Cloning of the gamma chain of the human IL-2 receptor. Science 257, 379-382.

Takeshita, T., Ohtani, K., Asao, H., Kumaki, S., Nakamura, M., & Sugamura, K. (1992b). An associated molecule, p64, with IL-2 receptor β chain. J. Immunol. 148, 2154-2158.

Takeuchi, Y., Tanaka, T., Hamamura, K., Sugimoto, T., Miyasaka, M., Yagita, H., & Okumura, K. (1992). Expression and role of interleukin-2 receptor beta chain on CD4-CD8- T cell receptor alpha beta+ cells. Eur. J. Immunol. 22, 2929-2935.

Talmadge, J.E., Wiltrout, R.H., Counts, D.F., Herberman, R.B., McDonald, T., & Ortaldo, J.R. (1986). Proliferation of human peripheral blood lymphocytes induced by recombinant human interleukin 2: Contribution of large granular lymphocytes and T lymphocytes. Cell Immunol. 102, 261-272.

Tanaka, N., Ohtani K., Nakamura, M., & Sugamura, K. (1993). A novel human tyrosine kinase gene inducible in T cells by interleukin 2. FEBS Lett. 324, 1-5.

Tanaka, N., Asao, H., Ohbo, K., Ishii, N., Takeshita, T., Nakamura, M., Sasaki, H., & Sugamura, K. (1994). Physical association of JAK1 and JAK2 tyrosine kinases with the interleukin 2 receptor beta and gamma chains. Proc. Nat. Acad. Sci. USA 91, 7271-7275.

Tanaka, T., Tsudo, M., Karashyama, H., Toyama, N., Hatakeyama, M., Taniguchi, T., & Miyasaka, M. (1991). Signal transduction through the human IL-2 receptor beta-chain expressed in IL-6-dependent mouse B cell hybridoma. Int. Immunol. 3, 105-108.

Tentori, L., Longo, D.L., Zuniga-Pflucker, J.C., Wing, C., & Kruisbeek, A.M. (1988). Essential role of the interleukin 2-interleukin 2 receptor pathway in thymocyte maturation *in vivo*. J. Exp. Med. 168, 1741-1747.

Terada, N., Lucas, J.J., Szepesi, A., Franklin, R.A., Takase, K., & Gelfand, E. (1992). Rapamycin inhibits the phosphorylation of p70 S6 kinase in IL-2 and mitogen-activated human T cells. Biochem. Biophys .Res. Commun. 186, 1315-1321.

Tigges, M.A., Casey, L.S., & Koshland, M.E. (1989). Mechanism of interleukin-2 signaling: Mediation of different outcomes by a single receptor and transduction pathway. Science 243, 781-786.

Trinchieri, G., Matsumoto-Kobayashi, M., Clark, S.C., Seehra, J., London, L., & Perussia, B. (1984). Response of resting human peripheral blood natural killer cells to interleukin 2. J. Exp. Med. 160, 1147-1169.

Truitt, K.E., Mills, G.B., Turck, C.W., & Imboden, J.B. (1994). SH2-dependent association of phosphatidylinositol 3'-kinase 85-kDa regulatory subunit with the interleukin-2 receptor β chain. J. Biol. Chem. 269, 5937-5943.

Tsudo, M., Kitamura, F., & Miyasaka, M. (1989a). Characterization of the interleukin 2 receptor beta chain using three distinct monoclonal antibodies. Proc. Natl. Acad. Sci. USA 86, 1982-1986.

Turner, B., Rapp, U., App. H., Greene, M., Dobashi, K., & Reed J. (1991). Interleukin 2 induces tyrosine phosphorylation and activation of p72-74 Raf-1 kinase in a T-cell line. Proc. Natl. Acad. Sci.USA 88, 1227-1231.

Uchiyama, T., Broder, S., & Waldmann, T.A. (1981). A monoclonal antibody (anti-Tac) reactive with activated and functionally mature human T cells. I. Production of anti-Tac monoclonal antibody and distribution of Tac (+) cells. J. Immunol. 126, 1393-1397.

Ullrich, A., & Schessinger, J. (1990). Schlessinger, Signal transduction by receptors with tyrosine kinase activity. Cell 61, 203-212.

Valius, M., & Kazlauskas, A. (1993). Phospholipase C-gamma 1 and phosphatidylinositol 3 kinase are the downstream mediators of the PDGF receptor's mitogenic signal. Cell 73, 321-334.

von Boehmer, H., Crisanti, A., Kisielow, P., & Haas, W. (1985). Absence of growth by most receptor-expressing fetal thymocytes in the presence of interleukin-2. Nature 314, 539-540.

Voss, S.D., Leary, T.P., Sondel, P.M., & Robb, R.J. (1993). Identification of a direct interaction between interleukin 2 and the p64 interleukin 2 receptor γ chain. Proc. Natl. Acad. Sci. USA 90, 2428-2432.

Voss, S.D., Sondel, P.M., & Robb, R.J. (1992). Characterization of the interleukin 2 receptors (IL-2R) expressed on human natural killer cells activated *in vivo* by IL-2: Association of the p64 IL-2R γ chain with the IL-2R β chain in functional intermediate-affinity IL-2R. J. Exp. Med. 176, 531-541.

Waldmann, T.A., Goldman, C.K., Robb, R.J., Depper, J.M., Leonard, W.J., Sharrow, S.O., Bongiovanni, K.F., Korsmeyer, S.J., & Greene, W.C. (1984). Expression of interleukin 2 receptors on activated human B cells. J. Exp. Med. 160, 1450-1466.

Wang, H.M., & Smith, K.A. (1987). The interleukin 2 receptor. Functional consequences of its bimolecular structure. J. Exp. Med. 166, 1055-1069.

Wang, L-M., Keegan, A.D., Paul, W.E., Heidaran, M.A., Gutkind, J.S., & Pierce, J.H. (1992). IL-4 activates a distinct signal transduction cascade from IL-3 in factor-dependent myeloid cells. EMBO J. 11, 4899-4908.

Watowich, S.S., Hilton, D.J., & Lodish, H.F. (1994). Activation and inhibition of erythropoietin receptor function: role of receptor dimerization. Molec. Cell Biol. 14, 3535-3549.

Watowich, S.S., Yoshimura, A., Longmore, G.D., Hilton, D.J., Yoshimura, Y., & Lodish, H.F. (1992). Homodimerization and constitutive activation of the erythropoietin receptor. Proc. Natl. Acad. Sci.USA 89, 2140-2144.

Weigel, U., Meyer, M., & Sebald, W. (1989). Mutant proteins of human interleukin 2. Renaturation yield, proliferative activity and receptor binding. Eur. J. Biochem. 180, 295-300.

Weiss, A. (1993). T cell antigen receptor signal transduction: a tale of tails and cytoplasmic protein-tyrosine kinases. Cell 73, 209-212.

Whal, S.N., McCartney-Francis, N., Hunt, D.A., Smith, P.D., Whal, L.M., & Katona, I.M. (1987). Monocyte interleukin-2 receptor gene expression and interleukin-2 augmentation of microbicidal activity. J. Immunol. 139.

Witthuhn, B.A., Silvennoinen, O., Miura, O., Lai, K.S., Cwik, C., Liu, E.T., & Ihle, J.N. (1994). Involvement of the Jak-3 Janus kinase in signalling by interleukins 2 and 4 in lymphoid and myeloid cells. Nature 370, 153-157.

Yoshida, M., Miyoshi, I., & Hinuma, Y. (1982). Isolation and characterization of retrovirus from cell lines of human adult T-cell leukemia and its implication in the disease. Proc. Natl. Acad. Sci. USA 70, 2031-2035.

Yoshimura, A., Longmore, G., & Lodish, H.F. (1990). Point mutation in the exoplasmic domain of the erythropoietin receptor resulting in hormone-independent activation and tumorigenicity. Nature 348, 647-649.

Zhu, X., Suen, K-L., Barbacid, M., Bolen, J.B., & Fargnoli, J. (1994). Interleukin-2-induced tyrosine phosphorylation of Shc proteins correlates with factor-dependent T cell proliferation. J. Biol. Chem. 269, 5518-5522.

Zmuidzinas, A., Mamon, H.J., Roberts, T.M., & Smith, K.A. (1991). Interleukin-2-triggered Raf-1 expression, phosphorylation, and associated kinase activity increase through G1 and S in CD3-stimulated primary human T cells. Mol. Cell .Biol. 11, 2794-803.

Zuniga-Pflucker, J.C., Smith, K.A., Tentori, L., Pardoll, D.M., Longo, D.L., & Kruisbeek, A.M. (1990). Are the IL-2 receptors expressed in the murine fetal thymus functional? Dev. Immunol. 1, 59-66.

Zurawski, S.M., & Zurawski, G. (1989). Mouse interleukin-2 structure-function studies: Substitutions in the first alpha-helix can specifically inactivate p70 receptor binding and mutations in the fifth alpha-helix can specifically inactivate p55 receptor binding. EMBO J. 8, 2583-2590.

Zurawski, S.M., Imler, J.L., & Zurawski, G. (1990). Partial agonist/antagonist mouse interleukin-2 proteins indicate that a third component of the receptor complex functions in signal transduction. EMBO J. 9, 3899-3905.

www.ingramcontent.com/pod-product-compliance
Ingram Content Group UK Ltd.
Pitfield, Milton Keynes, MK11 3LW, UK
UKHW020230290926
15470UKWH00002B/455

9 780762 30117